THE SUN AND THE HELIOSPHERE AS AN INTEGRATED SYSTEM

ASTROPHYSICS AND SPACE SCIENCE LIBRARY

VOLUME 317

THE SUN AND THE HELIOPSPHERE AS AN INTEGRATED SYSTEM

Edited by

GIANNINA POLETTO

INAF – Osservatorio di Arcetri,
Firenze, Italy

and

STEVEN T. SUESS

NASA Marshall Space Flight Center,
National Space & Technology Center/Solar Physics,
Huntsville, AL, U.S.A.

KLUWER ACADEMIC PUBLISHERS

DORDRECHT / BOSTON / LONDON

A C.I.P. Catalogue record for this book is available from the Library of Congress.

ISBN 1-4020-2830-X (HB)
ISBN 1-4020-2831-8 (e-book)

Published by Kluwer Academic Publishers,
P.O. Box 17, 3300 AA Dordrecht, The Netherlands.

Sold and distributed in North, Central and South America
by Kluwer Academic Publishers,
101 Philip Drive, Norwell, MA 02061, U.S.A.

In all other countries, sold and distributed
by Kluwer Academic Publishers,
P.O. Box 322, 3300 AH Dordrecht, The Netherlands.

Printed on acid-free paper

Printed in the Netherlands.

Contents

Preface

Spacecraft observations of the Sun and the heliosphere are passing from the exploration phases to explicit studies of physical processes. As this change takes place, we learn about the natural connections that exist between various parts of the system. Wonderful examples of this are now known. One is the entry of interstellar matter into the heliosphere and how this depends both on sources of the solar wind at the Sun and the explicit manner in which the interplanetary magnetic field connects back to the Sun. Understanding these connections permits using composition measurements in the solar wind to analyze interstellar matter on the one hand and acceleration of particles to relativistic energies at shocks on the other hand. Furthermore, understanding of the system has enabled us to use that knowledge to detect the equivalent of heliospheres - astrospheres - around other stars producing solar-type winds. The circumstance which is producing this understanding is the simultaneous existence of spacecraft observing the Sun (SOHO, Yohkoh, TRACE, RHESSI) and spacecraft observing the interplanetary medium (Voyagers, Ulysses, ACE, IMP, CLUSTER, SMEI, WIND). These missions are both difficult and expensive, explaining why it has taken a long time for understanding of the system to come. Now, with some understanding, we have learned many ways to extend and expand the studies. This is leading to many new and proposed missions. Some of those which will be launched within the next 10 years include STEREO, Solar-B, SDO, SENTINELS, and SO. But, there are many more exciting missions on the drawing boards, including those that will actually go to the Sun, exit the solar system, or again pass over the poles of the Sun.

Here we introduce this subject by discussing the topics as they appear in the book. The order is, roughly, from the outer heliosphere inwards, in an attempt to move from the extreme outreach of our solar system and the ensuing global scenario to finer details within it. Work on the outer heliospheric boundary has been boosted recently by the discovery of the so called "hydrogen walls": that region of neutral interstellar hydrogen which is located at the interface of the heliosphere, or astrospheres, and the interstellar medium, in the upstream direction. J. Linsky's contribution shows that we can now study the environment of other (G and K) stars, on the basis of our understanding of the heliosphere, and, being able to widen our view beyond the Sun, we get new information also on the history of our own solar system. New observations of the far reaches of the heliosphere are being made by by Voyager, which may have touched the termination shock in 2003, at distances on the order of about 85 AU. Another means to gain insight into these remote regions is by analyzing radio emission from the heliosphere interface: I. Cairns focusses on modeling radio data from Voyager observations of the outer heliosphere and the physics of the heliosphere interface. In order to understand some of these phenomena we must take into account variations brought about by solar cycle variations, demonstrating that only a global vision of the Sun and the heliosphere is adequate at such large distances. The new data that are being acquired by spacecraft allow us to test the theoretical models of the SW/LIC (solar wind/local interstellar cloud) interaction against observations, a capability only recently acquired (see V. Izmodenov contribution).

On a smaller scale, our knowledge of the inner heliosphere, out to distances of ≈ 5 AU, has been enormously improved by the Ulysses mission, the first spacecraft to pass above the poles of the Sun and to give us a view of regions at high latitudes. In addition, the long life of the mission, in its 14th year of operations at the time of this writing, allowed us to study variations over time in the heliosphere, as Ulysses has flown over the solar poles at times close to solar minimum (1994/1995) and at solar maximum (year 2000). Highlights of the mission are given by Marsden: we may notice here that changes in the heliosphere caused by the solar activity cycle, which we mentioned as an explanation for outer heliospheric phenomena, are obviously observed in the inner heliosphere as well, and, taking the dynamic solar wind pressure as a reference parameter, amount to a factor two increase, independent of latitude. We still have to find out whether this is a typical figure or whether changes may occur from one to the next solar cycle and how large those may possibly be. *In situ* measurements are not the only means to get some insight into the heliosphere: Interplanetary Scintillation Measurements (IPS), although affected by integration effects - nowadays overcome by sophisticated techniques to analyze data - may provide important data with the advantage, with respect to in situ measurements, of requiring not as much time as a spacecraft needs to acquire data over a large span of latitudes (Kojima et al.).

Ulysses polar passes allowed, for the first time ever, observations of energetic particles at high latitudes (Sanderson): these events, whose

properties differ less than expected from those of low latitude events, give
new information about the latitudinal transport of particles, thus about
polar fields. However, observations do not yet allow us to answer questions about the magnetic field configuration very near to the Sun, as it
is not clear whether particles diffuse across the field, or whether the field
is distorted. Energetic particles and anomalous cosmic rays provide new
information as well on fieldlines which connect to very large distances,
reaching out to the heliospheric termination shock. Footpoint motion at
the Sun, coupled with solar wind speed variations, may solve open problems both in the physics of anomalous cosmic rays (ACR) and cosmic
ray transport, once more highlighting how heliospheric events cannot be
explained unless they are linked to the near-Sun physics (Schwadron).
We want to point out that although ACRs, as well as pick-up ions, have
been providing us with new information, we know very little, if anything,
about species which are mostly ionized and on their isotopic composition: knowledge of the latter might be used to constrain cosmological
and nucleosynthesis models contributing to the global picture of the solar
system and its evolution whose first steps are described in this volume.

Apart from the changes in the heliospheric configuration caused by
the solar activity (11 year) and magnetic activity (22 year) cycles, short
time variations are caused by the propagation of Coronal Mass Ejections
through the heliosphere. It is well known that interplanetary shocks
and magnetic clouds of CMEs affect particle propagation, but they have
other effects as well. We have not reached a comprehensive view of their
role. Do they, for instance, scatter galactic cosmic rays, and, in case, how
large is it their impact on the solar cycle modulation of galactic rays?
how large is the contribution from the changing topology of CME field
lines to the magnetic flux balance of the heliosphere? Our understanding
of CMEs has increased enormously after SOHO and Ulysses, and the
state of the art is reviewed by N. Gopalswamy. This is a field where
we may reasonably hope for further progress coming from future space
missions: STEREO, with its twin spacecraft providing stereo images and
its *in situ* data acquisition, and the Solar Mass Ejection Imager (SMEI),
with its camera imaging CMEs as they propagate through the interstellar
medium, may strongly contribute to broaden our understanding of these
events and of their influence on heliospheric phenomena.

The unperturbed solar wind physical parameters show the random
fluctuations characteristic of turbulence and it is well known that the
best place to study the solar wind alfvénic turbulence is in fast streams.
How does turbulence evolve with distance from the Sun is not well known
and, before the advent of Ulysses, knowledge was limited to the low
latitude behavior. Ulysses polar passes provided the solar/heliospheric

community with the first opportunity to analyze the evolution of turbulence with distance for the polar wind: results from these new data show a behavior which is consistent with what might have been predicted, being similar, although slower than in ecliptic turbulence evolution. While this is a nice achievement, we should keep in mind that we have yet to disentangle distance from latitude effects, to be able to refine the picture. The need for disentangling spatial and temporal effects is especially felt when interpreting intermittency phenomena whose origin is still debated (Bavassano et al.). Quite obviously, waves/turbulence may be generated in the solar corona and only those on open coronal fieldlines (as opposed to those confined in loop structures) will reach out to the interplanetary medium. What is the nature of the waves, how is the solar corona heated and the solar wind accelerated, are long-standing problems still waiting for a non controversial solution. SOHO data have revealed some new features - how heavy ions are accelerated over the first few solar radii as opposed to protons - which provide valuable information in identifying the nature of the waves. Nevertheless, further data on the behavior of heavy ions, such as Helium, are needed to pinpoint the nature of waves (Marsch). Hopefully, future missions such as SO, Solar Probe, or the proposed ASCE mission will provide the necessary additional information.

The hot corona also transfers energy downwards, towards the transition region (TR) and chromosphere. Quite interestingly, this downwards transport affects the solar wind mass flux - another example, on a more local basis, of how the physics of any region is intertwined with the physics of adjacent and sometimes more distant regions. The feedback between downward energy transport and outflowing solar wind occurs because the location of the TR-chromosphere interface changes as a function of the energy being transferred downwards and the pressure of the corona changes accordingly. Lie-Svendsen gives a beautiful example of how the wind mass flux, which is measured by *in situ* experiments, can't be accounted for, unless the energy supplied to the TR reaches a threshold value.

Analogously, the elemental composition of the solar wind depends on chromospheric processes (Raymond). Different regions in the solar corona and solar wind have quite different abundances and these differ from photospheric abundances as well. This variation is ascribed to processes, seemingly occurring at chromospheric levels, which appear to depend on the First Ionization Potential (FIP effect) of the elements and are not yet unambiguously identified. In stars, we have an even more baffling situation, as there is some evidence for an "inverse" FIP effect, for which we have no explanation. Hence the capabilities of abundance studies as a means for gathering further information on the mechanisms by which the corona is heated and solar wind accelerated could not, so far, been fully exploited.

Down at the very end of this brief compendium of studies, tools, questions, new discoveries, which are the theme of the present volume, we reach the actual core of the problem: magnetic fields. We would not have a heliosphere, shaped as it is, or astrospheres, without a magnetic field which is structuring, or being structured by, cosmic plasmas. Solanki's contribution illustrates the magnetic field behavior as it passes through different regimes, from the convection zone out to the heliosphere. However, we would not have the variety of phenomena that we are observing, unless magnetic field reconnection would not provide us with the capability of magnetic field restructuring. Priest's conclusive chapter gives us basic information on this process as well as examples of reconfiguring magnetic fields. This brings us back to the astro/heliospheres we started with, closing the circle of our integrated system.

Cover: The acronyms on the cover refer to the following missions and/or techniques: HST - Hubble Space Telescope, IPS - Interplanetary Scintillations, IMP - Interplanetary Monitoring Platform, ACE - Advanced Composition Explorer, SOHO - SOlar Heliospheric Observatory, WIND - not an acronym, SMEI - Solar Mass Ejection Imager, TRACE - Transition Region And Coronal Explorer, RHESSI - Ramaty High Energy Solar Spectroscopic Imager.

Chapter 1

HYDROGEN WALLS: MASS LOSS OF DWARF STARS AND THE YOUNG SUN

Jeffrey L. Linsky

JILA, University of Colorado and NIST
Boulder, CO 80309-0440 USA
jlinsky@jila.colorado.edu

Brian E. Wood

JILA, University of Colorado and NIST
Boulder, CO 80309-0440 USA
woodb@origins.colorado.edu

Abstract   Charge-exchange reactions between a fully ionized stellar wind and the partially ionized warm gas in the interstellar medium create a compressed region of hot neutral hydrogen atoms that are decelerated relative to the inflowing interstellar gas. This "hydrogen wall" produces a blue-shifted absorption component in the stellar Lyman-α emission line at many viewing angles that has now been detected in *Hubble Space Telescope* spectra of eight dwarf stars. Comparisons of the observed Lyman-α line profiles with theoretical models led to the first sensitive measurements of mass-loss rates as small as 4×10^{-15} $M_\odot$ yr^{-1} for solar-like dwarfs. Our study of astrospheres provides the first observational data (other than for the Sun) with which to test theories for dwarf star winds. We find an empirical correlation of stellar mass-loss rate with X-ray surface flux that allows us to predict the mass-loss rates of other stars and to infer the solar wind mass flux at earlier times, when the solar wind may have been as much as 1000 times stronger than at present. We mention some important ramifications for the history of planetary atmospheres in our solar system, that of Mars in particular, and for exoplanets.

G. Poletto and S.T. Suess (eds.), The Sun and The Heliosphere as an Integrated System, 1–22.
© 2004 *Kluwer Academic Publishers. Printed in the Netherlands.*

1.   Is the Solar Wind Unique?

The presence of blue-shifted absorption features in spectral line profiles convinced stellar spectroscopists in the early twentieth century that luminous stars must lose mass. The classic P Cygni line profile, which is typically seen in hot star spectra, consists of blue-shifted absorption and red-shifted emission. Observations of this profile provide clear evidence that a star is losing mass and that the expanding outer atmosphere is comparable to or larger than the stellar photosphere. Novae and supernovae demonstrate that mass loss can be very rapid at times. In a classic study of the α Her triple system, Deutsch (1956) measured the mass-loss rate of $\dot{M} \geq 3 \times 10^{-8} M_\odot$ yr^{-1} from the M5 giant in the system by analysis of absorption features superimposed on the spectrum of the G5 dwarf lying behind the wind of the M5 giant. This mass-loss rate, which is low compared to typical M supergiants, is nonetheless at least a factor of 10^6 times larger than the mean solar mass-loss rate of $2 \times 10^{-14} M_\odot$ yr^{-1}. Lamers & Cassinelli (1999) have reviewed the observational evidence and theoretical models proposed to explain the mass loss from luminous hot and cool stars.

Stellar mass loss plays important roles in astrophysics including: (i) chemically enriching the interstellar medium out of which the next generation of stars emerges, (ii) exposing to view the chemically processed material in the inner layers of rapidly mass-losing stars (e.g., Wolf-Rayet stars), and (iii) changing the evolution of stars that lose a large portion of their mass either when young or in their post-main sequence phases, especially in the asymptotic giant branch phase of their evolution.

Biermann (1951) first showed that the Sun loses mass based on observations that comet plasma tails always face away from the Sun. Parker (1958, 1960) predicted that the Sun should have a high-velocity outflow, which is driven by the thermal gas pressure of the hot corona. To describe this outflow he coined the term "solar wind". Subsequent measurements of solar wind properties *in situ* using instruments on many satellites, most recently by *Ulysses* in its journey over the solar poles, demonstrated that Parker was basically right. However, additional acceleration and heating by magnetic waves may play important roles. For example, the coronal magnetic field provides the structure and controls the speed and density of the solar wind. Transient events (e.g., coronal mass ejections) also play a role.

The *Ulysses* space probe is providing the first detailed three-dimensional picture of the solar wind from *in situ* measurements obtained over a wide range of solar latitudes and solar-cycle phases. The solar wind has at least three identified components, which are correlated with the local magnetic field structure. The solar wind properties obtained from analysis of the *Ulysses* data are summarized in Table 1.1 (Phillips et al. 1995 with mass-flux values updated from the Ulysses/SWOOPS website[1]). For each coronal region the table lists the wind speed (v_∞) at the Earth, the ion density (n_{ion}) at the Earth, the mass-loss rate ($\dot{M}_{total}$) if the entire Sun consisted of this region, and the percent of the solar surface typically covered by this region at solar-cycle maximum and minimum. An important point is that there is less than a factor of two difference in the solar mass-loss rate between regions of open magnetic fields in coronal holes characterized by a high-speed but low-density wind and the largely (but not entirely) closed magnetic-field regions with their slow-speed but higher-density wind. The time-averaged contribution of coronal mass ejections to the solar wind is not well known but is likely of order 10%. Thus a sensible value for the mean solar wind mass-loss rate is about 2×10^{-14} solar masses per year, nearly independent of solar cycle phase. This may be compared with the mass-loss rates of 10^{-5} to 10^{-10} $\dot{M}_\odot$ yr^{-1} often cited for luminous cool stars. While the solar wind mass-flux rate is tiny by comparison with M supergiants, it can still play important roles in the evolution of solar-like dwarf stars as we shall see.

Table 1.1. Typical Parameters for the Three Components of the Solar Wind.

Region	v_∞ (km s^{-1})	n_{ion} (cm^{-3})	$\dot{M}_{total}$ ($10^{-14} M_\odot$ yr^{-1})	Coverage at solar (max, min)
Open magnetic fields	700–800	2.7	1.5	20%, few %
Closed magnetic fields	≈ 400	7.0	2.6	$\approx 80\%$, $\approx 95\%$
Coronal mass ejections	≤ 1000	0.2	?	about 10%

If a middle-aged, relatively inactive star like the Sun has a wind, surely other dwarf stars with convective photospheres and presumably complex magnetic fields should also have winds. Until recently, proof of this assertion was considered impossible as there were no tools available to perform the measurements. There are no space probes available to make *in situ* measurements of plasma properties inside the winds of nearby stars, comets in other stellar systems are not visible, and blue-shifted absorption features are not seen in dwarf star spectra, presumably because their mass-loss rates are not very different from the Sun and thus are too small to provide significant opacity in stellar emission lines. High resolution spectra of the Fe XXI 1354 Å and Fe XII 1242, 1349 Å coronal emission lines observed in the UV spectra of solar-like dwarf stars by the *Goddard High Resolution Spectrograph (GHRS)* and *Space Telescope Imaging Spectrograph (STIS)* instruments on the *Hubble Space*

Telescope (HST) (Ayres et al. 2003) show no Doppler shifts that would indicate outflowing coronal gas. High-resolution spectra of the coronal Fe XVIII 977 Å line observed with the *FUSE* satellite by Redfield et al. (2003) also show no Doppler shifts. Furthermore, *Chandra* observations of coronal emission lines have yet to provide any evidence for systematic Doppler shifts (relative to the stellar radial velocity) for solar-like dwarf stars (e.g., Ayres et al. 2001), indicating that the observed coronal gas is confined by strong closed magnetic fields rather than located in open field regions. Given these nondetections of stellar wind indicators, most astronomers would conclude that the winds of solar-like dwarf stars are unobservable.

If there were any chance of finding evidence for dwarf star winds, one might guess that the most optically-thick spectral line of the most abundant element might provide the clue. The Lyman-α line of atomic hydrogen, therefore, should be the place to look. It turns out that the Lyman-α line is the right place to look, but the path that led to the first evidence for dwarf star winds was not straightforward. In fact, nobody studied the Lyman-α line to search for stellar winds, but analyses of this line for a very different purpose revealed an important discrepancy that led to the discovery of winds by a fortuitous coincidence.

Resonance lines of atoms and ions that are abundant in the interstellar medium (ISM) contain absorption features produced by the ISM gas. Important examples include the H I Lyman-α, C II 1334 Å, O I 1302 Å, and Mg II 2796, 2803 Å lines in the ultraviolet, which could be observed with the high resolution echelle gratings on *GHRS* and can now be observed with *STIS*. Linsky et al. (1993, 1995) analyzed high resolution *GHRS* spectra to derive column densities of important interstellar ions located along the lines of sight to the nearby stars Capella and Procyon. The prime objective of this program was to infer the D/H ratio in the local ISM. When analyzing the next data set in this program, Linsky and Wood (1996) found a major discrepancy in the α Cen A and α Cen B data. For these two adjacent lines of sight, the central velocity of the H I Lyman-α absorption is -15.7 km s^{-1}, whereas the central velocities of the interstellar D I Lyman-α, Mg II 2796, 2803 Å and Fe II 2599 Å lines are -18.1 km s^{-1} within measurement errors (see Table 1.2). This 2.4 km s^{-1} difference is much larger than the expected errors in the *GHRS* wavelength scale. Also, the H I and D I Lyman-α lines are only 82 km s^{-1} apart and were observed at the same time. Thus the wavelength discrepancy cannot be instrumental in origin. Incidently, Landsman et al. (1984) also noticed a velocity discrepancy between the H I and D I Lyman-α lines in their *Copernicus* observations of α Cen A, but they did not have an explanation for this discrepancy.

We were able to fit the observed H I Lyman-α absorption feature with two components: one centered at the velocity of the other ISM lines with thermal and turbulent broadening consistent with them and the other componant redshifted by about 4 km s^{-1} from the ISM velocity. The second component has a column density a factor of 1,000 smaller than the ISM component and a broad profile indicating a temperature of about 27,000 K. We initially had no idea as to the origin of this second component, but its existence is nearly certain.

Table 1.2. Interstellar Parameters for α Cen Lines of Sight.

Star	Ion	Line	Velocity (km s^{-1})	log N_{HI}	τ
α Cen A	Fe II	2599.396	-18.2 ± 0.1	12.441 ± 0.004	1.64
α Cen A	Fe II	2599.396	-17.7 ± 0.1	12.445 ± 0.022	1.39
α Cen A	Mg II	2795.528	-18.1 ± 0.1	12.698 ± 0.017	5.50
α Cen A	Mg II	2802.705	-17.8 ± 0.1	12.691 ± 0.003	2.76
α Cen A	Mg II	2802.705	-17.8 ± 0.1	12.776 ± 0.010	3.35
α Cen B	Mg II	2795.528	-18.2 ± 0.1	12.640 ± 0.003	4.48
α Cen B	Mg II	2802.705	-18.0 ± 0.1	12.802 ± 0.003	3.61
α Cen A	D I	1215.339	-18.2 ± 0.1	12.779 ± 0.015	0.72
α Cen B	D I	1215.339	-18.2 ± 0.1	12.775 ± 0.009	0.65
α Cen A	H I	1215.670	-15.9 ± 0.1	18.031 ± 0.003	68700
α Cen B	H I	1215.670	-15.6 ± 0.1	18.027 ± 0.002	68100

Serendipity provided the missing explanation for the origin of the second component in the Lyman-α line. On July 11, 1995, a number of interesting papers were presented at Symposium 404 "The Local Interstellar Cloud and the Boundary of the Heliosphere", which was a part of the IAGA General Assembly in Boulder Colorado. In oral presentations at this meeting, H.L. Pauls, G.P. Zank, and V.B. Baranov presented theoretical models for the interaction of inflowing ISM gas with the outflowing solar wind, including charge exchange reactions between protons and hydrogen atoms. In these models the interaction of the solar wind with the ISM gas produces enhanced hydrogen densities near and somewhat beyond the heliopause, producing the "hydrogen wall". The neutral hydrogen in the wall is decelerated by a few km s^{-1} relative to the undisturbed ISM flow, is heated above the ISM temperature, and has a column density of a few times 10^{14} cm^{-2}. These are the same parameters that Linsky and Wood (1996) had inferred empirically for the second component in the Lyman-α line without knowing the physical origin of the absorption. Our realization at the meeting that the hydrogen wall can simply explain the extra absorption component provided the tool that would eventually lead to the first measurements of mass-loss rates from dwarf stars and to sensible estimates of mass-loss rates for the young Sun as a function of age.

2.   Hydrogen Walls: A New Tool for Measuring Mass-Loss Rates

Figure 1.1 shows the Lyman-α emission line of α Cen B, the K2 V secondary star in the nearest stellar system to the Sun, which was observed by Linsky & Wood (1996) using the echelle-A grating of *GHRS* with a resolution of 3.6 km s^{-1}. Superimposed on the stellar emission line are two absorption features: the broad, saturated absorption feature centered near 1215.6 Å, which is produced mostly by interstellar neutral hydrogen (H I) and a narrow absorption feature centered near 1215.25 Å, which is produced by interstellar deuterium (D I). The upper solid line in the figure is our estimate of the intrinsic stellar Lyman-α emission line obtained by scaling the observed solar Lyman-α line profile and matching the far wings of the observed stellar emission line. The dashed line in Figure 1.1 is the Lyman-α emission remaining after interstellar absorption alone is considered, where the ISM H I absorption is forced to have the thermal broadening and centroid velocity consistent with that of D I. Excess absorption is clearly seen on both sides of the H I absorption feature. Linsky & Wood (1996) proposed that this excess absorption is produced by interstellar H I gas in the outer heliosphere that has been heated and Doppler shifted by its interaction with the solar wind.

This interpretation of the excess H I absorption has strong theoretical support from recent models of the heliosphere that include both the ionized plasma and neutral H atoms in a self-consistent manner. The first such self-consistent models were those of Baranov & Malama (1993, 1995). These were followed by other models such as those of Pauls et al. (1995) and Zank et al. (1996). The basic structure of the heliosphere is illustrated in Figure 1.2. The solar wind is decelerated to subsonic speeds at the termination shock, the ISM plasma is decelerated to subsonic speeds at the bow shock, and the heliopause between these two shocks separates the plasma flows of the ISM and the solar wind. The theoretical models show that charge exchange between protons and hydrogen atoms in the outer heliosphere lead to a population of hot H I gas that permiates the entire heliosphere, as shown in Figure 1.3. This hot H I is especially prominent in the high-density "hydrogen wall" region between the heliopause and the bow shock (the region in white in the left panel of Figure 1.3).

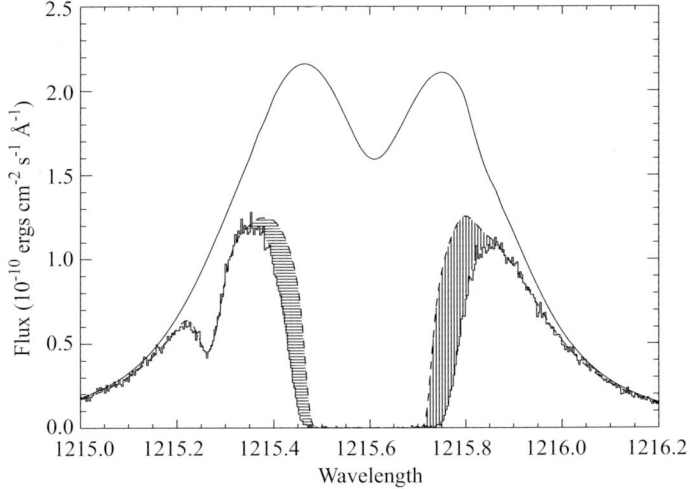

Figure 1.1. HST/GHRS Lyman-α line profile of α Cen B, showing broad H I absorption centered at 1215.6 Å and D I absorption centered at 1215.25 Å. The upper solid line is the assumed stellar emission line profile, and the dashed line is the effect of ISM absorption alone. The excess absorption is due to heliospheric H I (vertical lines) and astrospheric H I (horizontal lines).

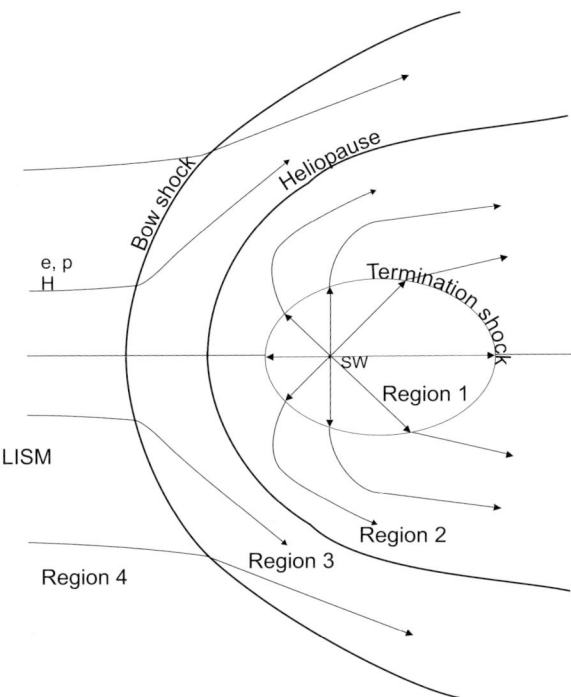

Figure 1.2 Schematic picture of the heliospheric interface, which can be divided into four regions, each with significantly different plasma properties as shown in the figure: (1) supersonic solar wind, (2) subsonic solar wind, (3) disturbed interstellar gas and plasma, and (4) undisturbed interstellar medium.

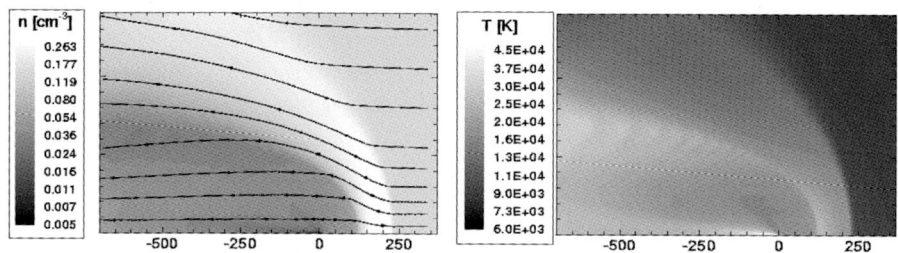

Figure 1.3. The H I density (left) and temperature (right) inside the heliosphere predicted by the Zank et al. (1996) hydrodynamic model of the ISM-solar wind interaction. The neutral hydrogen streamlines are shown in the left panel. The heliopause and bow shock are both visible at about 120 AU and 230 AU, respectively, in the upwind direction (i.e. on the right).

Gayley et al. (1997) used the H I absorption for the α Cen line of sight predicted by the models of Zank et al. (1996) to demonstrate that heliospheric H I absorption could successfully account for the excess H I absorption on the red side of the Lyman-α line (vertical lines in Figure 1.1) but not the excess absorption on the blue side (horizontal lines in Figure 1.1). Gayley et al. proposed that the blue side excess is due to "astrospheric" material surrounding α Cen, analogous to the heliospheric H I detected on the red side of the line. The various theoretical models predict that the heliospheric absorption should be redshifted relative to the ISM absorption in all directions, mostly because of the deceleration and deflection of ISM gas as it crosses the bow shock, although other factors are at work in downwind directions (e.g., Izmodenov et al. 1999; Wood et al. 2000b). Conversely, astrospheric absorption should be blueshifted with respect to the ISM absorption for most observing angles, since we are observing the neutral hydrogen decelerated relative to the star from outside the astrosphere rather than from inside. Thus excess absorption on the red side of Lyman-α is best interpreted as heliospheric, while excess absorption on the blue side is most likely astrospheric.

Heliospheric absorption has now been reported for four lines of sight: α Cen (Linsky & Wood 1996), Sirius (Izmodenov et al. 1999), 36 Oph (Wood et al. 2000a), and Proxima Cen (Wood et al. 2001). Three other lines of sight (31 Com, β Cas, and ϵ Eri) observed by HST sample different orientation angles through the heliosphere and provide upper limits for the amount of heliospheric absorption that could be present in those directions (Wood et al. 2000b; Izmodenov et al. 2002).

We show in Figure 1.4 the Lyman-α absorption profiles observed toward six of the seven stars mentioned above. We focus on the red sides of

Table 1.3.   Published Astrospheric Detections and Mass-Loss Measurements.

Star	Spectral Type	d (pc)	V_{ISM} (km s^{-1})	$\dot{M}$ ($\dot{M}_\odot$)	Sources
α Cen	G2 V+K0 V	1.3	25	2	a,b,c
Prox Cen	M5.5 V	1.3	25	< 0.2	c
ϵ Eri	K1 V	3.2	27	30	d,j
61 Cyg A	K5 V	3.5	86	0.5	e,j
ϵ Ind	K5 V	3.6	68	0.5	f,g,h
36 Oph	K1 V+K1 V	5.5	40	15	i,j
ξ Boo	G8 V+K4 V	6.7	32	1	k
λ And[1]	G8 IV-III+	25.8	53	5	f,g,h

[1] Uncertain detection.

Sources: (a) Linsky & Wood 1996. (b) Gayley et. al. 1997. (c) Wood et al. 2001. (d) Dring et al. 1997. (e) Wood & Linsky 1998. (f) Wood et al. 1996. (g) Müller et. al. 2001a. (h) Müller et al. 2001b. (i) Wood et al. 2000a. (j) Wood et al. 2002. (k) Preliminary value.

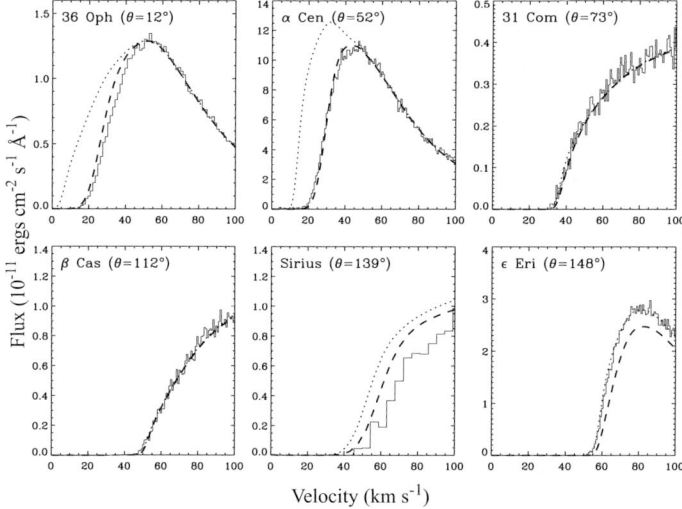

Figure 1.4.   Comparison of the heliospheric H I absorption predicted by a four-fluid heliospheric model (dashed lines) and the observations, where the model heliospheric absorption is shown after having been added to the ISM absorption (dotted lines). Reasonably good agreement is observed, although there is a slight underprediction of absorption towards 36 Oph and Sirius, and a slight overprediction towards ϵ Eri. From Wood et al. (2000b).

the profiles where the heliospheric absorption should be located. The θ angles given in the figure are the angles between the line of sight and the upwind direction of the interstellar flow. The six lines of sight sample a diverse range of orientation angles, from the nearly upwind direction toward 36 Oph ($\theta = 12°$) to the nearly downwind line of sight toward ϵ Eri ($\theta = 148°$). The dotted lines are the predicted Lyman-α profiles with only ISM absorption, for which the velocity and broadening parameters were obtained by fitting the D I absorption. Only 36 Oph, α Cen, and Sirius show excesses that reveal the presence of heliospheric absorption. A successful heliospheric model should reproduce accurately the amount of heliospheric absorption detected towards these three stars, while at the same time predicting no detectable absorption towards the other three stars. However, we have only begun to explore the potential that the HST data provide for testing the accuracy of heliospheric models, and for constraining the input parameters for these models that are poorly known observationally.

The predicted heliospheric H I absorption is very sensitive to the neutral hydrogen velocity distributions within the heliosphere. Unfortunately, the charge exchange processes in the outer heliosphere result in the hydrogen atoms being far from thermal equilibrium, and the velocity distributions are, therefore, far from Maxwellian (cf. Izmodenov et al. 2001). Ideally, a full kinetic treatment is required to accurately reproduce the distributions in such a situation. The models of Baranov & Malama (1993, 1995) and Müller et al. (2000) are examples of models that use a kinetic code to determine the neutral H properties. Wood et al. (2000b) used models computed with the Müller et al. (2000) code to predict absorption for comparison with the data in Figure 1.4. Despite assuming a range of values for the properties of the ISM surrounding the Sun, they were unable to find a model that is consistent with all of these lines of sight, mainly because the models predict too much absorption, especially in the downwind directions. Izmodenov et al. (2002) had somewhat better success with the kinetic models of Baranov & Malama (1993, 1995), but like Wood et al. (2000b) they also had problems with overpredicting the absorption in downwind directions.

Wood et al. (2000b) were able to find a so-called four-fluid model of the type described by Zank et al. (1996) that fits the heliospheric absorption data reasonably well. The model assumes the following ISM input parameters: $n(H\ I) = 0.14$ cm^{-3}, $n(H^+) = 0.1$ cm^{-3}, and $T = 8000$ K. The predicted absorption from this model is also shown in Figure 1.4. Despite a slight overprediction of absorption towards ϵ Eri and slight underpredictions for 36 Oph and Sirius, Wood et al. (2000b) argue that this four-fluid model fits the data acceptably well. The inclusion of magnetic fields in the solar wind and in the LISM can alter the heliosphere's shape (e.g., Linde et al. 1998) and thus the hydrogen wall absorption, perhaps leading to an explanation of the remaining discrepancies,

The four-fluid code of Zank et al. (1996) assumes that the hydrogen atoms can be treated as the sum of three distinct hydrodynamic fluids. (The fourth fluid is for the protons.) One would not expect that this approximation would lead to neutral H velocity distributions as accurate as those computed with codes that provide a full kinetic treatment for the neutrals, which make no assumptions about the shapes of H velocity distributions. Thus, the success of the four-fluid code in Figure 1.3 and the apparent failure of the kinetic codes are somewhat of a surprise. It is important to understand why the current kinetic codes have not been able to match successfully the observed heliospheric H I absorption, since kinetic codes should, in principle, produce more accurate velocity distributions. Further development of kinetic codes is required to determine whether or not more sophisticated kinetic codes can improve agreement with the data and lead to better estimates of ISM properties around the Sun.

Previous studies that used 2D models to predict heliospheric absorption assumed that the heliosphere is axisymmetric and the line of sight absorption is specified uniquely by the angle between the observed star and the upwind direction (θ in Figure 1.4). This approach is simplistic, because there are many effects that can change this conclusion, including latitudinal solar wind variations or a skewed orientation of the ISM magnetic field with respect to the ISM flow. The unstable jet sheet predicted by the models of Opher et al. (2003) is another example of a 3D effect that new generations of heliospheric models should include. The heliospheric absorption might provide important diagnostics for these effects. For example, if the magnetic field of the ISM is not perpendicular or parallel to the ISM flow, then the effective nose of the heliosphere could be shifted from the upwind direction (e.g., Ratkiewicz et al. 1998). It is possible that this shift would lead to changes in the distribution of heliospheric hydrogen that would be detectable in the Lyman-α absorption observations. However, in order for the HST data to be useful for studying 3D effects, many more lines of sight must be considered than the six in Figure 1.4, and more detections of heliospheric absorption must be found.

In our search of the HST archives conducted a little over a year ago, we found more than 40 lines of sight to stars within 100 pc of the Sun for which the observed Lyman-α line profiles had not yet been analyzed. (More distant lines of sight likely will have too much ISM absorption for heliospheric or astrospheric absorption to be detectable.) We have just begun to plow through the archival data, but we have already found

Map of HST Lines of Sight (Ecliptic Coordinates)

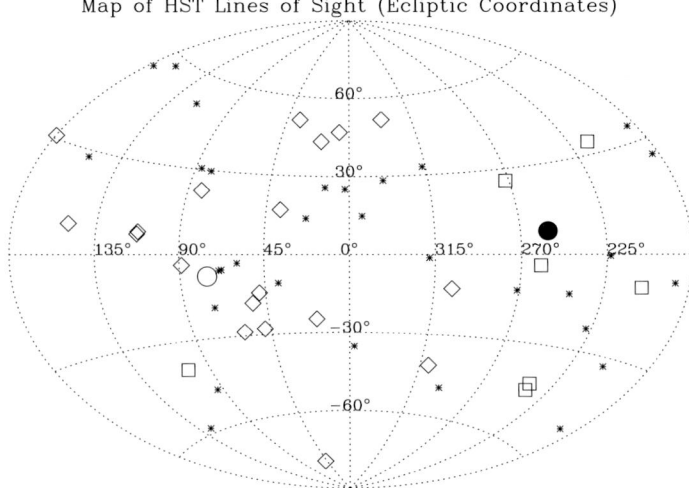

Figure 1.5. Sky map of HST-observed lines of sight in ecliptic coordinates, which have either yielded detected heliospheric Lyman-α absorption (boxes), nondetections of heliospheric absorption (diamonds), or have not yet been analyzed (asterisks). The filled and open circles are the upwind and downwind directions of the ISM flow as seen by the Sun.

three new heliospheric absorption detections and four new astrospheric detections. We show in Figure 1.5 a sky map of interesting HST-observed lines of sight in ecliptic coordinates. Boxes show lines of sight that have yielded detectable heliospheric absorption, while diamonds show nondetections. Six of the seven detections are near the upwind direction (identified by the filled circle). The heliospheric absorption is more easily detected in the upwind direction due to the substantial deceleration of hydrogen wall material, which redshifts the heliospheric absorption well away from the ISM absorption. The asterisks in the figure are lines of sight that will be analyzed in the near future.

2.1   Stellar Astrospheres

Not only has heliospheric H I been detected in HST data, but analogous astrospheric H I surrounding the stars has also been detected for some stars, as shown in Figure 1.1. The astrospheric absorption seen by HST represents the first detection, albeit indirect, of winds around other solar-like stars. Furthermore, the amount of absorption is a diagnostic for the stellar mass-loss rate, because the larger the mass-loss rate, the larger the astrosphere will be and the more Lyman-α absorption it will produce. Thus, mass-loss rates can be estimated from the astrospheric absorption, but as we will show below, the measurement process requires the computation of theoretical models analogous to the heliospheric models. Measurements of stellar mass-loss rates for stars of different ages and activity levels provides us with our very first opportunity to establish observationally how mass-loss rates vary with time for solar-like stars, including the Sun. Thus, measuring mass-loss rates from astrospheric absorption will allow us to determine the properties of the solar wind when the Sun was young.

The use of the four-fluid code of Zank et al. (1996) has already allowed significant progress in the estimation of stellar mass-loss rates from model fits to stellar Lyman-α line profiles. The first analysis was that of the α Cen astrospheric absorption (Wood et al. 2001), where the successful heliospheric four-fluid model from Wood et al. (2000b), as shown in Figure 1.6, was used in fitting the observed α Cen Lyman-α profile. To estimate the stellar mass-loss rate, one must first estimate the ISM wind vector seen by the star, by using the known stellar motion vector and the LISM flow vector. This vector is assumed to be the same as for the Local Interstellar Cloud (e.g., Lallement et al. 1995). These computations suggest that α Cen sees an interstellar wind speed of $V_{ISM} = 25$ km s^{-1}, almost identical to the 26 km s^{-1} velocity seen by the Sun (Witte et al. 1993). Our line of sight to α Cen is about $\theta = 79°$ from the upwind direction as seen by the star, so we are observing a sidewind portion of its astrosphere. In order to extrapolate from the successful heliospheric model mentioned above to an astrospheric model, we simply recompute the model assuming the stellar V_{ISM} value instead of the solar value while using the same values of the LISM density and hydrogen ionization. We then compute the astrospheric Lyman-α absorption for the line of sight described by the stellar value of θ. In order to experiment with different stellar mass-loss rates, we change the assumed stellar wind density.

The comparison of the Lyman-α profiles for Proxima Cen and α Cen B (see Figure 1.6) by Wood et al. (2001) provides an excellent example of heliospheric and astrospheric absorption. Since the two stars are only 2.2° apart in the sky and 1.3 pc distant from the Sun, the interstellar absorption (predicted from the deuterium and metal lines) and the heliospheric hydrogen wall absorption (on the red side of the interstellar absorption) should be the same. Indeed, the observations verify that this is the case.

On the other hand, the blue side of the Lyman-α absorption in the spectra of the two stars is different. Since this part of the absorption is due only to the astrospheric hydrogen wall absorption, the stellar winds

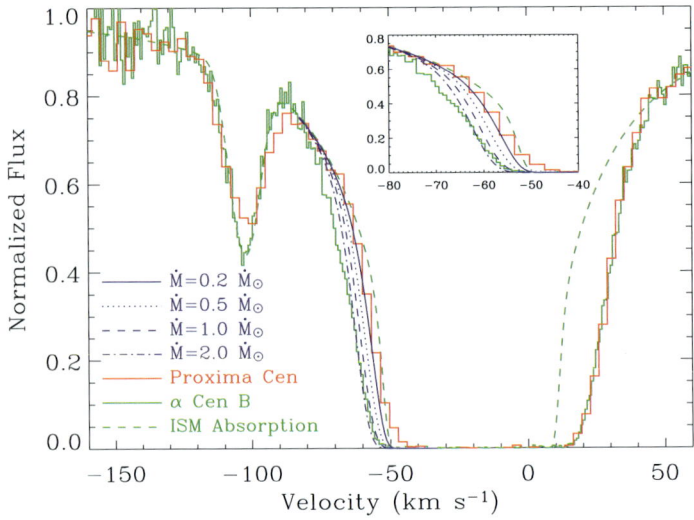

Figure 1.6. The α Cen B spectrum (thin solid line) and inferred ISM absorption (dotted line) from Figure 1.1. The dashed lines show the blue-side excess Lyman-α absorption predicted by four models of the α Cen astrosphere, assuming four different mass-loss rates. The 2.0 $\dot{M}_\odot$ model fits the α Cen spectrum reasonably well. The 0.2 $\dot{M}_\odot$ model is an upper limit for Proxima Cen. The insert shows an expanded view of the region of astrospheric absorption. From Wood et al. (2001).

of the two stars must be different. Comparison of such observations with theoretical models leads to estimates of the stellar mass-loss rate.

Figure 1.6 shows the α Cen B data compared with the astrospheric absorption predicted by four different models, with assumed mass-loss rates in the range $\dot{M} = 0.5$ to 2.0 $\dot{M}_\odot$. Since the $\dot{M} = 2\dot{M}_\odot$ model appears to fit the data best, we conclude that α Cen has about twice the mass-loss rate as the Sun. Since the α Cen binary separation is much smaller than the distance from the center of mass to the termination shock, both stars share the same astrosphere and the measured mass-loss rate represents the combined mass-loss rates of both α Cen A and α Cen B. Proxima Cen is separated from α Cen AB by at least 10^4 AU, so its astrosphere lies well outside of the astrosphere of α Cen AB and the two astrospheres are unlikely to interact. Note that the uncertainty in the inferred mass-loss rate due to possible errors in the assumed model parameters is minimized by extrapolating the astrospheric models from heliospheric models that fit the heliospheric Lyman-α absorption. In effect, the heliospheric absorption calibrates the models before they are applied to modeling astrospheres, and the mass-loss measurement technique is partially semi-empirical rather than purely theoretical. Fig-

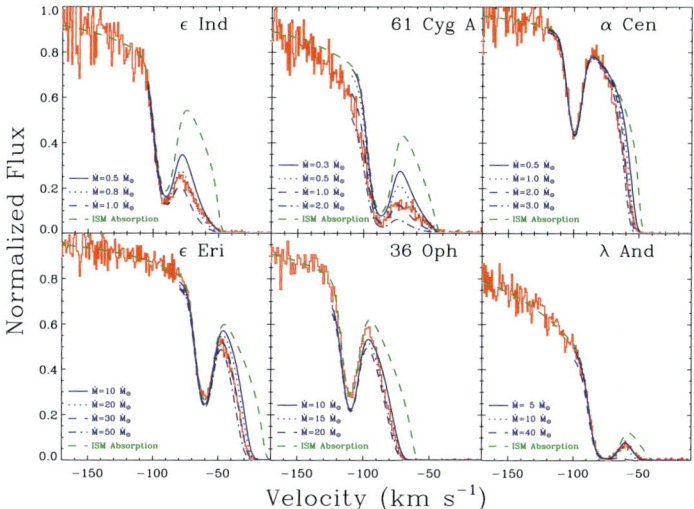

Figure 1.7. Closeups of the blue side of the H I Lyman-α absorption lines for all stars with detected astrospheric absorption, plotted on a heliocentric velocity scale. Narrow D I ISM absorption is visible in all the spectra just blueward of the saturated H I absorption. Green dashed lines indicate the interstellar absorption alone, and blue lines show the additional astrospheric absorption predicted by hydrodynamic models of astrospheres assuming various mass-loss rates.

ure 1.7 shows examples of our technique for estimating mass-loss rates for other stars by fitting the blue edge of the Lyman-α line absorption.

Table 1.3 lists the stars with published detections of astrospheric absorption. Mass-loss rates have been estimated for these stars using the method just described. The references in the table are to the papers that first reported the astrospheric detections and first presented mass-loss rate estimates. In Figure 1.8, we plot the mass-loss rates divided by the stellar surface area vs. the corresponding X-ray surface flux F_x (from Wood et al. 2002). For the solar-like G and K dwarfs, the data suggest that more active stars have higher mass-loss rates. The GK dwarf stars determine the mass-loss/activity relation, but the M dwarf Proxima Cen and the RS CVn system λ And (G8 IV-III+M V) are inconsistent with this relation. We had previously assumed that this was due to M dwarfs and RS CVn binaries having very different coronal structures than the GK dwarfs (Wood et al. 2002), but we have recently studied the astrosphere of the ξ Boo binary (G8 V + K4 V) by modelling the Lyman-α line in the *HST* archives. The preliminary mass-loss rate for ξ Boo (see Table 1.3) is consistent with the low mass-loss rates previously found for Proxima Cen and λ And. This new result calls into question our initial interpretation.

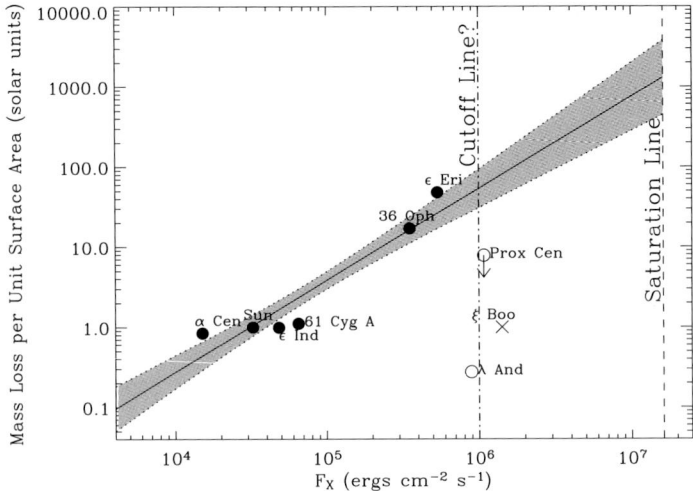

Figure 1.8.   Measured mass-loss rates (per unit surface area) plotted versus X-ray surface flux (F_x). A power law has been fitted to the solar-like G and K dwarfs (filled circles) that excludes Prox Cen (M5.5 Ve) and λ And (G8 IV-III+M V). The preliminary ξ Boo (G8 V + K4 V) data point is inconsistent with the mass-loss/activity relation but is roughly consistent with Prox Cen and λ And, which suggests that the mass-loss/activity relation defined by the lower activity GK dwarfs may have a high activity cutoff (dot-dashed line). The saturation line represents the maximum F_x value observed for very active stars.

A common feature of ξ Boo, Prox Cen, and λ And is that they are all very active stars with X-ray surface fluxes near 10^6 ergs cm^{-2} s^{-1}, about a factor of 30 larger than the Sun. These three active stars all lie well below the GK star mass-loss/activity relation in Figure 1.8. Whereas we had previously assumed that Prox Cen and λ And were different types of stars than the GK dwarfs, both stars in the ξ Boo system are normal G or K dwarfs. Thus the explanation for why ξ Boo, Prox Cen, and λ And lie below the previously determined relation for GK dwarfs could be their high activity. We suggest that there may be a high-activity cutoff to the mass-loss/activity relation (see Figure 1.8). Clearly more stars must be studied, especially the more active stars, to determine whether or not the apparent cutoff line near X-ray surface flux 10^6 ergs cm^{-2} s^{-1} is real and what the mass-loss/activity relation might be for even more active stars. If the cutoff is real, then it may indicate a difference in the coronal magnetic field configuration of highly active stars compared to less active stars. We are presently searching the *HST* archive for suitable stars to analyze but new observations are clearly needed to better determine the mass-loss/activity relation for stars with F_x both below and above the apparent cutoff near 10^6 ergs cm^{-2} s^{-1}.

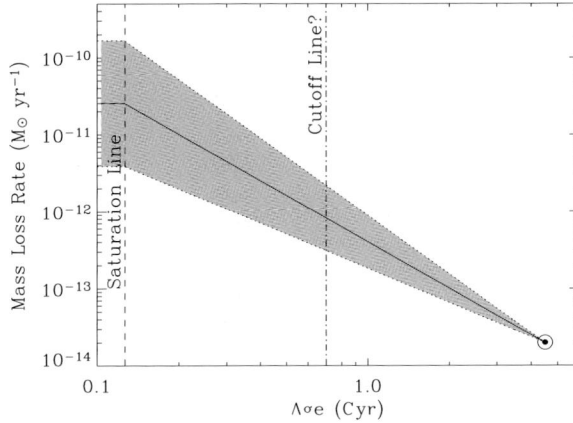

Figure 1.9 The mass-loss history of the Sun suggested by the power law relation (Equation 1.1) and the age/activity relation from Ayres (1997). If the high activity cutoff line is real, then the relation only applies for $t > 0.7$ Gyr.

Schrijver and Title (2001) have computed surface magnetic field structures for solar-like stars that assume the same magnetic flux injection and surface distribution properties as the present day Sun but higher rates for the emergence of magnetic fields that would correspond to a faster rotating Sun with an enhanced magnetic dynamo. They find that for a solar-like star with a rotation period of 6 days (rotating about 4 times faster than the Sun), the total unsigned magnetic flux on the stellar surface is 10 times that of the Sun at solar maximum. This star develops a large magnetic polar spot with an opening angle of 25°, and the astrosphere has a predominately magnetic dipole geometry. Active stars of all types, including RS CVn systems (like λ And), young GK stars, M dwarfs, and T Tauri stars, show evidence for polar spots (cf. Schrijver et al. 2003). Since the X-ray surface flux is roughly proportional to the magnetic flux density (cf. Schrijver et al. 2003) and the X-ray luminosity of the active Sun in the soft X-ray ROSAT band is about 4.7×10^{27} erg cm^{-2} s^{-1} (Peres et al. 2000), the model active star with a 6-day rotational period would have a surface X-ray flux $F_x \approx 7.6 \times 10^5$ erg cm^{-2} s^{-1}. Thus the proposed cutoff line in Figure 1.8 occurs at about the same X-ray surface flux as the appearance (both in observations and in the models) of large polar spots and strong astrospheric magnetic fields with a dipole geometry that could suppress mass loss over most of the stellar surface.

2.2 The Mass-Loss History of the Sun

The empirical relation between the mass-loss rate and the X-ray surface flux (F_x) shown in Figure 1.8 for $F_x \leq 10^6$ ergs cm^{-2} s^{-1} is

$$\dot{M} \propto F_x^{1.15 \pm 0.20}. \tag{1.1}$$

We can now use the known relation between stellar age and activity (e.g., Ayres 1997) to infer the dependence of mass-loss rate on stellar age. As stars age, their rotation rates (V_{rot}) slow due to magnetic braking and their X-ray fluxes decrease. There are many empirical relations decribing this evolution of stellar activity. We adopt the empirical relations of Ayres (1997),

$$V_{rot} \propto t^{-0.6\pm0.1},$$ (1.2)

and

$$F_x \propto V_{rot}^{2.9\pm0.3}.$$ (1.3)

Combining these three relations (see Wood et al. 2002), we obtain

$$\dot{M} \propto t^{-2.00\pm0.52}.$$ (1.4)

This relation predicts that when the young Sun ($t \approx 10^8$ years) had saturated X-ray emission ($L_x/L_{bol} = 10^{-4}$), its mass-loss rate was 200–10,000 times larger than at present. The cumulative mass loss from age 10^8 years to the present is $\leq 0.03 M_\odot$ (see Figure 1.10).

The mass-loss history of the Sun (Figure 1.9) predicts that the wind of the young Sun was roughly 1000 times stronger than at present. In addition to its obvious importance for solar and stellar astronomy, this result may have very important ramifications for planetary science as well, since solar wind erosion may have played an important role in the evolution of planetary atmospheres, especially that of Mars (e.g., Kass & Yung 1995). However, the existence of a high-activity cutoff in Figure 1.8 would mean that the solar wind evolution law shown in Figure 1.9 would only apply for $t > 0.7$ Gyr, and that at earlier times the solar wind may have actually been very weak.

Given the importance of these results to both solar and stellar astronomy and to planetary science, we believe it is crucial to test and refine these results with new astrospheric detections. The mass-loss rates measured from the detected astrospheric Lyman-α absorption should also be verified using astrospheric models other than the presently used four-fluid models, preferably kinetic models such as those of Baranov & Malama (1993, 1995). Once we have a heliospheric model that can successfully match the HST heliospheric data, then we can extrapolate from it astrospheric models assuming different mass-loss rates, compare these results with the astrospheric Lyman-α data, and determine whether or not the mass-loss rates that we derive are consistent with those listed in Table 1.3 from Wood et al. (2002).

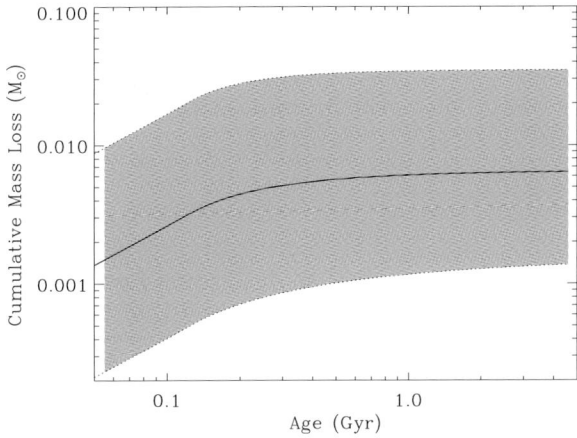

Figure 1.10 Cumulative mass loss for the Sun as a function of time, based on the mass-loss history of the solar wind in Figure 1.9.

3. Influence of Stellar Winds on Planets in the Solar System and Beyond

Equation (1.4) provides us with the ability to make quantitative estimates of the solar wind flux that planets received during the early solar system. This is important for addressing such questions as (i) why the young Earth had running water when the Sun was 25% fainter than today and (ii) whether solar wind erosion removed water from the young Martian atmosphere. Exposure to a stellar wind, together with UV and X-ray irradiation, can ionize and erode planetary atmospheres, and the higher mass-loss rates suggested for the young Sun and young stars with high X-ray fluxes would exacerbate these effects. Solar wind sputtering processes have been proposed to have important effects on the atmospheres of Venus and Titan, but Mars may be the most interesting case of solar wind erosion, since the history of its atmosphere is linked to whether or not water, and perhaps life, once existed on the surface.

Unlike Earth, the Martian atmosphere is not currently protected from the solar wind by a strong magnetosphere. There is evidence that Mars once had a magnetic field, which disappeared at least 3.9 Gyr ago (Acuña et al. 1999). At that time, the Martian atmosphere would have been directly exposed to a solar wind about 40 times stronger than the current wind. This would have had a dramatic effect on its atmosphere through ion-pickup and sputtering processes. Mars appears to have had running water on its surface in the distant past and a thicker atmosphere that could have supported a climate consistent with the existence of surface water (e.g., Jakosky & Phillips 2001). Solar wind erosion and XUV radiation are leading candidates for the cause of its present thin atmosphere and lack of surface water (e.g., Lundin 2001; Lammer et al. 2003). Planets around other stars could have similar histories.

4.   Conclusions

- Analysis of the astrospheric absorption features seen in the Lyman-α lines toward nearby stars provides the first quantitative measurements of mass-loss rates for G and K dwarf stars.

- We find that the mass-loss rates increase with activity, as measured by the X-ray surface flux, and thus with decreasing stellar age. More *HST/STIS* spectra are needed to determine whether an important change in mass-loss rates occures near $F_x \approx 10^6$ ergs cm^{-2} s^{-1}.

- The mass-loss rate for the Sun at 10^8 years is predicted to be 200–10,000 times larger than at present and likely played a major role in the evolution of planetary atmospheres.

We thank NASA and the Space Telescope Science Institute for grants to the University of Colorado and NSF for its support. We also thank Dr. Hans-Reinhard Müller and Prof. Gary Zank for their assistance in computing the astrosphere models. We also than Dr. Steven Suess and the referee for their careful reading of the manuscript and their helpful suggestions.

Notes

1. http://swoops.lanl.gov/

References

Acuña, M.J. et al. 1999, Science, 284, 790.

Ayres, T.R. 1997, JGR, 102, 1641.

Ayres, T.R., Brown, A., Osten, R.A., Huenemoerder, D.P., Drake, J.J., Brickhouse, N.S., and Linsky, J.L. 2001, ApJ, 549, 544.

Ayres, T.R., Brown, A., Harper, G.M., Osten, R.A., Linsky, J.L., Wood, B.E., and Redfield, S. 2003, ApJ, 583, 963.

Baranov, V.B. & Malama, Y.G. 1993, JGR, 98, 15157.

Baranov, V.B. & Malama, Y.G. 1995, JGR, 100, 14755.

Biermann, L. 1951, Zeitschrift für Astrophysik, 29, 274.

Deutsch, A.J. 1956, ApJ, 123, 210.

Dring, A. R., Linsky. J., Murthy, J., Henry, R. C., Moos, W., Vidal-Madjar, .A., Audouze, J., & Landsman, W. 1997, ApJ, 488, 760.

Jakosky, B.M. & Phillips, R.J., 2001, Nature, 412, 237.

Gayley, K. G., Zank, G. P., Pauls, H. L., Frisch, P. C., & Welty, D. E. 1997, ApJ, 487, 259.

Izmodenov, V. V., Gruntman, M., & Malama, Y. G. 2001, JGR, 106, 10681.

Izmodenov, V. V., Lallement, R., & Malama, Y. G. 1999, A&A, 342, 113.

Izmodenov, V. V., Wood, B. E., & Lallement, R. 2002, JGR, 107, 1308.

Kass, D. M., & Yung, Y. L. 1995, Science, 268, 697.

Lamers, H.J.G.L.M. & Cassinelli, J.P. 1999, *Introduction to Stellar Winds.* Cambridge: Cambridge University Press.

Lammer, H. et al. 2003, Icarus, 165, 9L.

Landsman, W.B., Henry, R.C., Moos, H.W., & Linsky, J.L. 1984, ApJ, 285, 801.

Lallement, R., Ferlet, R., Lagrange, A. M., Lemoine, M., & Vidal-Madjar, A. 1995, A&A, 304, 461.

Linde, T., Gombosi, T. I., Roe, P. L., Powell, K. G., & DeZeeuw, D. L. 1998, JGR, 103, 1889.

Linsky, J.L. et al. 1993, ApJ, 402, 694.

Linsky, J.L. et al. 1995, ApJ, 451, 335.

Linsky, J.L. & Wood, B.E. 1996, ApJ, 463, 254.

Lundin, R. 2001, Science, 291, 1909.

Müller, H. -R., Zank, G. P., & Lipatov, A. S. 2000, JGR, 105, 27419.

Müller,. H. -R., Zank, G. P., & Wood, B. E. 2001a, ApJ, 551, 495.

Müller,. H. -R., Zank, G. P., & Wood, B. E. 2001b, in The Outer Heliosphere: The Next Frontiers, ed. K. Scherer et al. (Amsterdam: Pergamon), 53.

Opher, M., Liewer, P. C, Gombosi, T. I., Manchester, W., DeZeeuw, D. L., Sokolov, I., & Toth, G. 2003, ApJ, 591, 161.

Parker, E.N. 1958, ApJ, 128, 664.

Parker, E.N. 1960, ApJ, 132, 821.

Peres, G., Orlando, S., Reale, F., Rosner, R., & Hudson, H. ApJ, 528, 537.

Phillips, J.L., Bame, S.J., Feldman, W.C., Goldstein, B.E., Gosling, J.T., Hammond, C.M., McComas, D.J., Neugebauer, M., Scime, E.E., & Suess, S.T. 1995, Science, 268, 1030.

Pauls, H. L., Zank, G. P., & Williams, L. L. 1995, JGR, 100, 21595.

Ratkiewicz, R., Barnes, A., Molvik, G. A., Spreiter, J. R., Stahara, S. S., Vinokur, M., & Venkateswaran, S. 1998, A&A, 335, 363.

Redfield, S., Ayres, T.R., Linsky, J.L., Ake, T.B., Dupree, A.K., Robinson, R.D., and Young, P.R. 2003, ApJ, 585, 993.

Schrijver, C.J., & Title, A.M., ApJ, 551, 1099.

Schrijver, C.J., DeRosa, M.L., & Title, A.M., ApJ, 590, 493.

Witte, M., Rosenbauer, H., Banaszkewicz, M., & Fahr, H. 1993, Adv. Space Res., 13, 121.

Wood, B.E. & Linsky, J.L. 1996, ApJ, 470, 1157.
Wood, B.E. & Linsky, J.L. 1998, ApJ, 492, 788.
Wood, B. E., Alexander, W. R., & Linsky, J. L. 1996, ApJ, 470, 1157.
Wood, B. E.., Linsky, J. L., & Zank, G. P. 2000a, ApJ, 537, 304
Wood, B. E., Muller, H. -R., & Zank, G. P. 2000b, ApJ, 542, 493.
Wood, B.E. et al. 2001, ApJ, 547, L49.
Wood, B.E. et al. 2002, ApJ, 574, 412.
Zank, G.P. et al. 1996, JGR, 101, 21639.

Chapter 2

THE HELIOSPHERIC INTERFACE: MODELS AND OBSERVATIONS

Vladislav V. Izmodenov

Moscow State University, Faculty of Mechanics and Matematics

Department of Aeromechanics and Gas Dynamics

izmod@ipmnet.ru

Abstract   The Sun is moving through a warm ($\sim$6500 K) and partly ionized local interstellar cloud (LIC) with velocity $\sim$26 km/s. The charged component of the interstellar medium interacts with the solar wind (SW), forming the heliospheric interface - the SW/LIC interaction region. Both the solar wind and interstellar gas have a multi-component nature that creates a complex behavior in the interaction region. The current state of art in the modeling of the heliospheric interface is reviewed in this paper. Modern models of the interface take into account the solar wind and interstellar plasma components (protons, electrons, pickup ions, interstellar helium ions, and solar wind alpha particles), the interstellar neutral component (H atoms), interstellar and heliospheric magnetic fields, galactic and anomalous cosmic rays, and latitudinal and solar cycle variations of the solar wind. Predictions of self-consistent, time-dependent, kinetic/gasdynamic modeling of the heliospheric interface are compared with available remote diagnostics of the heliospheric interface - backscattered solar Lyman-alpha radiation, pickup ions, the deceleration of the solar wind at large heliocentric distances measured by Voyager 2, heliospheric absorption of stellar light, anomalous cosmic rays (ACRs), and heliospheric neutral atoms (ENAs).

Keywords: Solar Wind, Local Interstellar Cloud, Interstellar H atoms, Termination Shock

1.   Introduction

The structure of the outer heliosphere and heliospheric boundary is determined by the interaction of the solar wind with the interstellar neighborhood of the Sun - the Local Interstellar Cloud (LIC). There is no

G. Poletto and S.T. Suess (eds.), The Sun and The Heliosphere as an Integrated System, 23–63.

doubt that the LIC is partly ionized and that the charged component of the LIC interacts with solar wind plasma. The interaction region, which is often called the *heliospheric interface*, is formed in this interaction (Figure 1). The heliospheric interface is a complex structure, where the solar wind and interstellar plasma, interplanetary and interstellar magnetic fields, interstellar atoms of hydrogen, galactic and anomalous cosmic rays (GCRs and ACRs) and pickup ions play roles.

Although a space mission into the Local Interstellar Cloud is becoming now more realizable and Voyager 1 is approaching the inner boundary of the heliospheric interface – the termination shock, there are as yet no direct observations inside the heliospheric interface. Therefore, at the present time the heliospheric interface structure and local interstellar parameters can only be explored with remote and indirect measurements. Currently, the major sources of information on the heliospheric interface structure and position of the termination shock are following: 1) direct measurements of interstellar pickup ions, which are interstellar atoms ionized by charge exchange and photoionization and measured by Ulysses and ACE spacecraft; 2) anomalous cosmic rays, which are those pickup ions that are accelerated to high energies and measured by Voyagers, Pioneers, Ulysses, ACE, SAMPEX and Wind; 3) backscattered (by interstellar atoms of hydrogen) solar Lyman-α radiation measured at 1 AU by SOHO and Hubble Space Telescope (HST) and in the outer heliosphere by Voyager and Pioneer spacecraft; 4) direct measurements of the solar wind at large heliocentric distances by Voyager 2 spacecraft. In addition, kHz emission detected on board of Voyagers can provide other constraints (Cairns, this volume). Recently, it was shown that study of Ly-α absorptions toward nearby stars can serve as remote diagnostics of similar interfaces and, in particular, the hydrogen wall around heliopause-equivalents (see, Linsky and Wood, 1996; Wood et al., 2000; Izmodenov et al., 1999a, 2002; Linsky, this volume). First detections of heliospheric energetic atoms (ENAs) by SOHO and IMAGE proved that the detailed imaging of the heliospheric interface in ENAs will be possible in the near future(Gruntman et al., 2001; planned NASA IBEX mission: http://ibex.swri.edu/).

To reconstruct the structure of the interface and the physical processes inside the interface using remote observations at one to several astronomical units, a theoretical model should be employed. Theoretical studies of the heliospheric interface have been performed for more than four decades, following the pioneering papers by Parker (1961) and Baranov et al. (1971). However, a complete theoretical model of the heliospheric interface has not yet been constructed. The difficulty in doing this is connected with the multi-component nature of both the LIC and the

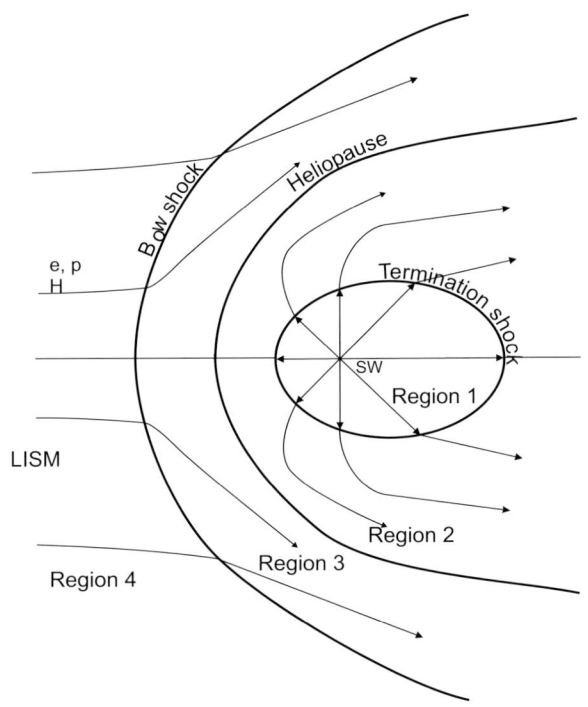

Figure 2.1 The heliospheric interface is the region of the solar wind interaction with LIC. The heliopause is a contact discontinuity, which separates the plasma of the solar wind from the interstellar plasma. The termination shock decelerates the supersonic solar wind. The bow shock may also exist in the supersonic interstellar wind. The heliospheric interface is thus divided into four regions: 1) supersonic solar wind; 2) subsonic solar wind in the region between the heliopause and termination shock; 3) disturbed interstellar plasma region (or "pile-up" region) around the heliopause; 4) undisturbed interstellar medium.

solar wind. The LIC consists of at least five components: plasma (electrons and protons), hydrogen atoms, interstellar magnetic field, galactic cosmic rays, and interstellar dust. The heliospheric plasma consists of original solar particles (protons, electrons, alpha particles, etc.), pickup ions and the anomalous cosmic ray component. Pickup ions modify the heliospheric plasma flow starting from ∼20-30 AU. ACRs may also modify the plasma flow upstream of the termination shock and in the heliosheath. Spectra of ACRs can serve as remote diagnostics of the termination shock (Stone, 2001). For a recent review on ACRs see Fichtner (2001).

Development of a theoretical model of the heliospheric interface requires choosing a specific approach for each of interstellar and solar wind component. Interstellar and solar wind protons and electrons can be described as fluids. At the same time the mean free path of interstellar H atoms is comparable with the size of the heliospheric interface. This requires kinetic description for the interstellar H atom flow in the interaction region. For the pickup ion and cosmic ray components the kinetic approach is also required.

This chapter focuses on 1) theoretical numerical models of the *global* heliospheric interface structure and 2) application of the models to interpreting different remote diagnostics of the SW/LIC interaction. Under global models I include those models that describe the entire interaction region, including the termination shock, the heliopause and possible bow shock. In this sense, this chapter should not be considered as a complete review of progress in the field. Many different approaches have been used to look into different aspects of the solar wind interaction with LIC connected with pickup ion transport and acceleration, with the termination shock structure under influence of ACRs, and pickup ions. For a more complete overview see recent reviews Zank (1999a), Fichtner (2001).

The structure of the chapter is the following: The next section briefly describes our current knowledge of the local interstellar and solar wind parameters. Section 3 discusses theoretical approaches to be used for the interstellar and solar wind components. Section 4 gives an overview of heliospheric interface models. Section 5 describes basic results of self-consistent two-component model of the heliospheric interface developed in Moscow and its recent developments. In section 6 we demonstrate possible analysis of space experiments on the basis of a theoretical model of the heliospheric interface. Section 7 gives a summary and underlines current problems in the modeling of the global heliosphere and discusses future perspectives.

2. Brief Summary of Observational Knowledge

Choice of an adequate theoretical model of the heliospheric interface depends on boundary conditions, i.e. on the undisturbed solar wind and interstellar properties.

2.1 Solar Wind Observations

At the Earth's orbit, the flux of the interstellar atoms is quite small and the solar wind can be considered undisturbed. Measurements of pickup ions and ACRs also show that these components do not have dynamical influences on the original solar wind particles at the Earth's orbit. Therefore, the Earth's orbit can be taken as inner boundary of the SW/LIC interaction problem.

Solar wind structure and behavior evolves over a solar cycle (e.g., Gazis, 1996; Neugebauer 1999; McComas et al. 2000, 2001, 2002). At solar minima, high-latitude regions on the Sun have well-developed coronal holes, which are sources of high-speed ($\sim$700 km/s), low-density solar wind. At low heliolatitudes the solar wind has low speed ($\sim$400 km/s) and high density. The dividing line between the fast and slow solar wind regimes is at about 20° heliolatitude. At solar maximum the slow, dense solar wind is present at all latitudes (McComas et al., 2002). Shortly after the solar maximum coronal holes and, therefore, the high-speed solar wind appear again. At this stage, both coronal holes the high-speed solar wind may appear even in the ecliptic, which results in increasing the average solar wind momentum flux at low latitudes shortly after solar maxima. Spacecraft near the ecliptic at 1 AU have detected variations of the solar wind momentum flux by a factor of two (Lazarus and McNutt, 1990; Gazis, 1996). Deep space probe data obtained with Pioneer and Voyager measurements in the distant solar wind also support this conclusion (Lazarus and McNutt, 1990; Gazis, 1996). A recent update of the Voyager 2 measurements of the solar wind can be found in Richardson et al. (2004). Apparently, over the past two solar cycles the momentum flux had a minimum value at solar maximum, then increased rapidly after solar maximum reaching a peak 1-2 years later. The pressures subsequently decreased until after the next solar maximum. It was unclear from the measurements in the ecliptic whether the variations of the momentum flux have global effect or limited to the ecliptic. Ulysses observations from its first full polar orbit showed that the momentum flux is diminished near the equator compared to higher latitudes. The effect is clearly evident in the period near solar minimum (May, 1995 - December, 1997) (McComas et al., 2000). Around solar maximum the three-dimensional structure of the solar wind is remarkably different from, and more complicated than, the simple, bimodal structure observed throughout much of the rest of the solar cycle. At maximum, the solar wind has the same properties at all latitudes (McComas et al., 2001, 2002).

Theoretical models predict that pickup and ACR components dynamically influence the solar wind at large heliocentric distances. Table 2.1 presents estimates of dynamic importance of the heliospheric plasma components at small and large heliocentric distances. The table shows that pickup ion thermal pressure can be up to 30-50 % of the dynamic pressure of solar wind.

2.2 Interstellar Parameters

Local interstellar temperature and velocity can be inferred from direct measurements of interstellar helium atoms by the Ulysses/GAS instrument (Witte et al., 1996; Witte, 2004). Atoms of interstellar helium penetrate the heliospheric interface undisturbed, because of the small strength of their coupling with interstellar and solar wind protons. Indeed, due to small cross sections of elastic collisions and charge exchange

Table 2.1.   Number Densities and Pressures of Solar Wind Components

| Component | 4-5 AU | | 80 AU | |
	Number Density cm^{-3}	Pressure eV/cm^{-3}	Number Density cm^{-3}	Pressure eV/cm^{-3}
Original solar wind protons	0.2-0.4	2.-4. (thermal) ~ 200 (dynamic)	$(7-14) \cdot 10^{-4}$	10^{-3} - 10^{-4} $\sim 0.5 - 1.$ (dynamic)
Pickup ions	$5.1 \cdot 10^{-4}$	0.5	$\sim 2 \cdot 10^{-4}$	~ 0.15
Anomalous cosmic rays				0.01 - 0.1

Table 2.2.   Local Interstellar Parameters

Parameter	Direct measurements/estimations
Sun/LIC relative velocity	25.3 ± 0.4 km s^{-1} (direct He atoms [1]) 25.7 km s^{-1} (Doppler-shifted absorption lines [2])
Local interstellar temperature	7000 ± 600 K (direct He atoms [1]) 6700 K (absorption lines [2])
LIC H atoms number density	0.2 ± 0.05 cm^{-3} (estimate based on pickup ion observations [3])
LIC proton number density	0.03 - 0.1 cm^{-3} (estimate based on pickup ion observations [3])
Local Interstellar magnetic field	Magnitude: 2-4 μG Direction: unknown

Pressure of low energetic part of cosmic rays$\sim$0.2 eV cm^{-3}

[1] Witte et al. (1996); [2] Lallement(1996); [3] Gloeckler (1996), Gloeckler et al. (1997)

with protons, the mean free path of these atoms is larger than the heliospheric interface. Independently, the velocity and temperature in the Local Interstellar Cloud can be deduced from analysis of absorption features in the stellar spectra (Lallement, 1996). However, this method provides mean values along lines of sight toward nearby stars in the LIC. A comparison of local interstellar temperatures and velocities derived from stellar absorption with those derived from direct measurements of interstellar helium shows quite good agreement (see Table 2.2).

Other local parameters of the interstellar medium, such as interstellar H atom and electron number densities, and strength and direction of the interstellar magnetic field, are not well known. In the models they can be considered as free parameters. However, indirect measurements of interstellar H atoms and direct measurements of their derivatives as pickup ions and ACRs provide important constraints on the local interstellar proton and atom densities and total interstellar pressure. The neutral H density in the inner heliosphere depends on filtration of the neutral H atoms in the heliospheric interface due to charge exchange. Since interstellar He is not perturbed in the interface, the local interstellar number density of H atoms can be estimated from the neutral hydrogen to the neutral helium ratio in the LIC, $R(HI/HeI)_{LIC}$: $n_{LIC}(HI) = R(HI/HeI)_{LIC}n_{LIC}(HeI)$. The neutral He number density in the heliosphere has been recently determined to be very likely around 0.015 ± 0.002 cm^{-3} (Gloeckler and Geiss, 2001). The interstellar ratio HI/HeI is likely in the range of 10-14. Therefore, expected interstellar H atom number densities are in the range of $0.13 - 0.25$ cm^{-3}. It was shown by modeling (Baranov and Malama, 1995; Izmodenov et al.,1999b) that the filtration factor, which is the ratio of neutral H density inside and outside the heliosphere, is a function of interstellar plasma number density. Therefore, the number density of interstellar protons (electrons) can be estimated from this filtration factor (Lallement, 1996). Independently, the electron number density in the LIC can be estimated from abundances ratios of ions of different ionization states (Lallement, 1996).

Note that there are other methods to estimate interstellar H atom density inside the heliosphere, based on their influence on the distant solar wind (Richardson, 2001) or from ACR spectra (Stone, 2001). Recent estimates of the location of the heliospheric termination shock using transient decreases of cosmic rays observed by Voyager 1 and 2 also provide constraints on the local interstellar parameters (Webber et al., 2001). However, simultaneous analysis of different types of observational constraints has not been done yet. Theoretical models should be employed to make such analysis. Table 2.2 presents a summary of our knowledge of local interstellar parameters. Using these parameters, we estimate local pressures of different interstellar components (Table 1.3). Although the dynamical pressure of interstellar H atoms is larger than all other pressures, all pressures have the same order of magnitude. This means that theoretical models should not neglect any of these interstellar components. A portion of the H atoms, ACRs and GCRs penetrates into the heliosphere, which makes their real dynamical influence on the heliospheric plasma interface difficult to estimate.

3.   Overview of Theoretical Approaches

In this section we consider theoretical approaches for components involved in the dynamical processes in the heliospheric interface.

Generally, any gas can be described on a kinetic or a hydrodynamic level. In the kinetic approach, macroscopic parameters of a gas of s-particles (or, briefly, s-gas) can be expressed through integrals of the ve-

Table 2.3. Local Pressures of Interstellar Components

Component	Pressure estimation, dyn cm^{-2}
Interstellar plasma component	
Thermal pressure	$(0.6 - 2.0) \cdot 10^{-13}$
Dynamic pressure	$(1.5 - 6) \cdot 10^{-13}$
H atoms	
Thermal pressure	$(0.6 - 2.0) \cdot 10^{-13}$
Dynamic pressure	$(4.0 - 9.0) \cdot 10^{-13}$
Interstellar magnetic field	$(1.0 - 5.0) \cdot 10^{-13}$
Low energy part of GCR	$(1.0 - 5.0) \cdot 10^{-13}$

locity distribution function $f_s(\vec{r}, \vec{w}, t)$: $n_s = \int f_s d\vec{w}$, $\vec{V}_s = (\int \vec{w} f_s d\vec{w})/n_s$, $P_{s,ij} = m_s \int (w_i - V_{s,i})(w_j - V_{s,j}) f_s d\vec{w}$, $\vec{q}_s = 0.5 m_s \int (\vec{w} - \vec{V}_s)^2 (\vec{w} - \vec{V}_s) f_s d\vec{w}$, where n_s is the number density of s-gas, $\vec{V}_s$ is the bulk velocity of s-gas, $P_{s,ij}$ are components of the stress tensor $\widehat{P}_s$, $\vec{q}_s$ is the thermal flux vector, and m_s is the mass of individual s-particle. In the hydrodynamic approach, some assumptions should be made to specify the stress tensor, $\widehat{P}_s$, and the thermal flux vector, $\vec{q}_s$ to make hydrodynamic system closed. For example, these values can be calculated by the Chapman-Enskog method, assuming $Kn = l/L << 1$, where l and L are the mean free path of the particles and characteristic size of the problem, respectively. The zero approximation of the Chapman-Enskog method gives local Maxwellian distribution, and the gas can be considered as an ideal gas, where the stress tensor reduces to scalar pressure P and $\vec{q} = 0$.

3.1 H Atoms

Interstellar atoms of hydrogen form the most abundant component in the circumsolar local interstellar medium (see, Table 2.2). These atoms penetrate deep into the heliosphere and interact with interstellar and solar wind protons. The cross sections of elastic H-H, H-p collisions are negligible as compared with the charge exchange cross section (Izmodenov et al., 2000). Charge exchange with solar wind/interstellar protons determines the properties of the H atom gas in the interface. Atoms, newly created by charge exchange, have the local properties of protons. Since plasma properties are different in the four regions of the heliospheric interface shown in Figure 1, the H atoms can be separated into four populations, each having significantly different properties. The strength of H atom-proton coupling can be estimated through the calculation of the mean free path of H atoms in the plasma. Generally, the mean free path (with respect to the momentum transfer) of an s-particle in a t-gas can be calculated by the formula: $l = m_s w_s^2 / (\delta M_{st}/\delta t)$. Here, w_s is the individual velocity of the s-particle, and $\delta M_{st}/\delta t$ is the individual s-particle momentum transfer rate in the t-gas.

Table 2.4 shows the mean free paths of H atoms with respect to charge exchange with protons. The mean free paths are calculated for typical atoms of different populations in different regions of the interface in the upwind direction. For every population of H atoms, there is at least one region in the interface where the Knudsen number $Kn \approx 0.5 - 1.0$. Therefore, the kinetic Boltzmann approach must be used to describe interstellar atoms in the heliospheric interface correctly.

Table 2.4. Mean free paths of H-atoms in the heliospheric interface with respect to charge exchange with protons, in AU.

Population	At TS	At HP	Between HP and BS	LISM
4 (primary interstellar)	150	100	110	870
3 (secondary interstellar)	66	40	58	190
2 (atoms originating in the heliosheath)	830	200	110	200
1 (neutralized solar wind)	16000	510	240	490

The velocity distribution of H atoms $f_H(\vec{r}, \vec{w}_H, t)$ may be calculated from the linear kinetic equation:

$$\frac{\partial f_H}{\partial t} + \vec{w}_H \cdot \frac{\partial f_H}{\partial \vec{r}} + \frac{\vec{F}}{m_H} \cdot \frac{\partial f_H}{\partial \vec{w}_H} = -f_H \int |\vec{w}_H - \vec{w}_p| \sigma_{ex}^{HP} f_p(\vec{r}, \vec{w}_p) d\vec{w}_p \quad (2.1)$$

$$+ f_p(\vec{r}, \vec{w}_H) \int |\vec{w}_H^* - \vec{w}_H| \sigma_{ex}^{HP} f_H(\vec{r}, \vec{w}_H^*) d\vec{w}_H^* - (\nu_{ph} + \nu_{impact}) f_H(\vec{r}, \vec{w}_H).$$

Here $f_H(\vec{r}, \vec{w}_H)$ is the distribution function of H atoms; $f_p(\vec{r}, \vec{w}_p)$ is the local distribution function of protons; $\vec{w}_p$ and $\vec{w}_H$ are the individual proton and H atom velocities, respectively; σ_{ex}^{HP} is the charge exchange cross section of an H atom with a proton; ν_{ph} is the photoionization rate; m_H is the atomic mass; ν_{impact} is the electron impact ionization rate; and $\vec{F}$ is the sum of the solar gravitational force and the solar radiation pressure force. The plasma and neutral components interact mainly by charge exchange. However, photoionization, solar gravitation, and radiation pressure, which are taken into account in equation (2.1), are important at small heliocentric distances. Electron impact ionization may be important in the heliosheath (region 2). The interaction of the plasma and H atom components leads to the mutual exchanges of mass, momentum and energy. These exchanges should be taken into account in the plasma equations through source terms, which are integrals of $f_H(\vec{r}, \vec{w}, t)$ specified later in equations (1.5)-(1.7).

3.2 Solar Wind and Interstellar Electron and Proton Components

Basic assumptions necessary to employ a hydrodynamic approach for space plasmas were reviewed by Baranov (2000). In particular, it was concluded there that interstellar and solar wind plasmas can be treated hydrodynamically. Indeed, the mean free path of charged particles in the local interstellar plasma is less than 1 AU, which is much smaller than the size of the heliospheric interface itself. Therefore, the local interstellar plasma is a collisional plasma, and a hydrodynamic approach can be used to describe it. Solar wind plasma is collisionless, because the mean free path of the solar wind particles with respect of coulomb collisions is much larger than the size of the heliopause. Therefore, the heliospheric termination shock (TS) is a collisionless shock. A hydrodynamic approach can be justified for collisionless plasmas when scattering of charged particles on plasma fluctuations is efficient ("collective plasma processes"). In this case, the mean free path l with respect to collisions is replaced by l_{coll}, the mean free path of collective processes, which is assumed to be less than the characteristic length of the problem L: $l_{coll} << L$. However, the integral of "collective collisions" is too complicated to be used to calculate the transport coefficient for collisionless plasmas.

A one-fluid description of heliospheric and interstellar plasmas is commonly used in the global models of the heliospheric interface. Governing equations of the one-fluid approach are mass, momentum and energy balance equations:

$$\frac{\partial \rho}{\partial t} + \nabla \cdot (\rho \vec{V}) = q_1, \tag{2.2}$$

$$\frac{\partial (\rho \vec{V})}{\partial t} + \nabla P + \nabla \cdot (\rho \vec{V} \otimes \vec{V}) - \frac{1}{4\pi}[rot\vec{B} \times \vec{B}] = -\nabla P_{cr} + \vec{q}_2 \tag{2.3}$$

$$\frac{\partial}{\partial t}\left(\frac{3}{2}P\right) + \nabla \cdot \left(\frac{5}{2}P\vec{V}\right) - \vec{V} \cdot \nabla P = q_3 - \vec{q}_2 \cdot \vec{V} - \vec{V} \cdot \nabla P_{cr} \tag{2.4}$$

Here $\rho = m_p(n_p + n_{pui})$; $P = P_e + P_p + P_{pui}$; P_e, P_p, P_{pui} and P_{cr} are pressures of electrons, protons, pickup ions and cosmic rays, respectively; q_1, $\vec{q}_2$ and q_3 are source terms in the plasma due to the charge exchange process with H atoms, photoionization and electron impact ionization:

$$q_1 = m_p n_H (\nu_{ph} + \nu_{impact}) \tag{2.5}$$

$$\vec{q_2} = m_p \int (\nu_{ph} + \nu_{impact}) \vec{v}_H f_H(\vec{v}_H) d\vec{v}_H +$$

$$m_p \int \int u\sigma_{ex}^{HP}(u)(\vec{w}_H - \vec{w}_p) f_H(\vec{w}_H) f_p(\vec{w}_p) d\vec{w}_H d\vec{w}_p + \quad (2.6)$$

$$m_p \int \int u\sigma_{ex}^{HP}(u)(\vec{w}_H - \vec{w}_i) f_H(\vec{w}_H) f_{pui}(\vec{w}_i) d\vec{w}_H d\vec{w}_i$$

$$q_3 = m_p \int (\nu_{ph} + \nu_{impact}) \frac{\vec{w}_H{}^2}{2} f_H(\vec{w}_H) f_p(\vec{w}_p) d\vec{w}_p d\vec{w}_H$$

$$+ m_p \int \int u\sigma_{ex}^{HP}(u) \frac{\vec{w}_H{}^2 - \vec{w}_p{}^2}{2} f_H(\vec{w}_H) f_p(\vec{w}_p) d\vec{w}_p d\vec{v}_H \quad (2.7)$$

$$+ m_p \int \int u\sigma_{ex}^{HP}(u) \frac{\vec{w}_H{}^2 - \vec{w}_i{}^2}{2} f_H(\vec{w}_H) f_{pui}(\vec{w}_i) d\vec{w}_i d\vec{w}_H$$

Here f_p is the Maxwellian velocity distribution of the solar wind and interstellar protons and f_{pui} is the velocity distribution of the pickup ions, which should be determined by a solution of the pickup ion kinetic transport equation or by assumption of complete assimilation of pickup ions into the solar wind plasma.

Faraday's equation

$$\frac{\partial \vec{B}}{\partial t} = rot[\vec{v} \times \vec{B}] \quad (2.8)$$

should be added to the system of governing equations (2.2)-(2.4) to make the system closed. This form of Faraday's equation follows from the classical form of the Ohm's law in the case of ideal conductivity: $\vec{E} = -\frac{1}{c}[\vec{V} \times \vec{B}]$.

Governing equations (2.2)-(2.8) for the one-fluid approach are obtained by summarizing mass, momentum and energy balance equations for electrons, protons and pickup ions under certain assumptions, which were recently discussed by Baranov and Fahr (2003a,b) and Florinsky et al. (2003). Ideal MHD equations with source terms q_1, $\vec{q_2}$, q_3 in the right-hand sides were solved self-consistently with equation (2.1) by Aleksashov et al. (2000) in the case when the IS magnetic field is parallel to the interstellar flow.

To derive the classical system of hydrodynamic equations applied for heliospheric interface in one-fluid models, one needs to ignore terms containing magnetic and electric fields in equations (2.3) and (2.4). In this case, equations (2.2)-(2.4) together with kinetic equation (2.1) for H atoms and expressions (1.5)-(1.7) for the source terms q_1, $\vec{q_2}$ and q_3 form a closed system of equations.

3.3   Pickup Ions

To study pickup ion evolution in the outer heliosphere and in the heliospheric interface, details of the process of charged particle assimilation into the magnetized plasma are needed. A newly created ion under the influence of the large scale solar wind electric and magnetic fields executes a cycloidal trajectory with the guiding center drifting at the bulk velocity of the solar wind. Assuming that the gyroradius is much smaller than the typical scale length, one can average the velocity distribution function over the gyratory motion. The initial ring-beam distribution of pickup ions is unstable. Basic processes that determine evolution of the pickup ion distribution are pitch-angle scattering, convection, adiabatic cooling in the expanding solar wind, injection of newly ionized particles, and energy diffusion in the wave field of both the solar wind and that generated by pickup ions and different kinds of instabilities of waves in the solar wind. The most general form of the relevant transport equation describing the evolution of gyrotropic velocity distribution function $f_{pui} = f_{pui}(t, \vec{r}, v, \mu)$ of pickup ions in a background plasma moving at a velocity $\vec{V}_{sw}$ was written by Isenberg (1997) and Chalov and Fahr (1998). f_{pui} is a function of the modulus of velocity in the solar wind rest frame, and μ is the cosine of pitch angle.

Theoretical models show (section 5) that the assumption of complete assimilation of pickup ions into the solar wind would lead to a great increase of plasma temperature with the heliocentric distance. Since such an increase is not observed, the solar wind and pickup protons represent two distinct proton populations up to the TS. Nevertheless, the radial temperature profile of protons measured by Voyager 2 shows a smaller decrease than predicted by adiabatic cooling. A fraction of the heating of solar wind protons may thus be connected with pickup generated waves (Williams et al.,1995; Smith et al., 2001). Many aspects of pickup ion evolution have been studied (e.g., Chalov and Fahr, 1999; for reviews, see Zank, 1999a; Fichtner, 2001). However, to date detailed models of the assimilation process of pickup ions into the solar wind have not been taken into account in the global models of the heliospheric interface structure.

3.4   Cosmic Rays

Cosmic rays are coupled to background flow via scattering by plasma waves. The net effect is that the cosmic rays tend to be convected along with the background plasma as they diffuse through the magnetic irregularities carried by the background plasma. Both galactic and anomalous cosmic rays can be treated as populations with negligible mass density but non-negligibe high energy density. At a hydrodynamical level, the cosmic rays may modify the wind flow via their pressure gradient ∇P_c, with the net energy transfer rate from fluid to the cosmic rays given by $\vec{V} \cdot \nabla P_c$. $P_c(\vec{r}, t) = \frac{4\pi}{3} \int_0^\infty f_c(\vec{r}, p, t) w p^3 dp$ is a cosmic ray pressure and $f_c(\vec{r}, p, t)$ is the isotropic velocity distribution of cosmic rays.

The transport equation of cosmic rays has the following form (e.g., Fichtner, 2001):

$$\frac{\partial f_c}{\partial t} = \frac{1}{p2} \frac{\partial}{\partial p} \left(p2 D \frac{\partial f_c}{\partial p} \right) + \nabla(\widehat{k} \nabla f_c) - \vec{V} \cdot \nabla f_c +$$
$$\frac{1}{3}(\nabla \cdot \vec{V}) \frac{\partial f_c}{\partial lnp} + S(\vec{r}, p, t) \qquad (2.9)$$

Here p is the modulus of the momentum of the particle; D is the diffusion coefficient in momentum space, often assumed to be zero; $\widehat{k}$ is the tensor of spatial diffusion; $\vec{V} = \vec{U} + \vec{V}_{drift}$ is the convection velocity; $\vec{U}$ is the plasma bulk velocity; $\vec{V}_{drift}$ is a drift velocity in the heliospheric or interstellar magnetic field; and $S(\vec{r}, p, t)$ is the source term.

At the hydrodynamic level, the transport equation of the cosmic rays is:

$$\frac{\partial P_c}{\partial t} = \nabla[\widehat{k} \nabla P_c - \gamma_c(\vec{U} + U_{dr}) P_c] + (\gamma_c - 1)\vec{U} \cdot \nabla P_c + Q_{acr,pui}(\vec{r}, t) \quad (2.10)$$

The last equation assumes that $D = 0$; U_{dr} is momentum-averaged drift velocity; γ is the polytropic index; and $Q_{acr,pui}$ is the energy injection rate describing energy gains of the ACRs from pickup ions. Chalov and Fahr (1996, 1997) suggested that $Q_{acr,pui} = -\alpha p_{pui} \nabla \cdot \vec{U}$, where α is a constant injection efficiency defined by the specific plasma properties (Chalov and Fahr, 1997). α is set to zero for GCRs since no injection occurs into the GCR component.

4. Overview of Heliospheric Interface Models

A complete model of the SW/LIC interaction has not yet been developed. However, in recent years, several groups have focused their efforts on theory and modeling of the heliospheric interface in order to understand influence of one (or several) components on the interface separately from others. In particular, the influence of the magnetic fields on the interface structure was studied in Fujimoto and Matsuda (1991), Baranov and Zaitsev (1995), Myasnikov (1997) for the two-dimensional case and in Ratkiewicz et al. (1998, 2000), Linde et al. (1998), Pogorelov and Matsuda (1998), Tanaka and Washimi (1999), Opher et al. (2003) for the three-dimensional case. Latitudinal variations of the solar wind have been considered in Pauls and Zank (1997). The influence of the solar cycle variations on the heliospheric interface was studied in the 2D case in Steinolfson (1994), Pogorelov (1995), Karmesin et al. (1995), Baranov and Zaitsev (1998), Wang and Belcher (1999), Zaitsev and Izmodenov (2001), and in Tanaka and Washimi (1999) for the 3D case. In spite of many interesting findings in these papers, these theoretical studies did not take into account interstellar H atoms or considered the population under greatly simplified assumptions, as it was done in Linde et al. (1998), where velocity and temperature of interstellar H atoms were assumed as constants in the entire interface.

Since most of the observational information on the heliospheric interface is connected with interstellar atoms and their derivatives as pickup ions and ACRs, we will focus on models which take into account the interstellar neutral component self-consistently together with the plasma component. These models can be separated into two types. Models of the first type (Table 2.5) use a simplified fluid (or multi-fluid) approach for interstellar H atoms. A kinetic approach was used in the models of the second type. Development of the fluid (or multi-fluid) models of H atoms was connected with the fact that fluid (or multi-fluid) approach is simpler for numerical realization. At the same time such an approach can lead to nonphysical results. Results of one of the most sophisticated multi-fluid models (Zank et al., 1996) were compared with the kinetic Baranov-Malama model in Baranov et al. (1998). The comparison shows qualitative and quantitative differences in distributions of H atoms. At the same time, it was concluded in Williams et al. (1997) that the two models agreed on the distances to the termination shock, heliopause and bow shock in the upwind direction, but not in positions of the termination shock in downwind direction.

One of the common features in the models by Wang et al. (1999), Zank et al. (1996), Liewer et al. (1995), Baranov and Malama (1993), Müller et al. (2000), Myasnikov et al. (2000a,b), Aleksashov et al. (2000, 2004), Izmodenov et al. (2003a,b) is that proton, electron and pickup ion components were considered as one fluid. The great advantage of this approach is that its equations are considerably simpler as compared with the kinetic approach for pickup ions. A key assumption of this approach is immediate assimilation of pickup protons into the solar plasma. In other words, it is assumed that immediately after ionization one cannot distinguish between original solar wind protons and pickup protons. Another important assumption is that electron and proton components have equal temperatures, $T_e = T_p$. For quasineutral plasma ($n_p + n_{pui} = n_e + o(n_e)$) this means that the pressure of the electrons is equal to half of the total plasma pressure $(P = n_e k T_e + (n_p + n_{pui}) k T_p \approx 2 n_e k T_e = 2 P_e)$.

For the solar wind, one-fluid models assume, essentially, that wave-particle interactions are sufficient for pickup ions to assimilate quickly into the solar wind, becoming indistinguishable from solar wind protons. However, as discussed above, Voyager observations have shown that this is probably not the case. Pickup ions are unlikely to be assimilated completely. Instead, two co-moving thermal populations can be expected. A model that distinguishes the pickup ions from the solar wind ions was suggested by Isenberg (1986). Electrons were considered as a third fluid. The key assumption in the model is that pickup ions and solar wind protons are co-moving $(V_p = V_{pui})$. It was also assumed that there is no exchange of thermal energy between solar wind protons and pickup ions. Isenberg's approach consists of two continuity equations for solar protons and pickup ions; one momentum equation and three energy equations for solar wind protons, electrons and pickup ions. Note that Isenberg used the simplified form of source terms suggested in Holzer (1972) and applied these equations to the spherically symmetric solar wind upstream of the termination shock only.

Another two-fluid model of the solar wind and pickup protons was developed recently in Fahr et al. (2000). This model also assumes that convection speed of pickup ions is identical to that of solar wind protons. The pressure of the pickup ions was calculated in the model under assumption of a rectangular shape for the pickup ion isotropic distribution function. In this case, the pressure can be expressed through the pickup ion density ρ_i and solar wind bulk velocity V_{sw} as

$$P_{pui} = \rho_{pui} V_{sw}^2 / 5. \tag{2.11}$$

The model also includes ACR and GCR components as two separate massless fluids. Therefore, the governing plasma equations in the Fahr et al. (2000) model are i) one-fluid equations for the mixture of solar protons, pickup ions and electrons; ii) a continuity equation for pickup ions; iii) two transport equations for ACRs and GCRs. The influence of cosmic ray components was taken as described by the terms $-\nabla(P_{ACR} + P_{GCR})$ and $-\vec{V} \cdot \nabla(P_{ACR} + P_{GCR}) - \alpha P_{pui} \nabla \cdot (\vec{V})$ in the right-hand side of the momentum and energy equations, respectively. This approach used by Fahr et al. (2000) is crude in the sense that the assumption made on the shape of the distribution function does not reflect such important physical processes as adiabatic cooling, stochastic acceleration, and charge-exchange process of pickup ions with interstellar H atoms in the heliosheath. In addition, Fahr et al. (2000) used a simple one-fluid approach to describe the flow of interstellar H atoms. Currently, our Moscow group is

Table 2.5. Models with multi-fluid or kinetic approaches for interstellar H atoms

Reference	GCR	ACR	IMF	HMF	Lat. SW asym.	Solar Cycle	Pickup SW Protons	H Atoms
MULTI-FLUID APPROACH								
Liewer et al., 1995	–	–	–	–	–	+	one-fluid	one-fluid
Zank et al. 1996	–	–	–	–	–	–	one-fluid	three-fluid
Pauls and Zank, 1997	–	–	–	–	+	–	one-fluid	one-fluid
Wang and Belcher, 1999	–	–	–	–	–	+	one-fluid	one-fluid
Fahr et al., 2000	+	+	–	–	–	–	two-fluid	one-fluid
Fahr and Scherer, 2003a,b	+	+	–	–	–	+	two-fluid	one-fluid
KINETIC APPROACH								
Osterbart and Fahr, 1992	–	–	–	–	–	–	No pickup ions	not self-consistent
Miller et al., 2000	–	–	–	–	–	–	one-fluid	particle mesh code
Moscow team models:								
Baranov and Malama, 1993	–	–	–	–	–	–	one- fluid	Monte Carlo with splitting
Myasnikov et al, 2000a,b	+	–	–	–	–	–	one- fluid	Monte Carlo with splitting
Aleksashov et al, 2000	–	–	+	–	–	–	one-fluid	Monte Carlo with splitting
Izmodenov et al., 2003a	–	–	–	–	–	+	one-fluid	Monte Carlo with splitting
Alexashov et al., 2004	–	+	–	–	–	–	one-fluid	Monte Carlo with splitting
Malama et al., 2004	–	–	–	–	–	–	two-fluid	Monte Carlo with splitting

developing a multi-component model which is free of these limitations (Malama et al., in preparation, 2004). The kinetic equation for the H atom component will be solved self-consistently with the total plasma mass, momentum and energy equations, and the kinetic equation for pickup ion component.

5. Self-Consistent Two-Component Model of the Heliospheric Interface and Recent Advancements of the Model

The first self-consistent model of the interaction of the two-component (plasma and H atoms) LIC with the solar wind was developed by Baranov and Malama (1993). The interstellar wind was assumed to be a uniform parallel flow. The solar wind was assumed to be spherically symmetric at the Earth's orbit. Under these assumptions, the heliospheric interface has an axisymmetric structure.

Plasma and neutral components interact mainly by charge exchange. However, photoionization, solar gravity and solar radiation pressure, which are especially important in the vicinity (< 10-15 AU) of the Sun, are also taken into account.

Kinetic and hydrodynamic approaches were used for the neutral and plasma components, respectively. The kinetic equation (2.1) for neutrals was solved together with the Euler equations for one-fluid plasma (2.2) - (2.4). The influence of the interstellar neutrals is taken into account in the right-hand side of the Euler equations that contain source terms q_1, $\vec{q_2}$, q_3, which are integrals (2.5)-(2.6) of the H atom distribution function $f_H(\vec{V_H})$ and can be calculated directly by a highly efficient Monte Carlo method with splitting of trajectories (Malama, 1991). The set of kinetic and Euler equations is solved by an iterative procedure, as suggested in Baranov et al. (1991). Supersonic boundary conditions were used for the unperturbed interstellar plasma and for the solar wind plasma at the Earth's orbit. The velocity distribution of interstellar atoms is assumed to be Maxwellian in the unperturbed LIC. Results of this model are discussed below, in this section.

5.1 Plasma

Interstellar atoms strongly influence the heliospheric interface structure. The heliospheric interface is much closer to the Sun in the case when H atoms are taken into account in the model, as compared to a pure gas dynamical case (Figure 2). The termination shock becomes more spherical. The Mach disk and the complicated tail shock structure,

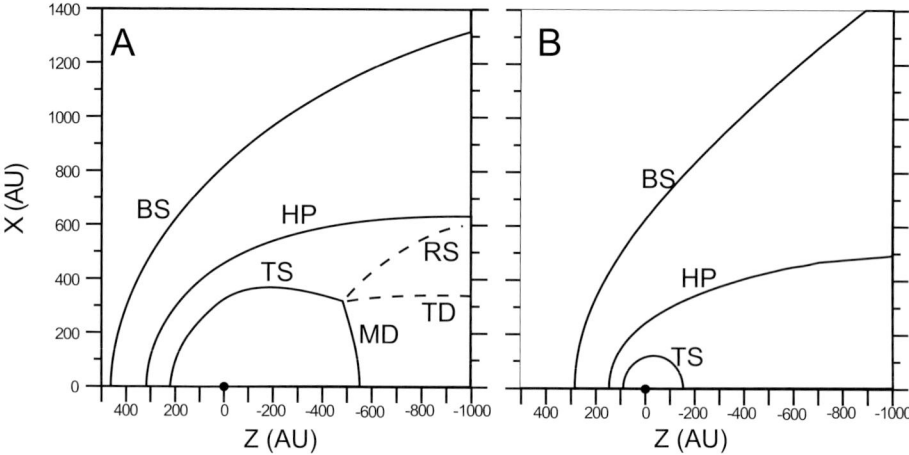

Figure 2.2. Effect of the interstellar neutrals on the size and structure of the interface structure. (a) The heliospheric interface pattern in the case of fully ionized local interstellar cloud (LIC), (b) the case of partly ionized LIC. BS is the bow shock. HP is the heliopause. TS is the termination shock. MD is the Mach disk. TD is the tangential discontinuity and RS is the reflected shock. (Izmodenov and Alexashov, 2003)

consisting of the reflected shock (RS) and the tangential discontinuity (TD), disappear.

The supersonic plasma flows upstream of the bow and termination shocks are disturbed. The supersonic solar wind flow is disturbed by charge exchange with the interstellar neutrals. The new protons created by charge exchange are picked up by the solar wind magnetic field. The Baranov-Malama model assumes immediate assimilation of pickup ions into the solar wind plasma. The solar wind protons and pickup ions are treated as one-fluid, called the solar wind. The number density, velocity, temperature, and Mach number of the solar wind are shown in Figure 3A. The effect of charge exchange on the solar wind is significant. By the time the solar wind flow reaches the termination shock, it is decelerated (15-30 %), strongly heated (5-8 times) and mass loaded (20-50 %) by the pickup ion component.

The interstellar plasma flow is disturbed upstream of the bow shock by charge exchange of the interstellar protons with secondary H atoms. These secondary atoms originate in the solar wind. This leads to heating (40-70 %) and deceleration (15-30 %) of the interstellar plasma before it reaches the bow shock. The Mach number decreases upstream of the BS and for a certain range of interstellar parameters ($n_{H,LIC} \gg n_{p,LIC}$)

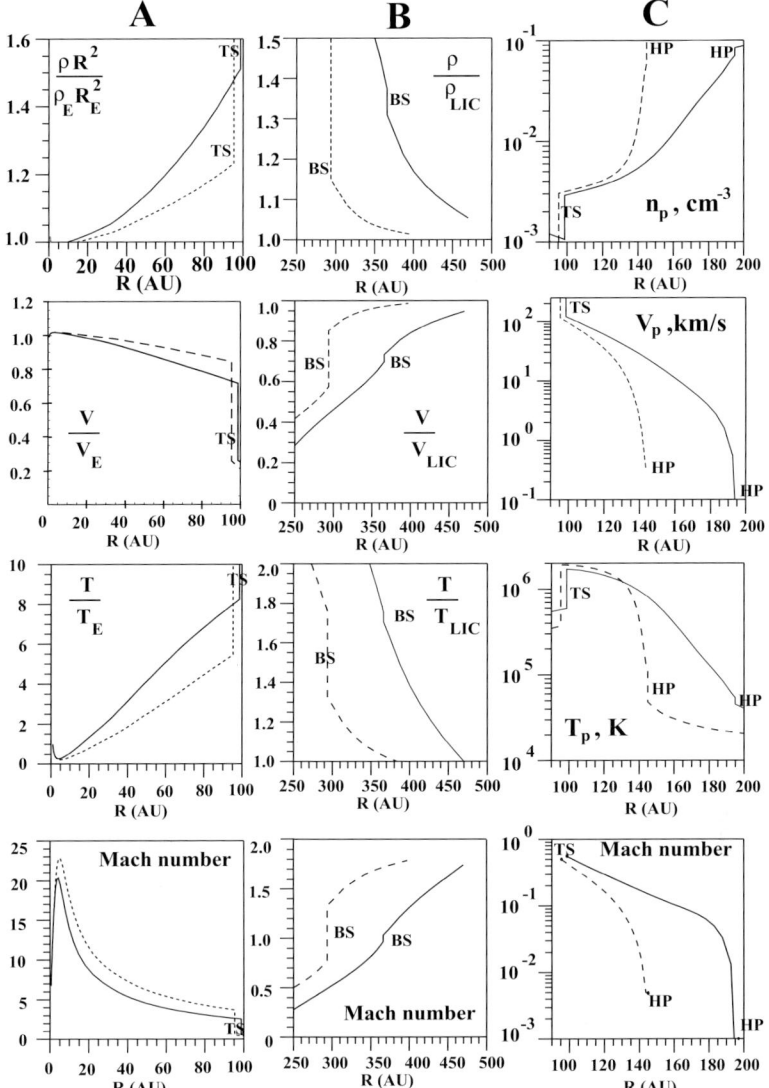

Figure 2.3. Plasma density, velocity, temperature and Mach number upstream of the termination shock (A), upstream of the bow shock (B), and in the heliosheath (C). The distributions are shown for the upwind direction. Solid curves correspond to $n_{H,LIC}$=0.2 cm^{-3}, $n_{p,LIC}$=0.04 cm^{-3}. Dashed curves correspond to $n_{H,LIC}$=0.14 cm^{-3}, $n_{p,LIC}$=0.10 cm^{-3}. V_{LIC}=25.6 km/s, T_{LIC}=7000 K. (Izmodenov (2000)

the bow shock may disappear. Solid curves in Figure 3B correspond to a small ionization degree in the LIC ($n_p/(n_p + n_H) = 1/6$) and the bow shock almost disappears in this case.

Interstellar neutrals also modify the plasma structure in the heliosheath. In a pure gas dynamic case (without neutrals) the density and temperature of the postshock plasma are nearly constant. However, the charge exchange process leads to a large increase in the plasma number density and decrease in its temperature (Figure 1.3C). The electron impact ionization process may influence the heliosheath plasma flow by increasing the gradient of the plasma density from the termination shock to the heliopause (Baranov and Malama, 1996). The influence of interstellar atoms on the heliosheath plasma flow is important, in particular, for the interpretations of kHz radio emission detected by Voyager (Gurnett et al., 1993; Gurnett and Kurth, 1996; Treumann et al. 1998) and possible future heliospheric imaging in energetic neutral atom (ENA) fluxes (Gruntman et al., 2001; IBEX mission).

5.2   H Atoms

Charge exchange significantly alters the interstellar atom flow. Atoms newly created by charge exchange have the velocity of their ion counterparts in charge exchange collisions. Therefore, the velocity distribution of these new atoms depends on the local plasma properties in the place of their origin. It is convenient to distinguish four different populations of atoms, depending on region in the heliospheric interface where the atoms were formed. Population 1 is the atoms created in the supersonic solar wind up to the TS. Population 2 is the atoms created in the inner heliosheath, the region between the TS and HP. Population 3 is the atoms created in the outer heliosheath - the region of disturbed interstellar wind between the HP and the BS. We will call original (or primary) interstellar atoms population 4. The number densities and mean velocities of these populations are shown in Figure 4 as functions of the heliocentric distance. The velocity distribution function of interstellar atoms $f_H(\vec{w}_H, \vec{r})$ can be represented as a sum of the distribution functions of these populations: $f_H = f_{H,1} + f_{H,2} + f_{H,3} + f_{H,4}$. The Monte Carlo method allows us to calculate these four distribution functions. The velocity distributions of the interstellar atoms at the 12 selected points in the heliospheric interface were shown by Izmodenov (2001), Izmodenov et al. (2001). As an example, the velocity distributions at the termination shock in the upwind direction are shown in Figure 5 for four introduced populations of H atoms. Note that velocity distributions of H atoms in the heliosphere were also presented by Müller et al. (2000). However, different populations of H atoms cannot be considered separately in the mesh particle simulations of H atoms (Lipatov et al., 1998) which were used in that paper.

Original (or primary) interstellar atoms are significantly filtered (i.e. their number density is reduced) before reaching the termination shock (Figure 4A). Since slow atoms have a small mean free path in comparison to fast atoms, they undergo more intensive charge exchange. This kinetic effect, called "selection", results in a deviation of the interstellar distribution function from Maxwellian (Figure 5A). The selection also results in ~10% increase in the primary atom mean velocity towards the termination shock (Figure 4C).

The secondary interstellar atoms are created in the disturbed interstellar medium by charge exchange of primary interstellar neutrals with protons decelerated by the bow shock. The secondary interstellar atoms collectively make up the "H wall", a density increase at the heliopause. The "H wall" has been predicted in Baranov et al. (1991) and detected in the direction of α Cen (Linsky and Wood, 1996). At the termination shock, the number density of secondary neutrals is comparable to the number density of the primary interstellar atoms (Figure 4A, dashed curve). The relative abundances of secondary and primary atoms entering the heliosphere vary with degree of interstellar ionization. It has been shown by Izmodenov et al. (1999b) that the relative abundance of the secondary interstellar atoms inside the termination shock is larger for the models with interstellar proton number density. The bulk velocity of population 3 is about -18 to -19 km/s. The sign "-" means that the population approaches the Sun. One can see that the velocity distribution of this population is not Maxwellian (Figure 5B). The reason for the abrupt behavior of the velocity distribution for $V_z > 0$ is that the particles with significant positive V_z velocities can reach the termination shock only from the downwind direction. The velocity distributions of different populations of H atoms were calculated in Izmodenov et al. (2001) for different directions from upwind. The fine structures of the velocity distribution of the primary and secondary interstellar populations vary with direction. These variations of the velocity distributions reflect the geometrical pattern of the heliospheric interface. The velocity distributions of the interstellar atoms can be a good diagnostics of the global structure of the heliospheric interface.

Another population of the heliospheric neutrals (population 2) is **the neutrals created in the inner heliosheath** from hot and compressed solar wind protons and pickup ions. The number density of this population is an order of magnitude smaller than the number densities of the primary and secondary interstellar atoms. This population has a minor importance for interpretation of Ly α and pickup ion measurements inside the heliosphere. However, it was recently pointed out by Chalov and Fahr (2003) that charge exchange of these atoms with solar

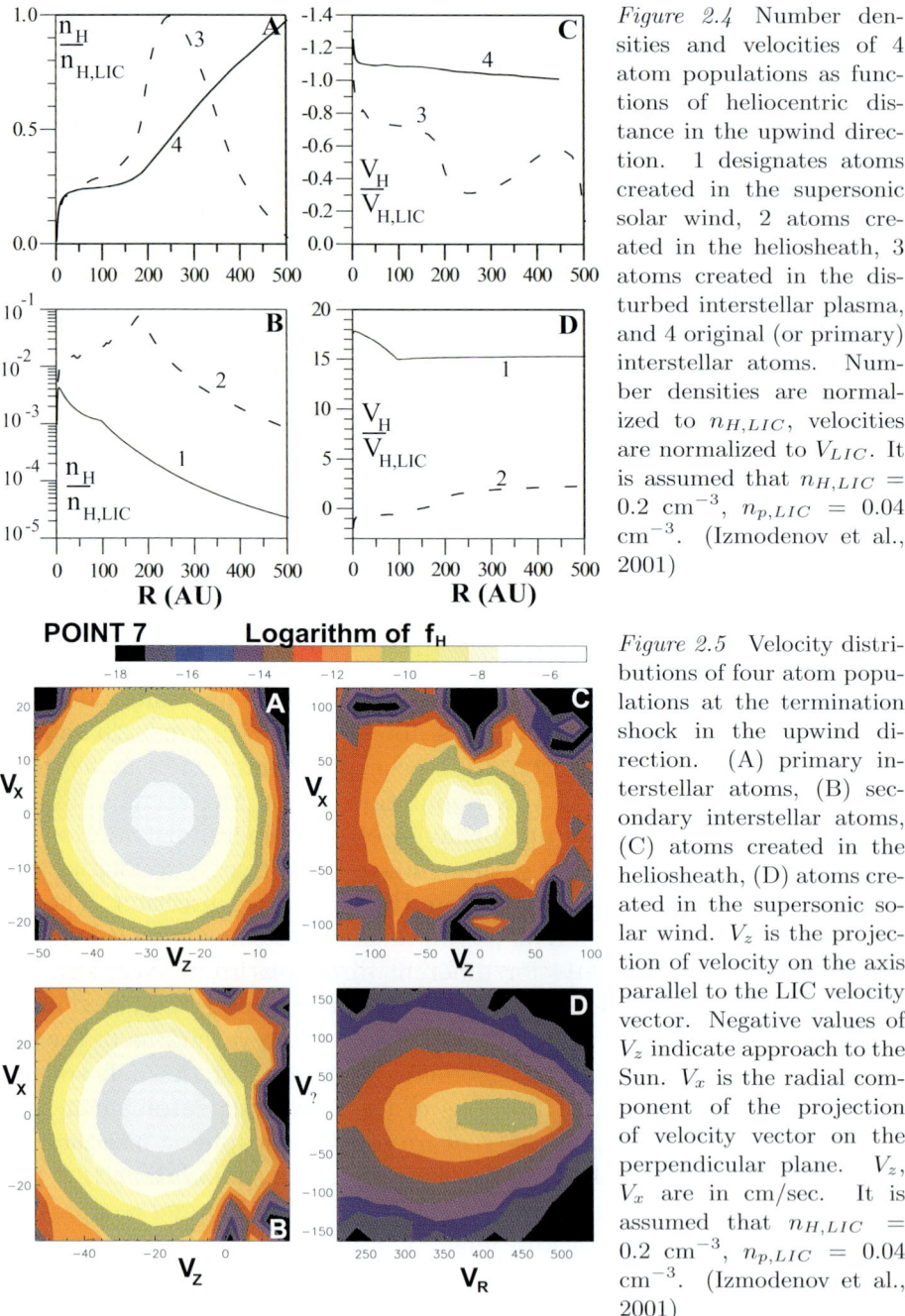

Figure 2.4 Number densities and velocities of 4 atom populations as functions of heliocentric distance in the upwind direction. 1 designates atoms created in the supersonic solar wind, 2 atoms created in the heliosheath, 3 atoms created in the disturbed interstellar plasma, and 4 original (or primary) interstellar atoms. Number densities are normalized to $n_{H,LIC}$, velocities are normalized to V_{LIC}. It is assumed that $n_{H,LIC} = 0.2$ cm^{-3}, $n_{p,LIC} = 0.04$ cm^{-3}. (Izmodenov et al., 2001)

Figure 2.5 Velocity distributions of four atom populations at the termination shock in the upwind direction. (A) primary interstellar atoms, (B) secondary interstellar atoms, (C) atoms created in the heliosheath, (D) atoms created in the supersonic solar wind. V_z is the projection of velocity on the axis parallel to the LIC velocity vector. Negative values of V_z indicate approach to the Sun. V_x is the radial component of the projection of velocity vector on the perpendicular plane. V_z, V_x are in cm/sec. It is assumed that $n_{H,LIC} = 0.2$ cm^{-3}, $n_{p,LIC} = 0.04$ cm^{-3}. (Izmodenov et al., 2001)

wind protons produces tails in the velocity distribution of pickup ions measured at one or several AU during quiet time periods. Some atoms of the population that may be detectable by Ly α hydrogen cell experiments due to their large Doppler shifts (Quemerais and Izmodenov, 2002). Due to their high energies and large mean free path, a portion of the atoms in this population penetrate upstream of the BS and disturb the pristine interstellar medium at large heliocentric distances. Inside the termination shock the atoms propagate freely. Thus, these atoms are a rich source of information on the plasma properties at the place of their birth, i.e. the inner heliosheath. There are plans to measure this population of atoms on future missions, including HIGO and a proposed Small Explorer called the Interstellar Boundary Explorer (IBEX) mission.

The last population of the heliospheric atoms is **the atoms created in the supersonic solar wind**. The number density of this atom population has a maximum at $\sim$5 AU. At this distance, the number density of the population is about two orders of magnitude smaller than the number density of interstellar atoms. Outside the termination shock the density decreases faster than $1/r2$, where r is the heliocentric distance (curve 1, Figure 4B). The mean velocity of population 1 is about 450 km/sec, which corresponds to the bulk velocity of the supersonic solar wind. The velocity distribution of this population is also not Maxwellian (Figure 5D). The extended "tail" in the distribution function is caused by the solar wind plasma deceleration upstream of the termination shock. This "supersonic" atom population penetrates the interface and charge exchanges with interstellar protons beyond the BS. The process of charge exchange leads to heating and deceleration upstream of the bow shock and, therefore, to the decrease of the Mach number ahead of the bow shock.

The Baranov-Malama (1993) model takes into account essentially two interstellar components: H atoms and charged particles. To apply this model to space experiments, one needs to evaluate how other possible components of the interstellar medium influence the results of this two-component model. Recently, several effects were taken into account in the frame of this axisymmetric model.

5.3 Effects of Interstellar and Solar Wind Ionized Helium

Recent measurements of interstellar helium atoms (Witte et al., 1996; Witte, 2004) and interstellar He pickup ions (Gloeckler and Geiss, 2001; Gloeckler et al., 2004) inside the heliosphere, as well as of the interstellar helium ionization (Wolff et al., 1999) allow us to estimate the number density of interstellar helium ions to be 0.008-0.01 cm^{-3}. Cur-

Table 2.6. Sets of model parameters and locations of the TS, HP and BS in the upwind direction

#	$n_{H,LIC}$	$n_{p,LIC}$	$\frac{n_{\alpha,sw}}{n_{e,sw}}$	HeII/(HeI+HeII)	R(TS)	R(HP)	R(BS)
	cm^{-3}	cm^{-3}	%		AU	AU	AU
1	0.18	0.06	0	0	95.6	170	320
2	0.18	0.06	0	0.375	88.7	152	270
3	0.18	0.06	2.5	0	100.7	176	330
4	0.18	0.06	2.5	0.150	97.5	168	310
5	0.18	0.06	2.5	0.375	93.3	157	283
6	0.18	0.06	4.5	0.375	97.0	166	291
7	0.20	0.04	0	0	95.0	183	340
8	0.20	0.04	2.5	0.375	93.0	171	290

rent estimates of the proton number density in the LIC fall in the range 0.04 - 0.07 cm^{-3}. Since helium ions are four times heavier than protons the dynamic pressure of the ionized helium component is comparable to the dynamic pressure of the ionized hydrogen component. Therefore, interstellar ionized helium cannot be ignored in the modeling of the heliospheric interface. For the first time, effects of interstellar ionized helium were studied by Izmodenov et al. (2003b). Simultaneously with interstellar ionized helium, the paper took into account solar wind alpha particles, which constitute 2.5 - 5 % of the solar wind and, therefore, produce 10 - 20 % of the solar wind dynamic pressure.

To evaluate possible effects of both interstellar ions of helium and solar wind alpha particles Izmodenov et al. (2003b) performed parametric model calculations with the eight sets of boundary conditions given in Table 2.6. Calculated locations in the upwind direction of the termination shock, the heliopause, and the bow shock are given for each model in the last three columns of Table 2.6. It is seen that the heliopause and the termination and bow shocks are closer to the Sun when the influence of interstellar helium ions is taken into account. This effect is partially compensated by additional solar wind alpha particle pressure that we also took into account in our model. The net result is as follows: the heliopause, termination and bow shocks are closer to the Sun by ∼12 AU, ∼ 2 AU, ∼ 30 AU, respectively in the model taking into account both interstellar helium ions and solar wind alpha particles (model 5) as compared to the model ignoring these ionized helium components (model 1). It was also found that both interstellar ionized helium and solar wind alpha particles do not influence the filtration of the interstellar H atoms through the heliospheric interface.

5.4   Effects of GCRs, ACRs and the Interstellar Magnetic Field

The influence of the galactic cosmic rays on the heliospheric interface structure was studied by Myasnikov et al. (2000a,b). The study was done in the frame of two-component (plasma and GCRs) and three-component (plasma, H atoms and GCRs) models. For the two-component case it was found that cosmic rays could considerably modify the shape and structure of the solar wind termination shock and the bow shock and change heliocentric distances to the heliopause and the bow shock. At the same time, for the three-component model it was shown that the GCR influence on the plasma flows is negligible when compared with the influence of H atoms. The exception is the bow shock, a structure that can be strongly modified by the cosmic rays. The dynamical influence of ACRs on the solar wind flow in the outer heliosphere and on the structure of the termination shock has been studied by Fahr et al. (2000) and Alexashov et al. (2004). Whereas earlier research on the cosmic-ray-modified heliosphere was devoted mainly to the dependence of the termination shock structure and position on the injection rate of ACRs, Alexashov et al. (2004) studied effects connected with changes in the value of the diffusion coefficient while keeping the injection rate fixed. Different values of the diffusion coefficient were considered for the reason that K is poorly known in the outer heliosphere and especially in the heliosheath, and in addition the value of the diffusion coefficient varies with the solar cycle. It was shown that: 1. The effect of ACRs on the solar wind flow near the termination shock leads to formation of a smooth precursor, followed by the subshock, and to a shift of the subshock towards larger distances in the upwind direction. This result is consistent with earlier findings based on one-dimensional spherically symmetric models. Both the intensity of the subshock and the magnitude of the shift depend on the value of the diffusion coefficient, with the largest shift (about 4 AU) occurred at medium values of K. 2. The precursor of the termination shock is rather pronounced except the case with large K. It has been shown by Berezhko (1986), Chalov (1988a, 1988b), and Zank et al. (1990) that the precursor of a cosmic-ray-modified shock is highly unstable with respect to magnetosonic disturbances if the cosmic-ray pressure gradient in the precursor is sufficiently large. The possible detection of oscillations in the magnetic field and solar wind speed by the Voyager spacecraft in the near future could be considered as evidence for the termination shock. The oscillations connected with the instability of the precursor have a distinctive feature: the magnetic field in more unstable modes oscillates in the longitudinal direction, while the solar wind speed oscillates in the direction perpendicular to the ecliptic plane (Chalov, 1990). 3. The postshock temperature of the solar wind plasma is lower in the case of the cosmic-ray-modified termination shock when compared with the shock without ACRs. The decrease in the temperature results in a decrease in the number density of hydrogen atoms originating in the region between the termination shock and heliopause. 4. The cosmic-ray pressure downstream of the termination shock is comparable to the thermal plasma pressure for small K when the diffusive length scale is much smaller than the distance to the shock. On the other hand, at large K the postshock cosmic-ray pressure is negligible when compared with the thermal plasma pressure. There is pronounced upwind-downwind asymmetry in the cosmic-ray energy distribution due to a difference in the amount of the energy injected into ACRs in the up- and downwind parts of the termination shock. This difference in the injected energy is connected with the fact that the thermal plasma pressure is lower in the downwind part of the shock when compared with the upwind part.

Effects of the interstellar magnetic field on the plasma flow and on distribution of H atoms in the interface were studied in Aleksashov et al. (2000) in the case of magnetic field parallel to the relative Sun/LIC velocity vector. In this case, the model remains axisymmetric. It was shown that effects of the the interstellar magnetic field on the positions of the termination and bow shocks and the heliopause are significantly smaller when compared to model with no H atoms (Baranov and Zaitsev, 1995). The calculations were performed with various Alfven Mach numbers in the undisturbed LIC. It was found that the bow shock straightens out with decreasing Alfven Mach number (increasing magnetic field strength in LIC). It approaches the Sun near the symmetry axis, but recedes from it on the flanks. By contrast, the nose of the heliopause recedes from the Sun due to tension of magnetic field lines, while the heliopause in its wings approaches the Sun under magnetic pressure. As a result, the region of compressed interstellar medium around the heliopause (or "pileup region") decreases by almost 30 %, as the magnetic field increases from zero to 3.5×10^{-6} Gauss. It was also shown in Aleksashov et al. (2000) that H atom filtration and heliospheric distributions of primary and secondary interstellar atoms are virtually unchanged over the entire assumed range of the interstellar magnetic field ($0 - 3.5 \cdot 10^{-6}$ Gauss). The magnetic field has the strongest effect on the density distribution of population 2 of H atoms, which increases by a factor of almost 1.5 as the interstellar magnetic field increases from zero to $3.5 \cdot 10^{-6}$ Gauss.

5.5   Effects of the Solar Cycle Variations of the Solar Wind

More than 30 years' (three solar cycles) observations of the solar wind show that its momentum flux varies by factor of ∼2 from solar maximum to solar minimum (Gazis, 1996; Richardson, 1997). It has been shown theoretically that such variations of the solar wind momentum flux strongly influence the structure of the heliospheric interface - the region of the solar wind interaction with the local interstellar medium (e.g., Karmesin et al., 1995; Wang and Belcher, 1999; Baranov and Zaitsev, 1998; Zank, 1999b; Zaitsev and Izmodenov, 2001; Scherer and Fahr, 2003; Zank and Müller, 2003).

Most global models of solar cycle effects ignored the interstellar H atom component or took this component into account by using simplified fluid approximations. These simplifications were done because it is difficult to solve 6D (time, two dimensions in space, and three dimensions in velocity-space) kinetic equation for the interstellar H atom component.

Recently we developed the non-stationary, self-consistent model of the heliospheric interface and used it to explore the solar cycle variations of the interface (Izmodenov et al., 2003a; Izmodenov et al., 2004b). We obtained the periodic solution of the system of Euler equations for plasma, and the kinetic equation for interstellar H atoms with the periodic boundary conditions for the solar wind at the Earth's orbit. Izmodenov et al. (2003a) presented results of the model, where the IMP 8 solar wind data were used as boundary conditions at the Earth's orbit. Detailed theoretical study of the solar cycle effects on the structure of the heliospheric interface was reported by Izmodenov et al. (2004b), where the solar wind dynamic pressure was sinusoidally varied by a factor of two over an 11-year period.

The basic results can be summarized as follows. The discontinuities vary sinusoidally with the 11-year period dictated by the inner boundary conditions. The termination shock varies within ±7 AU over the solar cycle in the upwind direction. Fluctuations of the TS become larger as we go from upwind to downwind. The variation of the TS in the downwind direction is 25 AU from it minimal to maximal value. The upwind fluctuation of the TS is nearly in anti-phase with its downwind fluctuations. The heliopause fluctuates with a smaller (∼3 AU) amplitude when compared to the TS. The fluctuation of the BS with the solar cycle is less than 0.1 AU in upwind. The mean solar-cycle value of the locations of the TS, HP and BS are very close to their values obtained in the stationary model with solar cycle averaged boundary conditions.

The strength of the TS has important consequences for anomalous cosmic rays (ACRs), because the velocity jump at the TS relates to the spectral index of ACRs, β, where the intensity of the cosmic rays j varies with energy E as $j \sim E^\beta$. Izmodenov et al. (2004b) have computed the solar cycle variation of the velocity jump at the TS. The upwind jump varies insignificantly, from 2.92 to 3.09, and from 2.92 to 3.17 downwind.

Plasma parameters perform 11-year fluctuations in the entire computation region. However, the wave-length of the plasma fluctuations in the solar wind is large when compared with the distances to the TS and HP in the upwind direction. Large-scale waves are also seen in plasma distributions of the post-shocked plasma in the downwind region. Amplitudes of the waves are much less than in the upwind direction and the wavelength is $\sim$200 AU. The situation is different in the outer heliosheath, the region between the HP and BS. Motion of the heliopause produces a number of shocks and rarefaction waves. The amplitudes of these shocks and rarefaction waves decreases while they propagate away from the Sun due to the increase of their surface areas and interaction between the shocks and rarefaction waves. The characteristic wavelength is $\sim$40 AU. It was shown also that the 11 year averaged distributions of plasma parameters practically coincides with plasma parameters obtained in stationary model with 11 year average boundary conditions.

For interstellar H atoms, Izmodenov et al. (2004b) obtained the following results. A clear 11-year periodicity is seen in the fluctuations of H atom densities. Deviations from exact 11-year sinusoidal periodicity are connected with errors of our statistical calculations, which are $\sim$2-3%. In the outer heliosphere and, in particular, at the TS the variations of densities are within $\pm$ 5% of their mean value for primary and secondary interstellar populations and for the populations of atoms created in the inner heliosheath. Closer to the Sun, for distances < 10 AU, the amplitude of fluctuations increases up to $\pm$ 15% around the mean value. The variation of the number density of H atoms created in the supersonic solar wind is $\pm$ 30% of the mean value. Variations of mean velocity and kinetic temperature of primary and secondary interstellar atoms are negligibly small for both interstellar populations, in comparison to statistical uncertainty. However, the mean velocity and kinetic temperature of atoms created in the inner heliosheath vary with the solar cycle by 10-12%. This is connected with the fact that the most of the H atoms of the latter population are created in the vicinity of the heliopause and they reflect the long wavelength plasma variations in this region.

It is important to note that number densities of the three components (primary and secondary interstellar populations, and the population of H atoms created in the inner heliosheath) fluctuate in phase. This coherent behavior of the fluctuations exists in all supersonic solar wind regions for the three populations and, also, in the inner heliosheath for the primary and secondary atoms. The reason of such coherent behavior of the variations of H atom densities becomes evident when the fluctuations are compared with plasma density variations (see, Izmodenov et al., 2004b). The two variations are almost in anti-phase. Apparently, such a correlation is only possible when temporal variations of the H atom densities are caused by variation of the local loss of the neutrals due to the charge exchange and ionization processes.

However, coherent fluctuations of different populations of H atoms disappear in the regions where the populations originate and the process of creation is dominant when compared with the losses. Indeed, in the inner heliosheath fluctuations of number density of H atoms created in this region are shifted as compared with coherent fluctuations of primary and secondary interstellar atom populations and are in phase with variations of proton number density near the heliopause. Variations of the secondary interstellar atom population are in anti-phase with variations of primary atoms in the outer heliosheath and almost in phase with plasma fluctuations in the region. Again, the creation process is dominant in the outer heliosheath for the population of secondary interstellar atoms.

It is important to note that the above described behavior of H atom populations in the heliospheric interface is kinetic in nature. Variations are determined basically by loss and creation processes, but not by convection and pressure gradient terms as it would be in fluid or multi-fluid approaches. The validity of fluid approach is determined by the simple criterium that the Knudsen number $Kn = l/L \ll 1$, where l and L are the mean free path of the particles and characteristic size of the problem, respectively. For stationary problems, the distance between the HP and BS, which is approximately 100 AU, can be chosen as characteristic size. The mean free path of H atoms in the region is ~ 50 AU (Izmodenov et al., 2000). Therefore, $Kn_{stationary} \approx 0.5$. As was described above, kinetic and fluid approaches were compared by Baranov et al. (1998) and Izmodenov et al. (2001), showing explicitly that the velocity distribution function of H atoms is not Maxwellian anywhere in the interface. For the time-dependent problem considered in this section, the characteristic size, L, is half of the wavelength in the plasma distributions. In the region between the HP and BS, $L \approx 20$ AU. Therefore, $Kn_{time} \approx 2.5$ and fluid approaches become even less appropriate when compared with stationary models. This fundamental point could be the main reason for the large difference between results presented by Izmodenov et al.

(2004b) and results obtained by Zank and Müller (2003) and Scherer and Fahr (2003a,b) who used multi-fluid approaches.

5.6   Heliotail

Plasma and H atom distributions in the tail of the LIC/SW interaction region were not of interest up until recently. However, modeling of the heliospheric interface gives answers to the two fundamental questions: 1. Where is the edge of the solar system plasma? 2. How far downstream can the influence of the solar wind be felt on the surrounding interstellar medium?

To supply an answer to the first question we need to define the solar plasma system boundary. It is natural to assume the heliospheric boundary is the heliopause that separates the solar wind and interstellar plasmas. This definition is not completely correct, because the heliopause is an open surface and, therefore, the heliosphere ends at infinity. To resolve the problem, and to address the second question, detailed specific modeling of the structure of the tail region up to 50000 AU was performed by Izmodenov and Alexashov (2003), Alexashov and Izmodenov (2003), Alexashov et al. (2004). It was shown that the charge exchange process qualitatively changes the solar wind - interstellar wind interaction in the tail region. The termination shock becomes more spherical and the Mach disk, reflected shock and tangential discontinuity disappear. This result was obtained previously by Baranov ana Malama (1993), who performed calculations up to 700 AU in the heliotail. In addition, Alexashov et al. (2004) found that the jumps of density and tangential velocity across the heliopause become smaller in the heliotail and disappear at about 3000 AU. Parameters of solar wind plasma and interstellar H atoms approach their interstellar values at large heliocentric distances. This allows an estimation of the influence of the solar wind, and, therefore, the solar system size in the downwind direction to about 20,000 - 40,000 AU. An illustration of the results is shown in Figure 6. The figure shows isolines of the Mach number up to 10,000 AU. The solar wind plasma has a velocity 100 km/s and a temperature of $1.5 \cdot 10^6$ K immediately after passing the TS. Then the velocity becomes smaller due to new protons injected by charge exchange and approaches the value of interstellar velocity. Since interstellar H atoms are effectively cooler when compared with postshocked protons, the solar wind also becomes cooler. This makes the Mach number increase. At distances of 4,000 AU the solar wind again becomes supersonic and the Mach number then approaches its interstellar value at 40,000 - 50,000 AU, where the solar wind gas dynamic parameters become undistin-

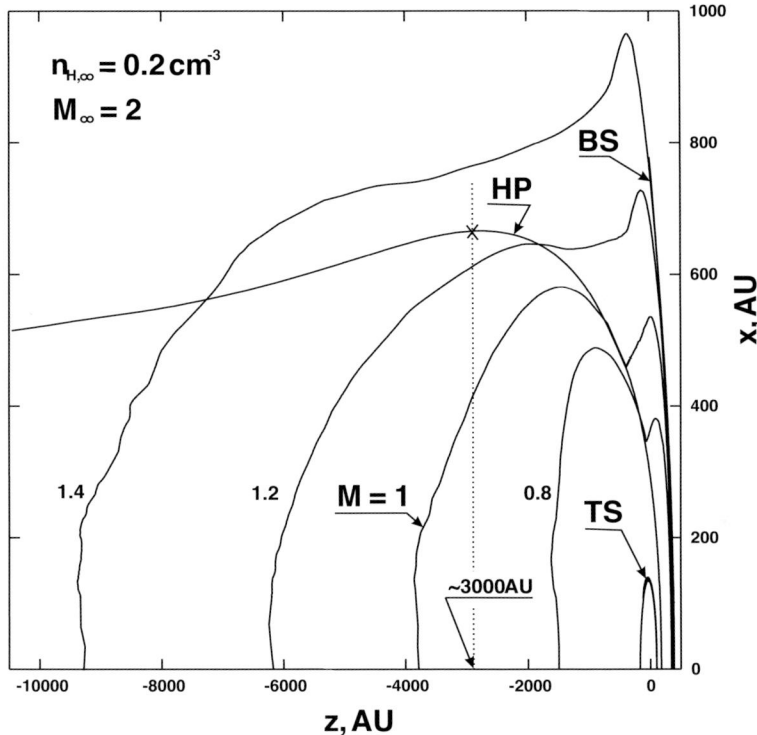

Figure 2.6. Isolines of Mach number of solar wind and interstellar plasma flows in the downwind direction (Alexashov et al., 2004).

guishable from undisturbed interstellar parameters. This result cannot be obtained in the absence of H atoms because the solar wind flow in the heliotail remains subsonic in that case.

6. Interpretations of Spacecraft Experiments Based on the Heliospheric Interface Model

As it was stated in Section 2, the Sun/LIC relative velocity and the LIC temperature are now well constrained (Witte et al., 1996; Witte, 2004; Lallement and Bertin, 1992; Linsky et al., 1993; Lallement et al., 1995, 1996; Moebius et al., 2004). To obtain other interstellar parameters one needs to use one or several remote diagnostics of the interface and a theoretical model.

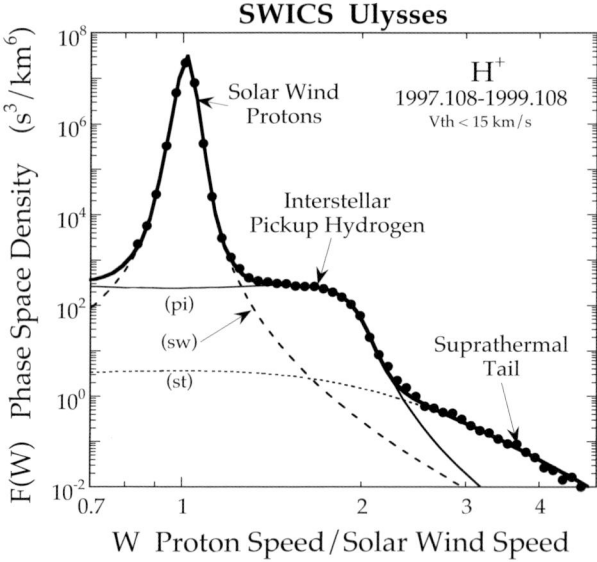

Figure 2.7. Two-year-averaged phase space density of H$^+$ versus normalized speed W (proton speed divided by solar wind speed) observed with the Solar Wind Ion Composition Spectrometer (SWICS) on Ulysses in the slow solar wind at ∼5 AU and low latitude. Only quiet-time periods, characterized by a low solar wind thermal speed (< 15 km/s) were used in this average. Interstellar pickup hydrogen is the dominant component in the flat portion of the spectrum between W ∼1.3 and ∼2.2. Solar wind protons, modeled by curve (sw), dominate below W ∼1.3 and accelerated protons form the suprathermal tail above W ∼2.2, modeled by curve (st). During quiet time periods used here, the solar wind distribution is sufficiently narrow and the suprathermal tail sufficiently weak, to reveal more fully the pickup ion component of the spectrum. The curve labeled (pi) is computed using model parameters given in the text. The atomic hydrogen density at the termination shock is found to be 0.100 ± 0.008 cm^{-3} (Izmodenov et al., 2003a).

6.1   Pickup ons

The charge exchange process in the heliospheric interface leads to a predictable reduction, or filtration, of interstellar atomic hydrogen that enters the heliosphere. The most accurate determination of the density of interstellar H atoms inside the heliosphere comes from measurements of pickup H$^+$. An example of a typical proton velocity distribution is given in Figure 7 (Figure 2, Izmodenov et al., 2003a; see also Gloeckler and Geiss, 2001). This spectrum was observed with the Solar Wind Interstellar Composition Spectrometer (SWICS) on Ulysses in the slow solar wind at ∼5 AU during quiet times at low latitudes.

The pickup ion distribution (labeled 'pi' in Figure 7) is modelled by using the "hot model" of Thomas (1978) for the spatial distribution of hydrogen atoms in the heliosphere and equations (9) and (10b) of Vasyliunas and Siscoe (1976) derived under the assumption of rapid pitch-angle scattering and hence isotropy the phase-space density of the resulting pickup protons. Model parameters, in particular the neutral hydrogen density at the termination shock, are adjusted until the best fit to the measured spectrum is obtained. The assumption of isotropic pickup ion distributions is justified in this case because at 5 AU in the ecliptic plane the average magnetic field direction is nearly perpendicular to the solar wind flow direction.

Heliospheric parameters affecting the model pickup ion distribution are known from direct measurements. During the 2-year time period of Figure 7 the average photoionization rate for hydrogen, derived from Lyman alpha measurements on SOHO was 0.8×10^{-7} s^{-1} (D. R. McMullin and D. L. Judge, priv. comm.). The ionization rate from charge exchange, a product of solar wind flux (measured with SWICS) and charge exchange cross section was 5.4×10^{-7} s^{-1}, giving a total loss rate of 6.2×10^{-7} s^{-1}. The total production rate of pickup hydrogen is somewhat smaller (5.1×10^{-7} s^{-1}) because the average solar wind flux for time periods of low thermal speed was measured to be lower than for the entire two year time period. With these values for the loss and production rates, an excellent fit to the measured pickup proton distribution is obtained and the number density of atomic H at the termination shock is determined to be 0.100 ± 0.005 cm^{-3}. The number density can be compared with the number density at the TS obtained in the theoretical models. Izmodenov et al. (2003b) used the heliospheric interface model, which was described above and takes into account influence of both interstellar and solar ionized helium. Parametric studies were performed in that paper by varying the interstellar proton and hydrogen atom number densities in the ranges of 0.03- 0.1 and 0.16 - 0.2 cm^{-3}, respectively. Figure 8 shows updated results compared with the calculations presented in Izmodenov et al. (2003a). Red curves are isolines of H atom number density at the TS. The dashed red area shows the range of possible pairs of ($n_{H,LIC}$, $n_{p,LIC}$), which are comparable to $n_{H,TS}$ obtained from pickup ions data. To reduce the range of possible interstellar atom and proton densities we need to employ other observational diagnostics. These could include, for example, the interstellar helium ionization ($\chi(He) = n_{He^+,LIC}/(n_{He^+,LIC} + n_{He,LIC})$) range of 0.3-0.4 derived from line-of-sight Extreme Ultraviolet Explorer measurements toward white dwarf stars in the local interstellar medium (Wolff et al. 1999). Using 1) ULYSSES/GAS measurements of the LIC

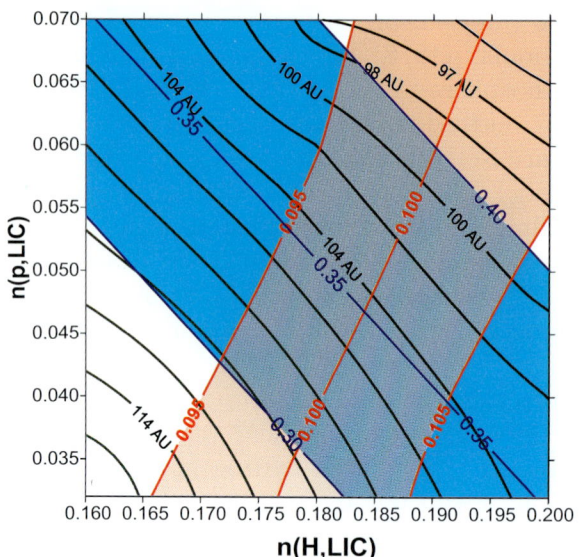

Figure 2.8 Contour plots of the interstellar H atom number density at the termination shock, the LIC helium ionization fraction, and the termination shock location in the Voyager direction.

atomic He density ($= 0.015$ cm^{-3} ; Gloeckler and Geiss 2001; Witte, priv. comm.), 2) the standard universal ratio of the total H to He, $(n_{p,LIC}+n_{H,LIC})/(n_{He^+,LIC}+n_{He,LIC})=10$, the relation between $n_{H,LIC}$ and $n_{p,LIC}$ can be established with

$$n_{p,LIC} + n_{H,LIC} = 0.15(1 - \chi(He^+))^{-1}, cm^{-3}.$$

Isolines of interstellar helium ionization $\chi(He^+)$ are shown in Figure 8. The intersection of the two shaded areas gives the most likely values of interstellar proton and atomic hydrogen number densities compatible with the observations: $n_{H,LIC} = 0.185 \pm 0.01$ cm^{-3} and $n_{p,LIC} = 0.05 \pm 0.015$ cm^{-3}. The interstellar hydrogen ionization fraction derived from this result is in agreement with recent calculations of the photoionization of interstellar matter within 5 pc of the Sun (Slavin and Frisch, 2002).

6.2 Location of the Termination Shock in the Direction of Voyager 1

Based on measurements of the low-energy particle fluxes, spectra, and composition by the Voyager 1 Low Energy Charged Particle instrument and of an indirect determination of the solar wind speed using particle anisotropy measurements, Krimigis et al. (2003) reported the probable crossing of the TS by Voyager 1 at 85 AU in the summer of 2002 and the return to the TS upstream region about 6 months later. McDonald et al. (2003) suggested another interpretation of the Voyager data arguing that the spacecraft remained in the supersonic wind, but in the precursor region. In any case, recent Voyager 1 data suggest that the TS was close to 85 AU in the Voyager 1 direction. To compare this evidence with model predictions we plot in Figure 8 isolines of the distance to the TS in the middle of 2002, computed in the Voyager direction. It is seen that for ($n_{H,LIC}$, $n_{p,LIC}$) comparable to the ionization range of interstellar helium of 0.3 - 0.4 the TS location is 104 ± 4 AU, which is ~20 AU farther from the Sun than Voyager 1. One possible solution to get the TS at ~85 AU in the model is to increase the interstellar atom and proton number densities. Our model-calculations show that for $n_{p,LIC} = 0.11 - 0.12$ cm^{-3} and $n_{H,LIC} \approx 0.22$ cm^{-3} the TS was at 85-86 AU in 2002 and the number density of H atoms at the TS is ~0.1 cm^{-3}. However, such a solution implies 55 % interstellar helium ionization.

6.3 Filtration of Interstellar Oxygen and Nitrogen

Additional constraints on the local interstellar properties and, in particular, local interstellar abundances can be obtained using measurements of pickup ions of heavier interstellar atoms made by Ulysses/SWCIS (Gloeckler and Geiss, 2001) and theoretical calculations of filtration of the atoms through the heliospheric interface.

Izmodenov et al. (2004a) performed comparative studies of the penetration of the interstellar atoms of H, O, N into the heliosphere through the heliospheric interface. We made a parametric study by varying local interstellar proton and atom number densities. It was found that 54 ± % of interstellar hydrogen atoms, 68 ± 3 % of interstellar oxygen and 78 ±2 % of interstellar nitrogen penetrate through the interaction region into the interface. In the case of a lower electron temperature in the heliosheath 81 ±2 % and 89 ±1 % of interstellar oxygen and nitrogen penetrate, respectively. Using our filtration coefficients and SWICS Ulysses pickup ion measurements we conclude that $n_{OI,LIC}$ =(7.8 ± 1.3)·10^5 cm^3 and $n_{NI,LIC}$ =(1.1 ±0.2)·10^5 cm^3. Finally, the local interstellar OI to HI and NI to OI ratios are (OI/HI)$_{LIC}$ =(4.3±0.5)·10^4 and (NI/OI)$_{LIC}$ =0.13±0.01. Our interstellar OI/HI ratio is slightly lower than the ratio (4.8±0.48)·10^4 determined by Linsky et al. (1995) from spectroscopic observations of stellar absorption.

7. Summary

The Local Interstellar Medium interacts with the solar wind and influences the outer heliosphere in a complicated way. Several particle populations and magnetic fields are involved in this interaction. From the interstellar side, the interacting populations are the plasma (electron and proton) component, H atom component, interstellar magnetic field, and galactic cosmic rays. Heliospheric plasma consists of the original solar wind protons, electrons, pickup protons, and the anomalous component of cosmic rays. A large effort has been made to study the theoretical physics of the interaction region. However, a complete, self-consistent model of the heliospheric interface has not yet been constructed, because of the difficulty incorporating both the multi-component nature of the heliosphere and the requirement for different theoretical approaches for different components of the interaction. Many aspects were studied and reported here. However, some aspects require additional theoretical exploration. Most of the theoretical models employ the one-fluid approach for solar wind and interstellar plasmas. It has been shown that several assumptions are needed to derive one-fluid approach equations. A key assumption that looks reasonable is that invoking a co-moving character for all components. Another assumption for a one-fluid plasma model is the immediate assimilation of the pickup ion component into the solar wind. As demonstrated by space experiments, this is not the case and it would be more natural to consider solar protons and pickup protons as separate co-moving populations. The electron component should also be treated as a distinct population. However, since the assumption of the co-moving character of these three heliospheric plasma populations looks reasonable to order one, the one-fluid approach gives us a reasonably accurate picture of the flow pattern (positions of the shocks and heliopause) and plasma velocity distributions. Theoretical kinetic models of the pickup ion transport and acceleration can be employed to determine the distribution of thermal energy between the solar wind and pickup proton components. A similar study should be done for electrons.

Finally, growing interest in heliospheric interface studies is connected with expectations that Voyager 1 recently crossed the termination shock or will cross the shock in the near future. Many predictions of the time of the termination shock being crossed by Voyager appeared in the literature. However, it seems that much more work should be done to explain and reconcile all available indirect observations of the heliospheric interface based on the unique model of the heliospheric interface. This work should be done especially because NASA plans to explore the interaction region remotely by ENA imaging (HIGO, or the proposed Small Explorer Mission known as IBEX, Web-page at http://ibex.swri.edu/) and to send a spacecraft (the Interstellar Probe) to a heliocentric distance of at least 200 AU with a flight-time of only 10 or 15 years (http://interstellar.jpl.nasa.gov/). Intensive theoretical study will help to optimize goals, instrumentation, and, finally, the scientific profit of the "interstellar" missions.

Acknowledgements. This work was supported in part by INTAS Award 2001-0270, RFBR grants 04-02-16559, 03-01-39004, 03-02-04020, 04-01-00594 and the International Space Science Institute in Bern. I thank Steve Suess for his numerous corrections and suggestions to improve style and English of the paper. I also thank Sergey Chalov for useful discussions.

References

Aleksashov, D., Baranov, V., Barsky, E., Myasnikov, A., Astronomy Letters, 26 (2000), 743-749.

Alexashov. D. and V. Izmodenov, in SOLAR WIND TEN: Proceedings of the Tenth International Solar Wind Conference, Eds. M. Velli and R. Bruno, AIP Conference Proceedings 679, pp. 218-221, 2003.

Alexashov, D., V.V. Izmodenov and S. Grzedzielski, Adv. Space Res., in press, 2004.

Alexashov, D., Chalov, S.V., Myasnikov, A., Izmodenov V., Kallenbach, R., Astron. Astrophys., in press, 2004.

Baranov, V.B., Krasnobaev, K.V., Kulikovksy, A.G., Sov. Phys. Dokl. 15 (1971), 791.

Baranov, V. B., Lebedev, M.,Malama Y., Astrophys. J. 375 (1991), 347-351.

Baranov, V., Malama, Y., J. Geophys. Res. 98 (1993), 15157.

Baranov, V. B., Malama, Y. G., J. Geophys. Res. 100 (1995), 14,755-14,762.

Baranov, V.B., Zaitsev, N.A., Astron. Astrophys. 304 (1995), 631.

Baranov, V., Zaitsev, N., Geophys. Res. Let. 25 (1998), 4051.

Baranov, V. B., Izmodenov, V., Malama, Y., J. Geophys. Res. 103 (1998), 9575-9586.

Baranov, V. B., Astrophys. Space Sci. 274 (2000), 3-16.

Baranov, V. B., and Y. G. Malama, Space Sci. Rev. 78 (1996), 305-316.

Baranov, V. B., and Fahr, H. J., J. Geophys. Res. (Space Physics), Vol. 108,A3, pp. SSH 4-1, CiteID 1110, DOI 10.1029/2001JA009221, 2003a.

Baranov, V. B., and H. J. Fahr, J. Geophys. Res., 108(A12), 1439, 10.1029/2003JA010118, 2003b.

Berezhko, E. G. 1986, Soviet Astron. Let., 12, 352

Chalov, S. V. 1988a, Soviet Astron. Let., 14, 114

Chalov, S. V. 1988b, Astrophys. Space Sci., 148, 175

Chalov, S. V. 1990, in Physics of the Outer Heliosphere, ed. S. Grzedzielski, & D. E. Page (Pergamon), 219.

Chalov, S. V., Fahr, H., Astron. Astrophys. 311 (1996), 317-328.

Chalov, S. V., Fahr, H., Astron. Astrophys. 326 (1997), 860-869.

Chalov, S. V., Fahr, H., Astron. Astrophys., 335 (1998), 746-756.

Chalov, S. V., Fahr, H., Solar Physics 187, 123-144 (1999).

Chalov, S. V., Fahr, H., Astron. Astrophys., 401 (2003), L1-L4.

Fahr, H., Kausch, T., Scherer, H., Astron. Astrophys. 357 (2000), 268-282.

Fichtner, H., Space Sci. Rev. 95 (2001), 639-754.

Florinski, V., and G. P. Zank, J. Geophys. Res., 108(A12), 1438, doi: 10.1029/2003JA009950, 2003.

Fujimoto Y., Matsuda, T., Preprint No. KUGD91-2, Kobe Univ., Japan, 1991.

Gazis, P.R., Reviews of Geophysics, 34,4, 379-402, 1996.

Gloeckler, Space Sci. Rev. 78 (1996), 335-346.

Gloeckler, G., Fisk, L.A., Geiss, J., Nature 386 (1997), 374-377.

Gloeckler, G., and Geiss, J., Joint SOHO/ACE workshop "Solar and Galactic Composition". Edited by Robert F. Wimmer-Schweingruber. Publisher: American Institute of Physics Conference proceedings vol. 598 location: Bern, Switzerland, March 6 - 9, 2001., p.281

Gloeckler, G., Moebius, E., Geiss, J., Astron. Astrophys., in press, 2004.

Gurnett, D. A., Kurth, W., Allendorf, S., Poynter, R., Science 262 (1993), 199-202.

Gurnett, D.,Kurth,W.,Space Sci. Rev. 78 (1996), 53-66.

Gruntman et al., J. Geophys. Res. 106, 15767-15782 (2001).

Holzer, J. Geophys. Res. 77 (1972), 5407.

Isenberg, P., J. Geophys. Res. 91 (1986), 9965.

Isenberg, P., J. Geophys. Res. 102 (1997), 4719-4724.

Izmodenov, V., Lallement, R., Malama, Y., Astron. & Astrophys, 342 (1999a), L13-L16.

Izmodenov V., Geiss, J., Lallement, R., et al., J. Geophys. Res. (1999b), 4731-4742.

Izmodenov, V., Astrophys. Space Sci. 274 (2000), 55-69.

Izmodenov, V., Malama, Y., Kalinin, A.,et al., Astrophys. Space Sci. 274 (2000), 71-76.

Izmodenov, V., Space Sci. Rev. 97 (2001): 385-388.

Izmodenov, V., Gruntman, M., Malama, Y., J. Geophys. Res. 106 (2001), 10681.

Izmodenov, V., Wood, B., and R. Lallement, J. Geophys. Res. 107(10), doi: 10.1029/2002JA009394, 2002.

Izmodenov, V., Malama, Y.G., Gloeckler, G., Geophys. Res. Let., 2003a.

Izmodenov, V., Malama, Y.G., Gloeckler, G., Geiss, J., Astrophys. J., 594:L59-L62, 2003b.

Izmodenov, V., and, D. Alexashov, Astronomy Letters, Vol. 29, No. 1, pp. 58-63, 2003.

Izmodenov, V., Malama, Y.G., Gloeckler, G., Geiss, J., Astron. Astrophys, 414, L29-L32, 2004a.

Izmodenov, V., Malama, Y.G., Ruderman, M.S., Astron. Astrophys, in press, 2004b.

Karmesin, S., Liewer, P, Brackbill, J., Geophys. Res. Let. 22 (1995), 1153-1163.

Krimigis, S. M., Decker, R. B., Hill, M. E., et al., Nature, 426, Issue 6962, pp. 45-48 (2003).

Lallement R., Bertin, Astron. Astrophys. 266 (1992), 479-485.

Lallement, R., Space Sci. Rev. 78 (1996), 361-374.

Lallement, R., Ferlet, A., et al., Astron. Astrophys. 304 (1995), 461-474.

Lallement, R., Linsky, J., Lequeux, J., Baranov, V., Space Sci. Rev. 78 (1996), 299-304.

Lazarus, A.J., and R.L. McNutt, Jr., in Physics of the Outer Heliosphere, Ed's S. Grzedzielski and D.E. Page, Pergamon (1990).

Liewer, P., Brackbill, J., Karmesin, S., International Solar Wind 8 Conference, p.33, 1995.

Linde, T., Gombosi, T., Roe, P., J. Geophys. Res. 103 (1998), 1889-1904.

Linsky, J., Wood, B., Astrophys. J. 463 (1996), 254.

Linsky, J., Brown A., Gayley, K., et al., Astron. J. 402 (1993), 694-709.

Linsky J.L., Dipas A., Wood B.E., et al., *Astrophys. J.* 476, 366, 1995.

Lipatov, A.S., Zank, G.P., Pauls, H.L., J. Geophys Res. 103, (1998), 20631-20642.

Malama, Y. G., Astrophys. Space Sci., 176 (1991), 21-46.

McComas, D.J., Barraclough, B.L., Funsten, H.O., Gosling, J., T., et al., J. Geophys. Res., 105, A5, 10419-10433, 2000.

McComas, D.J., et al., Space Science Rev., 97, 189, 2001.

McComas, D.J., Elliot, H.A., Gosling, J.T., et al., Geophys. Res. Letters, 29, 9, doi:10.1029/2001GL014164, 2002.

McDonald, F. B., Stone, E. C., Cummings, A. C., et al., Nature, 426, Issue 6962, pp. 48-51 (2003).

Myasnikov, A., Preprint No. 585, Institute for Problems in Mechanics, Russian Academy of Sciences, 1997.

Myasnikov, Izmodenov, V., Alexashov, D., Chalov, S., J. Geophys. Res. 105 (2000a), 5179.

Myasnikov, Alexashov, D., Izmodenov, V., Chalov, S., J. Geophys. Res. 105 (2000b), 5167.

Moebius, E., Space Sci. Rev. 78 (1996), 375-386.

Müller et al., J. Geophys. Res., 27,419-27,438 (2000).

Neugebauer, M., Rev. Geophys. 37, 1, 107-126, 1999.

Osterbart, R., and H. Fahr, Astron. Astrophys. 264 (1992), 260-269.

Opher, M., Liewer, P.C., Gombosi, T.I., et al., Astrophys. J. 591 (2003) L61-L65.

Parker, E. N., Astrophys. J. 134 (1961), 20-27.

Pauls, H. and G. Zank, J. Geophys. Res. 102 (1997), 19779-19788.

Pogorelov, N., Astron. Astrophys. 297 (1995), 835.

Pogorelov, N., Semenov, A., Astron. Astrophys. 321 (1997), 330.

Pogorelov, N., Matsuda, T., J. Geophys. Res. 103 (1998), 237-245.

Quemerais, E. and V. Izmodenov, Astronomy and Astrophysics, v.396, p.269-281, 2002.

Ratkiewicz, R., Barnes, A, et al., Astron. Astrophys. 335 (1998), 363.

Ratkiewicz, R., Barnes, A., J. Spreiter, J. Geophys. Res. 105 (2000), 25,021-25,031.

Richardson, J.D., Geophys. Res. Lett. 24 (1997), 2889-2892.

Richardson, J.D., The Outer Heliosphere: The Next Frontiers, Edited by K. Scherer, H. Fichtner, H. Fahr, and E. Marsch, COSPAR Colloquiua Series, 11. Amsterdam: Pergamon Press (2001), 301-310.

Richardson, J.D., Wang, C., Burlaga, L.F., Advances in Space Research, in press, 2004.

Slavin, J. D., and Frisch, P.C., Astrophys. J. 565, 364-379, 2002.

Smith, C. W., Matthaeus, W. H., Zank, G. P., et al., J. Geophys. Res. 106 (2001), 8253-8272.

Steinolfson, R.S., J. Geophys. Res. 99 (1994), 13,307-13,314.

Scherer, K., Fahr, H. J., Geophys. Res. Lett. 30 (2003), doi: 10.1029/ 2002GL016073, 2003a.

Scherer, K., and Fahr, H. J., Annales Geophys., 21 (2003b), 1303-1313.

Stone, E.C., *Science, 293, 55-56*, 2001.

Tanaka, T., Washimi, H., J. Geophys. Res. 104 (1999), 12605.

Thomas, G. E., *Ann. Rev. Earth Planet. Sci.* 6, 173, 1978.

Treumann, R., Macek, W., Izmodenov, V, Astron. Astrophys. 336 (1998), L45.

Vasyliunas, V. M. and. Siscoe, G. L., system, *J. Geophys. Res.* 81, 1,247-1,252, 1976.

Webber, W.R., Lockwood, J., McDonald, F., Heikkila, B., J. Geophys. Res. 106 (2001), 253-260.

Williams, L., Zank, G., Matthaeus, W., J. Geophys. Res. 100 (1995), 17059-17068.

Wang, C., Belcher, J, J. Geophys. Res. 104 (1999), 549-556.

Williams, L., Hall, D. T., Pauls, H. L., Zank, G. P., Astrophys. J. 476 (1997), 366.

Witte, M., Banaszkiewicz, M.; Rosenbauer, H., Space Sci. Rev. 78 (1996), 289-296.

Witte, M., Astron. Astrophys., in press, 2004.

Wood, B. E., Muller, H.; Zank, G. P., Astrophys. J. 542 (2000), 493-503.

Wolff, B., Koester, D., and Lallement, R., *Astron. Astrophys.*, 346, 969-978,1999.

Zaitsev, N., Izmodenov V., in The Outer Heliosphere: The Next Frontiers, Edited by K. Scherer, H. Fichtner, H. Fahr, and E. Marsch, COSPAR Colloquiua Series, 11. Amsterdam: Pergamon Press (2001), 65-69.

Zank, G. P., Axford, W.I., & McKenzie, J. F. 1990, Astron. Astrophys., 233, 275.

Zank, G., Pauls, H., Williams, L., Hall, D., J. Geophys. Res. 101 (1996), 21639-21656.

Zank, G., Space Sci. Rev. 89 (1999a), 413-688.

Zank, G., in Solar Wind Nine, edited by S.R. Habbal et al., AIP (1999b), 783-786.

Zank, G. P., Müller, H.-R., The dynamical heliosphere, J. Geophys. Res., 108,pp. SSH 7-1, CiteID 1240, DOI 10.1029/2002JA009689, 2003.

Chapter 3

RADIO EMISSION FROM THE OUTER HELIOSPHERE AND BEYOND

Iver H. Cairns

School of Physics, University of Sydney, NSW 2006, Australia.

cairns@physics.usyd.edu.au

Abstract   The Voyager spacecraft have observed episodic bursts of radio emission near 2-3 kHz that are generated outside the inner heliosphere, arguably when shock waves driven by global merged interaction regions (GMIRs) reach the vicinity of the heliopause. Voyager observations and theories for the source region and generation of the radiation are reviewed in this paper. Special foci are the GMIR model for interpreting the data, and the successes of the new Priming/GMIR theory which can explain the turn-on and generation of the radiation in the outer heliosheath, near the heliopause nose. Unresolved issues and directions for future work are also summarized, including the direction of the local interstellar magnetic field, timing of radio events, and propagation of the radiation into the inner heliosphere.

Keywords: radio emissions, outer heliosphere, GMIR, shock, pickup ions, electron acceleration, Langmuir waves, lower hybrid waves, radiation processes, nonlinear wave-wave interactions, propagation effects.

1.   Introduction

In 1983 the two Voyager spacecraft were beyond the orbit of Saturn and separated by several astronomical units (AU) when they observed radio emissions at frequencies $f \approx 2 - 3$ kHz with very similar dynamic spectra Kurth et al., 1984. The observations were interpreted in terms of emissions originating in the outer heliosphere and, most likely, a signature of the solar wind's interaction with the very local interstellar medium (VLISM). The basic rationale is that propagating radiation must have $f \geq f_p$, the electron plasma frequency, and f_p exceeds $2 - 3$ kHz interior to Saturn's orbit for nominal solar wind conditions.

G. Poletto and S.T. Suess (eds.), The Sun and The Heliosphere as an Integrated System, 65–90.
© 2004 *Kluwer Academic Publishers. Printed in the Netherlands.*

Kurth et al.'s (1984) basic interpretation remains favored today, over 20 years later, with the bursts now associated with global merged interaction regions (GMIRs) engendered by solar activity Gurnett et al., 1993; Gurnett and Kurth, 1996; Cairns and Zank, 2001; Zank et al., 2001. The phenomenon and its interpretation continue to attract major scientific interest due to them simultaneously combining space/solar system physics and astrophysics, with significant implications for both fields, similar to the related phenomena of cosmic rays and interstellar pick-up ions.

Figure 1 summarizes the plasma regions and discontinuities expected in the outer heliosphere. More details can be found elsewhere Zank, 1999; Izmodenov, 2004; Linsky, 2004. The superalvénic, supersonic solar wind plasma undergoes a shock transition at the termination shock, probably at a (solar cycle dependent) heliocentric distance $\approx 85 - 100$ AU. The inner heliosheath contains shocked solar wind plasma, which is slowed to speeds ≈ 100 km s^{-1}, compressed by a factor $\approx 2 - 4$, heated to temperatures $\approx 10^6$ K, deflected in direction, and the magnetic field amplified and rotated, at the termination shock. Analogous to the terrestrial magnetopause, the heliopause is a rotational discontinuity between the inner and outer heliosheaths which separates the shocked solar wind plasma from interstellar plasma. Finally, if the solar system moves superalfvénically or supersonically then the VLISM plasma will be shocked at a bow shock, and modified similarly to the solar wind at the termination shock. The outer heliosheath contains the VLISM plasma processed by the bow shock (or a bow wave if the flow is subsonic and subalfvénic).

Neutral particles are free to move across these plasma boundaries and to undergo charge-exchange collisions, resulting in pickup ions and associated momentum and energy changes to the plasma flows. So-called "region 1", "2" and "3" neutrals are formed in the VLISM and outer heliosheath, inner heliosheath, and solar wind, respectively. For instance, charge-exchange interactions between interstellar hydrogen atoms (region 1 neutrals) and shocked solar wind protons leads to a population of hot, slow neutrals moving outwards across the heliopause from the inner heliosheath (region 2 neutrals) and a concomitant reduction of interstellar neutrals entering the solar wind (the filtering factor is $\gtrsim 90\%$ for hydrogen) Holzer, 1989; Zank, 1999; Izmodenov, 2004. Ordinary collisions with region 1 neutrals, and multiple charge-exchange collisions, then lead to a region of enhanced neutral density and temperature called the "hydrogen wall", that is typically between the heliopause and bow shock. Of direct interest here is the secondary charge-exchange between region 2 neutrals and interstellar protons in the outer heliosheath (in the

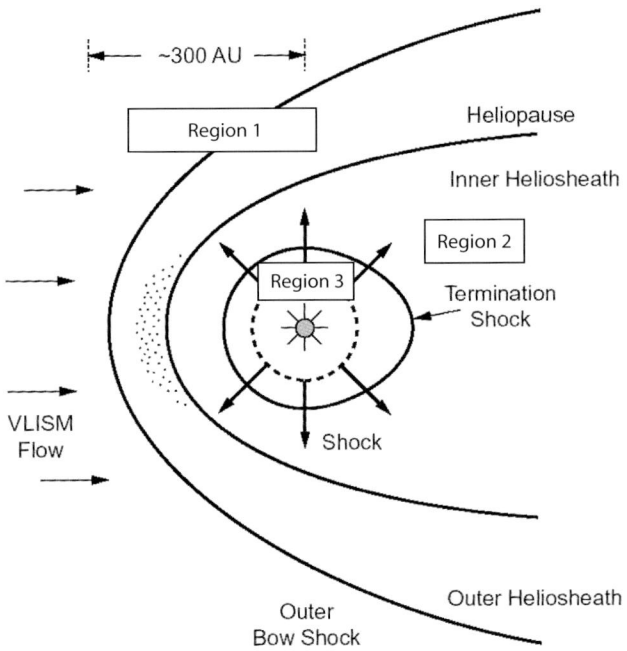

Figure 3.1. Major plasma regions and boundaries expected for the solar wind - VLISM interaction. The dashed line denotes a GMIR shock moving outwards. The dotted region shows the source region predicted for the Voyager emissions by the GMIR model and Priming/GMIR theory.

hydrogen wall region). This charge-exchange is predicted to prime a region of the outer heliosheath near the heliopause nose with an enhanced population of superthermal electrons (via pickup ions, the growth of lower hybrid waves, and resonant electron acceleration by the lower hybrid waves) and the Voyager radio events are produced when suitable shocks enter this source region Cairns and Zank, 2001; Cairns and Zank, 2002; Cairns et al., 2004; Mitchell et al., 2004. The dotted region in Figure 1 shows this predicted source region.

The aim of this paper is to review the status of observations and theories for the 2-3 kHz emissions, concentrating on the overall picture and recent progress. Other reviews exist Gurnett and Kurth, 1995; Cairns and Zank, 2001; Czechowski, 2004 and the detailed history of the subject is left to them and the original literature. Since the theory has advanced further than the observations in the last 5 years, the paper's primary focus is on reviewing recent theoretical progress. The remainder of the paper is organised as follows. Section 2 describes the observational status of the emissions, emphasizing the exciting progress made recently on the source location. The primary theoretical issues and previous theoretical models are summarized in Section 3. Section 4 describes the Priming/GMIR theory for the radiation and recent theoretical results are summarized in Section 5. Remaining theoretical and observational issues are discussed in Section 6, which also contains a brief summary.

2. Current Observational Status

Figure 2 is a recent dynamic spectrum of the emissions from 1982 until June 2003, showing the emission frequency versus time with the intensity color coded. It was obtained from the Voyager Plasma Wave Subsystem (PWS) site http://www-pw.physics.uiowa.edu/wsk/vgr/recent.html, courtesy of D.A. Gurnett and W.S. Kurth, and is similar to figures in other publications Kurth et al., 1987; Gurnett et al., 1993; Gurnett et al., 2004. The intense and continuous red band near 2.4 kHz is interference from Voyager 1's power supply system. The signals below 1 kHz at all times, and the diffuse light blue signals below 2 kHz from 1982 to mid-1992, are dominated by interference from other spacecraft instruments and systems. The signals of interest here are the three relatively intense, longlasting episodes of emission from $\approx 1.8 - 3.6$ kHz (1983-84, 1992-95, and 2003-2004), together with weaker events (e.g., early 1986, late 1989, and late 1991) in the same frequency range. Note that these major episodes are separated by an approximately 9-10 year period, reminiscent of the solar cycle, and that the third outburst started in mid-2003 and was relatively short-lived.

Calculations of the source power, assuming isotropic emission, are $\gtrsim 10^{13}$ W Gurnett et al., 1993. This power greatly exceeds those of planetary radio emissions ($\lesssim 10^{11}$ W for auroral emissions from Earth and Jupiter) and type II and III solar radio bursts. The outer heliospheric emissions therefore have the largest power of all solar system radio emissions Gurnett et al., 1993.

The Voyager emissions are presently categorized in two classes Kurth et al., 1987; Cairns et al., 1992; Gurnett et al., 1993: (1) the "2 kHz component", which remains in the frequency range $1.8 - 2.6$ kHz, is longer lasting (≈ 3 years), and does not drift significantly in frequency; (2) "transient" or "drifting" emissions which drift up in frequency, have a range of starting and ending frequencies within the domain $1.8 - 3.6$ kHz, frequency drift rates in the range $\approx 1 - 3$ kHz/year, and last for $\approx 100 - 300$ days. Clear evidence exists for frequency fine structures, particularly for the transient emissions. Figures 2 and 3 show that the transient emissions during 1994 often occurred as pairs of signals with

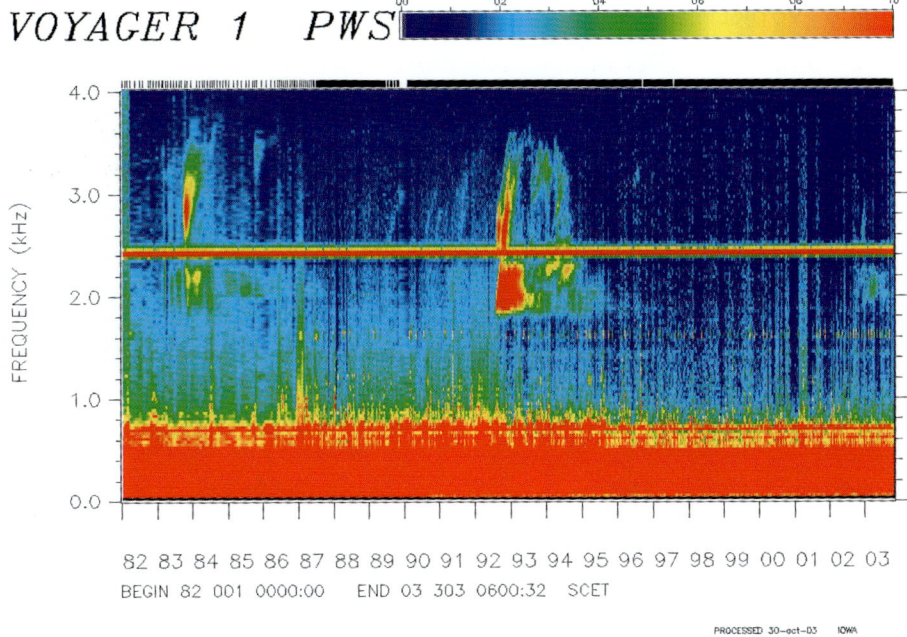

Figure 3.2. Voyager 1 dynamic spectrum for the period 1982-2003. Major episodic radiation events are visible, as well as weak drifting events. Two classes of radiation event are discernible: the "2 kHz component" and drifting "transient emissions". See the text for more details.

very similar drifts that are offset in frequency. This "pairing" characteristic is not understood but is reminiscent of split-band and multiple-lane
type II solar radio bursts Wild et al., 1963; Nelson and Melrose, 1985.
Possible interpretations include the shock moving across 2 regions with
slightly different densities or the splitting being an intrinsic feature of the
emission process; however, splitting at the electron cyclotron frequency
Cairns, 1994 yields unrealistically large magnetic fields $\gtrsim$ 10 nT.

Following earlier work on event triggers McNutt, 1988; Grzedzielski
and Lazarus, 1993, Gurnett et al. (1993) postulated that the episodic
bursts are produced when global merged interaction regions (GMIRs)
crossed the heliopause. GMIRs are formed by the merging of multiple
interacting coronal mass ejections (CMEs) and other fast plasma flows
produced by solar activity into a global disturbance of the plasma density, magnetic field and flow speed that propagate outwards faster than
the ambient solar wind. Figure 4 shows when multiple spacecraft ob-

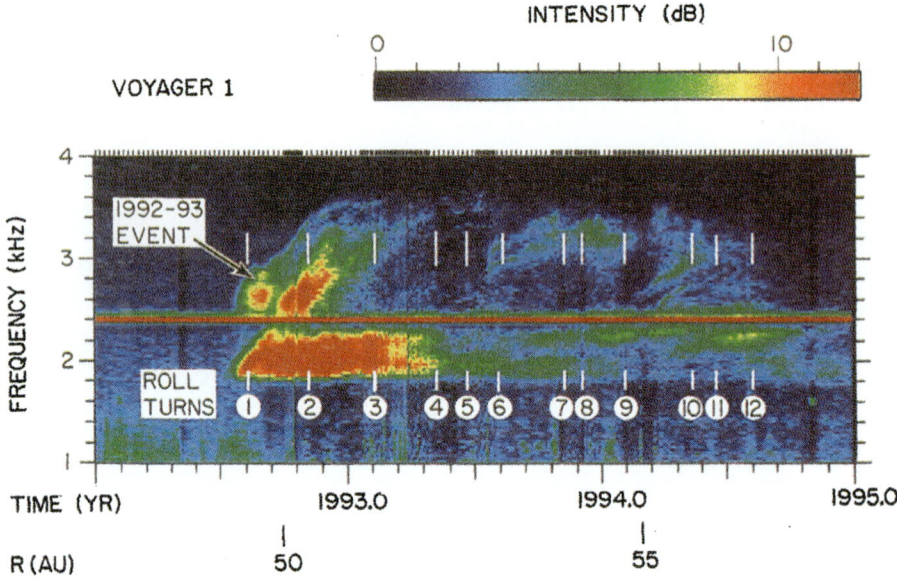

Figure 3.3. Closeup of transient emissions and the 2 kHz component for the 1992-94 event Gurnett et al., 1998. Numbers show the times of spacecraft rolls and associated direction-finding analyses discussed below.

served the shock driven by the GMIR associated with the 1992-1994 radiation event, as well as the Forbush decreases in cosmic ray flux caused by cosmic rays being reflected and scattered by the enhanced and disturbed magnetic field of the GMIR material. Detection of these signatures by the widely separated Voyager and Pioneer spacecraft confirms that the disturbance was truly global.

Figure 5 confirms the association between large GMIRs and the major radio outbursts Gurnett and Kurth, 1995: the 1983-84 and 1992-94 outbursts are associated with the two largest Forbush decreases observed in the Climax Neutron Monitor data. Moreover, the time-lag between the Forbush decreases at Earth and the radio onsets at Voyager are consistent, at 415 ± 4 days and the two GMIR propagation speeds to Voyager were consistent at $\approx 830 \pm 20$ km s^{-1}. Using these speeds and time-lags, together with plausible estimates for shock slowing, Gurnett et al. (1993) estimated the source to be at a radial distance $R \approx 140 - 190$ AU. These distances are plausible for the heliopause.

Recently Kurth and Gurnett, 2003 have combined GMIR time-of-flight effects with two other analyses in order to constrain further the distance and direction to the radiation source, extending earlier work

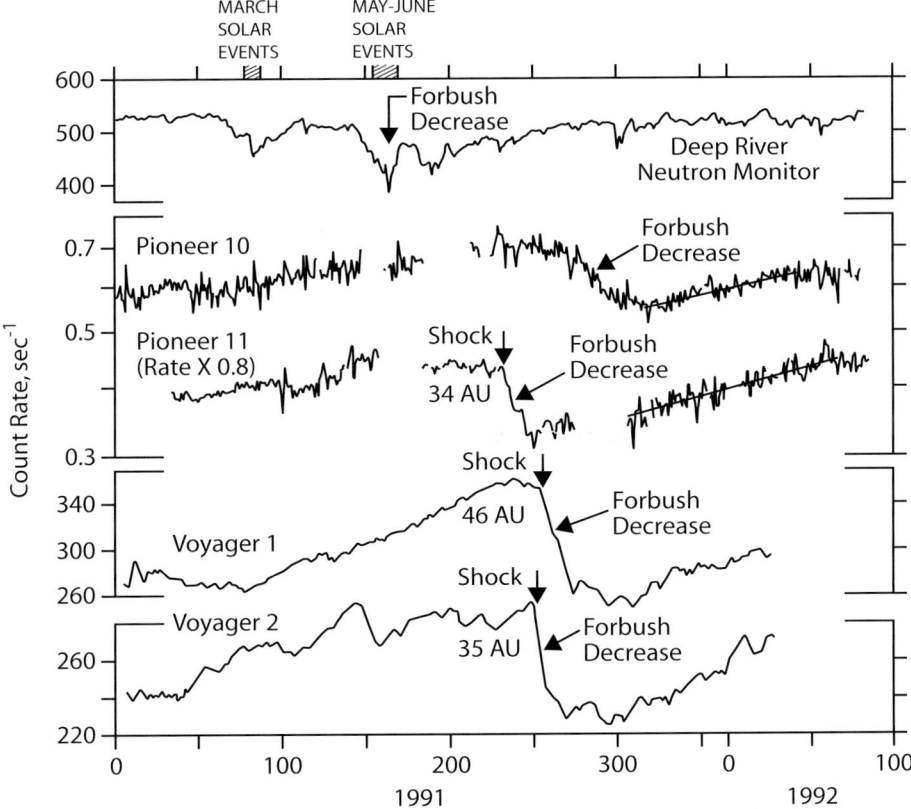

Figure 3.4. Forbush decreases in cosmic rays as functions of time at Earth and various widely separated spacecraft in the outer heliosphere Gurnett et al., 1993. These show the global nature of the GMIR associated with the 1992-94 radio event.

(Gurnett et al., 1993, 1998). One analysis involves the fact that the radiation has different amplitudes for the transient emissions but otherwise very similar dynamic spectra when observed by Voyagers 1 and 2 (e.g., Figure 6 Kurth and Gurnett, 2003): using the known locations of the spacecraft and the falloff of the radiation flux with inverse distance squared from its source, the relative amplitudes define a 3-D locus. The second analysis involves roll modulation patterns: Voyager is rolled about axis of the high-gain communications antenna (typically directed towards Earth) and the direction to the source in the spin plane is determined from minima in the measured radiation intensity (e.g., Gurnett et al., 1998).

The three techniques yield simultaneous solutions for multiple observations of the transient emissions during the 1992-94 event, as shown

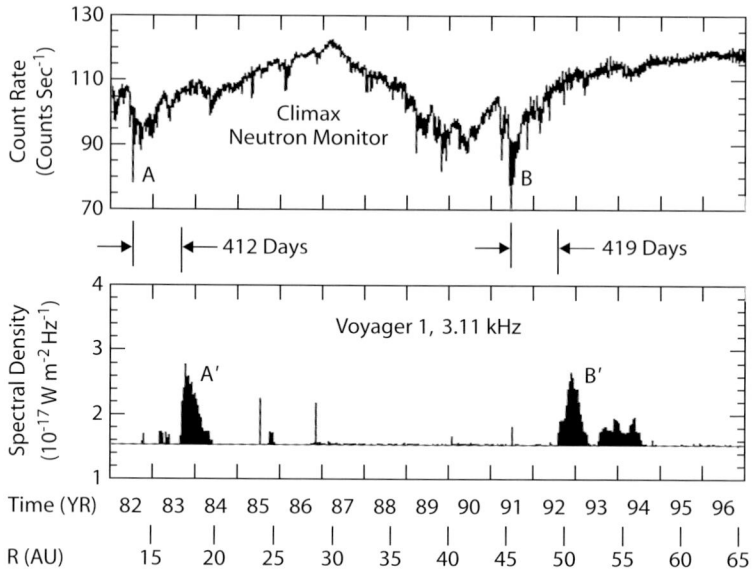

Figure 3.5. Association between the largest Forbush decreases at Earth (A and B) and the major 2-3 kHz radio events (A' and B') Gurnett and Kurth, 1996.

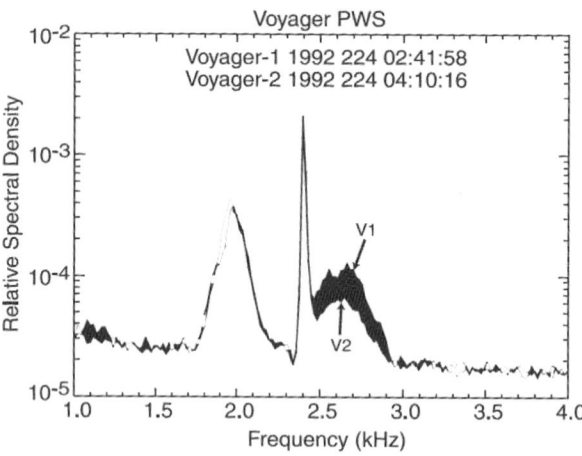

Figure 3.6 Relative spectral densities for the widely separated Voyager 1 and 2 spaceccraft on day 224, 1992. The different amplitudes of the transient emissions above ≈ 2.5 kHz constrain the source location Kurth and Gurnett, 2003.

in Figure 7 as a function of galactic latitude and longitude Kurth and Gurnett, 2003. The results are clear. First, the source of the transient emissions is initially very close to the nose direction for the heliopause. Second, the source direction changes with time but lies along a line closely parallel to the galactic plane. Third, the source generally moves

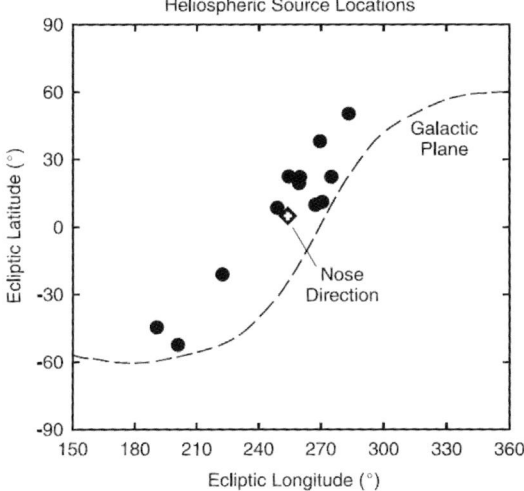

Heliospheric Source Locations

Figure 3.7 Source directions of transient emissions as functions of [1950] ecliptic latitude and longitude derived from analyses of roll modulations, relative amplitudes between Voyagers 1 and 2, and GMIR-radio time-lags Kurth and Gurnett, 2003. Source start near the heliopause nose and typically move away with increasing time.

away from the nose direction and the modulus of its ecliptic latitude typically increases with time. The first and third of these results confirmed earlier analyses (Gurnett et al., 1993, 1998). A detailed physical explanation for these results was not attempted. However, it was speculated that the second result arises because the apparent source axis is aligned with **B** in the outer heliosheath, implying that **B** is parallel to the galactic plane. A physical interpretation is suggested for this below based on the lower hybrid priming mechanism and magnetic draping at the heliopause: the source elongation defines the direction, in the plane of the sky, of **B** in the outer heliosheath.

3. Basic Theoretical Issues

The foregoing observational data argue compellingly that the 2-3 kHz emissions are produced when a GMIR shock crosses the heliopause and enters the outer heliosheath Gurnett et al., 1993. By analogy with solar and interplanetary type II radio bursts and radiation from Earth's foreshock Wild et al., 1963; Gurnett, 1975; Knock et al., 2001; Kuncic et al., 2002a, all of which are associated with shock waves travelling through the solar wind, the 2-3 kHz emissons are most likely produced at multiples of f_p. Nevertheless, a number of fundamental issues must be resolved for this GMIR model Cairns and Zank, 2001. They include:

1 Why and where does the radiation turn on?

2 Why does the GMIR shock not emit observable radiation in the solar wind and inner heliosheath?

3 Can a detailed theory be developed for the observed radiation levels?

4 Why is the observed radiation at frequencies of order plausible values of f_p in the outer heliosheath, with no evidence for $2f_p$ radiation?

The theory reviewed in Sections 4 and 5 can answer these questions semi-quantitatively. Before proceeding there, however, reasons are briefly sketched for why other mechanisms and interpretations for the radiation are not currently considered relevant.

First, could the radiation originate in the magnetosphere of Jupiter? Reasons against this include Gurnett et al., 1993; Gurnett and Kurth, 1994: (1) the absence of a significant correlation between the 2-3 kHz events and increases in Jovian emissions, (2) the observed correlation and time-lag between large GMIRs and the large radiation events, and (3) the 1992-94 event being more intense than the 1983-84 event despite the Voyagers being much more distant. Reason (2) also argues against Saturn or another planet being relevant.

Second, could the radiation be produced at multiples of f_p in the foreshock region sunwards of the termination shock Macek et al., 1991; Cairns and Gurnett, 1992; Cairns et al., 1992? The problem here is that the termination shock is likely near $80 - 100$ AU, where f_p is expected to be ≈ 300 Hz and so much less than the observed radiation frequencies. The interpretation is thus implausible Gurnett et al., 1993; Cairns and Zank, 2001. However, existing calculations predict that significant levels of radiation should be produced in this foreshock region Macek et al., 1991; Cairns and Gurnett, 1992. Voyager observations in the range 200 - 600 Hz, unfortunately contaminated by spacecraft interference (e.g., Figures 2 and 3), may detect this radiation as the Voyagers approach the termination shock.

Third, synchrotron emission and cyclotron maser emission are strongly implausible Cairns and Zank, 2001. This conclusion is based on likely values of the electron cyclotron frequency f_{ce} and f_p beyond the heliopause, the resulting ratios $f/f_{ce} \approx 10^3$ and $f_p/f_{ce} \gg 10$, and conditions on the mechanisms to produce observable narrowband radiation.

4. The Priming/GMIR Theory

The Priming/GMIR theory involves combining Gurnett et al.'s (1993) GMIR model for the radiation with a theory for priming/triggering of the radiation (Cairns and Zank, 2001, 2002) and a theory developed for type II solar radio bursts Knock et al., 2001. The resulting theory Cairns et al., 2004; Mitchell et al., 2004 provides a quantitative theoretical basis for Gurnett et al.'s GMIR model.

The theory involves the following primary concepts Cairns and Zank, 2001; Cairns and Zank, 2002; Cairns et al., 2004; Mitchell et al., 2004. (1) The emission is f_p radiation produced in foreshock regions upstream of a rippled GMIR shock. (2) The radiation turns on (or is triggered) when the GMIR shock enters a region primed with an enhanced superthermal electron tail just beyond and near (within ≈ 50 AU) of the heliopause nose. (3) The priming mechanism involves pickup ions driving lower hybrid (LH) waves which then resonantly accelerate the electron tail by a process called lower hybrid drive (LHD). (4) The pickup ions result from charge-exchange in the outer heliosheath of Region 2 neutrals produced originally in the inner heliosheath. (5) Constraints on LHD localize the priming to the outer heliosheath near the magnetic draping region. These concepts are now reviewed in more detail.

Neutrals from the inner heliosheath (formerly shocked solar wind protons which charge-exchanged with interstellar neutrals) move outwards across the heliopause at speeds ≈ 100 km s^{-1} and undergo another charge-exchange in the outer heliosheath (e.g., Zank et al., 1996; Zank, 1999; Izmodenov, 2004). The newly charge-exchanged ions are then picked-up by the plasma's convection electric field, developing a gyromotion and a ring distribution in velocity space (Figure 8), perpendicular to **B**. (If the neutral's initial $v_{\parallel} \neq 0$, then the distribution is more properly a ring-beam distribution, but this detail is ignored below.) The ring speed $v_r \approx 100$ km s^{-1}. When the ring is sufficiently narrow in velocity space and the plasma β (equal to the ratio of the thermal pressure to the magnetic pressure) is sufficiently low, then the ring is unstable to the growth of LH waves McBride et al., 1972; Omelchenko et al., 1989.

LH waves are low frequency ($f \approx f_{LH} = (f_{ce}f_{ci})^{1/2}$), primarily electrostatic waves that propagate almost perpendicular to **B** (with $\mathbf{k} \approx \mathbf{k}_{\perp}$ and $k_{\perp}/k_{\parallel} \approx (m_i/m_e)^{1/2}$. Their phase fronts then move very fast along **B**, and the associated parallel electric fields can accelerate electrons to large speeds via the Cerenkov resonance. This latter process is sometimes called lower-hybrid drive (LHD). Pickup ion distributions can effectively drive LH waves that resonate with both the ion and electron distributions, with $\omega \approx \omega_{LH} \approx k_{\perp}v_r \approx k_{\parallel}v_{\parallel,e}$. Under these conditions the LH waves accelerate a superthermal tail out of the thermal electron distribution McBride et al., 1972; Omelchenko et al., 1989; McClements et al., 1993; Shapiro et al., 1998; Cairns and Zank, 2001; Cairns and Zank, 2002, with maximum speed

$$v_m = (m_i/m_e)^{1/2}v_r \, , \tag{3.1}$$

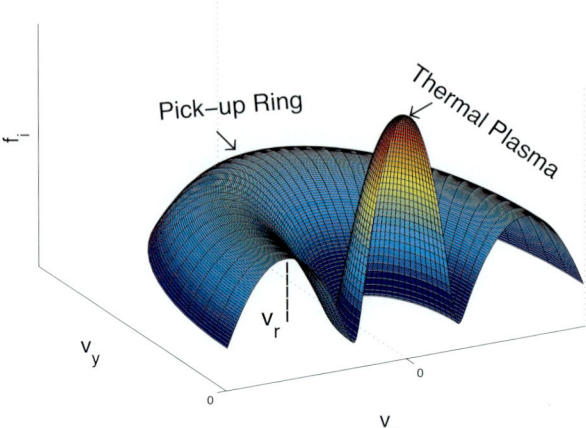

Figure 3.8 Superposition of a pickup ion ring on a Maxwellian distribution of background thermal ions. Here **B** is parallel to the z axis.

as illustrated in Figure 9. Estimates for LHD in the outer heliosheath have $v_m \approx 10V_e \approx 10^7$ m s^{-1}, where $V_e \approx 10^6$ m s^{-1} is the electron thermal speed, and a total tail fraction $\approx 10^{-5} - 10^{-4}$ of the background electron number density Cairns and Zank, 2002.

Due to constraints on LHD the priming mechanism is localized to the outer heliosheath in the vicinity of the heliopause nose, the presumed site of magnetic draping. Specifically: (1) previous simulations show that LHD is only efficient when $v_r/V_A \lesssim 5$ Omelchenko et al., 1989; Shapiro et al., 1998, where V_A is the Alfvénen speed; (2) the LH waves must have minimal damping by thermal ions and electrons. Figure 10 shows estimates of the ratio v_r/V_A Cairns and Zank, 2002 along the Sun-heliopause nose axis, calculated using values of n_e obtained from a plasma-neutral simulation Zank et al., 1996 and values for B calculated using the convected field approximation for $\mathbf{B}_{VLISM}$ (assumed perpendicular to the solar system's velocity relative to the VLISM plasma). Magnetic draping and flow stagnation at the heliopause lead to v_r/V_A decreasing from values ≈ 10 in the VLISM to values $\lesssim 5$ close to the heliopause nose, before increasing again in the solar wind. Enhanced

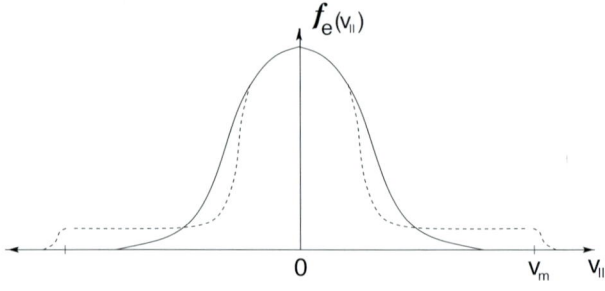

Figure 3.9 LHD causes a superthermal electron tail (dashed curve) to be drawn out of an initial thermal distribution (solid curve).

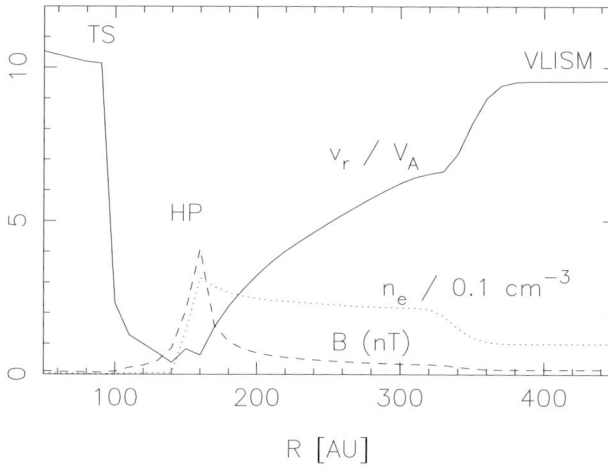

Figure 3.10 Ratio of v_r/V_A, B, and n_e based on plasma-neutral simulations and the convected field approximation Cairns and Zank, 2002.

lower hybrid damping precludes effective LHD occurring in the inner heliosheath. The reason is that $T_i \gtrsim 10^6$ K and $T_e \gtrsim 10^5$ K there, due to heating at the termination shock, whence $v_r \lesssim V_i$ and $v_m \lesssim 2V_e$ so that growth of LH waves is quenched. Thus the LH ring instability and enhanced superthermal electron tail produced by LHD are limited to the magnetic field draping region of the outer heliosheath, presumably within $\lesssim 50$ AU of the heliopause nose.

The theory used here for f_p and $2f_p$ radiation from the foreshock region upstream of a shock wave was developed for interplanetary type II radio bursts (Knock et al., 2001, 2003) and emission upstream from Earth's bow shock Kuncic et al., 2002a. The theory is analytic, quantitative, and has four main elements (Figure 11). First, electrons are reflected by the shock's magnetic mirror and undergo shock-drift acceleration. This is treated analytically using conservation of the magnetic moment and energy (including the electrostatic cross-shock potential ϕ_{cs}) in the de Hoffman-Teller frame, with the magnetic and density jumps calculated using the Rankine-Hugoniot conditions and ϕ_{cs} using a theoretical model Kuncic et al., 2002b. Second, electron beams develop in the foreshock via time-of-flight effects: an initial dispersion in $v_{\parallel}$ causes spatial dispersion, because the convection electric field $\mathbf{E}$ causes all electrons to $\mathbf{E} \times \mathbf{B}$ drift downstream perpendicular to $\mathbf{B}$ with the same speed $v_d = v_{E \times B}$, so that a minimum $v_{\parallel}$ exists for reflected electrons to reach a given foreshock location. Reduced electron distribution functions $f(v_{\parallel}, \mathbf{r}) = 2\pi \int dv_{\perp} v_{\perp} f(v_{\perp}, v_{\parallel})$ are calculated throughout the foreshock using Liouville's Theorem, by tracing particle paths back to the shock and using the reflection analysis to write this in terms of the particle distribution function in the undisturbed upstream plasma. Third, these beams drive Langmuir waves, whose growth causes quasilinear relaxation of the beam towards a state of marginal stability with a plateaued electron distribution. Assuming marginal stability on average, consistent with stochastic growth theory Robinson et al., 1993; Cairns and Robinson, 1999, the energy flux $dW_L(\mathbf{r})/dt$ into the Langmuir waves is given by the convective derivative of the available free energy in the beam (the difference between the initial beam kinetic energy and the plateaued beam) times the quasilinear efficiency predicted by standard theory (e.g., Melrose, 1986). Then

$$\frac{d}{dt}W_L(\mathbf{r}) = \mathbf{v}.\frac{\partial}{\partial\mathbf{r}}\left(\frac{n_b v_b \Delta v_b}{3}\right) \approx \frac{n_b v_b^2 \Delta v_b}{3r} , \qquad (3.2)$$

where n_b, v_b, and Δv_b are the number density, speed, and spread in speed of the beam, and r is the distance from the tangent point to the location $\mathbf{r}$ Knock et al., 2001; Cairns et al., 2004; Mitchell et al., 2004. Finally, the known efficiencies with which Langmuir energy is converted into f_p and $2f_p$ radiation for specific nonlinear processes are multiplied by $dW_L(\mathbf{r})/dt$ to predict the volume emissivities of f_p and $2f_p$ radiation, $j_F(\mathbf{r}$ and $j_H(\mathbf{r})$, respectively Knock et al., 2001. These quantities are the power flux into f_p or $2f_p$ radiation per unit volume and solid angle. For completeness, the processes used are the electrostatic decay $L \rightarrow L' + S$, which stimulates the electromagnetic decay $L \rightarrow T(f_p) + S'$ that produces f_p radiation, and provides the scattered L' waves for the coalescence $L + L' \rightarrow T(2f_p)$ that produces $2f_p$ radiation Robinson et al., 1994; Knock et al., 2001; Kuncic et al., 2002a. Here L and L' denote Langmuir waves, S and S' ion sound waves, and T radio waves.

The predictions of the combined Priming/GMIR theory follow in several steps. First, use the priming theory to predict the existence and properties of the superthermal electron tail superposed onto the background electron distribution for a given location of the shock. Second, use the shock emission theory to predict the results for reflection of this superposed distribution by the GMIR shock. The theory presently predicts five quantities: (1) $f(v_{\parallel}, \mathbf{r})$ in the foreshock, (2) $j_F(\mathbf{r})$ and $j_H(\mathbf{r})$ in the foreshock, and (3) the fundamental and harmonic radiation fluxes $F_F(\mathbf{r}_{ob})$ and $F_H(\mathbf{r}_{ob})$ at an observer location $\mathbf{r}_{ob}$, respectively, calculated by integrating the volume emissivities throughout the 3-D source taking into account the $|\mathbf{r} - \mathbf{r}_{ob}|^{-2}$ falloff.

A qualitative question is answered now before proceding to review the Priming/GMIR theory's predictions: Why does the radiation turn on when the GMIR shock enters the primed region? The reasons are at least three-fold. (1) The reflection and shock-drift acceleration of the tail electrons leads to much denser, fast electron beams – which are predicted to produce the bulk of the emission. (2) These enhanced beams

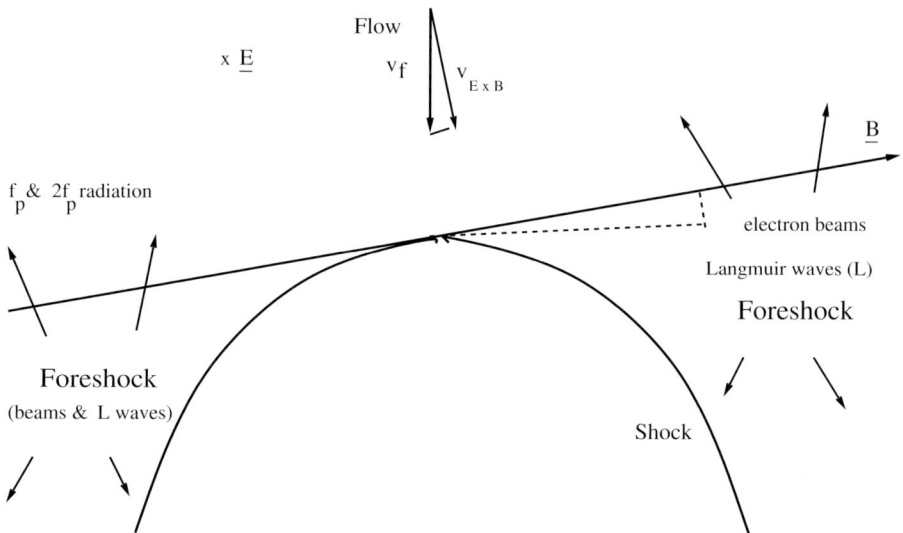

Figure 3.11. Schematic figure of the foreshock upstream of the GMIR ripple, showing the convection electric field and associated $\mathbf{E} \times \mathbf{B}$ drift, the formation of electron beams by time-of-flight effects (see dotted lines), and the regions where electron beams, Langmuir waves, and radiation are generated.

drive Langmuir waves with much higher electric fields E_L, since quasi-linear relxation predicts $E_L^2 \propto n_b v_b \Delta v_b$. (3) The rates of the nonlinear Langmuir processes scale at least as fast as E_L^2, thereby predicting higher radiation fluxes for higher E_L.

5.   Recent Theoretical Results

Theoretical results are presented here for a single paraboloidal ripple on the global GMIR shock, first for a given set of outer heliosheath plasma and magnetic parameters and second for the shock propagating from 10 AU to beyond the heliopause. (Alternatively, the shock parameters can be assumed to correspond to a single paraboloidal shock rather than to a single ripple on a larger shock.) Table 3.1 lists the shock speed U relative to the plasma and the shock's radius of curvature R_c at the nose. The background electron distribution before priming is assumed to be a generalized Lorentzian distribution with $f_\kappa(v_\parallel) \propto (v_\parallel^2 + V_e^2)^{-\kappa/2}$. Values for κ, n_e, T_e, and B are specified in Table 3.1. The axis of the ripple is assumed parallel to $\mathbf{U}$ and perpendicular to $\mathbf{B}$. The properties of the LHD tail are given by (3.1) and the ratio $n_T/n_e = 10^{-5}$ pre-

Table 3.1. Nominal shock and plasma parameters for the outer heliosheath.

U	R_c	T_e	n_e	κ	B
600 km s^{-1}	0.42 AU	8000 K	0.1 cm^{-3}	5	0.1 nT

dicted assuming the charge-exchanging pickup ions have number density $n_{ce} = 10^{-4}$ cm^{-3}. Finally, note that $f_p = 2.6$ kHz for $n_e = 0.1$ cm^{-3}.

Theoretical calculations show major enhancements of the foreshock electron beams for the case of LH priming Mitchell et al., 2004. Figure 12 shows the electron beams predicted for two foreshock locations under three sets of conditions (J.J. Mitchell, personal communication, 2004): (1) LH priming and shock acceleration (solid curves), (2) LH priming but no shock acceleration (dashed curves), and (3) shock acceleration of the background electrons with no LH priming (dotted curves). The left and right panels shows $f(v_\parallel)$ at the locations $(x, R) = (15, 100)$ Gm and $(3, 50)$ Gm, respectively, in the plane defined by $\mathbf{U}$ and $\mathbf{B}$. Here R and x are the coordinates along and perpendicular to $\mathbf{B}$, respectively, and the origin is the magnetic tangent point on the shock. It is clear that LH priming greatly increases the number of reflected electrons (compare the solid and dotted curves) and that the shock accelerates electrons significantly (compare the solid and dashed curves). Since reflected electrons require larger $v_\parallel \gtrsim v_d R/x$ to reach the location $(x, R) = (3, 50)$ Gm than $(15, 100)$ Gm, the larger number of fast seed electrons available in the primed case at large $v_\parallel$ means that priming increases the relative beam number density more proportionately at larger $v_\parallel$ (and so closer to the foreshock boundary) than at smaller $v_\parallel$. This favors f_p radiation over $2f_p$ radiation Cairns and Zank, 2001, as shown below.

Figure 13 shows predictions for the fundamental volume emissivity $j_F(\mathbf{r})$ with and without LHD priming. Including the superthermal LHD tail leads to the maximum values of j_F increasing by a factor $\approx 10^4$ and the development of two regions with significant j_F in the foreshock: the region with large v_b near the foreshock boundary corresponds to shock-accelerated tail electrons while the other corresponds to accelerated background electrons. Figure 14 shows similar results to Figure 13 but for harmonic emission. While priming is clearly vital again, it is also found that the maximum volume emissivity of $2f_p$ radiation is of order $3 - 4$ orders of magnitude less than for f_p radiation. Analyses for other plasma parameters shows that the lack of $2f_p$ radiation is primarily

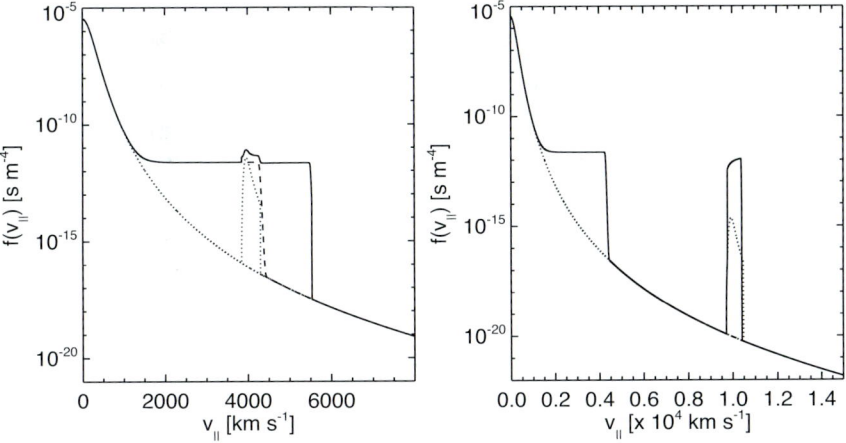

Figure 3.12. Distributions $f(v_\parallel)$ predicted at two foreshock locations: [left)] $(x, R) = (15, 100)$ Gm and [right] $(3, 50)$ Gm. See the text for more details.

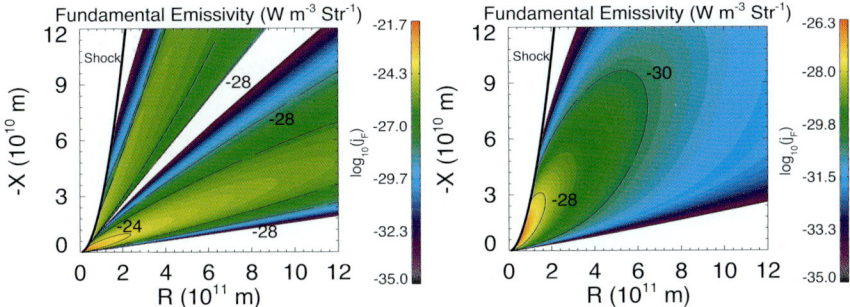

Figure 3.13. Volume emissivity $j_F(R, x)$ for f_p radiation (left) with and (right) without the superthermal tail predicted for LHD and the shock and plasma parameters in Table 3.1 Cairns et al., 2004. The solid black line shows the shock, x and R are the foreshock coordinates parallel to $\mathbf{U}$ and $\mathbf{B}$, respectively, the origin is the tangent point on the shock, and the calculation is for the plane defined by $\mathbf{U}$ and $\mathbf{B}$. The color bar shows $\log_{10} j_F(R, x)$.

due to the low value of $T_e = 8000$ K assumed for the outer heliosphere Mitchell et al., 2004.

The fluxes predicted for a remote observer are calculated by integrating the predictions for $j_F(R, x)$ and $j_H(R, x)$ over the 3-D foreshock volume taking into account the inverse falloff with distance squared (Table 3.2). These estimates assume that the predictions in Figures 13 and 14 remain correct in all foreshock planes defined by $\mathbf{U}$ and $\mathbf{B}$. Several important results are clear. (1) The flux of f_p radiation is predicted to be over 4 orders of magnitude larger than that of $2f_p$ radiation. This means

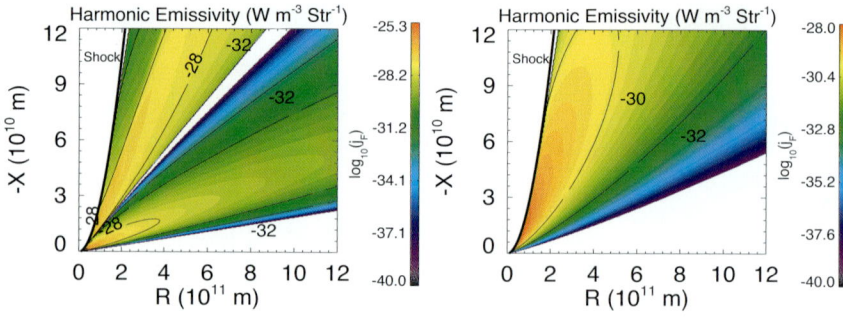

Figure 3.14. Volume emissivity $j_H(R, x)$ for $2f_p$ radiation (left) with and (right) without the superthermal tail predicted for LHD, in the same format and with the same shock and plasma parameters as Figure 13 Mitchell et al., 2004.

Table 3.2. Fluxes predicted for an observer in the ecliptic plane at 50 AU for the shock and plasma parameters in Table 3.1.

	f_p flux $(\mathrm{Wm^{-2}Hz^{-1}})$	$2f_p$ flux $(\mathrm{Wm^{-2}Hz^{-1}})$
Tail	3×10^{-17}	1×10^{-22}
No Tail	8×10^{-22}	7×10^{-25}

that no harmonic structrure is likely to be observed for the $2 - 3$ kHz radiation and that the observed radiation is almost certainly f_p radiation. (2) Priming is vital: it increases the predicted flux of f_p radiation by a factor $\approx 4 \times 10^4$ for these parameters and can plausibly account for the radiation turning on beyond the heliopause once the GMIR shock enters the primed region. (3) The predicted f_p flux is of order that observed by the Voyager spacecraft, $\approx 8 \times 10^{-17}$ $\mathrm{Wm^{-2}Hz^{-1}}$ Gurnett et al., 1993, for these parameters. These results confirm predictions made previously Cairns and Zank, 2001; Cairns and Zank, 2002. Accordingly, these calculations provide an underlying and semi-quantitative theoretical basis for Gurnett et al.'s (1993) GMIR model for the radiation Cairns et al., 2004; Mitchell et al., 2004.

What radio emissions are predicted to occur as the GMIR shock travels through the solar wind and inner heliosheath before entering the primed region and eventually the VLISM? Figure 15 answers this question by predicting the dynamic spectrum for a GMIR shock moving through the spatial profiles in density, flow speed, and ion and electron temperature given by the two-shock, cylindrically-symmetric, plasma-

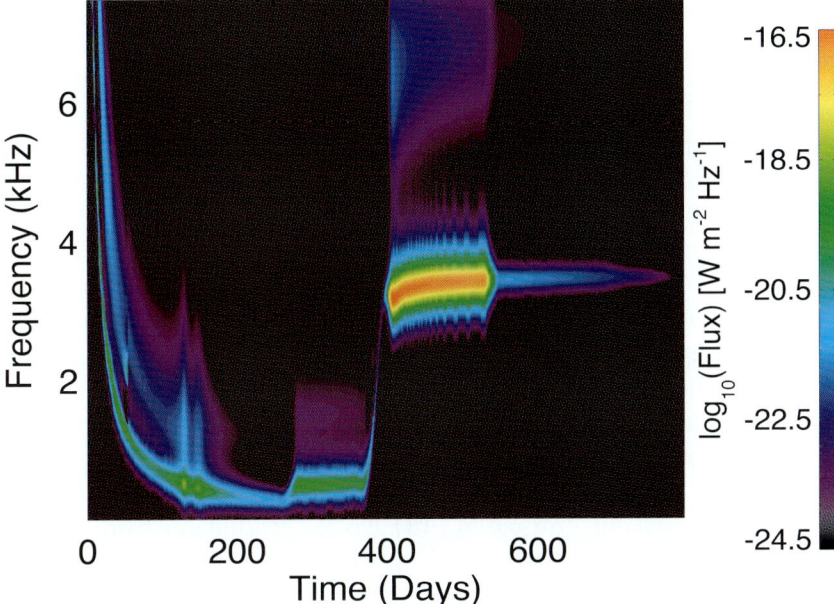

Figure 3.15. Dynamic spectrum predicted for the GMIR shock described in the text, with the logarithim of the flux color-coded for an observer located 50 AU from the Sun and just off the Sun-heliopause nose axis (Mitchell et al., 2004). The vertical ribbing from days 400 to 550 is a plotting artifact.

neutral, 4-fluid simulation of Zank et al. (1996). For simplicity, the shock is assumed to have the values U and R_c specified in Table 3.1 and **B** is always perpendicular to **U**, which is directed along the simulation's symmetry axis, the direction between the Sun and the nose of the heliopause. The background electron distribution is assumed to be a generalized Lorentzian with $\kappa = 5$ and the temperature predicted by the simulation. The Cairns and Zank (2001, 2002) model for the LHD superthermal tail is superposed onto the background distribution in the outer heliosheath, where LHD is predicted to occur.

In Figure 15 the frequency bands drifting downwards from 8 kHz to about 200 Hz, from early times until about day 300, are f_p (primarily) and $2f_p$ radiation produced when the GMIR is in the solar wind. The time origin corresponds to the GMIR leaving the Sun. (The enhancement near day 150 is when the GMIR passes near the observer.) The radiation produced in the solar wind is predicted to be very weak, typically well below about $10^{-18.5}$ Wm^{-2}Hz^{-1}, whereas the Voyager threshold is near 10^{-17} Wm^{-2}Hz^{-1}. The GMIR is in the inner heliosheath from about day 250 to day 400, primarily producing f_p radiation, after which the rapidly rising tone corresponds to f_p emission from the heliopause density ramp. The f_p radiation is predicted to increase in intensity by ≈ 4 orders of magnitude, to values in excess of the Voyager threshold, when the GMIR is in the outer heliosheath. Moreover, this emission remains almost constant in frequency and lasts for approximately 150 days, starting shortly after day 400. The GMIR enters the VLISM near day 550 and the emission then becomes very weak.

The following results follow from Figure 15. (1) The GMIR radiation is predicted to be unobservable by the Voyager spacecraft when the GMIR is in the solar wind, the inner heliosheath, and the VLISM. (2) The radiation is predicted to turn on when the GMIR enters the primed outer heliosheath, because the radiation flux increases above the Voyager threshold. (3) The f_p radiation predicted for the primed outer heliosheath closely resembles the 2 kHz component. Clearly, however, at present the theory cannot account for the existence and frequency properties of transient emissions. (4) No $2f_p$ radiation from the outer heliosheath or any other region is expected to be observable above the Voyager threshold. (5) The timing and frequency of the radio event predicted above the Voyager threshold are semi-quantitatively consistent with the observations. (In this simulation the radiation frequency, and so n_e in the outer heliosheath, are slightly high.)

The predicted fluxes depend on the shock, plasma, and tail parameters Cairns et al., 2004; Mitchell et al., 2004. Figure 16 shows that the flux depends strongly on the shock speed U for $U \lesssim 100$ km s^{-1} Mitchell et al., 2004. Since the 2-3 kHz emissions observed to date are at most a factor $\lesssim 3$ above the Voyager detection thresholds, it is clear that U is an important parameter that could determine whether a particular GMIR would produce observable emission. These and analagous calculations for other shock, plasma and tail parameters show that increased emission is expected for faster and larger shock ripples (specifically the flux increases as R_c^2), larger plasma temperatures, lower κ, larger pickup ion number densities, and larger ring speeds Mitchell et al., 2004.

Figure 16 gives a possible reason why the 2003-2004 emission event is weaker than the 1983-84 and 1992-94 events. Specifically, the Figure predicts that the lower GMIR speed inferred for the 2003-2004 radiation event (≈ 560 km s^{-1}) compared with the 1983-84 and 1992-94 events (≈ 850 km s^{-1}) should lead to a smaller radiation intensity by a factor ≈ 2. This prediction is not inconsistent with the radio observations in Figure 2.

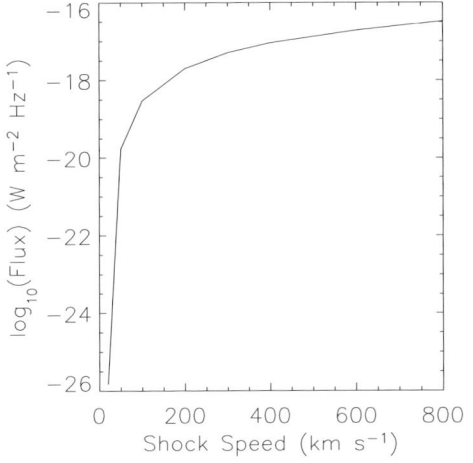

Figure 3.16 Predicted flux as a function of shock speed U for an observer located 50 AU from the Sun and just off the Sun-heliopause nose axis Mitchell et al., 2004. The other ripple, plasma, and tail parameters are as in Table 3.1.

6.   Discussion and Conclusions

The GMIR model of Gurnett et al. (1993), which involves the 2-3 kHz emissions being generated at multiples of f_p when a GMIR shock reaches the vicinity of the heliopause, is very attractive. Specifically, it can account for the correlation between GMIRs and radio events, the associated time-lags, and recent analyses of the source location which show the radiation starting in the close vicinity of the heliopause nose (based on direction finding, relative amplitudes, and inferred source distances of order $130 - 160$ AU) Gurnett et al., 1993; Gurnett and Kurth, 1996; Kurth and Gurnett, 2003. Major issues that the model does not explain include why and where the radiation turns on, how to explain the levels of radiation, why the radiation is at frequencies likely of order f_p in the outer heliosheath, why harmonic structure is not observed, why the radiation source maps out a primarily linear structure parallel to the galactic plan, and why two classes of radiation events exist (the 2 kHz component and the transient events).

The necessity for having two triggers for the radiation, a GMIR shock and a region primed to emit radiation when the GMIR shock enters, is clear from the event-GMIR correlation and the absence of detectable emission from the GMIR in the solar wind. Recently a Priming/GMIR theory has been developed Cairns and Zank, 2001; Cairns and Zank, 2002; Cairns et al., 2004; Mitchell et al., 2004. The theory is semi-quantitative and analytic: it combines the detailed plasma microphysics of two theories into a macroscopic theory for the radiation, first a theory for f_p and $2f_p$ emission from the foreshock regions of shock waves Knock et al., 2001; Knock et al., 2003; Kuncic et al., 2002a and, second, a theory for priming the outer heliosheath (Cairns and Zank, 2001, 2002). The magnetic draping region of the outer heliosheath, near the heliopause nose, is primed with an enhanced superthermal electron tail that results from resonant acceleration by lower hybrid waves driven by pickup ion ring instabilities (LHD). The pickup ions result from charge-exchange of region 2 neutrals in the outer heliosheath and the priming is localized to the predicted region by the physics of the lower hybrid-drive. The radiation turns on when the GMIR shock enters the primed region and accelerates the electron tail, strongly enhancing the levels of f_p and $2f_p$ radiation produced in the GMIR foreshock (and the properties of the electron beams and associated Langmuir waves). Semi-quantitative theoretical calculations show that the levels of f_p radiation dominate those of $2f_p$ radiation (by a factor $\geq 10^3$) and are of order those observed by the Voyager spacecraft for nominal GMIR, plasma, and neutral parameters. Moreover, the radiation does turn on at levels above the Voyager threshold when the GMIR reaches the primed region, with radiation fluxes predicted to be lower by about 4 orders of magnitude in the solar wind, inner heliosheath and VLISM. Finally, the predicted dynamic spectrum closely resembles that of the 2 kHz component.

Thus the Priming/GMIR theory can resolve five of the seven issues raised two paragraphs above and provide a detailed theoretical basis for most of the observations available and for Gurnett et al.'s (1993) GMIR model. Of course, further work is required on a number of areas.

First, why does the radiation source map out a linear band on the sky with time Kurth and Gurnett, 2003 and can the direction of $\mathbf{B}_{VLISM}$ be inferred? A qualitative answer is that the priming region is where magnetic draping occurs, and that draping results in a region with enhanced magnetic field which should be extended parallel to $\mathbf{B}$ and narrower transverse to $\mathbf{B}$ and the plasma flow. Thus, when viewed from the inner heliosphere, the draping region should take the form of a linear band aligned with $\mathbf{B}$ in the outer helioseath. Interpreting the radio data in this way, $\mathbf{B}$ in the outer heliosheath is parallel to the galactic plane in the plane of the sky. With the outer bow shock expected to be barely supersonic and superalfvénic, if it exists at all, the plane of the sky component of $\mathbf{B}$ in the outer heliosheath is expected to be close to that of $\mathbf{B}_{VLISM}$.

Second, how does the radiation reach the inner heliosphere if it is produced near f_p upstream of a shock, thereby being prevented from immediately propagating Sunwards? A qualitative answer is given by Cairns and Zank (2001): scattering by density irregularities diffuses the f_p radiation around the sides of the GMIR shock until it reaches locations where the GMIR shock has not reached the heliopause, thereafter allowing immediate propagation across the shock (since $f \ll f_p$ locally). Quantitative ray-tracing calculations are needed to test this idea.

Third, how will shock slowing and the existence of multiple simultaneously active shock ripples affect the predictions? Shock slowing is shown in Figure 16 to be potentially important, perhaps even explaining why the 2003-2004 radiation event is weaker than the previous major outbursts. Shock slowing beyond the termination shock is likely to be of order 30% per plasma discontinuity Gurnett et al., 1993; Story and Zank, 1997; Zank et al., 2001, thereby reducing the speed by a factor ≈ 2 from the distant solar wind to the outer heliosheath, and so the radiation flux by about a factor $\gtrsim 3$. Another point is that having N identical, simultaneously active ripples on the GMIR shock would increase the predicted radiation flux by a factor of N. Since $N \approx (20\text{AU}/0.4\text{AU})^2 \approx 2500$, for a GMIR shock with a radius of curvature of 20 AU, this increase is large. However, propagation losses into the inner heliosphere (see above) are expected to be large, likely reducing the emission by a factor $\gtrsim 100$. Accordingly, while the calculations for a single ripple yield reasonable agreement with the Voyager data, the preliminary discussion by Cairns et al. (2004) on the effects of multiple shock ripples, shock slowing, and propagation losses needs to be developed further and quantified.

Fourth, how can transient emissions be explained? Recent simulations show that solar cycle variations of the solar wind ram pressure can inject density waves into the outer heliosheath Zank and Muller, 2003; Scherer and Fahr, 2003. As the GMIR shock rides up these density ramps, or others caused by interactions of the termination shock with interplanetary shocks and other solar wind structures Zank et al., 1994, emissions with increasing frequency will be produced.

Fifth, why are the major episodes of radio activity produced a few years after solar maximum? One possibility is that solar cycle variations in the solar wind ram pressure causes not only temporal variations in the location of the termination shock and heliopause and injection of density waves into the outer heliosheath Zank and Muller, 2003; Scherer and Fahr, 2003, but also spatio-temporal changes in the neutrals and locations of maximum charge-exchange in the outer heliosheath. If the rate of charge-exchange is relatively small in the magnetic draping region of the outer heliosheath at some phases of the solar cycle, then priming will be ineffective at these times. Preliminary calculations along these lines are encouraging (J.J. Mitchell, personal communication, 2004) and need to be extended.

In conclusion, a detailed semi-quantitative theory exists for the 2-3 kHz emissions, based on combining multiple pieces of plasma microphysics into a macroscopic theory. This Priming/GMIR theory can account for many aspects of the observations, especially for the 2 kHz component, and provides a detailed theoretical basis for Gurnett et al.'s GMIR model. Further work remains to be done, however, including a detailed examination of propagation effects, multiple ripples, shock slowing, solar cycle effects, and the drifting pairwise nature of transient emissions.

Acknowledgments

Funding is acknowledged from the Australian Research Council, J.J. Mitchell is thanked for providing Figure 12, and J.J. Mitchell, G.P. Zank, S.A. Knock, and H. Mueller are thanked for helpful discussions.

References

Cairns, I. H. (1994). *J. Geophys. Res.*, , 99:23,505-23,513.

Cairns, I. H., and Gurnett, D. A. (1992). *J. Geophys. Res.*, 97:6235-6244.

Cairns, I. H., and Robinson, P. A. (1999) *Phys. Rev. Lett.*, 82:3066–3069.

Cairns, I. H. and Zank, G. P. (2001). in *The Outer Heliosphere: The Next Frontiers*, eds K. Scherer, H. Fichtner, et al., pp. 253-262, Pergamon, Amsterdam.

Cairns, I. H. and Zank, G. P. (2002). *Geophys. Res. Lett.*, 29(7), DOI 10.1029 2001GL014112, 47-1–4.

Cairns, I. H., Kurth, W. S., and Gurnett, D. A. (1992). *J. Geophys. Res.*, 97:6245-6259.

Cairns, I. H., Mitchell, J. J., Knock, S. A., and Robinson, P. A. (2004). *Adv. Space Res.*, in press, 2004.

Czechowski, A. (2004). *Adv. Space Res.*, in press, 2004.

Grzedzielski, S., and Lazarus, A. J. (1993). *J. Geophys. Res.*, 98:5551-5558.

Gurnett, D. A. (1975). *J. Geophys. Res.*, 80:2751-2763.

Gurnett, D. A., and Kurth, W. S. (1994) *Geophys. Res. Lett.*, 21:1571-1574.

Gurnett, D. A., and Kurth, W. S. (1995) *Adv. Sp. Res.*, 16(9):279-290, 1995.

Gurnett, D. A., and Kurth, W. S. (1996) *Space Sc. Rev.*, 78:53-66, 1996.

Gurnett, D. A., Kurth, W. S., Allendorf, S. C., and Poynter, R. L. (1993). *Science*, 262:199-203.

Gurnett, D. A., Allendorf, S. C., and Kurth, W. S. (1998). *Geophys. Res. Lett*, 25:4433-4436.

Gurnett, D. A., Kurth, W. S., and Stone, E. C. (2004). *Geophys. Res. Lett*, 30(23): SSC8-1-4.

Holzer, T. E. (1989). *Ann. Rev. Astron. Astrophys.*, 27:199–234.

Izmodenov, V. (2004). This book.

Knock, S., Cairns, I. H., Robinson, P. A., and Kuncic, Z. (2001). *J. Geophys. Res.*, 106:25,041–25052.

Knock, S., Cairns, I. H., Robinson, P. A., and Kuncic, Z. (2003). *J. Geophys. Res.*, 108(3):SSH 6-1–12, DOI 10.1029 2002JA009508.

Kuncic, Z.; Cairns, I. H., Knock, S., and Robinson, P. A. (2002a). Geophys. Res. Lett., 29(8):2-1–4, DOI 10.1029 2001GL014524.

Kuncic, Z., Cairns, I. H., and Knock, S. A. (2002b). *J. Geophys. Res.*, 107(8):SSH 11-1–10, DOI 10.1029 2001JA000250.

Kurth, W. S., and Gurnett, D. A. (2003). *J. Geophys. Res.*, 108(10):LIS 2-1-14, DOI 10.1029/2003JA009860.

Kurth, W. S., Gurnett, D. A., Scarf, F. L., and Poynter, R. L. (1984). *Nature*, 312:27-29.

Kurth, W. S., Gurnett, D. A., Scarf, F. L., and Poynter, R. L. (1987). *Geophys. Res. Lett.*, 14:49-52.

Linsky, J. (2004). This book.

Macek, W. M., Cairns, I. H., Kurth, W. S., and Gurnett, D. A. (1991). *Geophys. Res. Lett.*, 18:357-360.

McNutt, R. L., Jr. (1988). *Geophys. Res. Lett.*, 15:1307-1310.

McBride, J. B., Ott, E., Boris, J. P., and Orens, J. H. (1972). *Phys. Fluids*, 15:2367–2383.

McClements, K. G., Bingham, R., Su, J. J., Dawson, J. M., and Spicer, D. S. (1993). *Astrophys. J.*, 409: 465–475.

Melrose, D. B., *Instabilities in Space and Laboratory Plasmas* (Cambridge and new York, Cambridge University Press, 1986).

Mitchell, J. J., Cairns, I. H., and Robinson, P. A. (2004). *J. Geophys. Res.*, in press, 2004.

Nelson, G. J., and Melrose, D. B. (1985). In *Solar radiophysics: Studies of emission from the sun at metre wavelengths* (Cambridge and New York, Cambridge University Press, 1985), p. 333-359.

Omelchenko, Y. A., Sagdeev, R. A., Shapiro, V. D., and Shevchenko, V. I. (1989). *Sov. J. Plasma Phys.*, 15:427–431.

Robinson, P. A., Cairns, I. H., and Gurnett, D. A. (1993). *Astrophys. J.*, 407:790–800.

Robinson, P. A., Cairns, I. H., and Willes, A. J. (1994). *Astrophys. J.*, 422:870–882.

Scherer, K., and Fahr, H. J. (2003). *Geophys. Res. Lett.*, 30(2):17-1–4

Shapiro, V. D., Bingham, R., Dawson, J. M., Dobe, Z., et al. (1998). *Physica Scr.*, T75:39–45.

Story, T. R., and Zank, G. P. (1997). *J. Geophys. Res.*, 102:17,381–17394.

Wild, J. P., Smerd, S. F., Weiss, A. A. (1963). *Ann. Rev. Astron. & Astrophys*, 1:291–366.

Zank, G. P. (1999). *Space Sc. Rev.*, 89(3-4):413–688.

Zank, G. P., and Muller, H.-R. (2003). *J. Geophys. Res.*, 108(6): SSH7-1–15.

Zank, G. P., Cairns, I. H.,Donohue, D. J. and Matthaeus, W. M. (1994). *J. Geophys. Res.*, 99, 14,729-14,735.

Zank, G. P., Pauls, H. L., Williams, L. L., and Hall, D. T. (1996). *J. Geophys. Res.*, 101:21,639–21656.

Zank, G. P., Rice, W. K. M., Cairns, I. H., et al. (2001) *J. Geophys. Res.*, 106:29,363–29372.

Chapter 4

ULYSSES AT SOLAR MAXIMUM

Richard G. Marsden

Research and Scientific Support Department of ESA

ESTEC, P.O. Box 299, 2200 AG Noordwijk

The Netherlands

Richard.Marsden@esa.int

Abstract   One of the key space missions to contribute to our knowledge of the Sun and heliosphere as an integrated system is undoubtedly the joint ESA-NASA mission Ulysses. The spacecraft's unique orbit, almost perpendicular to the ecliptic plane, and the broad range of measurements being made, make it ideal for studying the global heliosphere in four dimensions: space and time. With high-latitude passes at solar minimum and solar maximum, this mission profile has enabled a picture of the heliosphere between 1 and 5 AU to be built up that covers the full range of heliographic latitudes, and a wide range of solar activity conditions. In this review, we focus on the results obtained from the second set of high-latitude passes that occurred during the maximum phase of solar cycle 23. Among the major findings were the relatively simple dipolar nature of the heliospheric magnetic field, even at solar maximum; the compelling evidence for efficient transport of solar energetic particles and cosmic rays in latitude and longitude, leading to low-energy particle reservoirs and minimal cosmic ray gradients; confirmation of a fundamental relationship between solar wind speed and coronal electron temperature, leading to a new model for the origin of the fast and slow solar wind.

Keywords: Ulysses; heliosphere; solar wind; magnetic field; cosmic rays; interstellar dust; energetic particles.

1.   Introduction

The joint ESA-NASA collaborative space mission Ulysses, the first ever to fly over the poles of the Sun, has literally changed the way we view the heliosphere. The mission's primary objective is to explore the

G. Poletto and S.T. Suess (eds.), The Sun and The Heliosphere as an Integrated System, 91–112.

heliosphere in four dimensions: three spatial dimensions, and time (e.g., Wenzel et al., 1992). The Ulysses spacecraft carries nine on-board investigations that utilise scientific instruments measuring the solar wind, the heliospheric magnetic field, natural radio emission and plasma waves, energetic particles and cosmic rays, interplanetary and interstellar dust, neutral interstellar helium atoms, and cosmic gamma-ray bursts (Wenzel et al., 1992; Balogh et al., 2001b). The Ulysses science team is international, with investigators from many European countries, the United States, and Canada. Ulysses was launched by the space shuttle Discovery on 6 October, 1990, using a combined IUS/PAM-S upper-stage to inject the probe into a direct Earth/Jupiter transfer orbit. Arriving at Jupiter in February 1992, the spacecraft executed a gravity-assist manoeuvre that placed it in its final Sun-centred, out-of-ecliptic orbit (Smith and Wenzel, 1993; Wenzel and Smith, 1993). With a period of 6.2 years, the orbit, shown in Figure 1, is inclined at 80.2° to the solar equator, has a perihelion distance of 1.34 AU, and aphelion at 5.40 AU.

Although Ulysses has made ground-breaking discoveries at many points along its unique trajectory, the segments above 70° heliographic (solar) latitude in either hemisphere (referred to as "polar passes") have attracted special interest. The first such polar passes occurred in 1994 (in the south) and 1995 (north), as the Sun's activity approached a minimum. The results from this phase of the mission (often referred to as the "First Solar Orbit") have been described extensively (Balogh et al., 2001a; Marsden, 1995; Smith et al., 1995; Marsden and Smith, 1996a,b; Marsden et al., 1996). Ulysses arrived over the Sun's south polar regions for the second time in November 2000, followed by the rapid transit from maximum southern to maximum northern helio-latitudes that was completed in October 2001. Solar activity reached its maximum in April 2000, so that Ulysses experienced a very different environment over the poles from the one it encountered during the first high-latitude passes. In addition to the general increase in the frequency of transient solar wind disturbances and particle events associated with flares and coronal mass ejections (CME) at solar maximum, the second set of high-latitude passes were characterised by the corresponding changes occurring in the polar coronal holes and the Sun's polar magnetic fields (Harvey and Recely, 2002). The timing of the polarity reversal of the polar fields, and the associated effects on the propagation of charged particles in the heliosphere, was of particular interest during this so-called "Ulysses Solar Maximum Mission". The results of these and other studies conducted during the second set of high-latitude passes have been summarised by Smith and Marsden (2003) and Smith et al. (2003). Preliminary findings are reported in Marsden (2001).

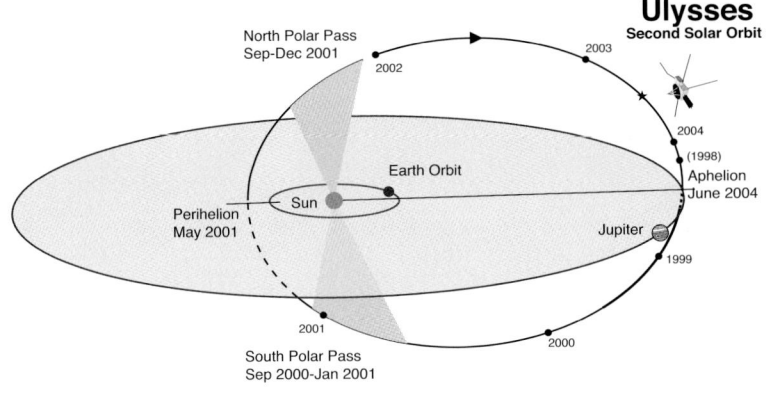

Figure 4.1. The Ulysses orbit viewed from 15 degrees above the ecliptic plane. Dots mark the start of each year. The position of the spacecraft on 1 July 2003 is shown (star).

In section 2 of this review, we focus on some of the scientific highlights from the Ulysses Solar Maximum Mission that are particularly relevant to the theme "the Sun and heliosphere as an integrated system". Section 3 provides an overview of the future plans for the mission.

2.   Scientific Highlights at Solar Maximum

It is beyond the scope of this overview to discuss all the new scientific insights that are being gained through the study of the unique data sets acquired by the Ulysses instruments. Instead, we will focus on a number of topics that are related to the large-scale structure of the heliosphere, its variation over the solar cycle, and its influence on the particle populations that form part of the "integrated system". Nonetheless, it should be noted that Ulysses is also making unique contributions in other areas, including studies of gamma-ray bursts (Hurley et al., 1995, 1999), a wide variety of natural radio emissions (McDowall and Kellogg, 2001), and the interstellar medium that surrounds the heliosphere (and its evolution since the formation of the solar system) (Gloeckler, Geiss and Fisk, 2001). Key aspects in the mission's success are the unprecedented level of data coverage (better than 95% on average throughout the 13-year mission to date), and the role that Ulysses plays in both the interplanetary network of gamma-ray detectors, and especially within the fleet of solar and solar-terrestrial missions that includes SOHO, ACE, Wind,

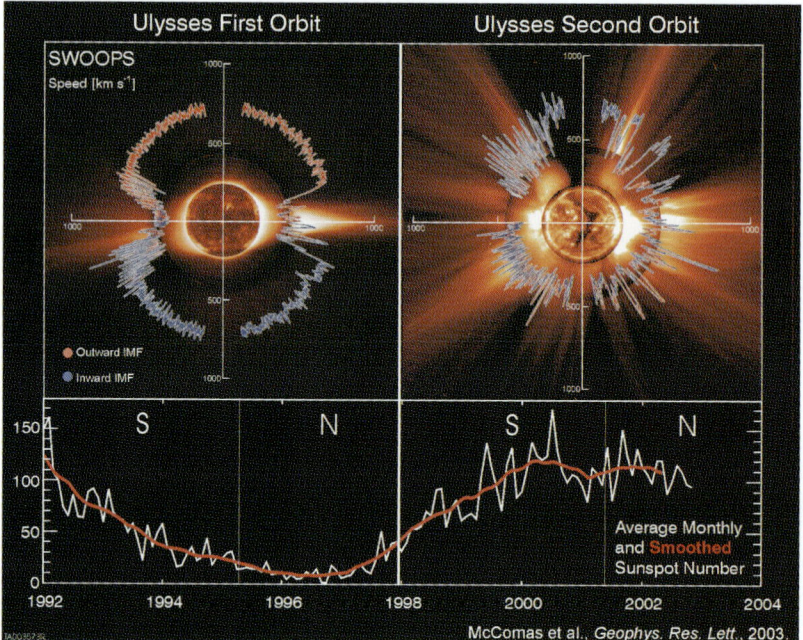

Figure 4.2. Solar wind speed plotted as a function of heliographic latitude for Ulysses' first orbit (top left panel), and second orbit (top right panel). Each trace runs anti-clockwise in time, and begins at aphelion (bottom left quadrant). Also shown (lower panels) is the averaged, smoothed Sunspot number. (From McComas et al. (2003). Reproduced by permission of AGU.)

and Voyager 1 and 2. Many multi-spacecraft studies rely on Ulysses' unique capabilities.

In order to provide context for some of the other results, we first discuss the solar wind and magnetic field observations.

2.1   Solar Wind

The exploration of the latitude dependence of solar wind characteristics (speed, temperature, composition) during maximum solar activity has revealed an entirely different configuration of the 3-dimensional heliosphere compared with that encountered near solar minimum. The variation in 3-dimensional structure over the solar cycle as observed by Ulysses is well illustrated in Figure 2, taken from McComas et al. (2003). At solar minimum, Ulysses found a heliosphere dominated by the fast wind from the southern and northern polar coronal holes. Slower, more variable, solar wind flows were confined to a relatively narrow latitudinal band (30°S to 30°N) around the equator (McComas et al., 1998; 2000).

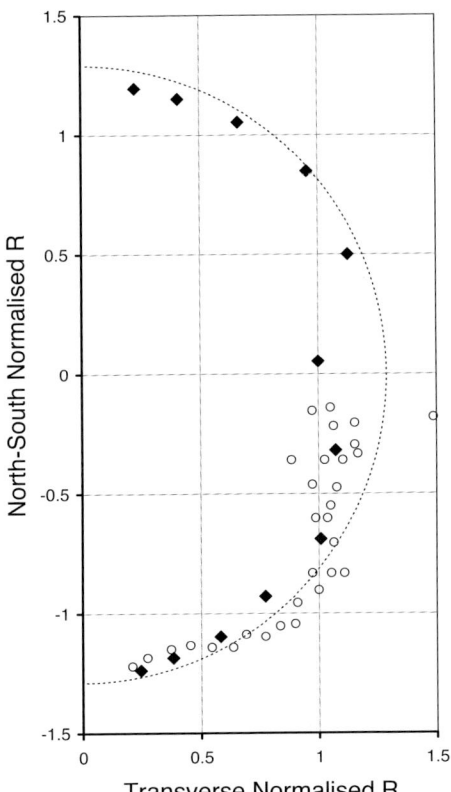

Figure 4.3 Radial distance to the heliopause (normalised) derived from Ulysses solar wind momentum flux measurements in a plane normal to the LISM flow. Diamonds and open circles indicate data acquired before and after the maximum southern latitude, respectively. The dashed line is a semi-circle fitted to the pre-polar pass data. After Phillips et al. (1996).

In contrast, during solar maximum, the large polar coronal holes had disappeared, and the heliosphere appeared much more symmetric. The solar wind flows measured throughout the south polar pass, and much of the rapid transit from south to north, showed no systematic dependence on latitude. The wind itself was generally slower and much more variable than at solar minimum at all latitudes. Nevertheless, when Ulysses reached high northern latitudes in late 2001, it witnessed the formation and growth of a new polar coronal hole. This clearly marked the start of a return to more stable conditions following activity maximum. The solar wind recorded at Ulysses became commensurately faster and more uniform, resembling the flows seen over the poles at solar minimum. The two large peaks between 40° and 70° north that were encountered in mid-2001 (the north-west quadrant in the polar plot) are temporary entries into the fast polar cap flow. Ulysses' soujourn in the relatively stable, fast solar wind at high latitudes was significantly shorter than at solar minimum, however. By the end of 2001, regular excursions into variable, slower wind were once again taking place, even though Ulysses was still poleward of 60° latitude. The explanation for these differences can be found in the fact that the Sun had only just begun its transition to more stable conditions during the second northern polar pass, whereas solar minimum-like conditions were already well established in 1994/95.

Of interest in studies of the global configuration of the heliosphere are the observed temporal and latitudinal variations in the solar wind dynamic pressure, or equivalently, momentum flux. In combination with the state of the local interstellar medium, the solar wind dynamic pressure largely determines the overall size and shape of the heliosphere. Data acquired during Ulysses' first rapid transit from high southern to high northern heliolatitudes, near solar minimum, led Phillips et al. (1996) to suggest that under these conditions, the heliopause would not be symmetric, but more like a figure-of-eight (illustrated in Figure 3). This conclusion was based on the finding that the measured momentum flux was lower during the northbound transit through low latitudes than at high southern or northern latitudes. Model calculations by Pauls and Zank (1996) supported this idea, showing a significant dependence of the shape of the heliopause on the equatorial ram pressure. The recent high-latitude observations, now at solar maximum, have allowed this finding to be placed in a broader context (McComas et al., 2003). As can be seen in the top panel of Figure 4, taken from McComas et al. (2003), the solar wind dynamic pressure (scaled to 1 AU) showed a rapid increase by a factor of $\sim$2 during 1991. This was followed by an overall decrease from 1991 to 2001 that is much larger than the latitudinal variation in 1995. A similar trend was also seen at IMP-8 and Voyager 2 (Richardson et al., 2001), and has been noted in previous solar cycles (Lazarus and McNutt, 1990). The implication is that, while the modest latitudinal variation seen by Ulysses during solar minimum is undoubtedly present, the internal pressure of the heliosphere undergoes a factor of 2 change over a solar cycle independent of latitude (Richardson et al., 2001). Interestingly, the most recent observations from Ulysses during the 2000 solar maximum (McComas et al., 2003) indicate that the dynamic pressure once again showed an increase, however only at the $\sim$50% level.

Solar wind observations from Ulysses have established a clear anticorrelation between the bulk flow speed and the coronal electron temperature derived from solar wind ionic charge states, as illustrated in Figure 5 (Gloeckler et al., 2003). The relationship is found to hold equally well for fast and slow solar wind streams, but not for Coronal Mass Ejection (CME)-related flows. As noted by Fisk (2003), the observed anticorrelation is surprising, since solar wind models that rely on proton heating predict a direct correlation between proton speed and coronal electron

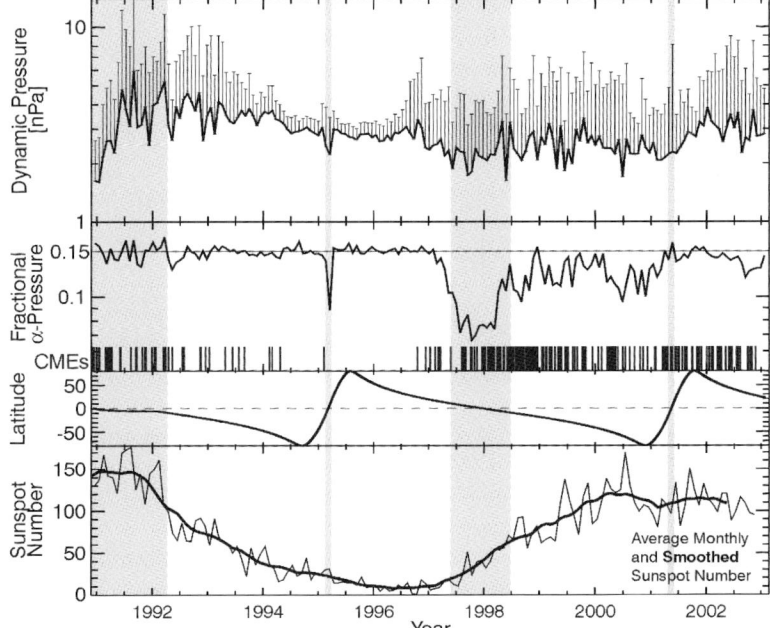

Figure 4.4. Mission plot of (from top to bottom): total dynamic pressure (momentum flux), fraction of alpha to proton pressure, spacecraft latitude, and Sunspot number. Vertical lines in the second panel mark the times of CMEs at Ulysses. (From McComas et al. (2003). Reproduced by permission of AGU.)

temperature (Hansteen et al., 1999). Furthermore, coronal electrons are not expected to play a direct role in the acceleration of the solar wind. In particular, their pressure is not adequate to accelerate the high-speed wind. The observed dependence can be explained, however, on the basis of a solar wind model put forward by Fisk (2003) that relies on reconnection between open magnetic field lines in the corona and closed coronal loops to produce both the fast and slow solar wind. In this model, the fast wind originates in smaller, low-temperature loops that are found in coronal holes, while the slow wind comes from larger, hotter loops outside coronal holes. The Fisk model predicts a specific relationship between solar wind speed and electron temperature, namely that $(V_{SW})^2$ is proportional to $1/T$. The data presented by Gloeckler et al. (2003) are fully consistent with this functional dependence, lending convincing support to the model. The breakdown of the speed-temperature relationship in the case of CMEs is not fully understood, but if confirmed, could be employed to routinely detect CME-related flows in the solar wind data.

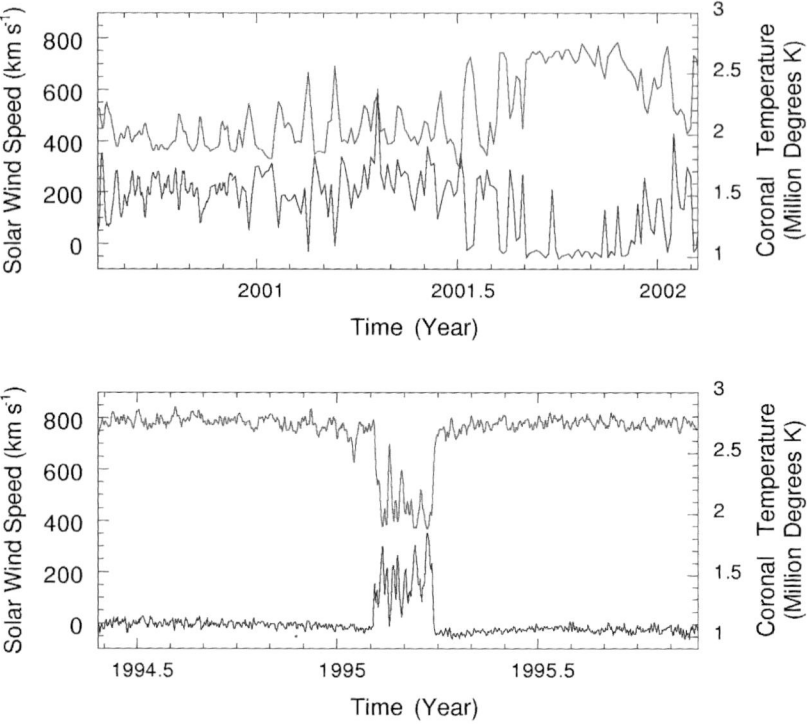

Figure 4.5. Plot showing the anticorrelation between solar wind speed (upper trace) and coronal electron temperature (lower trace, as derived from Ulysses ion composition data) at solar maximum (top panel) and solar minimum (lower panel). (Courtesy of G. Gloeckler.)

Although CMEs are known to be confined mainly to low- and middle solar latitudes, the occurrence of CMEs over the poles was one of the topics of interest during the recent polar passes at solar maximum. Indeed, Ulysses encountered CMEs up to 80° heliographic latitude in the southern hemisphere, all of which had similar properties to low-latitude events (Reisenfeld et al., 2003). Five CMEs were encountered above 70° latitude in the northern hemisphere. In a study of these events, all of which occurred in fast solar wind flow from the newly formed northern polar coronal hole, Reisenfeld et al. (2003) found that two CMEs were of the over-expanding type first observed by Ulysses during the high-latitude pass in 1994 (Gosling et al., 1998, and references therein). Of the remainder, two were fast CMEs (>800 kms^{-1}) that drove strong shocks. These CMEs were also observed in the ecliptic at 1 AU, both in situ, and as halo events by the LASCO coronagraph on SOHO. One of these events originated close to terrestrial subsolar point, implying significant lateral expansion (Reisenfeld et al., 2003). A CME-driven shock wave that passed over Ulysses during the rapid transit from south to north in May 2001 was responsible for the most intense interplanetary magnetic field, and highest solar wind density, observed by the spacecraft to date.

2.2 Magnetic Field

The development of the global solar wind characteristics discussed above was reflected in the Ulysses magnetic field observations. One of the key questions to be answered was how the high solar activity level would affect the structure and dynamics of the heliospheric magnetic field at high latitudes. As noted by Smith et al. (1997), the magnetic field near solar minimum as measured at Ulysses, possessed a relatively simple, dipole-like configuration. Above latitudes of ~30° in both hemispheres, the heliospheric fields were unipolar, the polarity coinciding with that of the respective polar cap field at the Sun's surface. In 1994/95, the fields in the north were positive (outward), and in the south, negative (inward). The heliospheric current sheet (HCS) separating the two polarities was relatively flat, and its axis was close to the solar rotation axis. Another striking feature of the heliospheric field during Ulysses' first orbit was that the open magnetic flux (given by $r^2 B_R$) was found to be independent of latitude (Smith et al., 1997). This was interpreted by Suess and Smith (1996) as implying a non-radial exansion of the solar wind at high latitudes near the solar surface, driven by excess magnetic pressure.

Surprisingly, at solar maximum the magnetic field at Ulysses (~1.5–2.5 AU from the Sun) maintained its dipole-like structure, even though the solar magnetic field, corona and solar wind were highly variable (Balogh and Smith, 2001; Smith et al., 2001; Jones and Balogh, 2003). In contrast to the situation at solar minimum, however, the equivalent magnetic poles were located at low latitudes rather than in the polar caps. As seen at Ulysses, the field in the southern hemisphere exhibited a 2-sector pattern almost up to the pole, with a single (negative) polarity seen only for one rotation at the highest latitude. This observation suggests that the HCS was highly inclined with respect to the solar equator, and was not flat (Smith et al., 2001). The same situation was encountered in the north, although now, consistent with the solar wind observations, a unipolar field originating in the newly formed polar coronal hole was observed for several rotations at latitudes poleward of ~70° (Jones and Balogh, 2003).

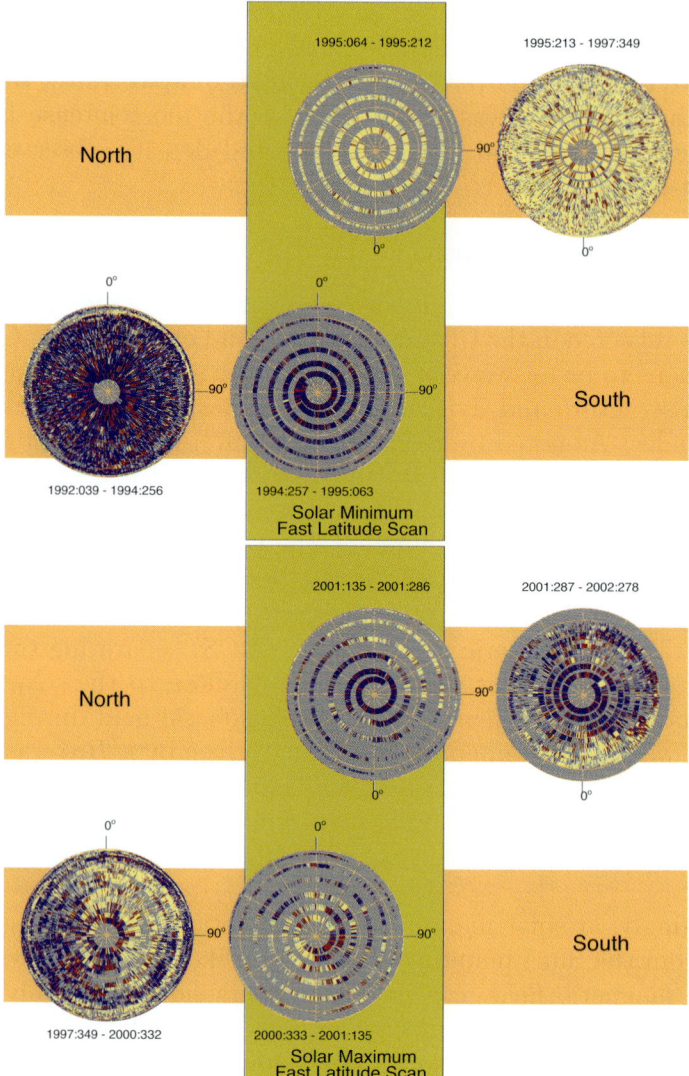

Figure 4.6. Heliospheric magnetic field polarity as measured by Ulysses, projected back to the Sun onto a spherical source surface at 2.5 Rs near solar minimum (top panel) and solar maximum (bottom panel). Views are shown from the northern and southern poles, as indicated. Time runs from bottom left to top right in each panel. Outward (positive) fields are shown in yellow, inward (negative fields in blue. Red indicates days for which the polarity was uncertain. (From Jones and Balogh (2003).)

The polar passes at solar maximum also offered a unique opportunity to observe the polarity reversal of the Sun's polar cap magnetic field from a high-latitude vantage point (Jones and Balogh, 2003; Smith and

Marsden, 2003; Smith et al., 2003). In a study of polar coronal holes during solar cycles 22 and 23 using ground-based observations, Harvey and Recely (2002) found that the southern polar cap fields reversed between 2002.31 and 2002.46 (i.e., between late April and mid-June, 2002), while the northern reversal occurred earlier, between 2001.19 and 2001.34 (mid-March and early May, 2001). The Ulysses data, shown in Figure 6, are consistent with this, in that the heliospheric field measured at high latitudes in the south between September 2000 and January 2001, still had the "old" (negative) polarity, whereas a year later, when Ulysses reached high northern latitudes in the second half of 2001, the heliospheric field had the "new" (also negative), polarity (Jones and Balogh, 2003). To zero order then, the Ulysses results support a model of the field reversal first proposed by Saito et al. (1978), in which the axis of the solar dipole simply rotates through 180° over the course of a solar cycle. As noted by Smith et al. (2003), however, the changes in the solar field are in reality much more complicated. Nevertheless, at solar maximum, the open fields that constitute the heliospheric magnetic field do exhibit a much simpler behaviour than the closed surface fields. As discussed above, during the solar minimum polar passes Ulysses found that the strength of the radial component of the heliospheric magnetic field, when corrected for radial distance effects, was independent of solar latitude. Perhaps surprisingly, as shown in Figure 7, this behaviour was found to be valid even in the more disturbed conditions at solar maximum (Smith and Balogh, 2003). Furthermore, the average value of $r^2 B_R$ was approximately the same in both phases of the solar cycle. This observed property of the distribution of open magnetic flux carried in the solar wind, its temporal invariance, has been explored in recent model by Fisk and Schwadron (2001). These authors argue that open flux of a given polarity can only be destroyed on the Sun by reconnection with open flux of the opposite polarity, a process that is rare given that open flux is ordered into regions of uniform polarity even at solar maximum. Effective invariance of open flux is thus expected. The model further treats the motion of open flux on the Sun as a diffusive process in which open flux reconnects with closed field loops, a process that can also explain the nature of the polarity reversal (Fisk and Schwadron, 2001). Here again, Ulysses observations have prompted a major step forward in our understanding of fundamental heliospheric processes.

2.3   Energetic Particles

One of the key discoveries made during the high-latitude passes at solar minimum was the unexpected ease with which energetic parti-

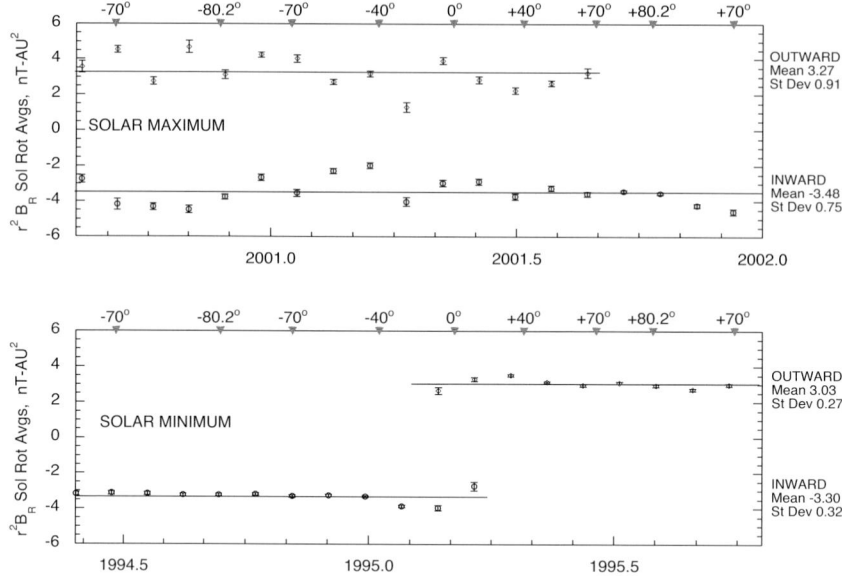

Figure 4.7. Plot of $r^2 B_R$ in outward (positive) and inward (negative) sectors averaged over successive solar rotations at solar maximum (top panel) and solar minimum (lower panel). Vertical bars show the standard error associated with each average. The solid lines show the means over the four transits between equator and pole. (From Smith et al. (2003). Reproduced with permission from AIP.)

cles, accelerated at low-to-middle latitudes, were able to gain access to the polar regions of the heliosphere (Kunow et al., 1999, and references therein). Recurrent increases in particle intensity, with a period of $\sim$26 days (i.e., close to the solar rotation period), were observed up to the highest latitudes even though the source of these particles, so-called Corotating Interaction Regions (CIRs), were confined to much lower latitudes (Sanderson et al., 1994, 1995; Simnett et al., 1994). This discovery prompted theorists to re-assess the role of perpendicular diffusion, and to reconsider the Parker model of the heliospheric magnetic field and departures caused by polar coronal holes and differential rotation. Modelling based on the latter effects led ultimately to new suggestions regarding the source of the solar wind itself (Fisk, 1996; Fisk and Jokipii, 1999; Fisk et al., 1999). An obvious question, then, when Ulysses returned to high latitudes at solar maximum, was: do energetic particles have the same easy access when the heliosphere is much more chaotic?

The observations provided an unequivocal answer: yes (McKibben et al., 2001). Indeed, the intensity profiles of energetic particles recorded at all latitudes were very similar to measurements made near the Earth,

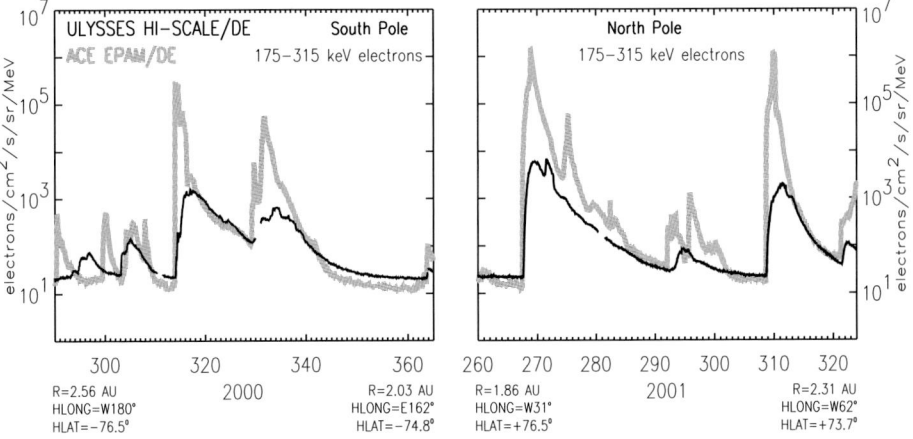

Figure 4.8. Comparison of 175-315 KeV electron intensities measured at Ulysses (HI-SCALE -black traces) and ACE (EPAM - grey traces) for the solar maximum polar passes in the south (left panel) and north (right panel). (From Lario et al. (2003). Reproduced with permission from Elsevier.)

close to the solar equator. Clearly, the source of energetic particles at solar maximum is not CIR-related, since the stable fast-slow solar wind stream structures that are responsible for CIRs do not exist during this phase of the solar cycle. As noted earlier, transient shock waves that are driven by fast CMEs give rise to the numerous large increases in particle flux that are characteristic of solar maximum. The Ulysses data from the recent high-latitude passes, in addition to confirming the presence of large fluxes of energetic particles over the poles, have revealed that the absolute intensity of these particles in the decay phase of many events is comparable to that measured simultaneously in the ecliptic near 1 AU, despite large separations in latitude and longitude (McKibben et al., 2001; Lario et al., 2003). An example is shown in Figure 8. This has led to the idea, originally proposed by Roelof et al. (1992), of the inner heliosphere near solar maximum acting as a "reservoir" for solar energetic particles. Further support for the existence of a particle reservoir has been presented by Patterson and Armstrong (2003), who found a persistent foreground population of energetic protons permeating the inner heliosphere. The measured proton energy spectra are complex in form, leading Patterson and Armstrong (2003) to suggest that multiple acceleration sources are involved.

The precise mechanism by which the particles are transported in latitude and longitude to fill this reservoir is still being debated. The process is certainly more complex than that in operation at solar minimum, where, as noted above, the prime candidates are enhanced perpendicular diffusion, and systematic changes to the underlying Parker spiral magnetic field configuration. The absence of long-lasting, stable structures in the corona at solar maximum tend to rule out any systematic motions of the field, and the evidence for perpendicular diffusion (Zhang et al., 2003) is not universally accepted (Sanderson et al., 2003). Alternatively, Neugebauer et al., (2002) have shown that open fields can connect back to active regions, and this may play an important role. These questions are discussed in more detail in the accompanying article by Sanderson (2004).

2.4   Cosmic Rays

Cosmic ray modulation is one of the central themes of the Ulysses mission. Galactic cosmic ray ions and electrons propagating through the heliosphere are affected by a combination of processes that include diffusion through irregularities in the heliospheric magnetic field, and gradient and curvature drifts resulting from the large-scale pattern of field lines (Parker, 1965; Fisk et al., 1998, and references therein). The direction of the drift motion is charge-dependent, and as such is also a function of the heliospheric magnetic field polarity. This in turn is related to the polarity of the open magnetic fields at the Sun, as discussed above. In this respect, the physics of cosmic ray modulation is a clear example of how the Sun and heliosphere form an integrated system. At the time of the Ulysses solar minimum high-latitude passes, the Sun's dipole field had positive polarity in the north. The predicted flow direction of positively charged cosmic ray nuclei entering the heliosphere was inward over the poles, and outward along the HCS. Conversely, electrons were expected to drift into the inner heliosphere mainly along the current sheet, and then outward via the polar regions. Indeed, Ulysses observed a positive latitudinal gradient in the flux of both galactic cosmic ray nuclei and the Anomalous Cosmic Ray (ACR) component, albeit smaller than predicted by most models (McKibben, 2001a,b). As in the case of the observations of lower energy particles accelerated at CIR and CME shocks, the cosmic ray data acquired near solar minimum suggested that transport across the average magnetic field is occurring on a much larger scale than previously thought.

Given the dependence on the heliospheric magnetic field polarity of the cosmic-ray drift patterns discussed above, the effects of the polar cap field reversals that occurred between 2001 and 2002 (Harvey and Recely, 2002) are clearly of interest. In particular, the presence of drift-

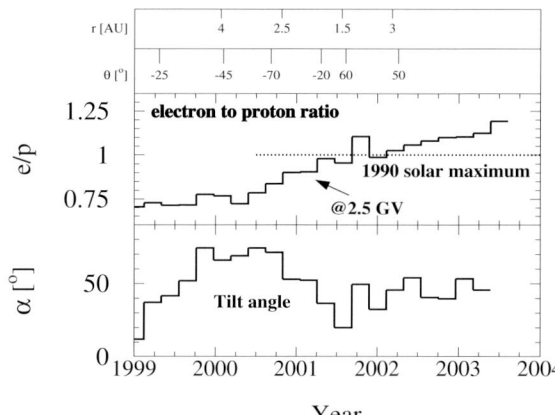

Figure 4.9 Electron-to-proton ratio at 2.5 GV measured along the Ulysses orbit from 1999-2003 (top panel). Also shown (lower panel) is the maximum extent of the heliospheric current sheet (a) shifted by 5 solar rotations to later times to account for the solar wind travel time to the heliopause. (From Heber et al. (2003). Reproduced by permission of AGU.)

dominated propagation at a given epoch should be indicated by (a) a different latitudinal dependence for oppositely charged species, and (b) different temporal behaviour of oppositely charged species as a result of the variations in the HCS during the solar cycle (e.g., Potgieter et al., 1997). The expectation then is that, once the new polarity configuration has become established throughout the heliosphere, the transport of positively charged cosmic-ray nuclei will be inward along the HCS, and outward via the poles. This in turn should lead to negative latitudinal gradients for these particles, with lower fluxes at high latitudes. As before, the opposite drift patterns should be observed in the case of electrons. The timing of the solar maximum polar passes was such that an observational test of this model prediction using Ulysses data proved to be difficult. The cosmic ray intensities recorded from pole to pole by the Ulysses instruments showed no measurable systematic variations with latitude (McKibben et al., 2003). This was presumably a result of the disordered state of the heliospheric magnetic field at that time, whereby a stable configuration with the new polarities had not yet been established. Data from the next solar minimum polar passes in 2006-2008, if available, will provide a better test.

As noted above, the drift motions of energetic electrons should be opposite to those of positively charged nuclei in a given magnetic polarity configuration. Measurements of cosmic ray electrons on board Ulysses have been invaluable in investigating these differences experimentally, thereby providing important tests of modulation theory (e.g., Potgieter,

1997). In 1994/1995, when Ulysses was at high latitudes near solar minimum, cosmic ray electrons did not show any significant latitudinal gradient, even though a clear proton gradient was present (Ferrando et al., 1996; Heber et al., 1999). The resulting variation in the electron-to-proton (e/p) ratio measured along the Ulysses trajectory was therefore consistent with the predictions of modulation models including drift effects, providing strong evidence for the importance of drifts in the modulation process at solar minimum. At solar maximum, the electron time profiles exhibited the same diffusion-dominated behaviour as the cosmic ray nuclei (Heber et al., 2003). As shown in Figure 8, between mid-2001 and mid-2002, the e/p ratio remained effectively constant, even though the polar cap fields were undergoing their polarity reversal, and the HCS tilt was decreasing. Positive latitudinal gradients in the electron fluxes, to be expected if drifts are also important at solar maximum, were not observed. This is further evidence that the global heliosphere had not yet adjusted to the new polarity, and that the predicted gradients, and corresponding increase in the e/p ratio, will only be observed once a stable configuration is achieved and drifts start to dominate again (Heber et al., 2003).

2.5   Interstellar Dust

The motion of the Sun through the Local Interstellar Cloud (LIC) enables a fraction of the gas and dust in the LIC to reach the inner heliosphere (Witte et al., 1993; Grün et al., 1993). Interstellar ions and dust grains smaller than ∼0.1 μm are excluded by the magnetised solar wind; however, larger grains, and neutral gas atoms, can flow into the heliosphere relatively unimpeded (Landgraf, 2000; Landgraf et al., 2003). As with other interstellar constituents, the flux of dust grains arriving at the heliospheric boundary is expected to remain relatively constant in time. Variations in the dust flux observed closer to the Sun, therefore, are assumed to be the result of processes occurring within the heliosphere. For example, the speed and direction of the grains are influenced by the Sun's gravity, and solar radiation pressure. In addition, the grains become positively charged as a result of electron emission under the influence of solar UV photons and the solar wind plasma (Mukai, 1981). As a result, particles of diameter less than ∼0.4 μm are strongly affected by the Lorentz force of the heliospheric magnetic field that varies with the 22-year magnetic cycle of the Sun. For larger grains, the effect is reduced owing to their smaller charge-to-mass ratios. Interstellar dust particles were measured for the first time in situ by the dust instrument on board Ulysses (Grün et al., 1993). Observations

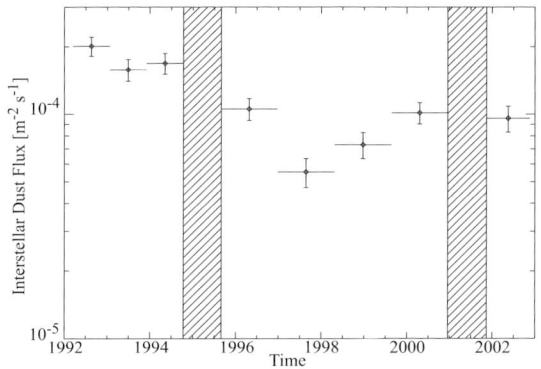

Figure 4.10 Interstellar dust flux as measured by the dust detector on board Ulysses. Shaded bars indicate the high-latitude passes at solar minimum and solar maximum. (From Landgraf et al. (2003). Reproduced by permission of AGU.)

acquired over the course of the mission to date have revealed how large-scale changes in the heliospheric magnetic field over the solar cycle, affect the flux of dust grains reaching the inner heliosphere (Landgraf, 2000; Landgraf et al., 2003).

The Ulysses dust measurements are shown in Figure 10. In mid-1996, a decrease in the interstellar dust flux was observed (Landgraf et al., 1999). This was attributed to the increased filtering effect (defocusing) of the heliospheric magnetic field configuration encountered by the dust grains during the last solar minimum (Landgraf, 2000). The model of Landgraf (2000) predicted that the flux would remain low until 2005, at which time the focusing effect of the polarity reversal that occurred in 2000-2002 would become apparent. Starting already in 2000, however, an increase in the interstellar dust flux has been observed (Landgraf et al., 2003). Recent modeling (Landgraf et al., 2003), in which a distribution of grain sizes ranging from 0.1 to 0.4 μm in radius was simulated, indicates that the Ulysses data covering the period 1992 to 2002 are best fit by a flux in which particles of effective radius 0.3 μm (having a charge-to-mass ratio of 0.6 Ckg^{-1}) form the dominant contribution, with a minor contribution from 0.4 μm grains. Even with this new model, the flux values measured between 1998 and 2000 are still higher than predicted, however. The cause of the earlier-than-expected rebound, therefore, remains to be understood.

3.   The Future of Ulysses

The Ulysses mission was originally foreseen to end in October, 1995. Owing to the outstanding scientific success, and equally, to the robustness of the spacecraft and its scientific payload, ESA and NASA recently agreed to continue the mission until March, 2008. This latest extension, the third in the history of the mission, will enable Ulysses to collect data during a third set of polar passes in 2007/2008. Conditions will be simi-

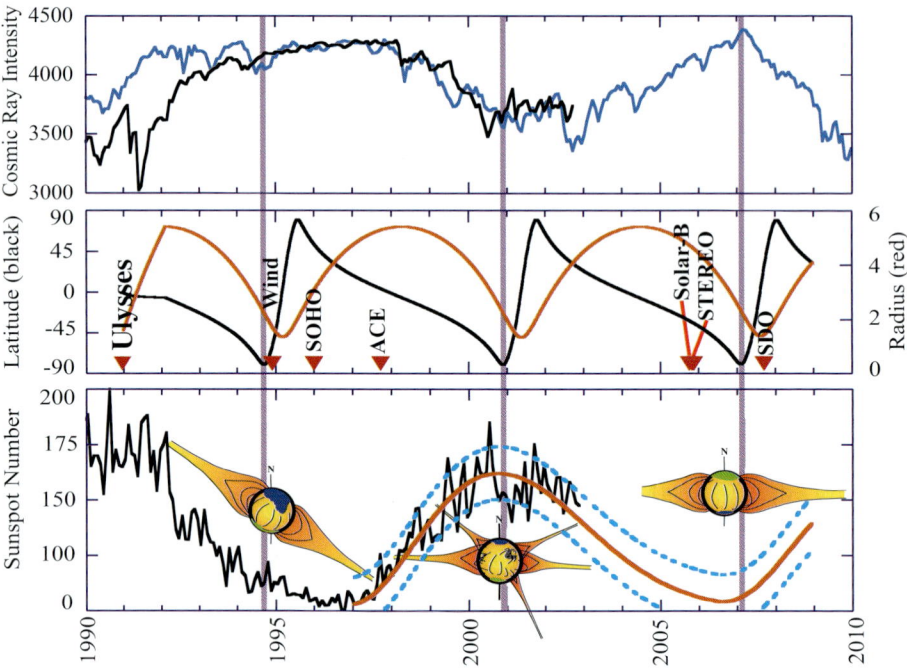

Figure 4.11. Top: Climax, Colorado cosmic ray intensity (black), and intensity shifted to the right by 22 years (blue). Middle: Ulysses orbital latitude (black) and radius (red) and launch dates for SOHO, Wind, ACE, Solar-B, STEREO, and SDO. Bottom: Solar Sunspot number and predicted Sunspot number through 2008. Schematics are of the typical corona at indicated times in the solar cycle. Vertical lavender bars mark the beginning of each Ulysses fast latitude scan. (Courtesy of S. Suess.)

lar to those in 1994/1995 when Ulysses first visited the Sun's poles. This time, however, the magnetic field of the Sun will be reversed in polarity, allowing Ulysses to search for differences in behaviour of the interplanetary medium related to the magnetic field reversal. Another difference between the solar minimum passes in 2007/2008 and those in 1994/1995, illustrated in Figure 11, is that Ulysses will have the additional benefit of being part of a fleet of solar and heliospheric spacecraft that includes ESA's SOHO and Cluster, and NASA's ACE and Cassini. New missions like NASA's dual-spacecraft STEREO, and the Solar Dynamics Observatory, will add a further dimension over the next few years.

    What of the era beyond Ulysses? Numerous questions will undoubtedly remain to be answered concerning the heliosphere and its many constituents. Plans for follow-up missions to Ulysses are currently being developed. One such mission is ESA's Solar Orbiter. With a likely launch date no earlier than 2014, Solar Orbiter will extend the range of out-of-ecliptic measurements to include imaging and spectroscopy of the Sun's polar regions, but from lower latitudes than Ulysses. On the NASA side, a mission called Telemachus (son of Ulysses) is being studied, that will also carry remote-sensing instruments, and will use the same Jupiter flyby technique as Ulysses to reach the highest latitudes, but with perihelion much closer to the Sun (0.2 AU). Even if approved, Telemachus is unlikely to be launched before Solar Orbiter. In the meantime, Ulysses remains the only spacecraft able to sample the Sun's environment away from the ecliptic plane.

Acknowledgments

The author wishes to thank the Ulysses PI teams for providing material used in this review. The helpful and insightful comments provided by Ed Smith, the NASA Project Scientist for Ulysses, and by one of the editors (Steve Suess), on an earlier version of this paper are gratefully acknowledged.

References

Balogh, A., R.G. Marsden, and E.J. Smith (eds.). (2001a). *The Heliosphere Near Solar Minimum: The Ulysses Perspective.* Springer-Praxis, Chichester.

Balogh, A., R.G. Marsden, and E.J. Smith. (2001b). In: A. Balogh, R.G. Marsden, and E.J. Smith (eds.): *The Heliosphere Near Solar Minimum: The Ulysses Perspective.* Springer-Praxis, Chichester, p. 1.

Balogh, A. and E.J. Smith. (2001). *Space Sci. Rev.* 97, 147.

Ferrando P., A. Raviart, L.J. Haasbroek, M.S. Potgieter, W. Dröge, B. Heber, H. Kunow, R. Müller-Mellin, H. Sierks, G. Wibberenz, and C. Paizis. (1996). *Astron. Astrophys.* 316, 528.

Fisk, L.A. (1996). *J. Geophys. Res.* 101, 15547.

Fisk, L. A. (2003). *J. Geophys. Res.* 108, 1157, 10.1029/2002JA009284.

Fisk, L.A., J.R. Jokipii, G.M. Simnett, R. von Steiger, and K.-P. Wenzel, (eds.). (1998). *Cosmic rays in the heliosphere.* Kluwer, Dordrecht.

Fisk, L.A., N.A. Schwadron, and T.H. Zurbuchen. (1999). *J. Geophys. Res.* 104, 19765.

Fisk, L.A. and J.R. Jokipii. (1999). In: A. Balogh, J.T. Gosling, J.R. Jokipii, R. Kallenbach, and H. Kunow (eds.): *Corotating Interaction Regions.* Kluwer, Dordrecht, p. 115.

Fisk, L.A. and N.A. Schwadron. (2001). *Astrophys. J.* 560, 425.

Gloeckler, G., J. Geiss, L.A. Fisk. (2001). In: A. Balogh, R.G. Marsden, and E.J. Smith (eds.): *The Heliosphere Near Solar Minimum: The Ulysses Perspective.* Springer-Praxis, Chichester, p. 287.

Gloeckler, G., T.H. Zurbuchen, and J. Geiss. (2003). *J. Geophys. Res.* 108, 1158, 10.1029/2002JA009286.

Gosling, J.T., P. Riley, D.J. McComas, and V.J. Pizzo. (1998). *J. Geophys. Res.* 103, 1941.

Grün, E., H.A. Zook, M. Baguhl, A. Balogh, S.J. Bame, et al. (1993). *Nature* 362, 428.

Hansteen, V. H., E. Leer, and T. E. Holzer. (1999). In: S. R. Habbal et al. (eds.): *Solar Wind Nine.* AIP Conf. Proc. 471, p. 17.

Harvey, K.L. and F. Recely. (2002). *Solar Phys.* 211, 31.

Heber, B., P. Ferrando, A. Raviart, G. Wibberenz, R. Müller-Mellin, H. Kunow, H. Sierks, V. Bothmer, A. Posner, C. Paizis, and M.S. Potgieter. (1999). *Geophys. Res. Lett.* 26, 2133.

Heber, B., J.M. Clem, R. Müller-Mellin, H. Kunow, S.E.S. Ferreira, and M.S. Potgieter. (2003). *Geophys. Res. Lett.* 30(19), 8032, doi: 10.1029/2003GL1017356.

Hurley, K. M. Sommer, T. Cline, M. Boer, and M. Niel. (1995). *Astrophys. Space Sci.* 231, 227.

Hurley, K., T. Cline, et al. (1999). *Nature* 397, 41.

Jones, G.H. and A. Balogh. (2003). *Ann. Geophysicae* 21, 1377.

Kunow, H., M.A. Lee, et al.. (1999). In: A. Balogh, J.T. Gosling, J.R. Jokipii, R. Kallenbach, and H. Kunow (eds.): *Corotating Interaction Regions.* Kluwer, Dordrecht, p. 221.

Landgraf, M., M. Müller, and E. Grün. (1999). *Planet. Space Sci.* 47, 1029.

Landgraf, M.. (2000). *J. Geophys. Res.* 105, 10303.

Landgraf, M. K., H. Krüger, N. Altobelli, and E. Grün. (2003). *J. Geophys. Res.* 108, 8030, doi: 10.1029/2003JA009872.

Lario, D., E.C. Roelof, R.B. Decker, and D.B. Reisenfeld. (2003). *Adv. Space Res.* 32(4), 579.

Lazarus, A.J. and R.L. McNutt, Jr. (1990). In: S. Grzedzielski and D.E. Page (eds.): *Physics of the Outer Heliosphere.* Pergamon Press, New York, p. 229.

MacDowall, R.J. and P.J. Kellogg. (2001). In: A. Balogh, R.G. Marsden, and E.J. Smith (eds.): *The Heliosphere Near Solar Minimum: The Ulysses Perspective.* Springer-Praxis, Chichester, p. 229.

Marsden, R.G. (ed.) (1995). *The High Latitude Heliosphere.* Kluwer, Dordrecht.

Marsden, R.G. (ed.) (2001). *The 3-D Heliosphere at Solar Maximum.* Kluwer, Dordrecht.

Marsden, R.G. and E.J. Smith. (1996a). *Adv. Space Res.* 17(4/5), 293.

Marsden R.G. and E.J. Smith. (1996b). *Il Nuovo Cimento* 19C, 909.

Marsden R.G., E.J. Smith, J.F. Cooper, and C. Tranquille. (1996). *Astron. Astrophys.* 316, 279.

McComas, D.J., A. Balogh, S.J. Bame, B.L. Barraclough, W.C. Feldman, R. Forsyth, B.E. Goldstein, J.T.Gosling, H.O. Funsten, M. Neugebauer, P. Riley, & R. Skoug (1998). *Geophys. Res. Lett.* 25, 1.

McComas, D.J., B.L.Barraclough, H.O.Funsten, J.T.Gosling, E. Santiago-Munoz, R.M. Skoug, B.E. Goldstein, M. Neugebauer, P. Riley, and A. Balogh. (2000). *J. Geophys. Res.* 105, 10419.

McComas, D.J., H.A. Elliott, N.A. Schwadron, J.T. Gosling, R.M. Skoug, and B.E. Goldstein. (2003). *Geophys. Res. Lett.* 30(19), 1517,doi: 10.1029/2003GL017136.

McKibben, R.B., C. Lopate, and M. Zhang. (2001). In: R.G. Marsden (ed.): *The 3-D Heliosphere at Solar Maximum.* Kluwer, Dordrecht, p. 257.

McKibben, R.B. (2001a). In: A. Balogh, R.G. Marsden, and E.J. Smith (eds.): *The Heliosphere Near Solar Minimum: The Ulysses Perspective.* Springer-Praxis, Chichester, p. 327.

McKibben, R.B. (2001b). *Proc. of ICRC 2001* 8, 3281.

McKibben, R.B., J.J. Connell, C. Lopate, M. Zhang, J.D. Anglin, A. Balogh, S. Dalla, et al. (2003). *Ann. Geophysicae* 21, 1217.

Mukai, T. (1981). *Astron. Astrophys.* 99, 1.

Neugebauer, M., P.C. Liewer, E.J. Smith, R.M. Skoug, and T.H. Zurbuchen. (2002). *J. Geophys. Res.* 107, 10.1029/2001JA000306.

Parker, E.N. (1965). *Planet. Space Sci.* 13, 9.

Patterson, J.D., and T.P. Armstrong. (2003). *Geophys. Res. Lett.* 30(19), 8037, doi: 10.1029/2003GL017154.

Pauls, H.L., and G.P. Zank. (1996). *J. Geophys. Res.* 101, 17081.

Phillips, J.L., S.J. Bame, W.C. Feldman, J.T. Gosling, D.J. McComas, B.E. Goldstein, M. Neugebauer, and C.M. Hammond. (1996). In: D. Winterhalter, J. Gosling, S. Habbal, W. Kurth, M. Neugebauer (eds.): *Solar Wind Eight.* AIP Conf. Proc. 382, p. 416.

Potgieter, M.S., L.J. Haasbroek, P. Ferrando, and B. Heber. (1997). *Adv. Space Res.* 19(6), 917.

Reisenfeld, D.B., J.T. Gosling, R.J. Forsyth, P. Riley, and O.C. St. Cyr. (2003). *Geophys. Res. Lett.* 30(19), 8031, 10.1029/2003GL017155.

Richardson, J.D., C. Wang, and K.I. Paularena. (2001). *Adv. Space Res.* 27(3), 471.

Roelof, E. C., R.E. Gold, G.M. Simnett, S.J. Tappin, T.P. Armstrong, and L.J. Lanzerotti. (1992). *Geophys. Res. Lett.* 19, 1243.

Saito, T., T. Sakurai, and K. Yumoto. (1978). *Planet. Space Sci.* 26, 413.

Sanderson, T.R. (2004). *this volume.*

Sanderson, T. R., R.G. Marsden, K.-P. Wenzel, A. Balogh, R.J. Forsyth, and B.E. Goldstein. (1994). *Geophys. Res. Lett.* 21, 1113.

Sanderson, T.R., R.G. Marsden, K.-P. Wenzel, A. Balogh, R.J. Forsyth, and B.E. Goldstein. (1995). *Space Sci. Rev.* 72, 291.

Sanderson, T.R., R.G. Marsden, C. Tranquille, S. Dalla, R.J. Forsyth, J.T. Gosling, and R.B. McKibben. (2003). *Geophys. Res. Lett.* 30(19), 8036, 10.1029/2003GL017306.

Simnett, G.M., K.A. Sayle, E.C. Roelof, and S.J. Tappin. (1994). *Geophys. Res. Lett.* 21, 1561.

Smith, E.J. and A. Balogh. (2003). In: M. Velli, R. Bruno, F. Malara (eds.): *10th International Conference on Solar Wind.* AIP Conf. Proc. 679, p. 67.

Smith, E. J. and K.-P. Wenzel. (1993). *J. Geophys. Res.* 98, 21111.

Smith, E.J., R.G. Marsden, and D.E. Page. (1995). *Science* 268, 1005.

Smith, E.J., A. Balogh, M.E. Burton, R. Forsyth, and R.P. Lepping. (1997). *Adv. Space Res.* 20(1), 47.

Smith, E.J., A. Balogh, R.J. Forsyth, and D.J. McComas. (2001). *Geophys. Res. Lett.* 28, 4159.

Smith, E.J. and R.G. Marsden. (2003). *Geophys. Res. Lett.* 30(19), 8027, 10.1029/
2003GL018223.

Smith, E. J., R.G. Marsden, A. Balogh, G. Gloeckler, J. Geiss, D.J. Mc-Comas, R.B. McKibben, R.J. MacDowall, L.J. Lanzerotti, N. Krupp, H. Krger, and M. Landgraf. (2003). *Science* 302(Nov 14), 1165.

Suess, S.T. and E.J. Smith. (1996). *Geophys. Res. Lett.* 23, 3267.

Wenzel, K.-P., R.G. Marsden, D.E. Page, and E.J. Smith. (1992). *Astron. Astrophys. Suppl.* 92(2), 207.

Wenzel, K.-P. and E.J. Smith. (1993). *Planet. Space Sci.* 41, 797.

Witte, M., H. Rosenbauer, M. Banaszkiewics, and H. Fahr. (1993). *Adv. Space Res.* 13(6), 121.

Zhang, M., J.R. Jokipii, and R.B. McKibben. (2003). *Astrophys. J.* 594, 493.

Chapter 5

PROPAGATION OF ENERGETIC PARTICLES TO HIGH LATITUDES

Ulysses' Second Northern Polar Pass

T. R. Sanderson

Research and Scientific Support Department of ESA

ESTEC, Noordwijk, The Netherlands

trevor.sanderson@esa.int

Abstract   We present observations of energetic particles in the energy range ∼1 MeV to ∼100 MeV made by the COSPIN instrument on board the Ulysses spacecraft during the recent second northern polar pass. For a short time during this pass the Ulysses spacecraft was at high heliographic latitude, above the current sheet, and immersed in high-speed solar-wind flow coming from the northern polar coronal hole. Four large solar energetic particle events were observed. We discuss the solar conditions prevalent during the Ulysses mission, and in particular the conditions during this polar pass. We discuss the rise to maximum of these events and examine the onset time and the anisotropy of the energetic particles. We find that during these events the particle angular distributions were almost isotropic, but with a net outward flow along the magnetic field lines. We conclude that particles reached these high latitudes travelling along the magnetic field lines.

Keywords: Ulysses, energetic particles, propagation, magnetic field, solar wind, coronal mass ejection, anisotropy

1.   Introduction

Ulysses was launched in 1990. It has now completed two orbits around the Sun. Conditions during both of these orbits were very different. This chapter summarises the conditions on the Sun and in the heliosphere which influenced the characteristics of the energetic particles observed. The particle observations of each orbit are discussed, and related to the

G. Poletto and S.T. Suess (eds.), The Sun and The Heliosphere as an Integrated System, 113–145.
© 2004 *Kluwer Academic Publishers. Printed in the Netherlands.*

conditions on the Sun, and then the unique observations of energetic particles over the pole of the Sun observed for the first time during the recent northern polar pass of the second orbit are presented and discussed.

In February 1992 Ulysses used an encounter with the planet Jupiter to begin its first out of the ecliptic orbit around the Sun, completing it in April 1998. This orbit began as the level of solar activity was falling. Most of the time when the spacecraft was over the south and then over the north pole of the Sun was close to the time of minimum of solar activity of solar cycle 22. At this time the axis of the dipolar component of the coronal magnetic field was nearly parallel to the Sun's spin axis, and the current sheet was almost flat, giving rise to a streamer belt which was more or less in the plane of the ecliptic. Large coronal holes were observed over the poles, giving rise to high-speed solar wind flow in the high latitude polar regions of the heliosphere. Figure 1a, after Suess et al. (1998) shows this configuration.

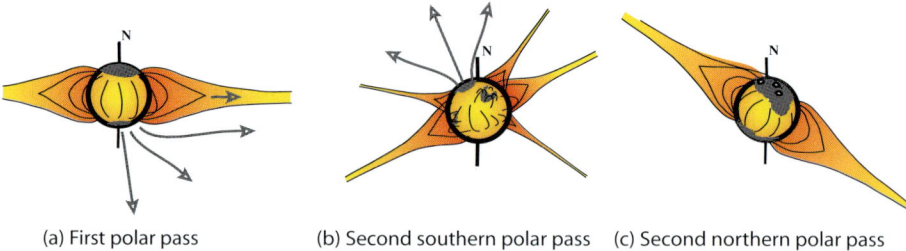

(a) First polar pass (b) Second southern polar pass (c) Second northern polar pass

Figure 5.1. Cartoon showing the coronal magnetic field observed during, from left to right, (a) the first polar passses, (b) the second southern polar pass and (c) the second northern polar pass (Suess et al., 1998).

As this was a period close to solar minimum, very few particle increases due to either Coronal Mass Ejections (CME) or Co-Rotating Interaction Regions (CIR) were observed in the polar regions during these passes. At mid-latitudes, significant increases due to CIRs were observed for a substantial fraction of the time that the spacecraft was above the current sheet, but again at the very highest latitudes very few, if any, increases due to CIRs or CMEs were observed.

The second orbit around the Sun began in April 1998 as the level of solar activity of solar cycle 23 was increasing. The second orbit was considerably different from the first. During most of the orbit, the heliosphere was dominated by the presence of CMEs as the level of solar activity increased, the spacecraft passing over the south and then the north poles of the Sun during the time around the maximum of solar activity of solar cycle 23. A summary of Ulysses observations during the maximum of cycle 23 can be found elsewhere in this volume (Marsden, 2004).

Conditions on the Sun were very different for the southern and the northern polar passes. During the southern pass, the dipole axis was oriented at around 135 degrees to the spin axis and the field had a significant quadrupole component. The current sheet reached up to very high latitudes, and there was no polar coronal hole over the southern pole, typical of that expected around solar maximum. So, Ulysses remained in slow solar wind flow as it passed over the pole. Figure 1b shows this configuration.

Over the northern pole the situation was quite different. The tilt of the dipole was similar. The dipole strength had increased, and the quadrupole term of the coronal field had diminished. The field was therefore much more dipolar, and so the current sheet only reached up to mid latitudes. A polar coronal hole had started to develop, and by the time Ulysses reached the highest northern latitudes it was immersed in fast solar wind. Figure 1c shows this configuration. Although the northern pass was still close to solar maximum, the configuration of the Sun was beginning to look more like that close to solar minimum.

Solar activity was still high, and so perhaps not surprisingly, CMEs and substantial particle increases were observed at the highest latitudes as the spacecraft passed over the northern pole of the Sun. Four large SEP events were observed at high latitude and in the fast solar wind. These events were unusual in that they were the only CME-related particle increases observed so far by Ulysses at high latitudes and in the fast solar wind. They were also unusual in that the onsets at high energies were delayed considerably, and the angular distributions at onset were almost isotropic.

In this chapter, we first discuss the solar conditions prevailing at the time of the Ulysses mission, and in particular at the time of the polar passes. We then discuss the propagation of energetic particles to high heliographic latitudes as observed when Ulysses was over the northern pole and immersed in the fast solar wind, and compare them with events observed in the slow solar wind over the southern pole.

2. Solar Conditions

2.1 Influence of the Sun on the Heliosphere

The coronal magnetic field controls the configuration of the heliosphere. In the corona, the current sheet winds its way around the positive and negative poles of the magnetic field, delineating the positive coronal magnetic field from the negative. The magnetic field propagates out into the heliosphere in the solar wind, and as it does so, the current sheet divides up the heliosphere into a region of positive magnetic field polarity (with outward pointing field) and a region of negative polarity (with inward pointing magnetic field). At solar minimum this is a simple configuration, but becomes more complicated at solar maximum.

The coronal holes follow the motion of the poles of the magnetic field as they move across the disc of the Sun in the course of the solar cycle. At solar minimum, large polar coronal holes exist over the poles. The coronal holes, the source of high speed flow, move equatorward during the course of the solar cycle, and break up into smaller mid-latitude coronal holes at solar maximum. At solar minimum, most of the high latitude region is filled with high speed solar wind flow. At solar maximum, only a few small low-latitude fast solar wind streams exist.

For the study of particle propagation, it is important to differentiate between propagation in the slow solar wind and the fast solar wind. The magnetic field in the fast solar wind is much more turbulent than in the slow solar wind, so propagation in the fast solar wind should be much more difficult.

However, the magnetic field in the slow solar wind tends to be full of discontinuities, channels, and other features, all of which, depending on the size and thickness of the discontinuities and the energy of the particles, affect propagation in different ways, whereas the field in the fast solar wind tends mainly to be homogeneous, and devoid of large scale discontinuities.

The slow solar wind more often than not contains magnetic field structures such as CIRs, and their associated forward and reverse shocks, and CMEs, with their associated interplanetary shocks, all of which again affect the propagation, and even acceleration of the particles.

All of the above has to be taken into account when considering the propagation of particles in the three-dimensional heliosphere.

2.2 Coronal Magnetic Field During a 22-year Solar Cycle

The starting point for the observations of the Sun presented here are two data sets based on daily observations of the Sun, one which covers the last ∼28 years, and the other which covers the last ∼15 years.

Figure 2 is an example from the first data set. This is a source surface map of the coronal magnetic field for one Carrington Rotation, computed using a potential field model. It is derived from daily large-scale photo-

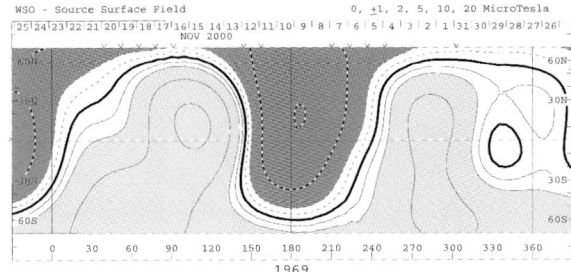

Figure 5.2 Wilcox Solar Observatory computed radial magnetic field at the source surface for November 2000, Carrington Rotation 1969.

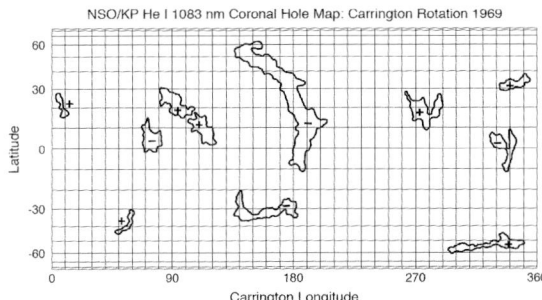

Figure 5.3 The boundaries of coronal holes derived from the Kitt Peak/NSO He I 1083 Coronal Hole spectroheliogram observations for November 2000, Carrington Rotation 1969.

spheric magnetic field observations made by the Wilcox Solar Observatory. A complete description of the method used to derive the parameters with which to describe the Sun's coronal field can be found in Hoeksema (1984) and at http://quake.stanford.edu/~wso/Description.ps.

The second data set is shown in Figure 3. This is a plot showing the location of the boundaries of the coronal holes derived from observations taken in the He I 1083nm line by the spectromagnetograph of the National Solar Observatory/Kitt Peak (Jones et al., 1992). A description of these observations can be found in Harvey and Recely (2002).

Figure 4 shows the long term behaviour of the Sun's coronal magnetic field as observed by the Wilcox Solar Observatory during the 28-year period starting in 1976, taken from Sanderson et al. (2003a). This corresponds to nearly three 11-year cycles, or one and one half 22-year cycles. This shows from bottom to top, the position of the Sun's source surface neutral line together with the spacecraft heliographic latitude, the sunspot number, dipole strength, quadrupole strength, and the angle between the dipole axis and the Sun's spin axis.

The position of the source surface neutral line is taken from the Carrington Rotation plots such as the one shown in Figure 2. The dipole and quadrupole strengths and the dipole axis direction are computed from the coefficients of the multipole expansion stored on the WSO home page. Here we used the WSO 9-order potential field model clas-

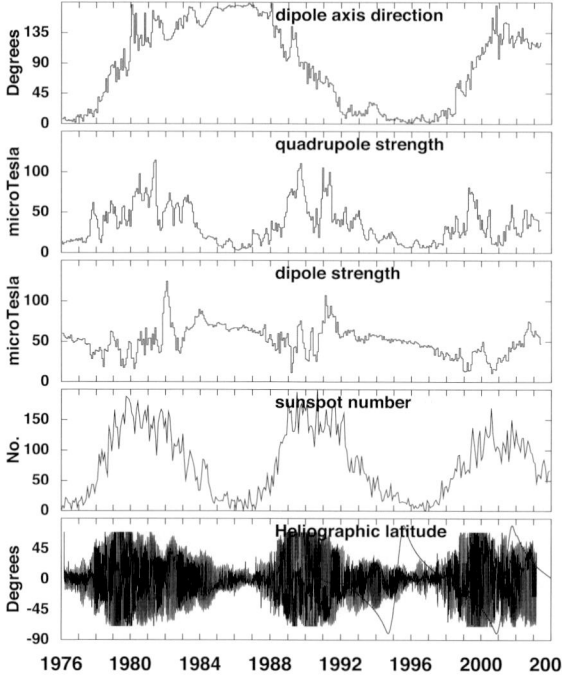

Figure 5.4 Latitude of the source surface neutral line together with the spacecraft heliographic latitude, sunspot number, dipole and quadruple strength and dipole axis direction for the 28 year period starting on 1 January 1976.

sical fit with a source surface at 2.5 Rs (Hoeksema, 1984). This gives the best overall agreement with the polarity pattern observed at Earth (http://quake.stanford.edu/∼wso/coronal.html).

In 1976 the sunspot number was at a minimum (cycle 20/21). The magnetic field was mainly dipolar, as the dipole strength was much higher than the quadrupole strength. The dipole axis was oriented along the Sun's spin axis, which meant that the current sheet was almost flat and in the ecliptic plane.

As the cycle progressed, the sunspot number increased, the quadrupole strength increased and the field became more quadrupolar. The dipole axis slowly tilted away from the Sun's axis until around solar maximum it was perpendicular to the spin axis, meaning that the current sheet was no longer flat, but was highly inclined, and reached up almost to the poles.

As the cycle progressed past maximum, the dipole rotated past the perpendicular and the polarity reversed. The sunspot number started to decrease, the quadrupole strength diminished and the field became more dipolar. As the dipole continued rotating, the current sheet became flatter until at the minimum in around 1986 the dipole axis was aligned anti-parallel to the spin axis, and the current sheet was flat and in the ecliptic plane again.

The whole process continued on until the minimum between cycle 22 and 23 in 1996 when the axis was parallel again, with the magnetic field positive again, so completing a complete rotation.

2.3   Coronal Magnetic Field and Coronal Holes During the Ulysses Mission

Figure 5 shows the same quantities again, together with the position of the spacecraft, but this time for the 14-year period starting in 1990, the year when Ulysses was launched. Vertical lines show the times of the four polar passes, the first southern, first northern, second southern and second northern passes.

The first southern and northern passes were close to the minimum in solar activity of cycle 22, with a very low sunspot number. The dipole strength was much higher than the quadrupole strength, so that the field was mainly dipolar. The axis of the dipolar component of the coronal magnetic field was nearly parallel to the Sun's spin axis, and the current sheet was more or less flat, giving rise to a streamer belt which was more or less in the plane of the ecliptic.

The second southern pass began close to the maximum phase of solar cycle 23, when the sunspot number was high. The quadrupole strength was higher than the dipole strength, implying that the field was some-what quadrupolar. The dipole axis was oriented at around 135 degrees to the spin axis and the current sheet reached up to high latitudes, but not as high as the highest latitude reached by Ulysses. This implied that the spacecraft would be in a uni-polar field, but this was not observed (Harvey and Recely, 2002, Jones et al., 2003), although one cannot expect observations of the Sun's current sheet at high latitudes to correspond exactly to conditions at Ulysses. There was no polar coronal hole over the southern pole, typical of that expected around solar maximum. So, Ulysses remained in slow-speed solar wind flow as it passed over the pole.

Over the northern pole the situation was quite different. The tilt of the dipole was similar, but the dipole strength had increased, and the quadrupole term of the coronal field had diminished. The field was therefore much more dipolar. The current sheet again reached up to moderately high latitudes. But now a large polar coronal hole had started to develop over the northern pole, and by the time Ulysses reached the highest northern latitudes it became immersed in the fast solar wind.

The overriding difference between the second southern and second northern polar passes was the presence of a large coronal hole over the

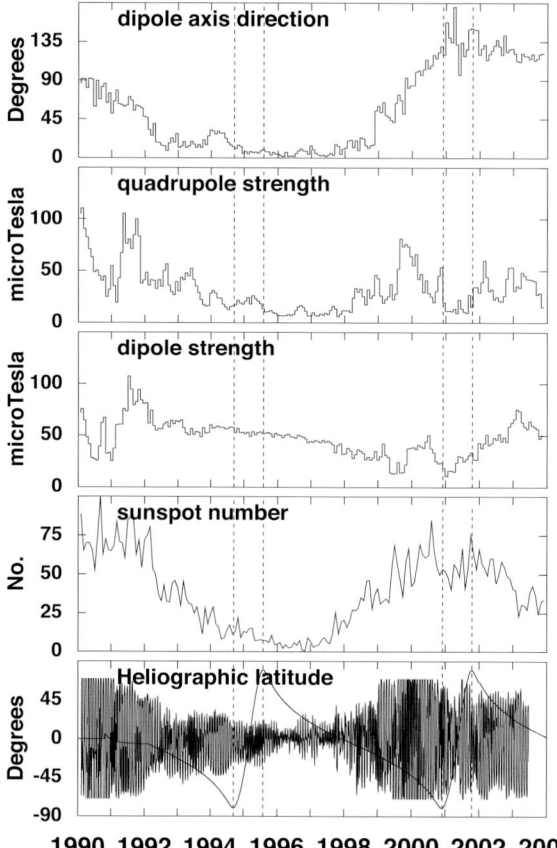

Figure 5.5 Latitude of the source surface neutral line together with the spacecraft heliographic latitude, sunspot number, dipole and quadruple strength and dipole axis direction for the 14 year period covering the Ulysses mission.

northern pole and the absence of a polar coronal hole over the southern pole.

Although the northern pass was still close to solar maximum, from the point of view of the dipole and quadrupole, conditions on the Sun were beginning to look more like conditions close to solar minimum. However, the Sun was still active, and during the period when Ulysses was over the northern pole, four large CMEs were observed whilst the spacecraft was in the fast solar wind (Reisenfeld et al., 2003)

3.   The First Orbit

In Figure 6 we show combined observations of the magnetic field and coronal hole configuration on the Sun at the time of the first northern and first southern polar passes (which corresponds to Carrington rotations 1887 and 1898), derived from data such as are shown in Figure 2 and Figure 3 (Sanderson et al. (2003a)).

At the top are synoptic plots of the coronal magnetic field similar to the ones produced by the Wilcox Solar Observatory, with the contours enhanced and with the location of the edges of the coronal holes super-imposed on top. At the bottom is the same data, but now plotted in a stereographic projection.

The panels at the left correspond to the first southern polar pass, whilst the right panels correspond to the first northern polar pass. During both passes, the axis of the dipolar component of the coronal magnetic field was nearly parallel to the Sun's spin axis. The current sheet was almost flat, giving rise to a streamer belt which was more or less in the plane of the ecliptic. Large coronal holes were observed over both the southern and northern poles, giving rise to fast solar wind in the high latitude polar regions of the heliosphere. During both the first southern and the first northern polar passes, the Ulysses spacecraft was immersed in the fast solar wind from these polar coronal holes.

In Figure 7 we show seven years of particle and plasma data covering the first out-of-the-ecliptic orbit of the Ulysses spacecraft, starting in 1992 and ending at the end of 1998 (The nominal start was February 1992 when the spacecraft passed Jupiter, and the nominal end was April 1998).

Particle observations presented here were made with the Low-Energy Telescope (LET) and the High-Energy Telescope (HET) of the COSPIN instrument (Simpson et al., 1992). Both the LET and HET are mounted perpendicular to the spacecraft spin axis. The spin axis always points to the Earth, so that the anisotropy observations are from a plane perpendicular to the Earth-spacecraft line. Magnetic field observations were made with the Ulysses magnetometer (Balogh et al., 1992), while solar wind measurements were made with the Ulysses Solar Wind experiment (Bame et al., 1992).

In Figure 7 we plot, starting from the bottom, position of the Sun's source surface neutral line together with the heliographic latitude of the spacecraft, the solar wind speed, the magnetic field magnitude, magnetic field azimuth, the 8.4 - 19.0 MeV/n alpha particle intensity and the 1.2 - 3.0 MeV proton intensity. The 8.4 - 19.0 MeV/n alpha particle intensity is reasonably representative of SEPs accelerated close to the Sun, either by the flare, or by the CME when close to the Sun, whereas the 1.2 - 3.0 MeV proton intensity is usually a mixture of SEP particles and particles accelerated locally, either by CIRs or by CMEs.

Already in 1990 and 1991, whilst the Ulysses spacecraft was at low latitudes and was still on its way to Jupiter there was a hint of the regular CIRs to come (Marsden et al., 1993) together with periods of elevated solar activity, culminating in the March 1991 and the July 1991

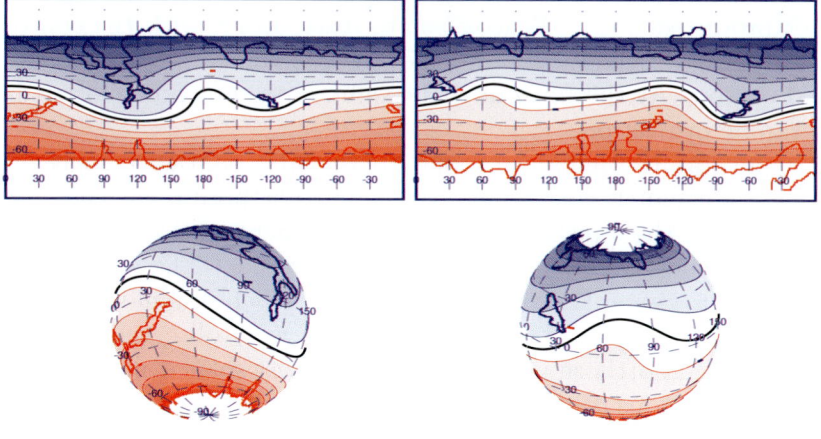

(a) First Southern Polar Pass, Carrington 1887   (b) First Northern Polar Pass, Carrington 1898

Figure 5.6. Contours of computed radial magnetic field at the source surface together with the boundaries of the coronal holes for (a) the first southern polar pass (Carrington Rotations 1887) and (b) the first northern polar pass (Carrington Rotation 1898).

series of events (Sanderson et al., 1992 and Roelof et al., 1992). The first out-of-the-ecliptic orbit began in 1992 with the spacecraft at low latitude at the position of Jupiter, immersed in the slow solar-wind. Within the slow solar wind there were many irregular compression regions and solar activity was moderately high, such that intensity increases at ~1 MeV were a mix of SEP and CIR accelerated particles.

In mid 1992, as the spacecraft started to climb to high southern latitudes, a series of regular particle increases (Sanderson et al., 1994, Simnett et al., 1994) due the presence of persistent pattern of CIR structures (Bame et al., 1993, Smith et al., 1993) were observed. Increases continued to be observed beyond mid-1993 when the spacecraft was continually immersed in high speed flow coming from the polar coronal hole, but only due to CIRs. From this time on until 1998 no solar energetic particle increases (such as are seen in the high energy alpha channel), were observed. Increases due to CIRs continued to be observed up to mid-1994, when the spacecraft was at a latitude of around 70 degrees south, well beyond the latitude that the forward-reverse shocks and the compression regions of the CIRs were observed.

Particle increases due to CIRs were again observed during the Fast Latitude Scan in early 1995 as the spacecraft quickly crossed through

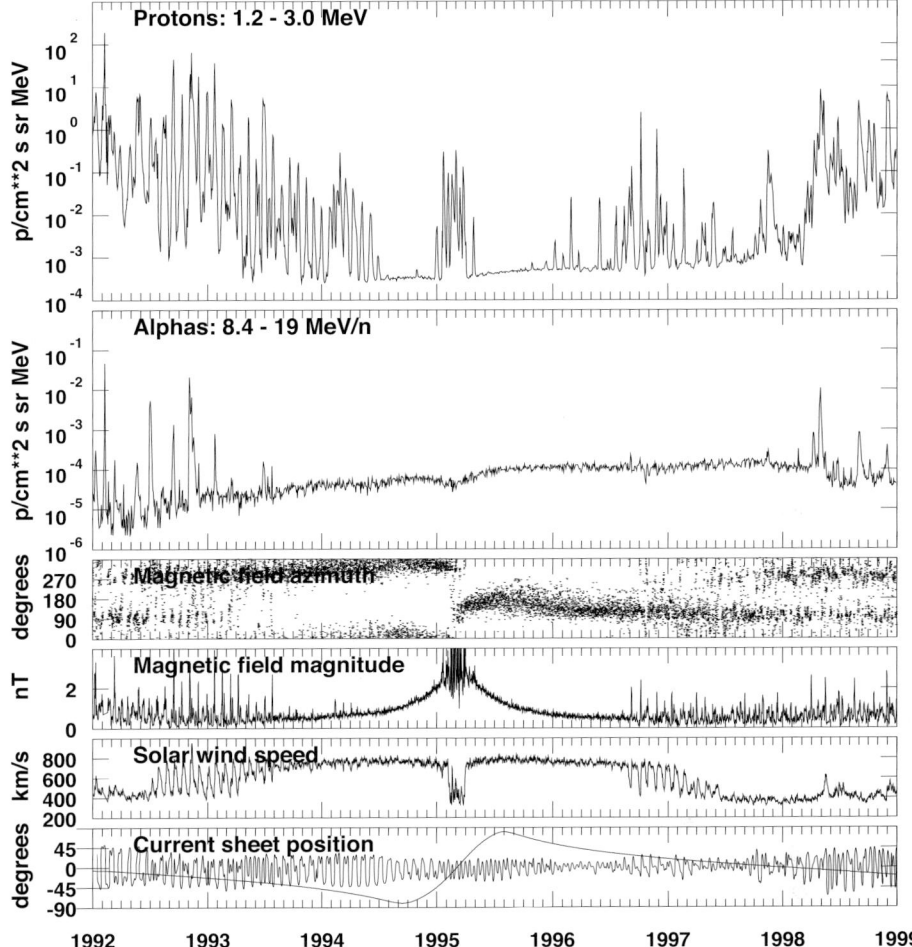

Figure 5.7. Seven-year plot starting on 1 January 1992 showing data for the first out-of-ecliptic orbit.

the ecliptic (Sanderson et al., 1995), but again, over the northern poles, no increases were observed.

Particle increases associated with CIRs (Roelof et al., 1997 and Sanderson et al., 1999) were again seen in 1996 as the spacecraft descended slowly down to low latitudes, though this time with not the same regularity as during the period in 1992 and 1993. Most of 1997 was dominated by CIR associated particle increases (Lario et al., 2000a), whilst most of 1998 was dominated by CME-associated particle increases (Lario et al.,

2000b) as the level of activity of the Sun began to increase as solar cycle 23 started to pick up again.

4.   The Second Orbit

In Figure 8 we show observations of the magnetic field and coronal hole configuration on the Sun at the time of the second northern and second southern polar passes (which corresponds to Carrington rotations 1969 and 1975 respectively), similar to Figure 6.

Conditions on the Sun were very different for the southern and the northern polar passes. During the southern pass (left panels), the dipole axis was oriented at around 135 degrees to the spin axis and the field had a significant quadrupole component. The current sheet reached up to high latitudes, and there were no polar coronal holes, typical of that expected around solar maximum. Ulysses never left the slow solar wind as it passed over the pole.

Over the northern pole the situation was quite different (right panels). The tilt of the dipole was similar. The dipole strength had increased, and the quadrupole term of the coronal field had diminished. The field was much more dipolar, and the current sheet still reached up to moderately high latitudes. A polar coronal hole had started to develop, and by the time Ulysses reached the highest northern latitudes it was immersed in high speed solar wind flow from this coronal hole. Although the northern pass was still close to solar maximum, the conditions on the Sun were beginning to look more like conditions close to solar minimum.

In Figure 9 we show particle and plasma data from the second orbit, starting on 1 January 1998. It can be seen immediately that the particle properties during this orbit differs considerably from the previous one. During the early part of the orbit, the high energy particle flux was dominated by the background cosmic ray flux, which slowly dropped during the course of the period 1998 to early 2000. From mid-2000 onwards, there were many intense increases in the 8.4 - 19 MeV/n Alpha particle intensity, the result of an increase in flare and CME activity on the Sun as solar maximum was approached. This activity peaked during 2001 and 2002. Activity increased again for a short period towards the end of 2003, coincident with the October/November 2003 events on the Sun.

Many more energetic particles, both high energy SEP particles and locally accelerated low-energy particles, were observed over the poles. Compare this with the lack of low-energy particles over the poles during the first orbit.

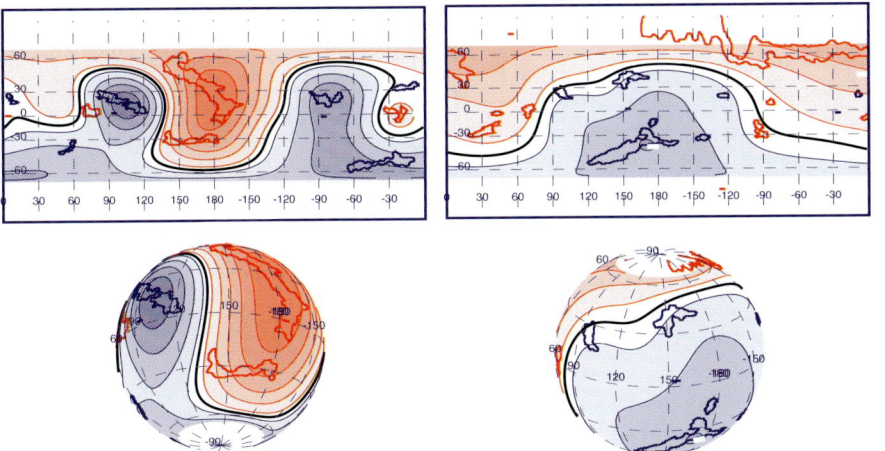

(a) Second Southern Polar Pass, Carrington 1969   (b) Second Northern Polar Pass, Carrington 1975

Figure 5.8.   Contours of computed radial magnetic field at the source surface together with the boundaries of the coronal holes for (a) the second southern polar pass, (Carrington Rotation 1969) and (b) the second northern polar pass, (Carrington Rotation 1975).

For most of the orbit, the solar wind speed was low, except for a short 3-4 month period at the end of 2001. Over the poles, the magnetic field was much more disturbed than during the first orbit.

4.1   The Second Polar Passes

In Figure 10, taken from Sanderson et al. (2003b), we show particle and plasma data taken during the second polar passes. This shows the second southern and northern polar passes, and the second fast latitude scan, with, from top to bottom the ∼1 MeV proton and ∼10 MeV/n alpha particle intensities, the magnetic field azimuth, solar-wind speed, and the position of the current sheet and the spacecraft for the 2-year period starting on 1 July 2000 and ending on 30 June 2002.

For the southern polar pass the current sheet was highly inclined, and although Ulysses was just above the projected position of the current sheet, the sector structure was still continually observed, and the polar coronal hole over the southern pole had disappeared. So, Ulysses never entered the fast solar wind, even if any was present over the pole.

The northern polar coronal hole was just beginning to re-form as Ulysses approached the northern pole. At the same time the inclination of the current sheet had decreased, so that as Ulysses reached the highest

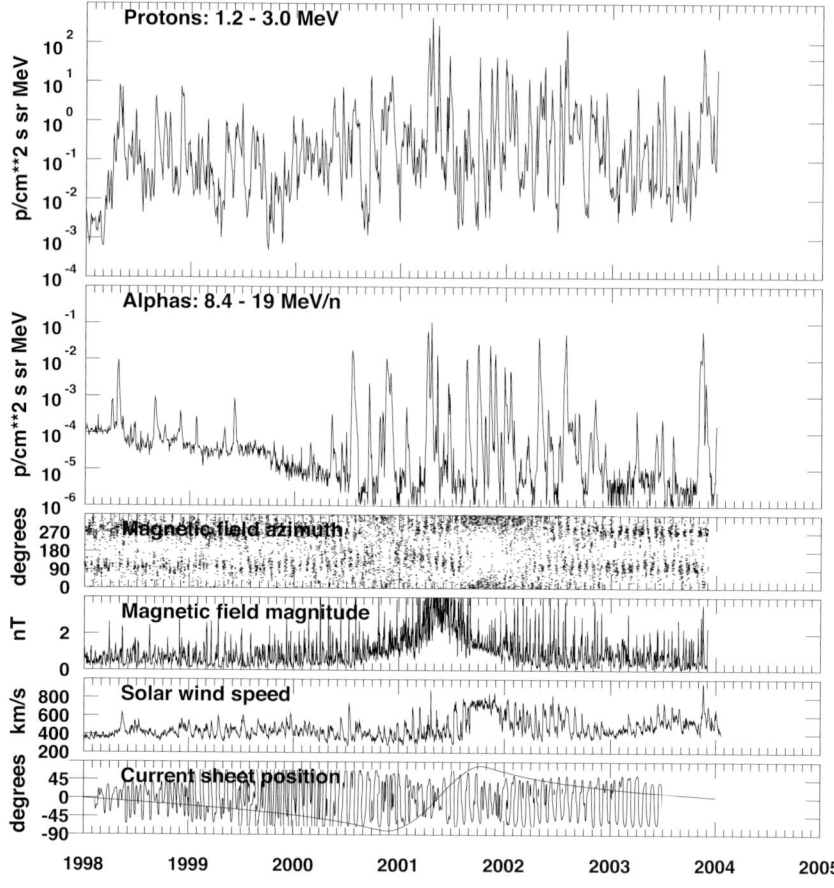

Figure 5.9. Data for the second out-of-ecliptic orbit. Seven year plot starting on 1 January 1998.

northern polar latitudes, it climbed for the first time during the second orbit into the region containing high-speed solar wind flow.

The alpha particle increases were mainly increases due to energetic particles which were accelerated at the Sun. The lower energy proton increases were due to either the passage of short-lived Stream Interaction Regions (SIR) or CIRs, lasting just two or three rotations or transient CMEs.

The largest increases seen during this period were due to the CMEs observed during the 3 month period around the middle of 2001, e.g. event 3 of Figure 10, when Ulysses was passing through the ecliptic. The events observed at high latitudes, which was also when Ulysses was

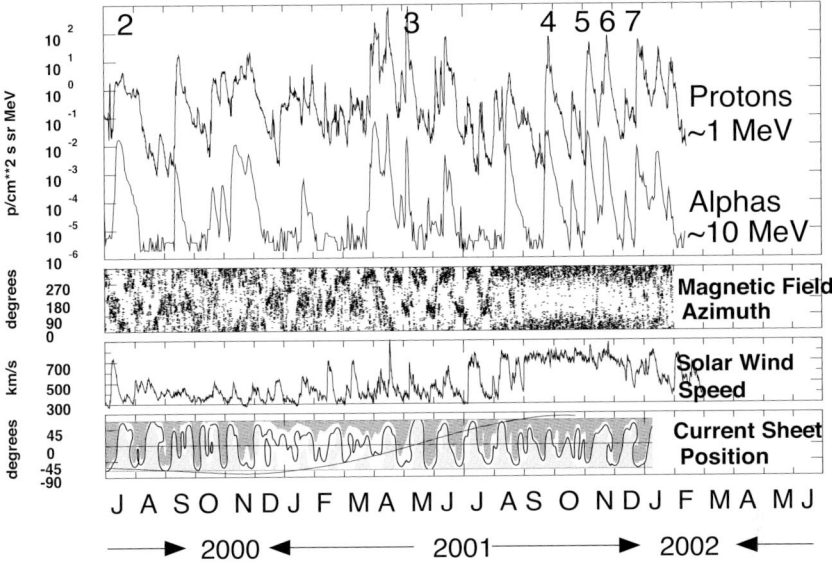

Figure 5.10. Summary plot of the second polar pass, from top to bottom, ∼1 MeV proton and ∼10 MeV/n alpha intensity, azimuth of the magnetic field, solar-wind speed, and the current sheet position on the Sun together with the spacecraft latitude.

at a greater radial distance, were approximately one order of magnitude less in both the proton and the alpha particle intensity.

During the 4-month period in late 2001 and early 2002 Ulysses was at high latitudes, above the current sheet (panel 4), immersed in high-speed solar-wind flow (panel 3) and in a unipolar magnetic field (panel 2). During this time four events, events 4 to 7 of Figure 10 were observed. These were the only events observed so far in the mission under these unique conditions. One of these, event 4, will be discussed later in more detail.

In Figure 11 we show a summary of the high energy particle data for the same period. This Figure, taken from McKibben et al. (2003) shows Ulysses high energy particle data (∼35-70 MeV and 70-95 MeV proton intensity), and the corresponding data from IMP-8 at 1 AU for the full polar pass, which includes the southern polar pass, the fast latitude scan, and the northern polar pass. According to this paper, all events observed at IMP-8 produced comparable intensity increases at Ulysses, independent of the connection footprint on the Sun of the field lines which passed through the spacecraft, from which the authors concluded that either the acceleration took place over a near global range of latitudes and longitudes, or that some mechanisms exists which transports

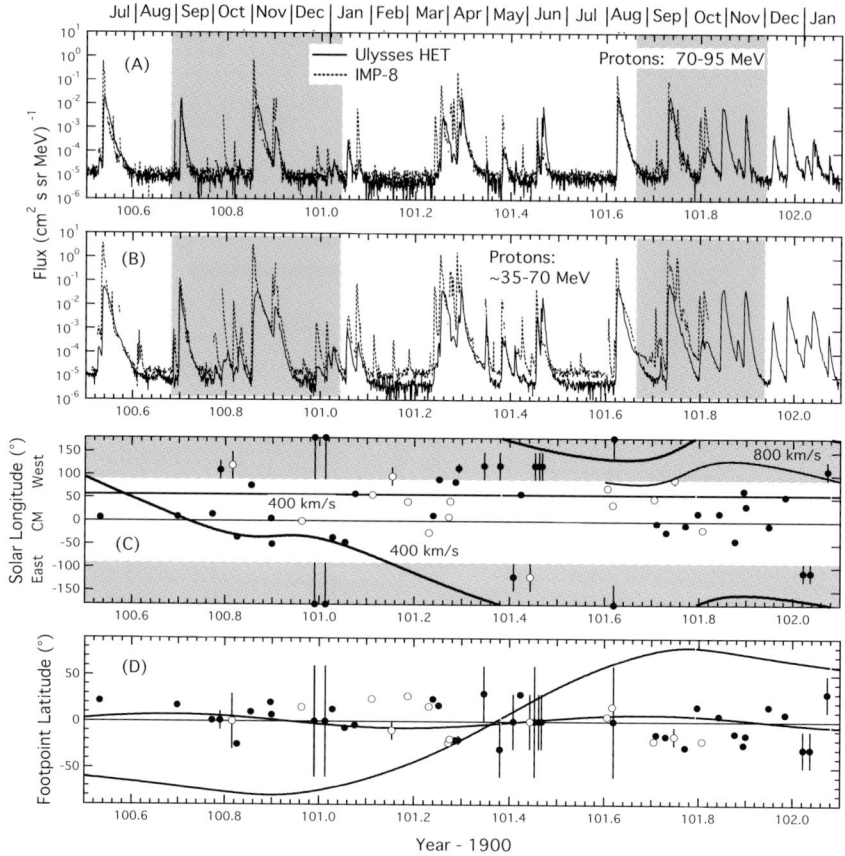

Figure 5.11. High energy particle intensity from IMP at 1 AU and Ulysses for the period early 2000 to early 2002 (Taken from McKibben et al. (2003)).

particles accelerated at a CME front throughout the heliosphere, both along and across mean magnetic field lines.

This is the so-called Reservoir Effect, first noted by McKibben (1972) and more recently discussed by many authors, e.g. Roelof et al. (1992). This effect is particularly noticeable during the decay phase. McKibben at al. (2003) concluded that within 3 - 4 days after almost every large event, the proton fluxes observed at energies between 10 and 100 MeV near Earth and at Ulysses remained nearly equal for the rest of the decay of the event. Note that at the onset of each event this is not so, the difference between IMP and Ulysses often being 1 - 2 orders of magnitude. This conclusion used the simultaneous observations at Ulysses and IMP-8. IMP-8 coverage stopped just after the observation of the first high latitude event seen over the northern pole, so that most of these conclusions were based on observations of Ulysses at high latitudes, but only within the slow-speed solar-wind flow.

5.   Discussion

5.1   The Second Northern Polar Pass

We begin by discussing the second northern polar pass. In Figure 12 we show low-energy particle data for the period when Ulysses was over the northern pole and in high speed solar wind flow. This plot, taken from Lario et al. (2004) shows data from the HISCALE/LEMS instrument on Ulysses.

This shows the intensity increases observed at Ulysses associated with the four large high latitude CMEs identified by Reisenfeld et al. (2003). At high energies, each particle event commenced with a rapid increase just a few hours after the time of release of the CME on the Sun. A small additional increase was observed a few days later, coincident with the arrival of the CME. Thereafter, the events decayed slowly to background level over a period of 10 - 20 days. At lower energies, the increases due to the CME were much larger, the largest intensities observed in the event being seen as the CME passed over.

Events 1, 3 and 5 were the largest, and were of similar magnitude. The profile of event 1 was much smoother, which according to Sanderson et al. (2003b) was due to the lack of any structure in the solar wind at Ulysses during the onset of the event. During events 3 and 5, small magnetic field structures were present during the onset of the event, modifying the time-intensity profiles of the particles.

In Figure 13 we show summary plots of the events number 2, 3 and 4 of Sanderson et al. (2003b), as numbered in Figure 10. (Event 4 corresponds to Event 1 of Lario et al. (2004) as shown in Figure 12). An additional event, Event 1, is included as a typical event observed close to the Earth.

Each plot covers 16 days. Each is split into three panels. In the bottom panel we show particle intensities from two electron channels ranging from bottom to top, from 10 to 1 MeV, in the next, from 5 proton channels ranging from 100 MeV to 20 MeV, and in the top panel, from 6 proton channels ranging from 20 MeV to 1 MeV.

Event 1 was observed shortly after Ulysses had been launched and was on its way to Jupiter, being at a radial distance of 1.07 AU and heliographic latitude of just 4° and is included here as an example of a 'typical' 1 AU event. Note the rapid onset, typical of events like this at 1 AU.

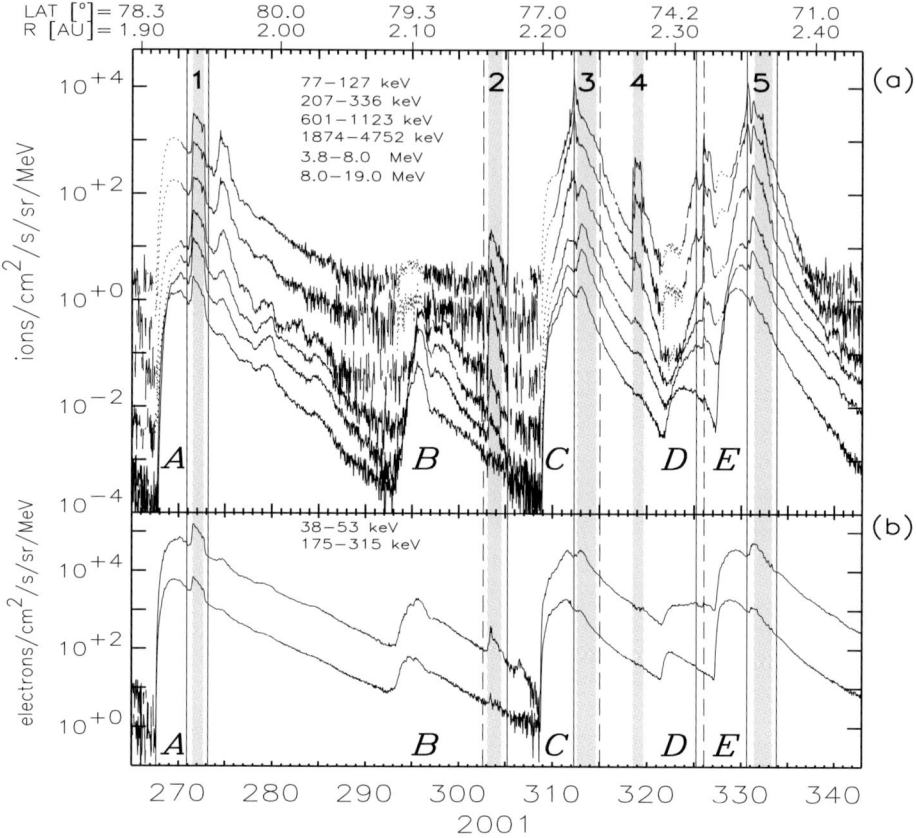

Figure 5.12. Summary of low-energy observations from Ulysses for the period day 265 2001 to day 343 day 2002 (Figure 4 of Lario et al. (2004)), showing low-energy data from the HI-SCALE/LEMS instrument. Vertical lines show the times of shock passage, and gray bars the times of CME passage.

Event 2 is the Bastille-day event, observed at mid-latitudes and in the slow solar wind. The onset and decay profiles in the electron channels and at high proton energies were relatively smooth (bottom and middle panel). A high background masked the onset of the event at low proton energies (top panel).

Event 3 is typical of the many low-latitude events, where the event occurs at the same time as a structure in the magnetic field such as a CIR or SIR passes the spacecraft during the onset. The time intensity profile during the first onset was disturbed considerably by the presence of the structure, parts of which acted as channels to allow the lower energy protons rapid access to the position of the spacecraft, and other

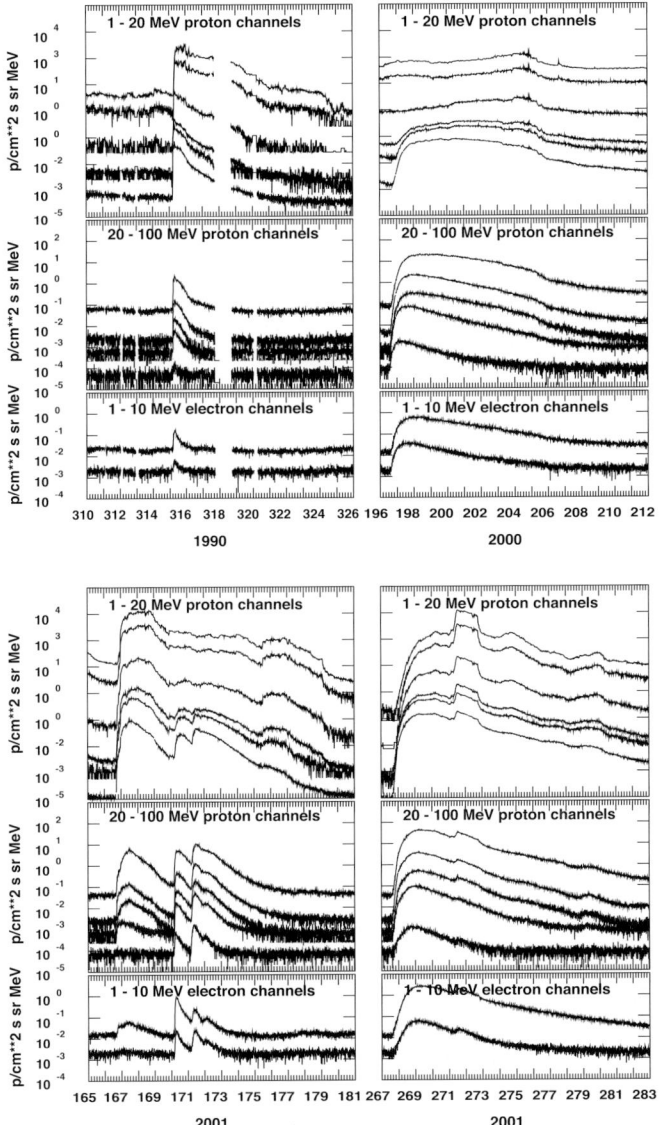

Figure 5.13. 16-day summary plots for events 1, 2, 3, and 4. The bottom panel of each plot shows particle intensities from 2 electron channels ranging from 10 to 1 MeV, in the next, from 5 proton channels ranging from 100 MeV to 20 MeV, and in the top panel, from 6 proton channels ranging from 20 MeV to 1MeV.

parts of which acted as sources of local acceleration. This gave rise to complicated intensity and anisotropy versus time profiles. This is the most often seen type of event at Ulysses.

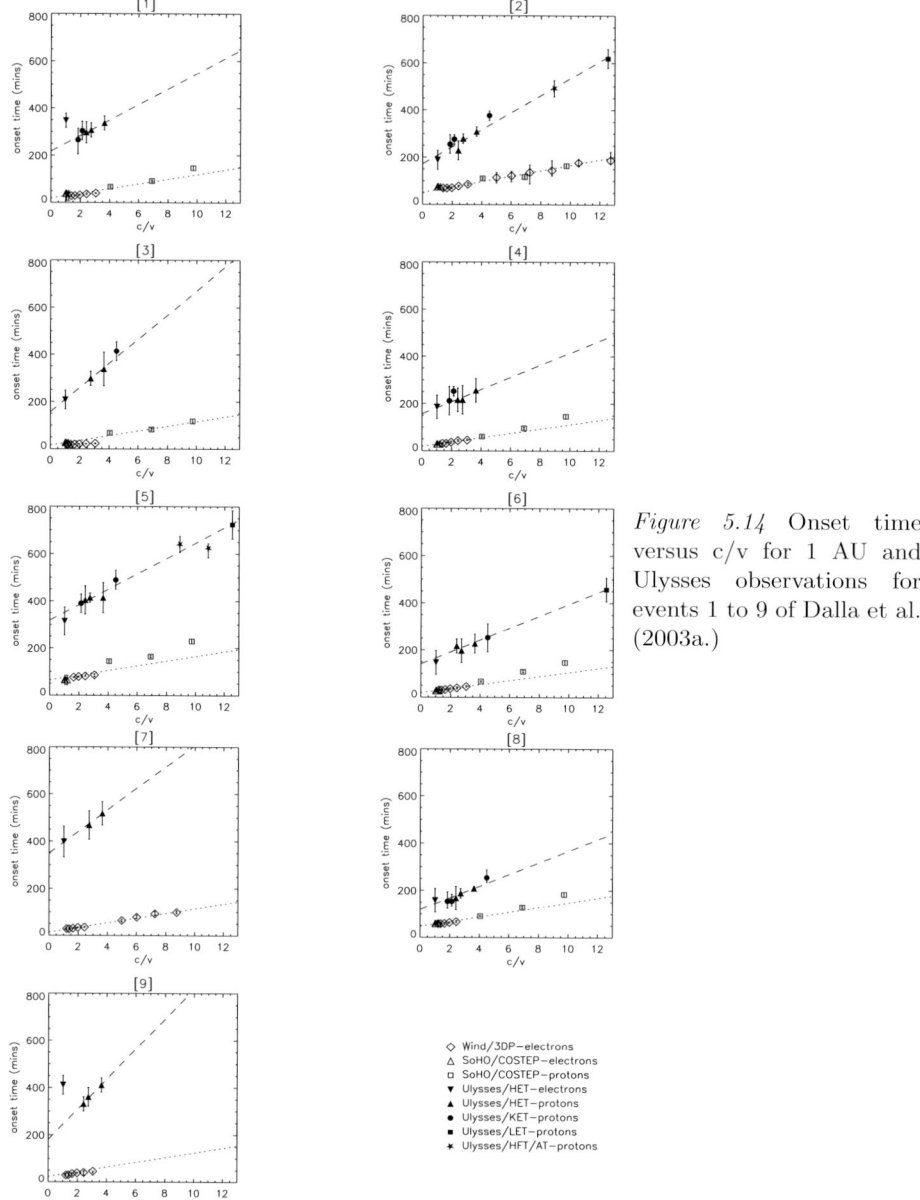

Figure 5.14 Onset time versus c/v for 1 AU and Ulysses observations for events 1 to 9 of Dalla et al. (2003a.)

Event 4 is one of only four large high-latitude events observed in the fast solar wind during the Ulysses mission, observed during the second high-latitude pass. This event had a relatively slow onset (compared with 1 AU in-ecliptic events) and during the first few days, a smooth time-intensity profile. The CME which followed a few days after the onset showed up as an increase lasting around 1 - 2 days in the low-energy particle intensity.

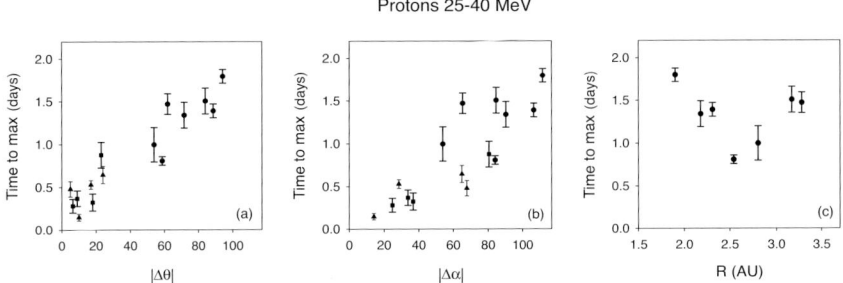

Figure 5.15. Time to maximum as a function of latitude, great circle angle and radial distance from the Ulysses to the Sun (Figure 3 of Dalla et al. (2003b).

Figure 14, reproduced from Dalla et al. (2003a) shows the onset time behaviour for 9 high-latitude events observed during the period 2000 to 2002. The first three of these were observed during the second polar pass over the southern pole, the next five were observed over the northern pole, whilst the remaining one was observed at a moderately high latitude as the spacecraft returned to low latitudes. Event 1 was the Bastille Day event. The first three and the last were observed while Ulysses was in low-speed flow, while the remaining five were observed in the fast solar wind. The top trace in each plot shows Ulysses onset times calculated for a wide range proton and electron velocities, while the lower trace shows observations at 1 AU taken from Soho and Wind. From these plots it is possible to calculate both the path length travelled by the particles, and the delay in releasing them at the Sun.

In Figure 15 we show results of another analysis by Dalla et al. (2003b) again using the same events, which show the correlation between times to maximum of the events versus latitude, great circle angle, and radial distance from Ulysses to the Sun.

Larger than expected delays in onset times were observed, corresponding typically to 120 to 350 minutes from the flare onset. These delay times and the path lengths were correlated against several variables. The best correlation was found with difference in latitude between the flare site and the latitude of Ulysses, this correlation being surprisingly better than the correlation with the angular separation between the site and Ulysses. This implies a very effective longitudinal transport of the particles, but a very inefficient transport latitudinally, which the authors concluded meant that cross field diffusion was the fundamental mechanism in getting the particles to high latitudes, in agreement with the suggestion by Zhang et al. (2003) for the Bastille Day event. However, they did not rule out the possibility that the delay was due to the time taken for the CME to reach the field lines connected to the spacecraft.

In Figures 16 and 17 we show particle intensity profiles and anisotropy parameters for two events, taken from Sanderson et al. (2003b). Event 1 is a 1 AU baseline event, and Event 4 is an event observed at high latitude and in the fast solar wind. Here we show, from top to bottom, the following: low-energy proton intensities, 1.8 - 3.8 MeV first order perpendicular anisotropy, first order parallel anisotropy, high energy proton intensities, 34 - 68 MeV first order perpendicular anisotropy, first order parallel anisotropy, magnetic field azimuth, elevation (in spacecraft coordinates), and magnitude. The particle instruments scan in a plane perpendicular to the spacecraft spin axis (z-axis), where the z-axis always points to the Earth. After rotating the x-axis to the direction of the magnetic field projected onto the scan plane, the sectored count rates in this frame of reference are Fourier analysed. In this way, the 2-dimensional parallel and perpendicular anisotropy amplitudes in the scan plane are derived. Anisotropies presented here are the ratios of the amplitude of the first order component (either parallel or perpendicular to the projected field direction) to the amplitude of the zero order component. The vertical line shows the arrival time of the shock.

Figure 16 is a 3-day plot showing the onset of event 1, observed on 11 November (day 315), 1990, the 1 AU baseline event. This event has a profile similar to profiles described in many models of propagation, but in fact was one of the rare occasions when we observed an onset without the disturbing effect of the presence of a magnetic field structure such as a CIR, SIR, or CME at the spacecraft.

Note how rapidly the onset took place, the higher energies arriving first, and all energies reaching maximum within only a few hours of each other.

The 34 - 68 MeV parallel anisotropy suddenly increased at the time the 34 - 68 MeV intensity started to rise, and dropped to zero within a few hours of the onset. Similarly, the 1.8 - 3.8 MeV anisotropy rose and fell, starting one or two hours after the 34 - 68 MeV particles. In both cases, during the rise to maximum, and for a few hours thereafter, there was a finite parallel anisotropy and a perpendicular component which was essentially zero, signifying that the particles were always field-aligned. The small fluctuations were mainly due to the limited counting statistics.

Compare this with the high-latitude, fast solar wind event in Figure 17. This shows in detail the September 2001 event. This is a 6-day plot, starting on 24 September (day 267), 2001. At high energies, the increase was about one order of magnitude less than the low latitude events. The event was most likely initiated by an X2.6 flare at 09:36

on 24 September, day 267, at S16 E23. This event has a moderately rapid increase at $\sim$50 MeV. Three and one half days after the onset, an over-expanding CME (start and stop times shown by the dashed lines) (Reisenfeld et al., 2003), preceded by a forward shock (shown by the solid line), passed the spacecraft, causing an additional increase of around one order of magnitude in the particle intensity at around $\sim$1 MeV. The event slowly decayed to background level after about 15 days.

The event was typical of the four large events seen in 2001 and 2002 when Ulysses was in high-speed solar-wind flow and at high northern latitudes. The events observed in the fast solar wind (events 4 to 7) had a much smoother profile than the events observed in the slow solar wind, due to the absence of magnetic field structures. Most low latitude particle events in the slow solar wind, such as event 3, had a much more ragged profile than the high-latitude high-speed flow events, because of the presence of an SIR, CIR or discontinuities.

One of the features of the events propagating in the high-latitude fast solar wind is that this medium is remarkably homogeneous and usually devoid of large scale structures such as CIRs, SIRs, CMEs and large scale discontinuities not related to the event being studied, which would otherwise add to the complication of the study of the propagation.

In general, the duration of the events observed in the slow solar-wind events were shorter than the fast solar wind events, a typical slow solar-wind event lasting 7 - 10 days, and a typical fast solar-wind event lasting 15 days. Onsets at high latitude and in the fast solar-wind tended to be smooth and rapid, lasting typically one day. Surprisingly the anisotropies associated with the onset were very small.

Comparing event 1, the baseline 1 AU event with event 4, the high-latitude, high-speed flow event, we see immediately the large difference between the two. The 1 AU event rose to a maximum in around 3 hours. The intensity during the high-latitude event rose in around 2 days, a factor $\sim$16 slower than the 1 AU event, which was located only a factor $\sim$2 closer to the Sun (measured along the field line).

The anisotropy in the high-latitude event was very small. Note that the scales of the anisotropy panels are the same. In the 34 - 68 MeV channel, the component of the anisotropy perpendicular to the field fluctuated back and forth around zero. There was a small but finite anisotropy which persisted for around one day after the onset in the 34 - 68 MeV proton channel, signifying that the particles were propagating outwards from the Sun (the field direction at this time was inward). This anisotropy persisted until around the time of maximum of the 34 - 68 MeV channel, and then remained around zero for the next couple

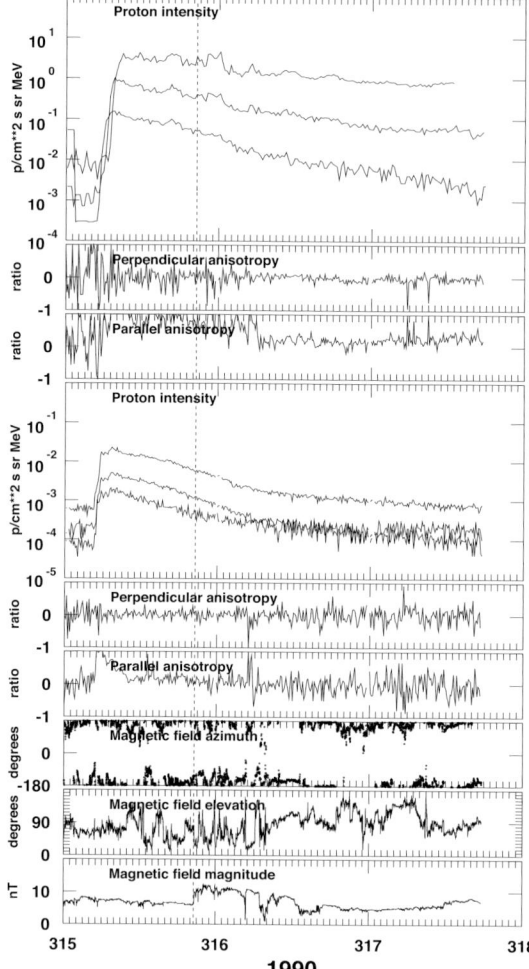

Figure 5.16 November 11 1990 event, from top to bottom, proton intensity from the 1.8 - 3.8, 3.8 - 8.0 and 8.0 - 19.0 MeV channels, 1.8 - 3.8 MeV first order anisotropy amplitude resolved perpendicular to the component of the magnetic field in the scan plane of the instrument, first order anisotropy amplitude resolved parallel to the component of the magnetic field in the scan plane of the instrument, proton intensity from the 24 - 31, 34 - 68 and 68 - 92 MeV channels, 34 - 68 MeV first order anisotropy amplitude resolved perpendicular to the component of the magnetic field in the scan plane of the instrument, first order anisotropy amplitude resolved parallel to the component of the magnetic field in the scan plane of the instrument, magnetic field azimuth, elevation (in spacecraft coordinates), and magnitude.

of days. All along, the perpendicular component remained around zero, indicating that there was no net flow across the field.

In the 1.8 - 8.0 MeV channel, the component of the anisotropy perpendicular to the field again fluctuated back and forth around zero. There was a larger negative parallel anisotropy which persisted for around half a day after the onset in the 1.8 - 3.8 MeV channel, again signifying that the particles were propagating outwards from the Sun. A small field aligned anisotropy persisted during the next couple of days. Again, all along, the perpendicular component remained around zero, indicating that the net flow was field aligned.

The measured anisotropies were very small. Within the limits of accuracy of measurement of the anisotropy, when there was a net flow,

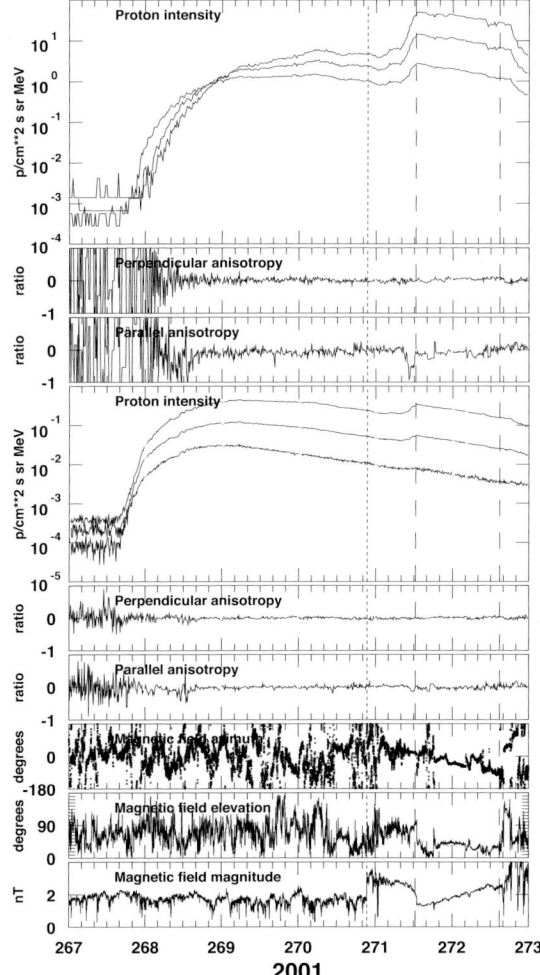

Figure 5.17 The September 2001 Event, plotted with the same parameters as in Figure 16).

the anisotropy directions were coincident with the magnetic field direction, projected onto the plane within which the anisotropy measurements are made. This meant that the flow was field aligned, and that particles reached high latitudes travelling along the field lines, and not by crossing over them. The particles were scattered significantly as they propagate outwards, which explains their relatively slow onset and their small anisotropy, but despite this, any net flow direction was still along the local magnetic field line direction.

In Figure 18 we show a drawing of the cross section in the meridian plane of an over-expanding CME. This plot, adapted from Gosling et al. (1994) and Forsyth and Gosling (2001), has been inverted so as to apply to the northern hemisphere. Drawn on top of this is a line

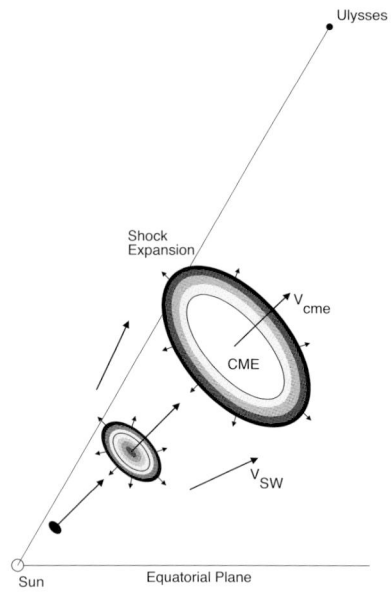

Figure 5.18 Cartoon of the configuration of the heliosphere during the September 2001 particle event, which was associated with an over-expanding CME (after Gosling at al., 1994 and Forsyth and Gosling (2001).

showing the latitude of the spacecraft. Since the original diagram did not quite extend to the latitude of observation, we have drawn the line at lower latitude. At the onset of this event, the CME responsible for accelerating the particles was still travelling on its way from the Sun, and still some considerable distance away. Ulysses was in the fast solar wind, so that the particles accelerated at the shock front closer to the Sun were travelling along the field lines in the fast solar wind. The CME which subsequently arrived at Ulysses was an over-expanding CME (Reisenfeld, 2003), preceded by a forward wave, and followed by a reverse shock, similar to the events observed by Gosling et al. (1994).

5.2   Comparison with the Second Southern Polar Pass

At high latitudes (70 - 80°N) and in the fast solar wind we found no evidence for any substantial net flow across the field lines, whereas at moderately high latitudes (62°S) in the slow solar wind Zhang et al. (2003) found evidence for cross-field flow.

In Figure 19 we show data from the Bastille-day event, using the same format as in Figure 16. This event has some similarities to the high-speed, high-latitude events, Events 4 to 7. However, the Bastille event was observed in the slow solar wind and is more like a low latitude event (such as shown in Figure 16) than a high latitude one. At Ulysses, the Bastille Day event was an unexceptional event. The increase discussed here was a secondary event sometime after the main event. It

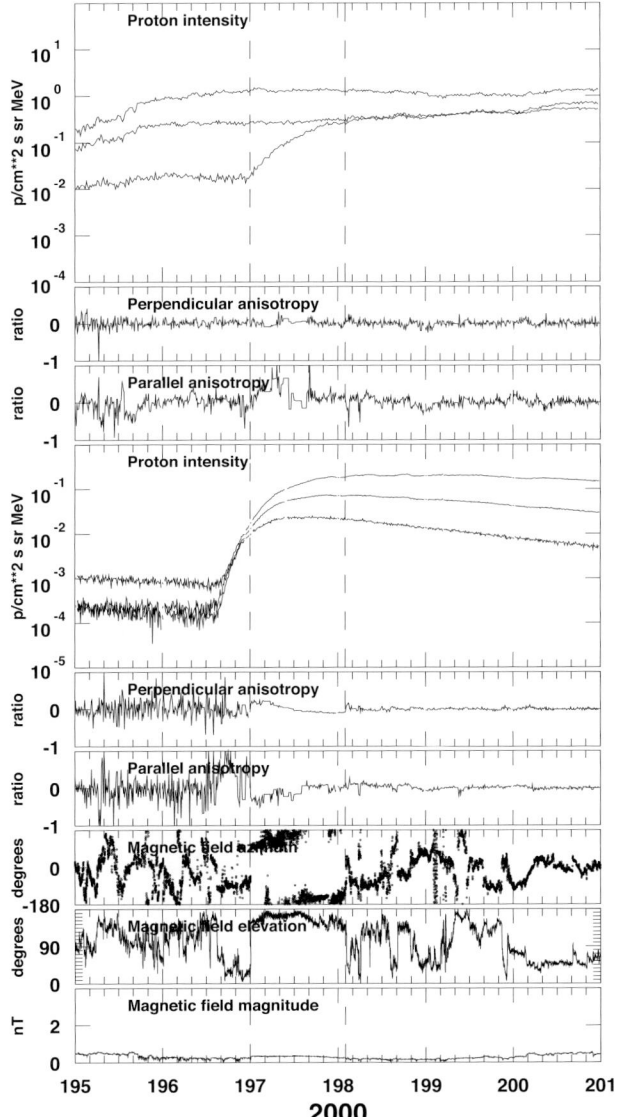

Figure 5.19 The Bastille day event, plotted with the same parameters as in Figure 15.

was observed at 3.17 AU, and at 62°S, which was around the same southern latitude as the southern-most extent of the current sheet, whereas Events 4, 5, 6 and 7 were observed at 2 AU and latitudes between 70° and 80°N, which at the time was ∼20° higher than the northern-most extent of the current sheet, as can be seen in Figure 10.

The main event began with an onset at high energies at Ulysses on day 193. The second, more substantial onset occurred at around 1600 UT on day 197. The anisotropy of the high energy particles suddenly increased at the time of onset, the particles streaming outwards along the field past the spacecraft. The anisotropy amplitude then started to decrease. At the beginning of day 197, a CME arrived at the position of Ulysses. The magnetic field direction in the spacecraft frame of reference changed as the spacecraft entered the CME, reversing the sign of the anisotropy amplitude. The anisotropy amplitude continued to decrease as the spacecraft entered the CME, whilst at the same time a comparable, but small, perpendicular component of the anisotropy (measured in the scan plane) was observed. Half way through the passage of the CME, the anisotropy had dropped essentially to zero. Although at a moderately high latitude, this event was observed in the slow solar wind, and has some similarities to the 1 AU event shown in Figure 16. Upstream of the CME was a region lasting around 6 hours within which the high-energy anisotropy was high. The perpendicular component was considerably smaller, which implies that the flow was essentially field aligned. Immediately after the onset, particles travelling along the field lines with a moderately substantial parallel anisotropy and essentially no perpendicular anisotropy were observed. The anisotropy amplitude then started to decay slowly, just like the 1 AU in-ecliptic event, except the time and magnitude scales were considerably different.

Approximately 10 hours after the onset, the spacecraft entered the previously-existing CME, quite unrelated to the CME, which was responsible for this event. The parallel anisotropy amplitude continued to decrease, but from this time on a perpendicular component of similar magnitude was observed. Both the perpendicular and parallel component amplitudes continued to decrease until both were zero at a time half way through the passage of the CME. This could imply that particles were crossing field lines within the CME, as suggested by Zhang et al. (2003) although this is probably unlikely, as the field is usually very quiet inside a CME, so scattering across the field lines is not to be expected. A more likely explanation is that inside the previously existing CME there was a substantial un-measurable field aligned component, as at this time the magnetic field in the spacecraft frame of reference was almost perpendicular to the scan plane. This signature could also possibly be due to the existence of a gradient, but the duration of this anisotropy is probably too long for this to be true.

In Figure 20 we show another drawing, this time based on Figure 1 of Gosling (1991), of the cross section of this pre-existing CME. This time the CME is immersed in slow solar wind. Drawn on this is a line showing the latitude of the spacecraft. Again, since the diagram does not extend up to any higher latitudes, we have drawn the line at a lower latitude.

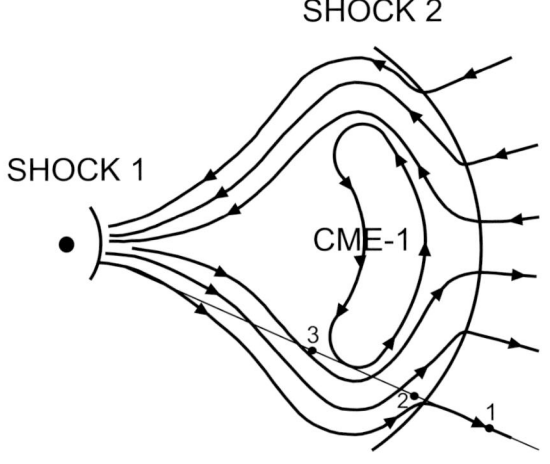

Figure 5.20 Cross section of a CME in the meridian plane (Adapted from Gosling (1991).

The CME which accelerated the particles which we were observing, was still much closer to the Sun.

At position 1, the onset of the event, the local magnetic field is unaware of the pending arrival of the pre-existing CME, which again, we repeat, is not the CME which accelerated the particles closer to the Sun. This CME was in between Ulysses and the acceleration site. Most likely, the particles travelled around the CME on their way to the spacecraft.

At position 2, just inside the CME, the field has deflected northward, and then at position 3, just before leaving the CME, the field is deflected southward. Within the CME, the particles should follow the field lines and continue to be field aligned. However, given sufficient scattering within the CME, the particles could cross the field lines. A more likely explanation is that there is a gradient within the CME. In this event, the draping of the field lines provided a distortion of the field lines as the CME passed over the spacecraft. There is obviously easy access in and out of the CME, as the particle intensities remained the same as the boundary of the CME passed the spacecraft.

6. Summary and Conclusions

The coronal magnetic field controls the configuration of the heliosphere. In the corona, the current sheet winds its way around and between the positive and negative poles of the magnetic field, delineating the positive coronal magnetic field from the negative.

The positive and negative polarity magnetic field propagates out into the heliosphere, frozen in the solar wind. As it does so, the current sheet divides up the heliosphere into a region of positive magnetic field polarity (with outward pointing field) and a region of negative polarity (with inward pointing magnetic field). At solar minimum this is a simple configuration, but at solar maximum it becomes much more complicated.

The coronal holes follow the motion of the poles of the magnetic field as they move across the disc of the Sun during the course of the solar cycle. At solar minimum, the poles of the dipole are located at the heliographic poles, and there is a large polar coronal holes at each pole. The coronal holes are the source of the fast solar wind. They move equatorward during the course of the solar cycle, following the dipole poles, and, as the quadrupole strength increases around solar maximum, break up into smaller mid-latitude coronal holes roughly co-located with the quadrupole poles.

At solar minimum, most of the high latitude region is filled with fast solar wind. At solar maximum, only a few small low-latitude fast solar wind streams exist.

For the study of particle propagation at high latitudes, it is important to differentiate between propagation in slow and fast solar wind, as the characteristics of the propagation differ considerably between the two.

During the highest northern-latitude parts of the second polar pass, Ulysses was immersed in the fast solar wind. The fast solar wind tends mainly to be homogeneous, and devoid of large scale discontinuities, but is much more turbulent than the slow solar wind, and so particles propagate to high latitudes in high-speed solar wind with some difficulty.

Energetic particle events observed during the part of the northern polar pass where Ulysses was at its highest latitude and in the fast solar wind had smooth time intensity profiles, near-isotropic particle angular distributions at all energies at the onset, flow directions during the rising phase of the events along the field, and no evidence for any net flow across the field lines. These particles propagated to the highest heliographic latitudes travelling along magnetic field lines and not across them.

Our observations do not allow us to draw conclusions about propagation closer to the Sun, but most likely, to reach the high latitudes, particles must either diffuse across field lines closer to the Sun, or else there was some large scale distortion of the magnetic field lines.

During the second southern polar pass, the spacecraft was continually in the slow solar wind. Most of the particle events observed at this time occurred at the same time as some other pre-existing and unrelated structure, such as a CME from a previous solar flare, or a CIR, passed over the spacecraft. Particle propagation was dominated by the presence of these structures, the frequent occurrence of which meant that it was quite rare to find an event where the event was unaffected by one.

The forward and reverse shocks and the stream interfaces of the CIRs, and interplanetary shocks and magnetic clouds of the CMEs all affected the particle propagation, sometimes even accelerating the lower energy particles locally. These structures tended to be full of discontinuities, which again affected the propagation, the effect depending on the size and thickness of the discontinuities and the energy of the particles.

During the southern polar pass, where structures were present we observed events with irregular time-intensity profiles, onset times and velocity dispersion modified by the presence or the lack of structures and discontinuities and field aligned flow. In between the shocks and discontinuities, the field tended to be relatively quiet and channelling could be observed. Occasionally, close to a boundary or interface, we observe a short period of non field-aligned flow.

On the rare occasions in low-speed flow when no structures were present, we observed a smoother time-intensity profile, a rapid onset with velocity dispersion and moderately high field aligned anisotropies at the onset, diminishing with time.

Finally we conclude with a summary table showing the energetic particle and the plasma characteristics of the two different high-latitude regions, the slow solar wind region observed during the second southern polar pass, and the fast solar wind region observed during second northern polar pass.

Table 5.1. Summary of observations during the second polar passes

Characteristic	*Southern Polar Pass*	*Northern Polar Pass*
Solar wind	Slow	Fast
Time intensity profiles	Irregular	Smooth
Structures at onset	Frequent	Rare
Event onset times	Rapid	Delayed
High latitude Propagation	Modified by draping	Direct along field
Anisotropy at onsets	Large	Nearly isotropic
Particles inside CME	Intensities depressed	Intensities elevated
Particle flow directions	Field aligned	Field aligned (non-field aligned near structures)

Acknowledgments

The NSO/Kitt Peak data on the coronal holes used here was prepared by the late Karen Harvey. The author acknowledges gratefully her contribution to this work, without which this study would not have been possible. It was a privilege working with her. The author gratefully acknowledges Todd Hoeksema for permission to use the Wilcox Solar Observatory data used here, A. Balogh, D. McComas, and R. B.

McKibben for permission to use Ulysses VHM/FGM, SWOOPS and COSPIN/HET data respectively, and S. Dalla, D. Lario, J. T. Gosling R. B. McKibben and R. J. Forsyth for permission to reproduce their diagrams. The author gratefully acknowledges R. G. Marsden and C. Tranquille for their contributions to this work.

References

Balogh, A., T. J. Beek, R. J. Forsyth, P. C. Hedgecock, R. J. Marquedant, E. J. Smith, D. J. Southwood, and B. T. Tsurutani, Astron. and Astrophys, 92, 221, 1992.

Bame, S. J. et al., Astron. and Astrophys, 92, 237, 1992.

Bame, S. J., B. E. Goldstein, J. T. Gosling, J. W. Harvey, D. J. McComas, M. Neugebauer, and J. L. Phillips, Geophys. Res. Letters, 20, 2323, 1993.

Dalla, S., A. Balogh, S. Krucker, A. Posner, R. Mueller-Mellin, J. D. Anglin, M. Y. Hofer, R. G. Marsden, T. R. Sanderson, B. Heber, M. Zhang, and R. B. McKibben, Annales Geophysicae, 21, 1367 - 1375, 2003a.

Dalla, S., A. Balogh, S. Krucker, A. Posner, R. Mueller-Mellin, J. D. Anglin, M. Y. Hofer, R. G. Marsden, T. R. Sanderson, C. Tranquille, B. Heber, M. Zhang, and R. B. McKibben, Geophys. Res. Lett., 30, No. 19, 8035, 2003b.

Forsyth, R. J. and Gosling, J. T., in "The Heliosphere Near Solar Minimum: The Ulysses Perspective", 107-166, Balogh, A., Marsden, R. G. and Smith, E. J., eds., Springer-Praxis, Chichester, UK, 2001.

Gosling, J. T., J. Geophys. Res., 96, No. A5, 7831-7839, 1991.

Gosling, J. T., D. J. McComas, J. L. Phillips, L. A. Weiss, V. J. Pizzo, B. E. Goldstein, and R. J. Forsyth, Geophys. Res. Lett., 21, 2271, 1994.

Harvey, K. L. and F. Recely, Solar Physics, 211,(31-52), 2002

Hoeksema, J. T., Ph. D. Thesis, Stanford University, Stanford, 1984.

Jones, H. P, et al., Solar Phys., 139, 211, 1992.

Jones, G. H., A Balogh, E. J. Smith, Geophys. Res. Lett., 2003, 30, (2-1), 2003.

Harvey, K. L., and F. Recely, Solar Physics, 211, 31-52, 2002.

Lario, D., et al., J. Geophys. Res., 105, 18235, 2000a.

Lario, D., et al., J. Geophys. Res., 105, 18251, 2000b.

Lario, D., R. B. Decker, E. C. Roelof, D. B. Reisenfeld, T. R. Sanderson, JGR, 109, A01108, 2004.

Marsden, R. G., This Volume, 2004.

Marsden, R. G., T. R. Sanderson, K. P. Wenzel, and S. J. Bame, Adv. Space Res., Vol. 13, No. 6, pp (6)95 - (6)98, 1993.

McKibben, R. B., Bulletin of the American Astronomical Society, Vol. 4, p.387, 1972.

McKibben, R. B., J. J. Connell, C. Lopate, M. Zhang, J. D. Anglin, A. Balogh, S. Dalla, T. R. Sanderson, R. G. Marsden, M. Y. Hofer, H. Kunow, A. Posner, and B. Heber, Ann. Geophys., 21, 1217, 2003.

Reisenfeld, D. B., J. T. Gosling, R. J. Forsyth, P. Riley, and O. C. St. Cyr, Geophys. Res. Lett., 30, No 19, 8031, 2003.

Roelof, E. C., R. E. Gold, G. M. Simnett, S. J. Tappin, T. P. Armstrong and J. L. Lanzerotti, Geophys. Res. Lett., 19, 1243, 1992.

Roelof, E. C., G. M. Simnett, R. B. Decker, L. J. Lanzerotti, C. G. Maclennan, T. P. Armstrong, R. E. Gold, J. Geophys. Res., 102, 11251, 1997.

Sanderson, T. R., et al., Geophys. Res. Lett., 21, 1113, 1994.

Sanderson, T. R., in "Proceedings of the Eight International Solar Wind Conference", Dana Point, CA, p411, 1995.

Sanderson, T. R., D. Lario, M. Maksimovic, R. G. Marsden, C. Tranquille, A. Balogh, R. J. Forsyth, B. E. Goldstein, Geophys. Res. Lett., 26, 1785, 1999.

Sanderson, T. R., T. Appourchaux, J. T. Hoeksema, and K. L. Harvey, J. Geophys. Res., No. A1, 1035, doi:10.1029/2002JA009388, 2003a.

Sanderson, T. R., R. G. Marsden, C. Tranquille, S. Dalla, R. J. Forsyth, J. T. Gosling, and R. B. McKibben, Geophys. Res. Lett., 30(19), doi:10.1029/2003GL017306, 2003b.

Simnett, G. M., et al., Geophys. Res. Letters, 21, 1561, 1994.

Simpson, J. A. et al., Astron. and Astrophys, 92, 365, 1992.

Smith, E. J., et al., Geophys. Res. Letters, 20, 2327, 1993.

Suess, S. T., J. L. Phillips, D. J. McComas, B. E. Goldstein, M. Neugebauer, and S. Nerney, Space Science Reviews, 83, 75-86, 1998.

Zhang, M., R. B. McKibben, C. Lopate, J. R. Jokipii, J. Giacalone, M.-B. Kallenrode, and H. K. Rassoul, J. Geophys. Res., 108, No. A4, 1154, doi:0.1029/2002JA009531, 2003

Chapter 6

SOLAR WIND PROPERTIES
FROM IPS OBSERVATIONS

Masayoshi Kojima, Ken-ichi Fujiki, Masaya Hirano, Munetoshi Tokumaru
Solar-Terrestrial Environment Laboratory, Nagoya University, Toyokawa 442-8507 Japan

Tomoaki Ohmi
ChudenCTI Co., Ltd., 1-27-2 Meiekiminami, Nakamura-ku, Nagoya 450-0003, Japan

Kazuyuki Hakamada
Department of Natural Science and Mathematics, Chubu University, Kasugai 487-8501, Japan

Abstract   Since *Hewish et al.* (1964) discovered the interplanetary scintillation (IPS) phenomena, the IPS method has been used as one of the few devices which can be used to observe solar wind in three-dimensional space. However because of the line-of-sight integration effect of IPS, solar wind had to be studied with blurred images. In the late 1990s new methods of IPS observation and analysis which can deconvolve the line-of-sight integration effect were developed independently by a group at University California at San Diego (*Grall et al.*, 1996) and a group at the Solar-Terrestrial Environment Laboratory, Nagoya University (*Asai et al.*, 1998; *Kojima et al.*, 1998; *Jackson et al.*, 1998). Today we can obtain unbiased solar wind images with high spatial resolution from IPS observations. The Ulysses spacecraft has been observing detailed structures of solar wind in three dimensions since its launch in 1990. However, Ulysses takes ten months even to make a rapid latitudinal scan from the south to north poles. IPS measurements have several advantages in comparison with spacecraft measurements. It can observe three-dimensional solar wind in a short time, and the observations can be carried out consistently over a solar cycle. Making use of these advantages of IPS, we have been studying several interesting solar wind features observed by Ulysses; namely, whether they are stable structures

147

G. Poletto and S.T. Suess (eds.), The Sun and The Heliosphere as an Integrated System, 147–178.
© 2004 *Kluwer Academic Publishers. Printed in the Netherlands.*

and how they depend on the solar cycle. We introduce these studies and propose a model to determine the solar wind velocity structure.

Keywords: Solar wind, interplanetary scintillation, tomography, fast wind, slow wind, latitudinal structure, bimodal structure, coronal magnetic field

1.   Introduction

The solar corona changes its structure dramatically with solar activity. As a consequence of this change, solar wind also changes dynamically. Solar wind has been observed in three dimensions with the interplanetary scintillation method (IPS). *Coles et al.* (1980) first showed how the latitudinal structure of solar wind changes over a solar cycle and compared it with the change of the polar coronal hole size. *Kojima and Kakinuma* (1987, 1990) introduced synoptic velocity maps with which it can be easily seen how the velocity structure of solar wind changes, and showed that low-speed wind tends to concentrate along a potential field neutral line on the source surface. *Rickett and Coles* (1991) analyzed solar wind structure changes over 16 years and compared this with coronal density and magnetic field structures. These IPS observations were confirmed by Ulysses' *in situ* observations (e.g., *Phillips et al.*, 1995; *Goldstein et al.*, 1996; *Woch et al.*, 1997; *McComas et al.*, 2000).

Ulysses observed that the velocity of high-latitude fast solar wind is in the range of 700-800 $\mathrm{km\,s^{-1}}$ and that, within a coronal hole, there is a small but noticeable gradual increase in velocity towards higher latitudes (*Woch et al.*, 1997). Velocity asymmetry was also observed in the high-latitude fast wind between the northern and southern hemispheres (*Goldstein et al.*, 1996). Ulysses (*Woch et al.*, 1997) confirmed the bimodal velocity structure which had been revealed by the IPS observations (*Kakinuma*, 1977). Since it took about ten months for Ulysses to observe these solar wind features at all the latitudes from the south to north poles, quicker measurements of solar wind in the full latitudinal range are both interesting and important in terms of finding out whether the solar wind structures observed by Ulysses are stable structures and how they change with solar activity. These solar wind properties have been studied using IPS observations by *Fujiki et al.* (2003a), and we introduce them in the following sections.

Since IPS measurements of solar wind properties integrate over the line of sight, the solar wind structures studied using the IPS method were sketchy except for those in the work of *Kakinuma* (1977), who analyzed the latitudinal velocity structure using a model fitting method to remove the line-of-sight integration effects. Nowadays, to study the three-dimensional solar wind structure we can use computer-assisted tomography (CAT) which was developed by *Asai et al.* (1998), *Jackson et al.* (1998), and *Kojima et al.* (1998) to analyze the IPS observations. In section 2 we introduce the IPS CAT method. In section 3 synoptic velocity maps are derived from the IPS CAT analysis, and the solar cycle dependence of the solar wind structures is briefly discussed. Section 3 also describes the IPS data used in the analyses and the analysis method to compare observations made at different heliocentric distances. Section 4 discusses the reliability and accuracy in retrieving solar wind properties from IPS measurements using the tomographic method. In section 5 we analyze the coronal hole size dependence of the wind velocity in polar regions. Since the coronal hole is a source of high-speed wind (≥ 750 km s^{-1}) as well as low-speed wind (≤ 400 km s^{-1}), in section 6 we also discuss the low-speed wind from a small coronal hole. In sections 7–9, the N-S asymmetry, the latitudinal velocity gradient and the bimodal structure observed by Ulysses are discussed in relation to their solar cycle dependence. In section 10, we summarize how the solar wind structure changes with the solar cycle.

Coronal holes play important role in determining the solar wind structure. A large-scale polar coronal hole is a source of fast solar wind, and medium and slow speed streams originate in smaller coronal holes (*Neugebauer et al.*, 1998). The velocity depends on the coronal hole scale size in a linear fashion (*Nolte et al.*, 1976). Not only the coronal hole scale size but also the flux expansion rate (*Wang and Sheeley*, 1990; *Sheeley et al.*, 1991) and energy supplied from the photosphere (*Fisk et al.*, 1999) determine the wind velocity. *Hirano et al.* (2003) found that a new physical parameter, which is the combination of the flux expansion rate and energy supply, can well determine the solar wind velocities from various kinds of coronal holes. This is illustrated in section 11.

2.   Interplanetary Scintillation Measurements

Radio waves from a compact radio source are scattered by electron density irregularities in solar wind, and the scattered radio waves interfere with each other as they propagate to the Earth, producing diffraction patterns on an observer's plane. This phenomenon is called interplanetary scintillation (*Hewish et al.*, 1964). We use IPS observations to study the three-dimensional solar wind structure, because they have several advantages over *in situ* spacecraft measurements. They can be used consistently for a long-term study of the solar cycle dependence of the solar wind structure. In addition, since a large number of IPS sources are available, vast regions of interplanetary space can be probed in a relatively short time.

Solar wind velocity is estimated by measuring the drift velocity of the diffraction pattern. It has been assumed that the velocity derived from the IPS observation represents the solar wind velocity around the point closest to the Sun on a line of sight (P-point assumption). Although the P-point assumption is approximately valid, especially for low-speed wind (*Watanabe and Kakinuma*, 1972; *Coles et al.*, 1978; *Coles and Kaufman*, 1978), the flow velocity obtained from the IPS measurement is a weighted integration along a line of sight of the velocity component perpendicular to the line of sight. Consequently, the IPS measurement leads to large underestimation of the velocity of fast solar wind and to blurred spatial resolution.

Two methods that can deconvolve the line-of-sight integration have been developed to improve the spatial resolution and retrieve the intrinsic velocity from IPS observations. One method uses remotely separated multi antennas with a baseline longer than a Fresnel radius, and the other employs the CAT method.

IPS observations made using widely separated antennas can resolve the line-of-sight integration into two components when bimodal solar wind is observed (*Grall et al.*, 1996). When the baseline is long enough, the cross-correlation function has two clearly separated peaks; one peak corresponds to high-speed wind, and the other to low-speed wind. However, use of this method is restricted to special observational conditions: the baseline length has to be a few times longer than the Fresnel radius, and the baseline should be parallel to the projected solar wind flow direction.

Another method which uses the CAT technique was developed by *Asai et al.* (1998), *Jackson et al.* (1998) and *Kojima et al.* (1998). With IPS observations made while the Sun rotates, both the solar rotation and solar wind outward motion give perspective views of three dimensional solar wind structures at various different view angles. These situations make it possible to use the CAT technique for the IPS analysis, and this technique can retrieve not only unbiased solar wind parameters but also provide high spatial resolution.

2.1   Tomographic Analysis of IPS Observations

In the tomographic analysis, an initial model of wind velocity distribution is first introduced on the reference sphere and then expanded radially outward with a constant velocity to make a three-dimensional solar wind model. IPS observations are simulated in this solar wind model for actually observed geometries of lines of sight and then compared with the observed IPS velocity (Figure 6.1). The discrepancy ΔV between the simulated and observed velocity is distributed on the reference sphere along a projected line of sight with a weighting factor to modify the velocity distribution model on the reference sphere. After completing the simulations for all observations, the initial solar wind model is modified and the IPS simulations are restarted. This process is iterated until the residuals become small enough. Usually this process converges after several iterations. It should be noted that the result from the tomographic analysis does not depend on the initial model given on the reference sphere to start the iteration process; that is, the tomographic iteration can be started from a structureless solar wind model. For a full description of the CAT analysis method, we refer the reader to *Asai et al.* (1998) and *Kojima et al.* (1998).

Figure 6.2 demonstrates how the IPS CAT can improve the velocity map. The upper map is derived from the IPS data obtained during Carrington rotations (CR) 1909–1913 (April – September 1996) using the previously used P-point assumption. Although the analyzed period is in the solar minimum phase, the velocities at high latitudes are lower than 700 km s^{-1}, and the separation between low-speed and high-speed regions is not sharp. The lower panel in the figure gives the results of the CAT analysis applied to the same IPS data used for the upper map. We can see that the map has a higher resolution and velocities higher than 700 km s^{-1} emanate from the high latitude region. These features agree

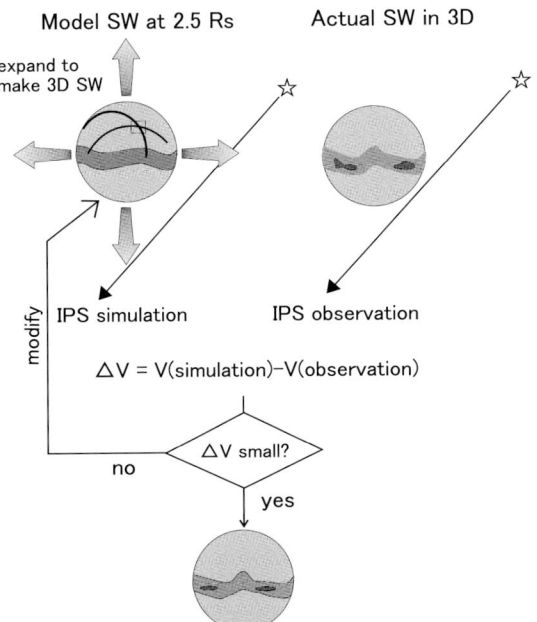

Figure 6.1 IPS CAT analysis method. IPS observations are simulated for the actual geometry of lines of sight in the solar wind model and compared with the observed IPS velocities. Using the residual between velocities obtained by simulation and observations, the solar wind model is modified.

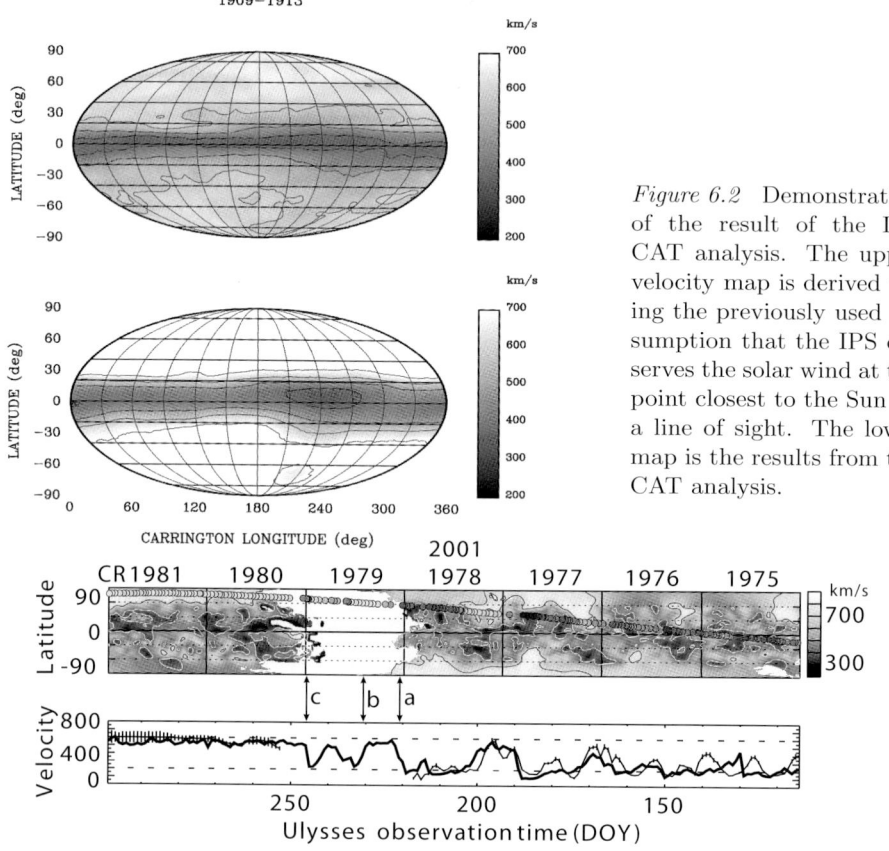

Figure 6.2 Demonstration of the result of the IPS CAT analysis. The upper velocity map is derived using the previously used assumption that the IPS observes the solar wind at the point closest to the Sun on a line of sight. The lower map is the results from the CAT analysis.

Figure 6.3. Comparison between the velocities derived by the IPS CAT analysis and the Ulysses observations. The velocity map is derived at a heliocentric distance 2.5 R_s by the IPS CAT, and the Ulysses trajectory is mapped on it. Velocities are sampled from the map along the Ulysses trajectory and are plotted with a thin line in the lower panel. Error bars ($\pm 1\sigma$) are shown on the velocity plot, but they are small except for those in the polar region where there were not enough IPS observations. The thick line is the Ulysses measurements. The structure between marks a and b is similar to the mesa-like steep structure of the fast stream observed by Helios around 0.3 AU (*Rosenbauer et al.*, 1977) which will be discussed in section 9. Ulysses entered into a polar high-speed region around DOY 245 (marked by c) with a steep velocity jump.

well with the solar wind structure obtained by the first Ulysses' rapid latitudinal scan in 1994 and 1995 (e.g., *Phillips et al.*, 1995; *Goldstein et al.*, 1996; *Woch et al.*, 1997).

Figure 6.3 compares IPS CAT velocities with Ulysses measurements in the solar maximum phase. The upper panel shows velocity distribution maps for CR1975–1981 (April – October 2001) derived by the IPS CAT analysis (*Fujiki et al.*, 2003b). In that year Ulysses made the second rapid latitudinal scan, and its trajectory is mapped back on the velocity map with gray colored circles. Velocities were sampled in the map along the Ulysses trajectory and compared with Ulysses observations in the lower panel. The thick line is the Ulysses observations and the thin line is the IPS observations. We find good agreement even though the observations were made in the solar maximum phase when the solar wind structure is not stable over a long time.

3. Synoptic Velocity Maps

From 13 years' observation during 1986–1998 we derived 21 solar wind velocity maps in heliographic latitude and Carrington longitude, and some of them are shown in Figure 6.4. These maps were derived from the IPS data obtained at the Solar-Terrestrial Environment Laboratory (STELab) using a four-antenna system operated at 327 MHz (*Asai et al.*, 1995). We used IPS data observed in an elongation range of 12°-64°; thus, the closest distance of a line of sight to the Sun is 0.2 to 0.9 AU where the solar wind has a steady cruising speed and radio scattering is weak at a frequency of 327 MHz. For this analysis, we selected those Carrington rotations over which the evolution of solar wind structures was minimal, and each map was derived from observations during two or three Carrington rotations so that there are a sufficient number of lines of sight traversing the polar regions.

In order to compare the synoptic velocity maps, which were derived from the IPS observations at 0.2–0.9 AU, with other observations at different heliocentric distances such as Ulysses data beyond 1.5 AU, those data are mapped back onto the source surface at 2.5 R_s based on an assumption of radial constant velocity. In order to examine coronal observations, such as magnetic filed, and interplanetary observations, interplanetary observations are mapped back onto the source surface, and coronal observations are traced on to the source surface along potential field magnetic field lines.

Schwenn et al. (1981) analyzed a solar wind acceleration rate at distances of 0.3– 1AU using Helios observations and obtained an acceleration rate of 7 ± 16 km s^{-1} AU^{-1} for fast wind and 52 ± 11 km s^{-1} AU^{-1} for slow wind. If we take into account these accelerations at distances where IPS observations were made, mapping back error in longitude is less than 5° in case of the slow wind. This acceleration may continue beyond 1 AU but may become less at further distances of Ulysses' orbit. Therefore, although the solar wind observed by Ulysses traveled a longer

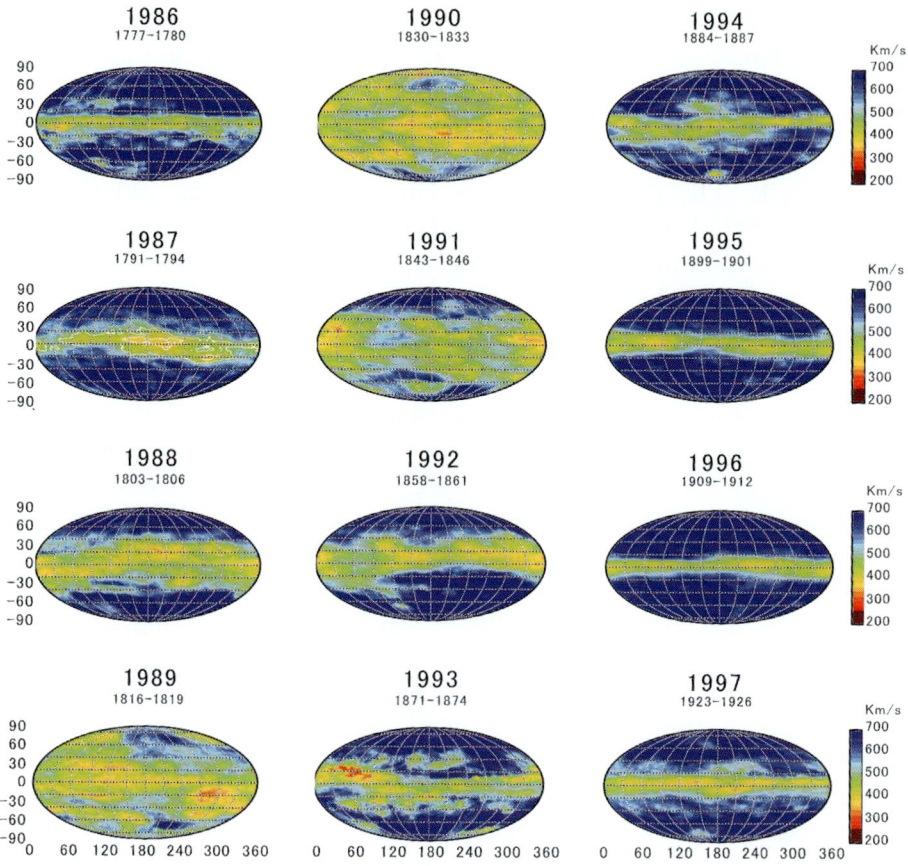

Figure 6.4. Velocity maps derived from the IPS CAT analysis. Each map was derived using data obtained during Carrington rotations shown below the year label.

distance under acceleration than the solar wind observed by the IPS, the mapping back error caused from the solar wind acceleration at distances beyond 0.3 AU is negligible. Even if the acceleration of the solar wind continues considerably beyond the source surface, this acceleration effect is cancelled in mapping back IPS and Ulysses observations because the solar winds observed at different distances equally experienced this acceleration.

3.1   Solar Cycle Dependence of Solar Wind Velocity Structure

Comparing Figure 6.4 with velocity maps which were derived earlier by *Kojima and Kakinuma* (1987, 1990) without using the CAT, we find that these new maps show sharp structures as well as several fine features such as compact slower speed regions. Here we briefly discuss the solar cycle dependence of the solar wind velocity structures, and detail structures are analyzed in the later sections.

In the solar minimum phase, the low-speed region distributes in a narrow belt along the solar equator separated from the high-speed region by a steep velocity gradient. As the solar activity increases, the low-speed region comes to have a wavy structure with a larger amplitude and wider distribution, while the high-speed region shrinks to the polar region and disappears at the maximum. Although the low-speed region tends to distribute along a potential field neutral line on the source surface (e.g., *Kojima and Kakinuma*; 1987, 1990), there is also a tendency for the lowest speed locus to deviate from the neutral line (*Crooker et al.*, 1997). This deviation has been explained by *Kojima et al.* (1999), and is illustrated in section 6. Using these velocity maps, we proceed to discuss the solar cycle dependence of the solar wind structure.

4.   Correction Factor for CAT Analysis Results

The CAT analysis is made using a limited number of line-of-sight data, and the data have some observational errors. *Kojima et al.* (1998, 1999, 2001) have shown that the IPS CAT has sufficient reliability and sensitivity in determining solar wind velocity structure. Here we make numerical evaluation of the accuracy of the CAT analysis and obtain numerical correction factors which will be used in the later sections. The evaluations are made as follows.

Velocity maps in the second row of Figure 6.5 are derived from actual IPS observations using the CAT analysis. From these maps we obtained the lower latitudinal boundary of the high-speed region, and their yearly variations are plotted in the top panel with a dotted line. Maps in the third row are used as models for observations. They have a steep velocity gradient between low ($400 \ \mathrm{km \, s^{-1}}$) and high ($750 \ \mathrm{km \, s^{-1}}$) speed regions. With these solar wind models we simulated IPS observations and then applied the CAT analysis to the synthetic IPS observations. Thus we obtained the CAT results that are shown at the bottom. By comparing the velocities of the model with the CAT results, we evaluated the accuracy of the CAT analysis for the high-latitude high- speed wind. Figure 6.6 shows the ratio of the model velocities to the CAT results for the high-latitude, high-speed wind. The abscissa gives the latitude of the lower boundary of the high speed region; the ratio of the high speed area to the surface area is also shown. The figure shows that CAT results are noticeably biased when the high- speed region shrinks to latitudes higher than 60°. Therefore, we exclude three years around the maximum phase from the analysis. The velocity ratio will be used as a speed reduction factor in the later analysis to correct the CAT results.

Another evaluation analysis was made for spatial resolution. The above solar wind models have a steep velocity gradient between the low-

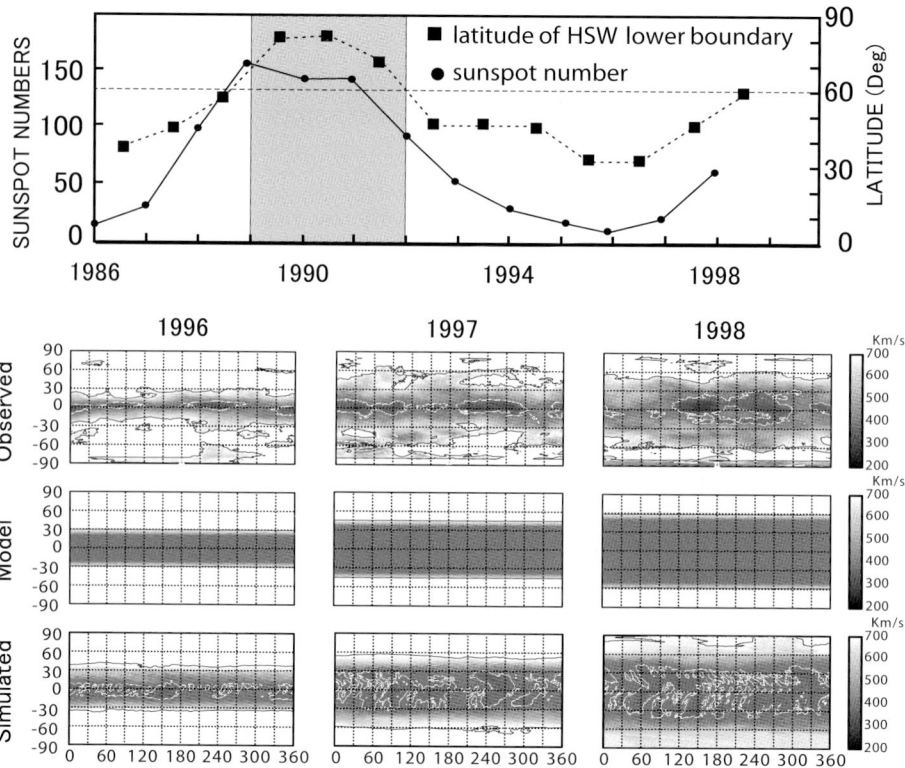

Figure 6.5. Evaluation of the IPS CAT analysis. The velocity maps in the second row are derived from the IPS observations using the CAT method. From these maps the lower latitude boundaries of the high-latitude high-speed region are obtained and plotted in the top panel with a dotted line. The maps in the third row are solar wind models simplified for the above observations. The bottom panels are derived using the CAT method from the synthetic IPS observations which were simulated in the above solar wind model.

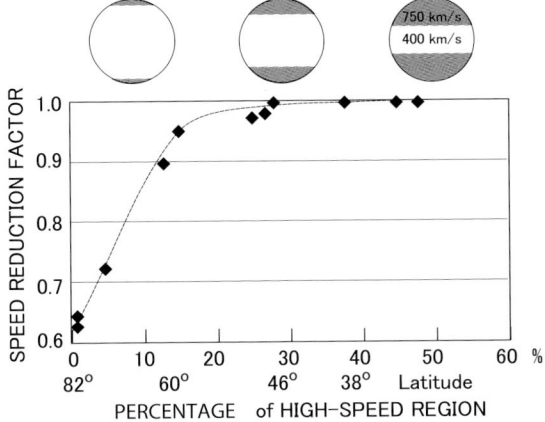
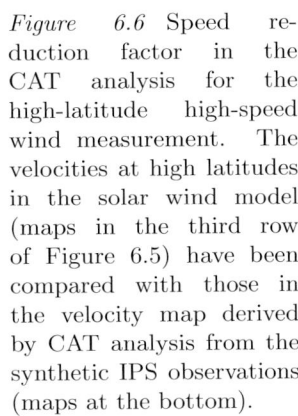

Figure 6.6 Speed reduction factor in the CAT analysis for the high-latitude high-speed wind measurement. The velocities at high latitudes in the solar wind model (maps in the third row of Figure 6.5) have been compared with those in the velocity map derived by CAT analysis from the synthetic IPS observations (maps at the bottom).

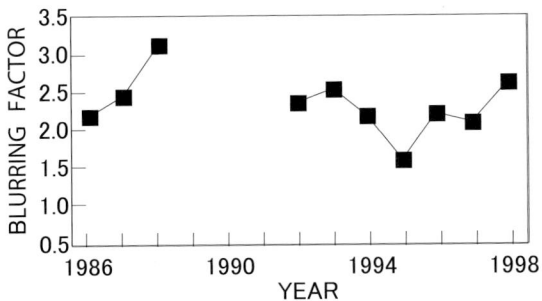

Figure 6.7 Blurring factor of spatial resolution of the CAT analysis. Area of medium-speed (450–700 $km\,s^{-1}$) region in the velocity map is calculated in both the solar wind model (maps at the third row in Figure 6.5) and the velocity map derived by CAT analysis from the synthetic IPS observations (maps at the bottom). The ratio of these two areas is used to evaluate the spatial resolution of the CAT analysis as a blurring factor.

speed and high-speed regions. We examine how this steep structure can be analyzed by the CAT method. The CAT results in the bottom row of Figure 6.5 show that the steep structure cannot be retrieved well when the high-speed region becomes small. To evaluate this numerically, the area occupied by the medium-speed (450–700 $km\,s^{-1}$) region in the map is calculated for the model and the CAT result, respectively, and the areas are compared by obtaining a ratio. The thus obtained ratio is shown as a blurring factor in Figure 6.7. This factor will be used in section 9 as a correction factor.

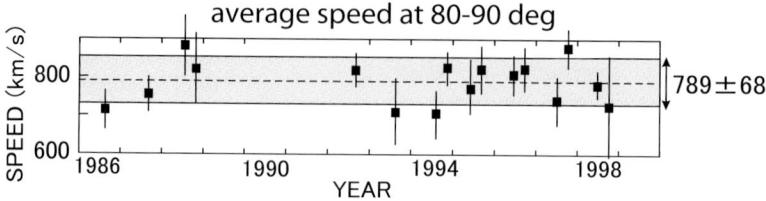

Figure 6.8. Mean velocities in the latitude range of 80°–90° in the velocity maps derived with the IPS CAT analysis. Velocities in the northern and southern polar regions were averaged. The total average through the analysis period is shown by the broken line with the gray belt having a range of ±1σ.

5. Coronal Hole Size Dependence of Solar Wind Velocity

From data analysis in the Skylab era, *Nolte et al.* (1976) reported that there is a high correlation between solar wind velocity and the size of a coronal hole; that is, slower solar winds originate from smaller coronal holes. Several works provide evidence that small coronal holes are sources of the slow (≤ 400 km s^{-1}) solar wind. They are reviewed in section 6. However, studies of the change of the polar wind speed as a function of the polar coronal hole size over a solar cycle are still missing.

We calculate a mean velocity in the latitude range of 80°–90° in the velocity maps derived with the IPS CAT analysis. The latitude of the lower boundary of the high-speed region changed from 30° to 60° during the time interval we analyzed, and the CAT analysis can retrieve the actual velocities with an accuracy of more than 90 % as shown in Figure 6.6. The velocities, which were corrected with the reduction factor, are shown in Figure 6.8 with a ±1σ error bar. The mean value calculated over the analysis period except for three years around the solar maximum is 789 ± 68 km s^{-1}, which agrees with the velocity estimated by extrapolating the Ulysses observations to the polar region. The figure shows no obvious systematic velocity change associated with the latitudinal change of the high-speed region.

During the time interval we examined the latitudinal boundary of the polar coronal hole varied in between 60° and 80°, and the coronal hole area varied in between 5×10^{10} and 40×10^{10} km^2. Coronal holes with these sizes were not analyzed by *Nolte et al.* (1976) because they investigated coronal holes within 10° of the ecliptic plane. The mean velocity obtained from this study is shown in Figure 6.9 together with the data of *Nolte et al.*. This figure indicates that the coronal hole has a critical scale size around 5×10^{10} km^2 beyond which the flow speed becomes independent of the coronal hole scale size. Since, however, this figure is the combination of the measurements for the equatorial and polar coronal holes, it is important to investigate whether the polar coronal hole turns to be the source of the slow solar wind when it becomes smaller than the critical scale size.

6.   Slow Solar Wind from a Small Coronal Hole

In velocity maps low-speed regions are often observed to cluster into a compact area associated with active regions (*Kojima et al.*, 1992; *Watanabe et al.*, 1995). From correlation analyses among an interplanetary parameter (density turbulence level with scale size around 100 km measured by the IPS), coronal parameters (Soft X-ray (SX) and Fe XIV intensities), and a source surface parameter (potential-field B_r at 2.5 R_s), *Hick and Jackson* (1995) and *Hick et al.* (1995) reported that enhanced density-turbulent regions tend to be mapped back to high-intensity regions of Fe XIV and SX rather than to a current sheet on the source surface. From the analysis of spacecraft observations and coronal magnetic field configuration, *Neugebauer et al.* (1998) suggested that medium and slow speed streams originate not only from open field regions just outside coronal hole boundaries but also from small coronal holes at low latitudes. *Arge et al.* (2003) also found that a narrow coronal hole observed in Yohkoh soft X-ray images was the source of slow solar wind. These low-speed regions tend to deviate from a potential-field neutral line on the source surface. In this section we examine the properties of two kinds of slow solar wind emanating from small coronal holes, one from a polar coronal hole and the other from an equatorial coronal hole.

Compact low-speed streams which appeared in the solar minimum phase near the solar equator were studied by *Kojima et al.* (1999). One

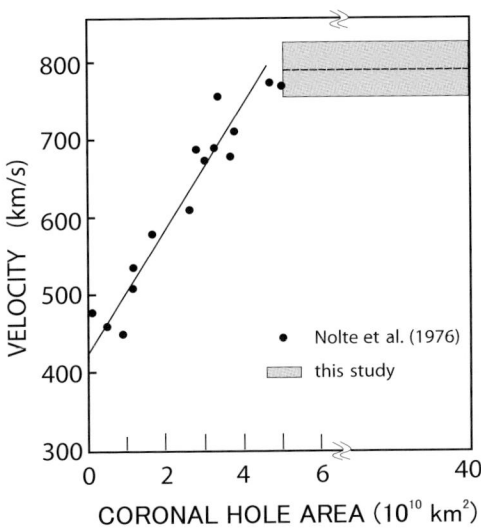

Figure 6.9 The relation between coronal hole size and velocity. Data plotted with dots are from *Nolte et al.* (1976), who analyzed equatorial coronal holes. The broken line and the gray colored square show the results of this study (789 ± 68 km s^{-1}) for the polar coronal hole.

of them is shown in Figure 6.10, which gives the velocity distribution on the source surface at 2.5 R_s, Kitt Peak photospheric magnetic field, and coronal potential magnetic field lines. The data is for CR1900 (August – September 1995). The field lines connecting the photosphere and the source surface are those with flux expansion factors larger than 2,000, which is defined by the ratio between the magnetic field intensity on the photosphere and that on the source surface, and their intersection with the source surface is labeled by white dots. Closed loops in the corona represent field lines with intensity ≥ 15 gauss on the photosphere. On the source surface, black areas are for velocities of ≤ 370 km s^{-1}. In this figure the slowest speed region is not located right above the closed loops (helmet structure), but is shifted away from them and connected to the open field regions (coronal hole) in the vicinity of the closed loops. Therefore, this compact stream is magnetically unipolar and consequently a neutral line does not traverse through it. This is the reason why the lowest speed locus tends to deviate from the neutral line.

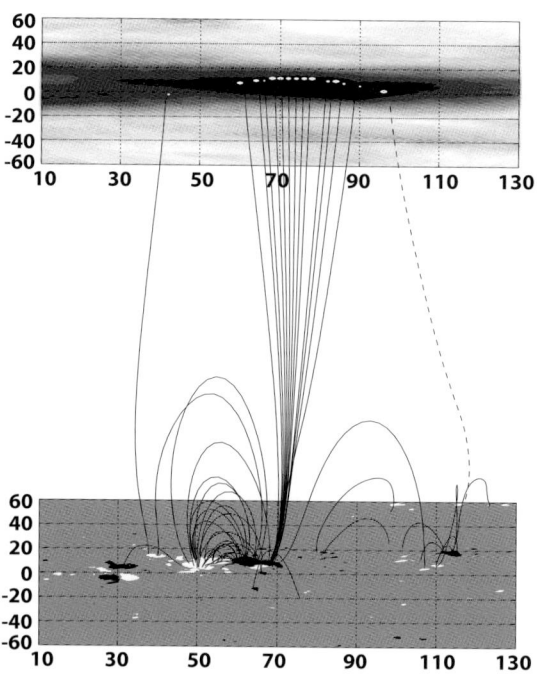

Figure 6.10 Magnetic flux tubes connecting the low-speed regions on the source surface and the coronal hole in the vicinity of active regions. Data are for CR1900. The magnetic field lines are for flux expansion factor larger than 2,000, which is defined by the ratio between the magnetic field intensity on the photosphere and that on the source surface, and the magnetic field in the closed loops shown in the figure is ≥ 15 gauss on the photosphere. The expansion factor was calculated for photospheric location with longitude and latitude binning size of $1° \times 1°$. On the source surface, black areas are for velocities of ≤ 370 km s^{-1}. White spots are the locations of the flux-expansion factors of $\geq 2,000$. Adapted from *Kojima et al.* (1999).

Table 6.1. Comparison of the physical properties of solar winds in a heliospheric plasma sheet (HPS), from a small equatorial coronal hole (seCH) and from a large equatorial coronal hole (leCH). Adapted from *Ohmi et al.* (2004).

	Slow (HPS)			Slow (seCH)			Fast (leCH)		
V_p [km s^{-1}]	343	±	12	323	±	9	673	±	14
N_p [cm^{-3}]	11.7	±	3.3	10.2	±	0.6	3.2	±	0.5
T_p [10^5K]	0.39	±	0.14	0.55	±	0.11	2.20	±	0.39
T_O [10^6K]	1.85	±	0.24	1.99	±	0.19	1.42	±	0.13
N_α/N_P	0.009	±	0.003	0.031	±	0.008	0.039	±	0.004

Physical properties of one of these low-speed streams were investigated by *Ohmi et al.* (2004) using WIND spacecraft observations. The top panel in Figure 6.11 (adapted from *Ohmi et al.* (2004)) shows the velocity distribution on a sphere at 1 AU, and other panels give parameters from WIND *in situ* observations. The bottom panel is an azimuthal angle of interplanetary magnetic field showing polarity change. Velocities measured by the IPS CAT analysis along the WIND trajectory are shown with a solid thick line in the second panel to evaluate the reliability of the IPS CAT analysis. WIND traversed a slow solar wind region at longitude 300°– 340° where the magnetic field did not change polarity (indicated by light gray). The physical properties of this slow solar wind (between two vertical lines) are listed in Table 6.1, and compared with those of a fast solar wind from a large equatorial coronal hole (leCH) observed around the longitude of 260° and a slow solar wind in a heliospheric plasma sheet (HPS) around the longitude of 0–120°. Velocity V_p, proton density N_p, proton temperature T_p, and freeze-in temperature T_o are all similar to those of the slow wind in an HPS. However, the helium to proton density ratio N_α/N_p is as large as the fast wind from a leCH, and, interestingly, the variance of N_p is as small as the leCH fast wind. The intensity of the photospheric magnetic field of this equatorial small coronal hole was 18 gauss, stronger than that in quiet regions.

Another source of slow wind was found in the polar region at the solar maximum in 1990 (CR 1829) and 1999 (CR 1955) by *Ohmi et al.* (2001, 2003). Figure 6.12 shows the velocity distribution on the source surface derived from the IPS CAT analysis and coronal potential magnetic fields. At the northern pole, a compact low-speed region was observed with a speed of 316±13 km s^{-1} for CR1829 and 290±39 km s^{-1} for CR1955. The magnetic neutral line (black solid line) does not traverse the slow speed region and open field lines connect the slow speed region to a small polar coronal hole. These open fields are surrounded by closed loops from mid latitudes, forming a large sea anemone structure. This structure is

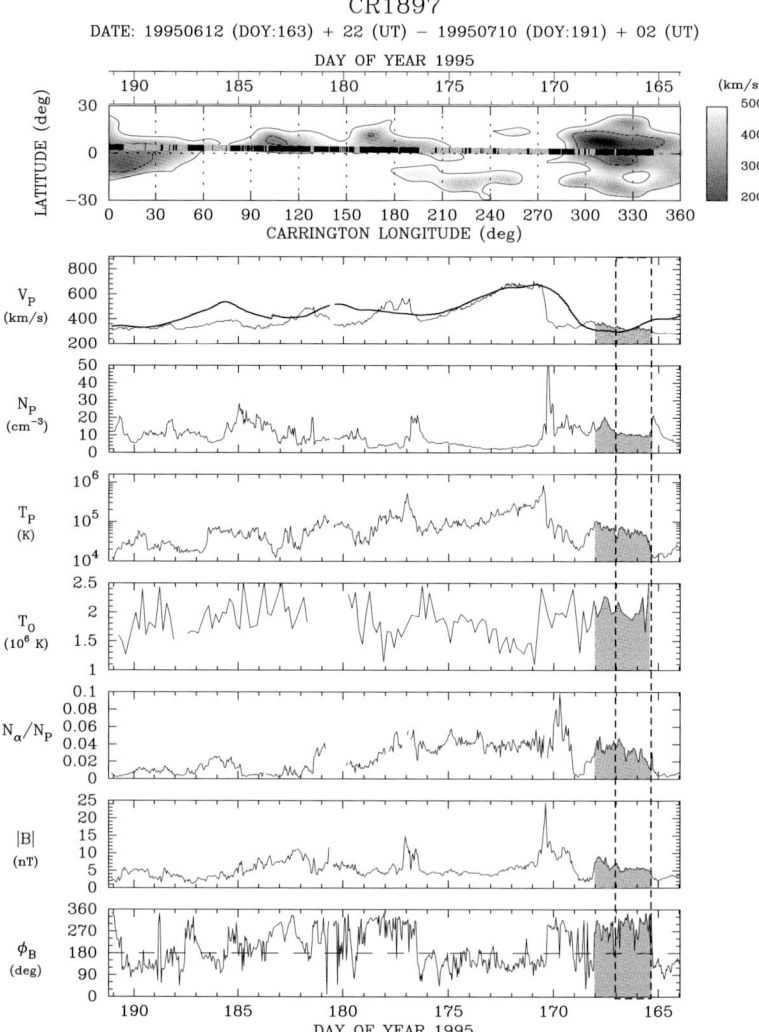

Figure 6.11. The top panel is the velocity distribution at 1AU derived from the IPS CAT analysis, and lower panels are measurements of the WIND spacecraft. The bottom panel is an azimuthal angle of interplanetary magnetic field showing polarity change. The trajectory of the WIND spacecraft is projected on the velocity map with a thick belt near the equator. In the second panel the velocities obtained by IPS CAT analysis (solid thick smooth line) are compared with WIND observations. WIND observed slow solar wind with a unipolar magnetic field in the gray colored region. The solar wind properties for the time interval between the vertical dashed lines are discussed in the text. Adapted from *Ohmi et al.* (2004).

similar to that of the open field region from a small coronal hole in the vicinity of active regions. The average magnetic field intensity of the small polar coronal hole observed at CR1955 is about 4–6 gauss, which is weaker than that in a large polar coronal hole (~10 gauss) observed in the solar minimum phase (*Wang and Sheeley*, 1995), because this is a polar coronal hole remnant about to disappear. This magnetic field intensity is also different from that in the equatorial small coronal hole. For the later discussion, these magnetic field intensities should be remembered. Although the small coronal holes at the pole and in the equatorial region have different magnetic field intensities, both were sources of low-speed wind.

7.   N-S Asymmetry of High-Latitude Fast Solar Wind

Ulysses observed in its first rapid latitudinal scan that the solar wind at mid-latitudes had a modest north-south asymmetry in velocity (*Goldstein et al.*, 1996); typically 15- 20 $\mathrm{km\,s^{-1}}$ faster velocities were observed in the northern hemisphere than in the southern hemisphere, and the velocity difference decreased with latitude from 24 $\mathrm{km\,s^{-1}}$ at 40° to 13 $\mathrm{km\,s^{-1}}$ at 80°. These velocity features were observed not only in the first latitudinal scan but also in its full polar orbit during the years of 1993–1997 (*McComas et al.*, 2000). Since, however, the Ulysses rapid

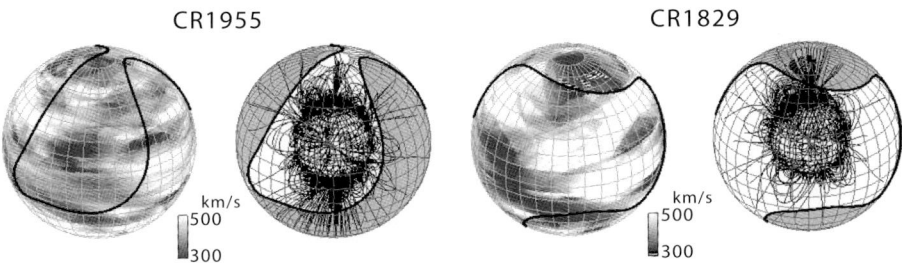

Figure 6.12.   Low-speed solar wind from a small polar coronal hole at the solar maximum. Solar wind velocity distribution is depicted on the sphere at 2.5 R_s. The black thick line is the magnetic neutral line from the potential field model. Coronal magnetic field lines are originating from 5° × 5° grids on the photosphere in the range of 5-250 gauss. The gray area is negative polarity and the white region is positive. The speeds in the polar low-speed region are 316±13 $\mathrm{km\,s^{-1}}$ for CR1829 and 290±39 $\mathrm{km\,s^{-1}}$ for CR1955. Figures for CR1829 and CR1955 are adapted from *Ohmi et al.* (2001) and (2003), respectively.

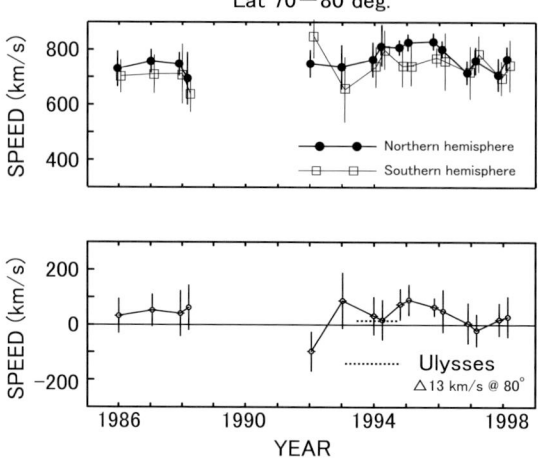

Figure 6.13 Average wind speed in the 70° to 80° latitude interval in the northern and southern hemisphere. The top panel shows velocity for each hemisphere separately, and the bottom panel gives the difference between the two hemispheres. North-south asymmetry observed by Ulysses in its first rapid latitudinal scan at latitude 80° is shown for reference with a dotted line (*Goldstein et al.*, 1996).

scan took about ten months to observe all the latitudes, the north-south asymmetry could be due to temporal variation of the solar wind. Therefore, we investigated whether this asymmetry was a stable structure lasting several years or much longer, and whether there was any solar cycle dependence. We made these investigations using the IPS CAT analysis.

We compared the velocities of northern and southern high latitudes. For this comparison we calculated the mean velocity over the range of latitudes of 70°–80°, for which there are more IPS observations than for polar regions. The top panel in Figure 6.13 shows the average velocities in the northern and southern hemispheres, and the lower panel indicates the difference between the two hemispheres. The velocity difference observed by Ulysses in 1994–1995 is shown by a dotted line for reference.

IPS and Ulysses observed the same asymmetry: that is, the speed was higher in the northern hemisphere than in the southern hemisphere. The asymmetry persisted over the solar cycle, apart from the 1992 and 1997 data points, and did not change when polarity of the solar dipole field reversed at the maximum.

It is important to note here that the IPS observed the asymmetry at distances within 0.9 AU. This means that the asymmetry was formed somewhere within 0.9 AU. *McComas et al.* (2000) suggested reverse waves from a stream-stream interaction region (*Gosling*, 1996) as an effect contributing to the asymmetry, though they did not consider it to be the major effect because no reverse waves were observed above 60° and the bulk solar wind parameters were nearly identical in the two hemispheres at the highest latitudes. We agree that the reverse waves are not the major cause, because the asymmetry is formed somewhere within 0.9 AU where the stream- stream interaction is not strongly developed. *McComas et al.* (2000) attributed the observed asymmetry to real temporal differences in solar wind such as that from the polar coronal hole, high-latitude CMEs, and reverse waves and shocks, because the observations in the southern hemisphere were made in the declining phase (1992–1995) while the northern hemisphere was observed in the minimum phase (1995–1996). However, the IPS observations have revealed that the asymmetry is a stable real structure which is not caused by temporal change of the solar wind condition.

Goldstein et al. (1996) and *Kojima et al.* (2001) compared the velocity asymmetry with that of a magnetic flux expansion factor in the corona. Goldstein et al. analyzed the Ulysses data observed in the first rapid latitudinal scan, and *Kojima et al.* analyzed the IPS data observed in 1996, one year after the Ulysses rapid latitudinal scan. They obtained qualitatively good agreement in terms of the empirical inverse relation between velocity and a magnetic flux tube expansion factor which was found by *Wang and Sheeley* (1990); that is, the expansion factor in the northern hemisphere is smaller than in the southern hemisphere.

8.   Velocity Gradient in High-Speed Region

In Ulysses' first rapid latitudinal scan, *Woch et al.* (1997) found that the solar wind from the polar coronal hole increased in velocity gradually with latitude, and *Kojima et al.* (2001) confirmed this with the IPS CAT analysis using data obtained during CR1894 to CR1896 (April – June 1995) (Figure 6.14). Mapping the solar wind velocity observed by Ulysses back to the source surface, *Neugebauer et al.* (1998) found that the solar wind from the inner most region in the polar hole was faster. *Goldstein et al.* (1996) and *Kojima et al.* (2001) pointed out that the

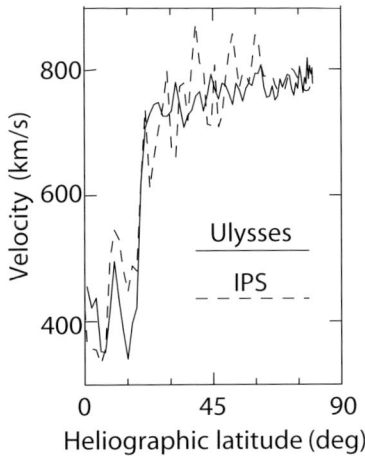

Figure 6.14 Velocity gradient observed by Ulysses (solid line) and the IPS CAT analysis (dotted line). The IPS velocities were sampled in the velocity map, which was derived by the IPS CAT analysis, along the Ulysses trajectory.

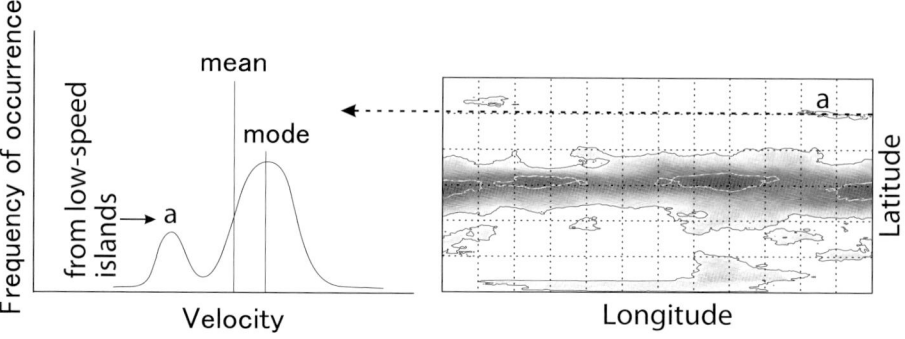

Figure 6.15. Process to analyze the latitudinal velocity gradient. First we analyze velocity distribution at each latitude along a longitude, and then calculate a mean value. Next step is to search the mode in a velocity distribution range larger than the mean value. If there is a low-speed region in the longitudinal scan, it will make a hump in the distribution function and cause the mean value to be underestimated.

latitudinal dependence of the velocity is a consequence of the empirical inverse relationship between speed and expansion factor.

Using velocity maps, we investigate how the velocity gradient changes with coronal hole scale size. Since the velocity maps are not uniform in high latitudes but have several fine structures such as low speed islands and abnormally high-speed regions, the analysis was made after removing these fine structures (Figure 6.15). First we obtain a velocity distribution function in longitude at each latitude, and then calculate a mean value. In the next step we search the mode in a distribution range larger than the mean value, which is the velocity obtained with the greatest frequency in the range. Since abnormally high-speed bins in the map contribute to a higher velocity tail in the distribution function and low speed islands cause the mean value to be underestimated, we do not use the mean but use the mode as the representative of the velocity at each latitude. The gradients of the velocity with latitude derived with this technique are plotted in Figure 6.16. The velocity gradient observed by Ulysses in 1994–1995 (*Woch et al.*, 1997) is shown by a dotted line for reference, with which the IPS observations show good agreement. The figure indicates that the velocity gradient tends to be small at the solar minimum and increases with the solar activity.

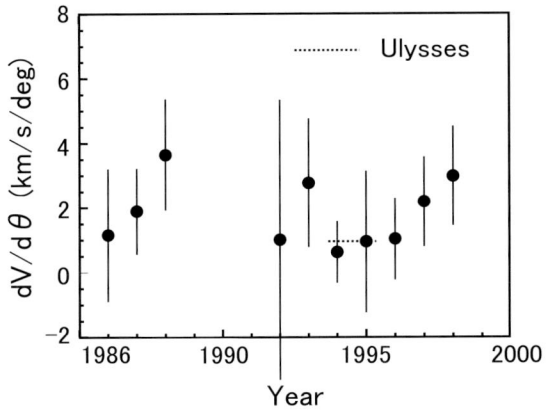

Figure 6.16 Yearly variation of the latitudinal velocity gradient of the high-speed wind. The gradient observed by Ulysses in its first rapid latitudinal scan is shown by a dotted line (*Goldstein et al.*, 1996) .

9.   Bimodal Structure of Solar Wind Velocity

Helios spacecraft revealed that fast streams have a mesa-like structure in longitude with steep edges at a heliocentric distance of 0.3 AU (*Rosenbauer et al.*, 1977). *Kakinuma* (1977) and *Kojima and Kakinuma* (1990) had revealed long before Ulysses' *in situ* observations that the solar wind in the minimum phase has a bimodal latitudinal distribution, in which the solar wind latitudinal structure mainly consists of a low-speed and a high-speed component separated by a steep gradient. For solar wind in 1974 *Kakinuma* (1977) derived a V vs. θ profile of

$$\begin{aligned} v &= 500\sin^2(2\theta) + 300 \quad \text{for } |\theta| < 45°, \\ &= 800 \quad \text{for } |\theta| \geq 45°, \end{aligned}$$

and for solar wind in 1985 *Kojima and Kakinuma* (1990) derived a much steeper gradient and N-S asymmetry of

$$\begin{aligned} v &= 400\sin^4(3(\theta + 2)) + 400 \quad \text{for } 28° > \theta > -2°, \\ &= 800 \quad \text{for } \theta \geq 28°, \\ v &= 350\sin^4(3(\theta + 2)) + 400 \quad \text{for } -2° \geq \theta \geq -32°, \\ &= 750 \quad \text{for } \theta < -32°. \end{aligned}$$

Thus, Helios and IPS observations revealed that the fast wind has steep velocity edges in longitude and latitude. *Newkirk and Fisk* (1985) analyzed the solar cycle dependence of the solar wind latitudinal structure by fitting a Gaussian function to spacecraft data. Their analysis shows that velocities of low-speed and high-speed winds do not change with the solar cycle, but the Gaussian half width does change. In the previous section we also showed that the polar high-speed wind did not change in speed when its lower latitudinal boundary changed from 30° to 60°. The results of these studies indicate that the velocities of slow and fast winds do not change with the solar activity. The question, then, is what the change in the Gaussian half width indicates. Does the boundary between low-speed and high-speed regions become less steep or does the low-speed region simply expand?

We analyzed how much of the fractional area in the map is occupied by low-, medium- and high-speed regions, respectively, and the results are shown as a histogram in Figure 6.17. The lower panel shows the histogram of the medium-speed region only. The low-speed (< 450 km s^{-1}) region increased in area from 10 % at the solar minimum to 60 % at the solar maximum, while the high-speed (> 700 km s^{-1}) region decreased in area from 50 % at the minimum phase to a few % at the maximum phase. There were no systematic changes in the medium-speed (450–700 km s^{-1}) area with the solar cycle. The variance of each velocity area throughout the analysis period except for the three years of 1989–1991 was calculated. Variances of the high- and low-speed areas are about 10 %, while variance of each medium-speed area is less than half of that.

Yearly change of the fractional area of the medium-speed region is shown in Figure 6.18. Square symbols are from the histogram, while diamonds show the values corrected using the blurring factor obtained in section 4. This figure shows that the medium-speed regions occupied a quite small area (< 25 %) consistently throughout the solar cycle, in other words, the solar wind structure was bimodal throughout the solar cycle.

In 2001 Ulysses reached the northern high latitude and its trajectory is mapped back on the velocity map in Figure 6.3 which has been built assuming a radial constant velocity. Although there is a data gap in IPS observations for one Carrington rotation, we can guess the velocity distribution in this rotation by interpolating it from observations in

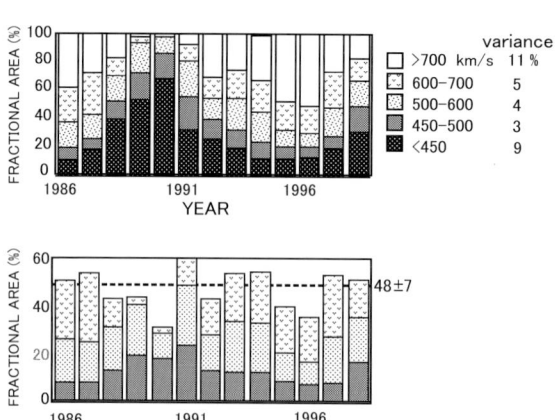

Figure 6.17 Fractional area in the velocity map occupied by each velocity bin for years 1986–1998. The lower panel shows only medium-speed region (450–700 km s^{-1}), and the mean fractional area is 48 ± 7%, shown by a broken line.

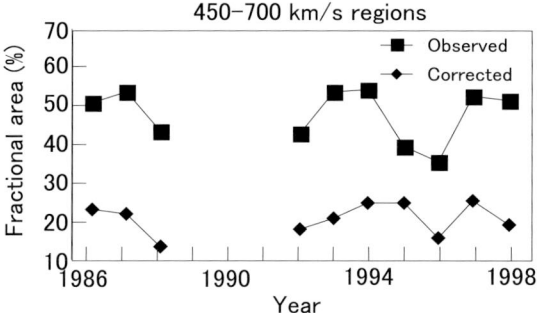

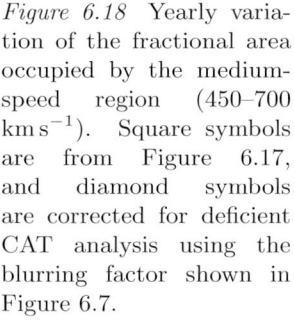

Figure 6.18 Yearly variation of the fractional area occupied by the medium-speed region (450–700 km s^{-1}). Square symbols are from Figure 6.17, and diamond symbols are corrected for deficient CAT analysis using the blurring factor shown in Figure 6.7.

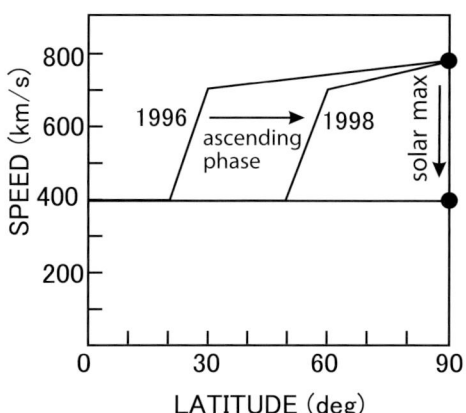

Figure 6.19 Solar cycle dependence of latitudinal velocity structure. In the ascending (from the year 1996 to 1998)/descending phase, the latitudinal width of the low-speed region extends/shrinks, and the velocity gradient in the high-latitude high-speed region becomes steeper/gradual. On the other hand, the velocities in the polar region and the velocity gradient between low-speed and high-speed regions do not change. When, at solar maximum, the polar coronal hole area shrinks and nearly vanishes, the velocity in the polar region decreases.

CR1978 and CR1980. Ulysses' repeated entrance and exit in/out of the high-speed bay structure extended to mid-latitudes from the polar region during days 190–240 (CR1978–1979). As we mentioned, the structure marked *a* and *b* in the figure is similar to the mesa-like steep structure of the fast stream observed by Helios around 0.3 AU (*Rosenbauer et al.*, 1977). Around day 245 (marked *c*) Ulysses entered the polar high-speed region with a steep velocity jump. These findings indicate that the polar high-speed stream has steep edges even in the solar maximum phase.

10.   Summary of Solar Cycle Dependence

We have studied the solar cycle dependence of solar wind properties using IPS data obtained at STELab during the years of 1986–1998. The mean velocity at the polar region was 789 ± 68 km s^{-1} throughout the analysis period except for the three years 1989–1991 around the solar maximum, and there was no systematic change in velocities when the latitudinal lower boundary of the polar coronal hole changed from 60° to 80°. This means that the speed of the polar solar wind is independent of the solar cycle except for the maximum phase. When the polar coronal hole shrank to a size smaller than a critical scale size of about 5×10^{10} km^2 in the solar maximum phase, it became the source of the slow solar wind. The fast and slow solar wind are separated by a steep velocity gradient, even if the scale of the high-speed region changes. The high-latitude fast wind shows a N-S asymmetry in velocity: this was a stable characteristic that lasted throughout the whole solar cycle we analyzed.

A resume of these results is given in Figure 6.19, which illustrates how the latitudinal structure changes with the solar activity. In the descending and ascending phases of solar activity, the velocities of low-speed wind and high-speed wind do not change, and the boundary between the low-speed and high-speed regions is always steep. What varies is the latitudinal width of the low-speed region and the latitudinal velocity gradient in the high-speed region. When a polar coronal hole shrinks and its size becomes very small at the solar maximum, the flow from it becomes slow.

11.   Solar Wind Velocity and Physical Condition in Corona

Coronal properties determine the solar wind velocity structure. One is the coronal hole scale size (see section 5). A coronal hole is not only the source of high- speed wind but also the source of low-speed wind if it is small, and there is a linear relation between solar wind velocity and coronal hole size (*Nolte et al.*, 1976). However this linear relation does not hold for a large-scale coronal hole. In this study we showed that there is a critical scale size in the coronal hole over which the wind velocity becomes independent of the scale size. Another coronal property is a flux expansion rate, which has good relation with wind velocity (*Wang and Sheeley*, 1990), and the third is the magnetic energy released by reconnection at the lower corona (*Fisk et al.*, 1999). In order to model the solar wind acceleration it is important to find a relation between the global properties of the solar wind and corona. However most coronal holes are at high latitudes where spacecraft cannot access, with the exception of the Ulysses. In this study, using the IPS CAT, which can derive an unbiased solar wind velocity map over all latitudinal ranges, we identify the relation between the wind velocity and the coronal magnetic condition.

11.1 Data

Three data sets are used in this analysis: 1) velocity maps derived on the source surface from the IPS CAT analysis using STELab data, 2) photospheric coronal hole structures inferred from the HeI absorption line measured at Kitt Peak National Solar Observatory, 3) coronal magnetic fields extrapolated via a potential field source surface model (*Hakamada and Kojima*, 1999) from photospheric fields measured at Kitt Peak National Solar Observatory. We analyze data obtained during a period of CR1896-1914 (May 1995 – October 1996) which were in the last solar minimum phase.

In the following coronal holes are identified by different labels; more precisely (Figure 6.20): (a and e) polar coronal holes, (b and c) distinct equatorial and mid-latitude coronal holes, and (d) low-latitude polar coronal hole extension. Although the polar coronal hole extended to latitudes ≤ 60°, we examine the coronal hole area at latitudes higher than 60° because the boundary region has a complex structure. If there is a extension from a polar coronal hole toward the equator beyond a latitude of 40°, it is treated as an independent mid-latitude coronal hole. Since a HeI coronal hole and an open magnetic field region do not exactly overlap, we analyze only the area in which the HeI coronal hole overlaps with the open magnetic field region (Figure 6.21). Thus, a total of 43 coronal holes have been analyzed.

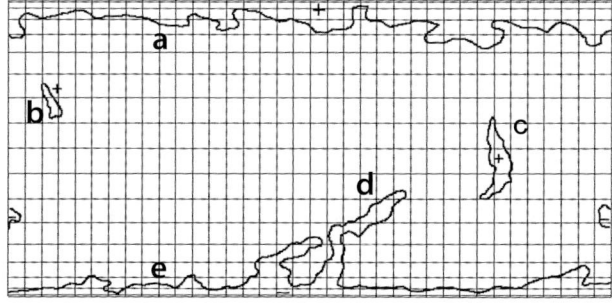

Figure 6.20. Coronal hole classification. (a, e) polar coronal hole, (b) mid-latitude coronal hole, (c) equatorial coronal hole, and (d) low-latitude polar coronal hole extension.

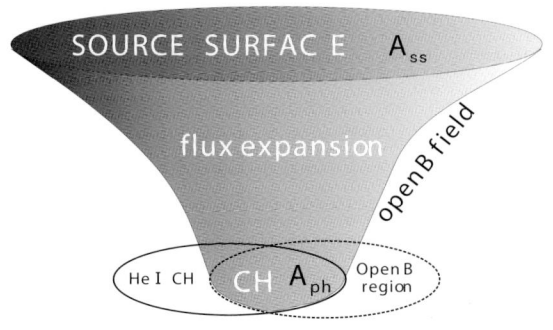

Figure 6.21 The area where open magnetic field region overlaps with a HeI coronal hole is analyzed. A flux expansion rate is defined by the area ratio of an open flux tube on the photosphere A_{ph} and on the source surface A_{ss} normalized by R_s^2/R_{ss}^2, where R_s is a solar radius and R_{ss} is a radius of the source surface.

Although a flux expansion rate f is generally defined by the ratio between the magnetic field intensity on the photosphere and that on the source surface, in this study we use the ratio of the area of an open flux tube on the photosphere A_{ph} and that on the source surface A_{ss} normalized by R_s^2/R_{ss}^2, where R_s is the solar radius and R_{ss} is the radius of the source surface. In the analysis we use a mean velocity V averaged over the area A_{ss} on the source surface and a mean photospheric magnetic field intensity B averaged over A_{ph} on the photosphere.

11.2 Cross-Correlation Analysis

The velocity V has been cross correlated with a) the inverse of an expansion rate $1/f$, b) photospheric magnetic field intensity B, and c) the ratio between these two parameters B/f. Results are given in Figures 6.22, 6.23 and 6.24. In these figures we focus on the following three kinds of coronal holes, which have different properties (Table 6.2): equatorial coronal holes (triangles, CH_{eq}) associated with active regions which have a large flux expansion rate and relatively strong magnetic field, isolated mid-latitude coronal holes (boxes, CH_{ml}) which have a medium expansion rate and weaker magnetic field, and polar coronal hole extensions (diamonds, CH_{ce}) which have a smaller expansion rate and weaker magnetic field.

Figure 6.22 is the correlation diagram for the flux expansion rate. Wind velocity V and the inverse of the flux expansion rate $1/f$ are well correlated with a correlation coefficient of 0.56. This verifies the empirical relation proposed by *Wang and Sheeley* (1990). However, there are four isolated data points which refer to two high- speed solar winds from polar coronal hole extensions (CH_{ce}-1, 2 (diamonds)) and two low-speed solar winds from isolated mid-latitude coronal holes (CH_{ml}-1, 2 (boxes)), which do not fit in the behavior of data points with a relatively small expansion rate.

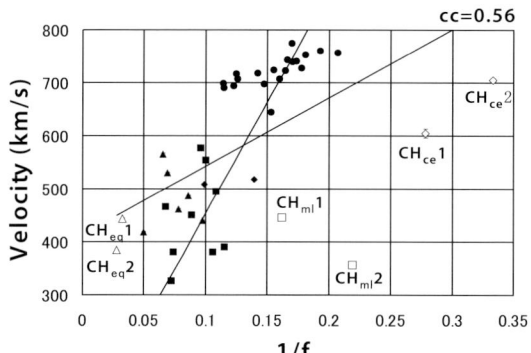

Figure 6.22 Correlation diagram between velocity and the inverse value of a flux expansion rate. CH_{eq} (triangle) is the equatorial coronal hole associated with active regions with a large flux expansion rate and relatively strong magnetic field, CH_{ml} (box) is the isolated mid-latitude coronal hole with a medium expansion rate and weaker magnetic field, and CH_{ce} (diamond) is a polar coronal hole extension with a smaller expansion rate and weaker magnetic field. The correlation coefficient is 0.56. Two lines in the diagram are regression lines for V from $1/f$ and for $1/f$ from V, respectively

Figure 6.23 gives the correlation between the velocity V and the photospheric magnetic field intensity B. This is a test of the model proposed by *Fisk et al.* (1999). Since we do not have direct measurements of magnetic field energy released by reconnection, we assume it is proportional to the power of magnetic field intensity B^β. We did not find a good correlation between V and B^β for any $0.5 \le \beta \le 2$. Data distribute around the magnetic field intensity of 5-10 gauss except for two low-speed winds from equatorial coronal holes (CH_{eq}-1, 2 (triangles)). These two regions have relatively strong magnetic field intensity, and we can find them in Figure 6.22, with a large flux expansion rate. In this correlation diagram

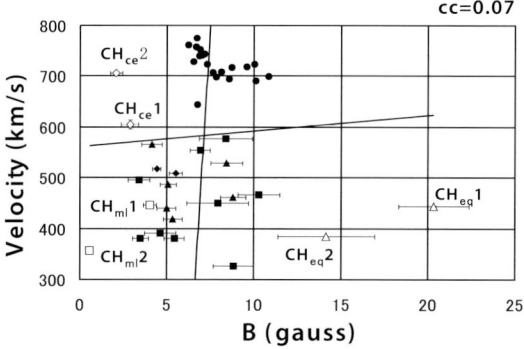

Figure 6.23 Correlation diagram between velocity and photospheric magnetic intensity. Coronal holes marked with CH_{eq}, CH_{ml} and CH_{ce} are the same coronal holes in Figure 6.22. The correlation coefficient is 0.07.

we find that the wind from the four regions (CH_{ce} (diamonds), CH_{ml} (boxes)), which deviated from the correlated data group in Figure 6.22, originated from weaker magnetic field regions.

In Figure 6.24 we tested the model which combines the flux expansion rate and the magnetic field intensity in the ratio B/f. This figure shows that the velocity V and B/f have a remarkably high correlation with a coefficient 0.88, and interestingly, three of the data points (CH_{eq} (triangle), CH_{ce} (diamond), CH_{ml} (box)) which deviated from the major data group in the correlation diagrams of other two models fall on the regression line, the only exception being the coronal hole $CH_{ce}2$.

Table 6.2. Properties of the coronal holes marked with CH_{eq}, CH_{ml} and CH_{ce} in Figures 6.22, 6.23, and 6.24.

Coronal hole	magnetic field intensity	flux expansion rate	speed
Equatorial coronal hole	strong	large	slow
Polar coronal hole extension	weak	small	fast
Mid-latitude coronal hole	weak	medium	slow

12.   Conclusion

We investigated the relation between the solar wind velocity and the coronal magnetic condition for various kinds of coronal holes, which have different magnetic field intensity B and flux expansion rate f. From cross-correlation analysis, we revealed that the parameter B/f is extremely well correlated with the wind velocity. It is interesting to note that this correlation holds for solar winds of various speeds from various kinds of coronal holes, as shown in Table 6.2.

The physical meaning of the parameter B/f can be understood by combining *Wang and Sheeley* (1990) and *Fisk et al.* (1999) models. The model of *Fisk et al.*, which employs B, is based on the energy

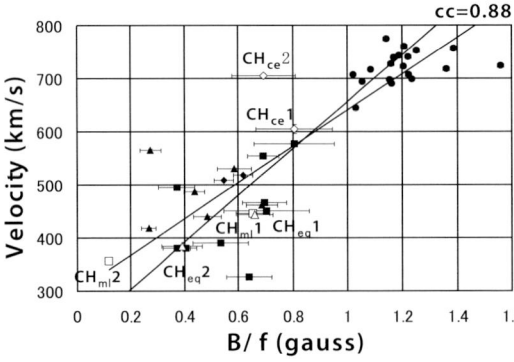

Figure 6.24 Correlation diagram between velocity and B/f. Coronal holes marked with CH_{eq}, CH_{ml} and CH_{ce} are the same coronal holes as in Figures 6.22 and 6.23. The correlation coefficient is 0.88.

supply to the solar wind from magnetic field reconnection in the low coronal, while *Wang and Sheeley* model, which employs $1/f$, shows how the reconnection energy is deposited efficiently in an expanding flux tube and accelerates the solar wind (*Wang and Sheeley*, 1991). The physics of this process is theoretically explained by *Suzuki* (2004) who invokes a dissipation mechanism of fast and slow MHD waves through MHD shock process.

At a different solar activity phase, coronal holes with different properties appear and different kinds of solar winds are produced. These form the three-dimensional solar wind structure which changes with solar activity. These solar wind properties will be explained by two physical parameters in the corona and on the photosphere, that is, B/f. Since the B/f parameter was tested in this study for solar wind in the solar minimum phase, it will be necessary to analyze the solar wind through the whole solar cycle in order to make this result universal.

Acknowledgments

The authors wish to thank the National Solar Observatory (NSO) at Kitt Peak for the use of their HeI coronal hole maps and magnetic field data, which are produced cooperatively by NSF/NOAO, NASA/GSFC, and NOAA/SEL. Our thanks go to Bernie Jackson, Paul Hick and Andrew Buffington at the University of California at San Diego for their collaboration in developing the CAT analysis method. We would like to acknowledge the engineering support from Y. Ishida, K. Maruyama and N. Yoshimi to keep the IPS observations running over two solar cycles at the STELab. Figure 6.11 and Table 6.1 are reprinted from a paper by *Ohmi et al.* (2004) with permission from Elsevier. This work was partially supported by the Japan Society for the Promotion of Science (grants 12440130, 15340162).

References

Arge, C. N., K. L. Harvey, H. S. Hudson, and S. W. Kahler, 2003, in *Solar Wind Ten*, AIP Conference Proceedings 679, AIP, New York, 1995, edited by M. Velli, R. Bruno and F. Malara, 202–205.

Asai, K., Y. Ishida, M. Kojima, K. Maruyama, H. Misawa, and N. Yoshimi, 1995, *J. Geomag. Geoelectr.*, **47**: 1107–1112.

Asai, K., M. Kojima, M. Tokumaru, A. Yokobe, B. V. Jackson, P. Hick, and P. K. Manoharan, 1998, wind, *J. Geophys. Res.*, **103**: 1991–2001.

Coles, W. A., J. K. Harmon, A. J. Lazarus, and J. D. Sullivan, 1978, *J. Geophys. Res.*, **83**: 3337–3341.

Coles, W. A., and J. J. Kaufman, 1978, *Radio Science*, **13**: 591–697.

Coles, W.A., B. J. Rickett, V. H. Rumsey, J. J. Kaufman, D. G. Turley, S. Ananthakrishnan, J. W. Armstrong, J. K. Harmons, S. L. Scott, and D. G. Sime, 1980, *Nature*, **286**: 239–241.

Crooker, N. U., A. J. Lazarus, J. L. Phillips, J. T. Steinberg, A. Szabo, R. P. Lepping, and E. J. Smith, 1997, *J. Geophys. Res.*, **102**: 4673–4679.

Fisk, L. A., N. A. Schwadron, and T. H. Zurbuchen, 1999, *J. Geophys. Res.*, **104**: 19,765- -19,772.

Fujiki, K., M. Kojima, M. Tokumaru, T. Ohmi, A. Yokobe, and K. Hayashi, 2003a, in *Solar Wind Ten*, AIP Conference Proceedings 679, AIP, New York, 1995, edited by M. Velli, R. Bruno and F. Malara, 141–143.

Fujiki, K., M. Kojima, M. Tokumaru, T. Ohmi, A. Yokobe, K. Hayashi, D. J. McComas, and H. A. Elliott, 2003b, *Ann. Geophy.*, **21**: 1257–1261.

Goldstein, B. E., M. Neugebauer, J. L. Phillips, S. Bame, J. T. Gosling, D. McComas, Y.-M. Wang, N. R. Sheeley, and S. T. Suess, 1996, *Astron. Astrophys.*, **316**: 296– 303.

Gosling, J. T., 1996, *Annu. Rev. Astron. Astrophys.*, **34**: 35–73.

Grall, R. R., W. A. Coles, M. T. Klinglesmith, A. R. Breen, P. J. S. Williams, J. Markkanen, and R. Esser, 1996, *Nature*, **379**: 429–432.

Hakamada, K., and M. Kojima, 1999, *Solar Phys.*, **187**: 115–122.

Hewish, A., P. F. Scott, and D. Wills, 1964, *Nature*, **203**: 1214–1217.

Hick, P. L., and B. V. Jackson, 1995, in *Solar Wind Eight*, AIP Conference Proceedings 382, AIP, New York, 1995, edited by D. Winterhalter, J.T. Gosling, S.R. Habbal, W.S. Kurth, and M. Neugebauer, 461–464.

Hick, P. L., B. V. Jackson, S. Rappaport, G. Woan, G. Slater, K. Strong, and Y. Uchida, 1995, *Geophys. Res. Lett.*, **22**: 643–646.

Hirano, M., M. Kojima, M. Tokumaru, K. Fujiki, T. Ohmi, M. Yamashita, K. Hakamada, and K. Hayashi, 2003, *Eos Trans. AGU*, **84**(46), Fall Meet. Suppl., Abstract SH21B-0164.

Jackson, B. V., P. L. Hick, M. Kojima, and A. Yokobe, 1998, *J. Geophys. Res.*, **103**: 12,049–12,067.

Kakinuma, T, 1977, in *Study of Traveling Interplanetary Phenomena*, edited by M. A. Shea, D. F. Smart, and S. T. Wu, pp. 101–118, D. Reidel, Norwell, Mass.

Kojima, M., and T. Kakinuma, 1987, *J. Geophys. Res.*, **92**: 7269–7279.

Kojima, M., and T. Kakinuma, 1990, *Space Sci. Rev.*, **53**: 173–222.

Kojima, M., H. Washimi, H. Misawa, and K. Hakamada, 1992, in *Solar Wind Seven*, Proc. 3rd COSPAR Colloquium, edited by E. Marsch and R. Schwenn, 201–204, Pergamon Press, Oxford.

Kojima, M., M. Tokumaru, H. Watanabe, A. Yokobe, K. Asai, B. V. Jackson, and P. L. Hick, 1998, *J. Geophys. Res.*, **103**: 1981–1989.

Kojima, M., K. Fujiki, T. Ohmi, M. Tokumaru, A. Yokobe, and K. Hakamada, 1999, *J. Geophys. Res.*, **104**: 16,993–17,003.

Kojima, M., K. Fujiki, T. Ohmi, M. Tokumaru, A. Yokobe, and K. Hakamada, 2001, *J. Geophys. Res.*, **106**: 15,677–15,686.

McComas, D. J., B. L. Barraclough, H. O. Funsten, J. T. Gosling, E. Santiago- Muñoz, R. M. Skoug, B. E. Goldstein, M. Neugebauer, P. Riley, and A. Balogh, 2000, *J. Geophys. Res.*, **105**: 10,419–10,422.

Neugebauer, M., R. J. Forsyth, A. B. Galvin, K. L. Harvey, J. T. Hoeksema, A. J. Lazarus, R. P. Lepping, J. A. Linker, Z. Mikic, J. T. Steiberg, R. von Steiger, Y.-M. Wang, and R. F. Wimmer-Schweingruber, 1998, *J. Geophys. Res.*, **103**: 14,587–14,599.

Newkirk, G., Jr., and L. A. Fisk, L. A, 1985, *J. Geophys. Res.*, **90**: 3391–3414.

Nolte, J. T., A. S. Krieger, A. F. Tomothy, R. E. Gold, E. C. Roelof, G. Vaiana, A. J. Lazarus, J. D. Sullivan, and P. S. McIntosh, 1976, *Sol. Phys.*, **46**: 303–322.

Ohmi, T., M. Kojima, A. Yokobe, M. Tokumaru, K. Fujiki, and K. Hakamada, 2001, *J. Geophys. Res.*, **106**: 24,923–24,936.

Ohmi, T., M. Kojima, K. Fujiki, K. Hayashi, M. Tokumaru, and K. Hakamada, 2003, *Geophys. Res. Let.*, **30**: 1409–1412.

Ohmi, T., M. Kojima, M. Tokumaru, K. Fujiki, and K. Hakamada, 2004, *Adv. Space Res.*, **33**: 689–695.

Phillips, J. L., S. J. Bame, A. Barnes, B. L. Barraclough, W. C. Feldman, B. E. Goldstein, J. T. Gosling, G. W. Hoogeveen, D. J. McComas, M. Neugebauer, and S. T. Suess, 1995, *Geophys. Res. Lett.*, **22**: 3301–3304.

Rickett, B. J., and W. A. Coles, 1991, *J. Geophys. Res.*, **96**: 1717–1736.

Rosenbauer, H., R. Schwenn, E. Marsch, B. Meyer, H. Miggenrieder, M. D. Montgomery, K. H. Muehlhaeuser, W. Pilipp, W. Voges, and S. M. Zink, 1977, *J. Geophys.*, **42**: 561–580.

Sheeley, N. R., Jr., E. T. Swanson, and Y.-M. Wang, 1991, *J. Geophys. Res.*, **96**: 13,861–13,868.

Suzuki, T. K., 2004, *Mon. Not. R. Astron. Soc.*, in press.

Schwenn, R., K.-H. Mülhäuser, E. Marsch, and H. Rosenbauer, 1981, in *Solar Wind Four*, Rep. MPAE-W-100-81- 31, edited by H. Rosenbauer, 126–130, Max-Planck-Institute für Aeron., Katlenburg- Lindau, Germany.

Wang, Y.-M., and N. R. Sheeley, Jr., 1990, *Astrophys. J.*, **355**: 726–732.

Wang, Y.-M., and N. R. Sheeley, Jr., 1991, *Astrophys. J.*, **372**: L45–L48.

Wang, Y.-M., and N. R. Sheeley Jr., 1995, *Astrophys. J.*, **447**: L143–L146.

Watanabe, T., and T. Kakinuma, 1972, *Publ. Astron. Soc. Japan.*, **24**: 459–467.

Watanabe, H., M. Kojima, Y. Kozuka, and Y. Yamauchi, 1995, in *Solar Wind Eight*, AIP Conference Proceedings 382, AIP, New York, 1995, edited by D. Winterhalter, J.T. Gosling, S.R. Habbal, W.S. Kurth, and M. Neugebauer, 117–120.

Woch, J., W. I. Axford, U. Mall, B. Wilken, S. Livi, J. Geiss, G. Gloeckler, and R. J. Forsyth, 1997, *Geophys. Res. Lett*, **24**: 2885–2888.

Chapter 7

COMPONENTS OF THE DYNAMICALLY COUPLED HELIOSPHERE:

New Pickup Ion and ACR Populations, Acceleration, and Magnetic Field Structures

Nathan Schwadron

Southwest Research Institute, San Antonio, TX, 78238

Abstract   The description of the heliosphere as a system has emerged recently due to a deeper understanding of the heliosphere's numerous interlinked components. The call for a system view is illustrated by several areas of discovery involving the sources and acceleration of energetic particles, the links between particle acceleration and magnetic field structure, and the dynamic solar sources of heliospheric magnetic field meso-scale and micro-scale structure. These research areas have seeded the discovery of the inner and outer pickup ion sources; the discovery of the outer source of anomalous cosmic rays; the discovery that nearly radial magnetic fields in co-rotating rarefaction regions and associated energetic particle dwells are caused by the motions of open magnetic field footpoints across coronal hole boundaries; the relevance of these footpoint motions for ubiquitous statistical acceleration in slow solar wind and for Favored Acceleration Locations at the Termination Shock (FALTS); and the Ulysses discoveries of multiple unplanned crossings of the distant cometary tails of Hyakutake and McNaught-Hartley. Here, we review these emerging areas of heliospheric research and discuss how they are leading to a deeper understanding of our intrinsically dynamic and interlinked heliosphere.

1.   Introduction

Both remote and in-situ observations reveal that the Sun's magnetic fields are inherently dynamic, which on heliospheric scales, causes continuous reorganization of the open magnetic fields that are carried out by the solar wind to fill the heliosphere. In more precise terms, the footpoints of open magnetic fieldlines most likely execute organized and

G. Poletto and S.T. Suess (eds.), The Sun and The Heliosphere as an Integrated System, 179–199.
© 2004 *Kluwer Academic Publishers. Printed in the Netherlands.*

systematic motions over the surface of the Sun (Fisk, 1996; Fisk et al., 1999; Fisk and Schwadron, 2001). As we will discuss in sections 4, 5 and 6, these organized footpoint motions coupled with the intrinsic speed gradients of the solar wind have important implications for the structure of the heliospheric magnetic field and for particle acceleration throughout the heliosphere.

Anomalous Cosmic Rays (ACRs) are among the most energetic particles generated within the heliosphere. For years they have thought to arise only from neutral atoms in the interstellar medium (Fisk et al., 1974) that drift freely into the heliosphere through a process that has four essential steps: first, there is a source of neutral particles, traditionally thought to be only interstellar neutral atoms that stream into the heliosphere (see sections 2, 3 and 4); second, the neutrals are converted into ions, called pickup ions since they are picked up and swept out by the solar wind; third, pickup ions are pre-accelerated by shocks and waves in the solar wind (see also sections 5, Schwadron et al., 1996); and finally, they are accelerated to their final energies at the termination shock (Pesses et al., 1981). This last step cannot occur by diffusive acceleration alone (see section 7); there is a lower energy threshold, or injection energy, above which particles are accelerated through diffusive acceleration, and below which the processes responsible for particle acceleration are a matter of debate. Easily ionized elements such as C, Si, and Fe are expected to be strongly depleted in ACRs since such elements are not neutral in the interstellar medium and therefore cannot drift into the heliosphere.

Due to instruments like SWICS on Ulysses, we have been able to detect pickup ions directly, and due to ongoing measurements by cosmic ray detectors on spacecraft such as Voyager and Wind, researchers have been able to detect non-traditional components ACRs (Reames, 1999; Mazur et al., 2000; Cummings et al., 2002). There is a growing understanding that, in addition to the traditional interstellar source, grains produce pickup ions throughout the heliosphere: grains near the Sun produce an "inner source" of pickup ions (section 2), and grains from the Kuiper Belt provide an "outer" source of pickup ions and anomalous cosmic rays (section 3).

Comets are also a source of pickup ions the heliosphere. Until recently it was thought that the only way to observe cometary pickup ions is to send spacecraft to sample cometary matter directly. However, Ulysses has now had two unplanned crossings of distant cometary tails (Gloeckler et al., 2000; Gloeckler et al., 2004). Comets were formed during the birth of our solar system and provide important primordial samples of matter. The two unplanned crossings of cometary tails by Ulysses sug-

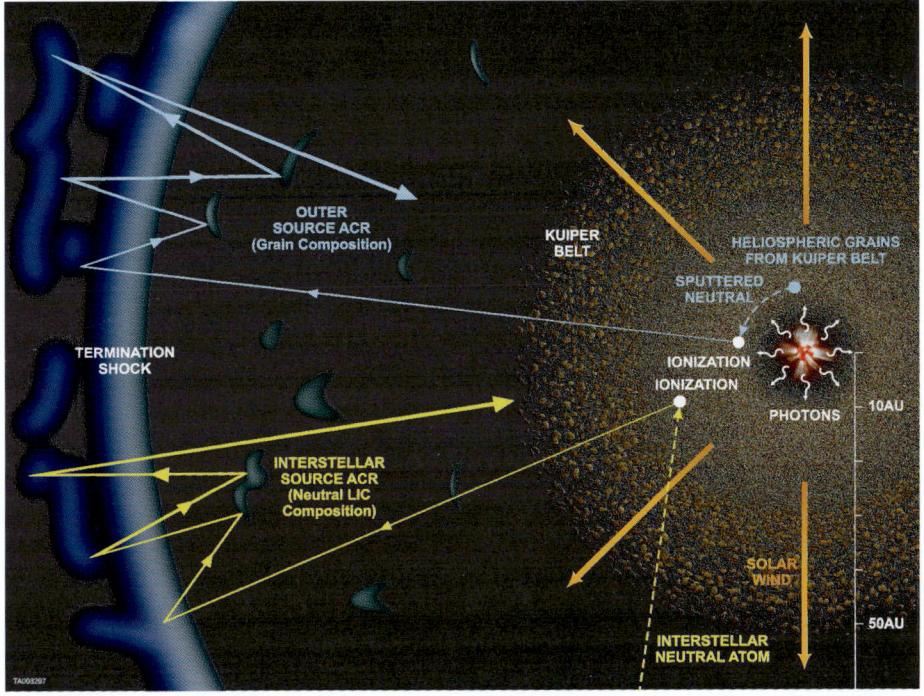

Figure 7.1. An illustration of ACR production (Schwadron et al., 2002). Yellow curves apply to the known interstellar source ACRs (adapted from Jokipii and Mc-Donald, 1995), while blue curves apply to the outer source, described later in the paper.

gest an entirely new way to observe comets through detection of distant cometary tails.

2.   Inner Source of Pickup Ions

The tell-tale signature of interstellar pickup ions, as seen for H^+ in Figure 7.2, is a cutoff in the distribution at ion speeds twice that of the solar wind in the spacecraft reference frame. This cutoff is exactly what we would expect since interstellar ions are initially nearly stationary in the spacecraft reference frame and subsequently change direction, but not energy, in the solar wind reference frame due to wave-particle interactions and gyration, thereby causing them to be distributed over a range of speeds between zero and twice the solar wind speed in the spacecraft frame.

By comparison to the interstellar pickup ion distributions, the C^+ and O^+ distributions in Figure 7.2 are puzzling. There is no cutoff at

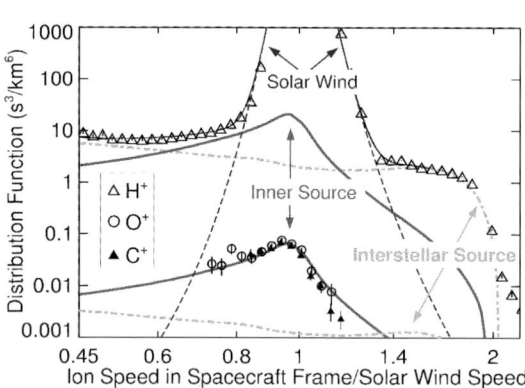

Figure 7.2 Observed distributions from Ulysses/SWICS of C^+ (solid triangles), O^+ (open circles), and H^+ (open triangles) vs. ion speed in the spacecraft frame normalized by the solar wind speed (Schwadron et al., 2000). The observations are compared with simulated distributions of solar wind H^+ (dashed black curve), interstellar H^+ (upper grey dash-dot curve), inner source H^+ (upper thick black line), inner source C^+, O^+ (lower thick black line), and interstellar O^+ (lower grey dash-dot curve).

twice the solar wind speed, and as opposed to being flat for ion speeds near the solar wind speed, there is a peak. The fact that the ions are singly charged precludes them from having been emitted directly by the Sun; if this were the case, they would be much more highly charged, e.g., C^{5+} and O^{6+} each common solar wind heavy ions. The C^+ and O^+ distributions in Figure 7.2 are consistent with a pickup ion source close to the Sun. As the ions travel out in the solar wind, they cool adiabatically in the solar wind reference frame, and as opposed to having a distribution that cuts off at twice the solar wind speed in the spacecraft reference frame, they have a peak near the solar wind speed. The solid curve that goes through the C^+ and O^+ data points is based on a transport model (Schwadron, 1998; Schwadron et al., 2000) for ions picked up close to the Sun, thereby proving the source.

Based on data like those presented in Figure 7.2, Geiss et al., 1995, Gloeckler and Geiss, 1998, Gloeckler et al., 2000 and Schwadron et al., 2000 concluded that there was evidence for a pickup ion source in addition to interstellar neutrals. Since the source was peaked near the Sun like interplanetary grains, it was concluded that grains were associated with the source.

Based on a grain source, a naive expectation was that the inner source composition would resemble that of grains, i.e., enhancements of carbon and oxygen and strong depletions of noble elements such as neon were expected. However, the composition strongly resembled that of the solar wind (Gloeckler et al., 2000). This conundrum was resolved by assuming a production mechanism whereby solar wind ions become embedded within grains and subsequently reemitted as neutrals. Remarkably, the existence of the inner source was hypothesized (Banks, 1971) many years prior to its discovery.

The concept that the inner source is generated due to neutralization of solar wind requires that sputtered atoms do not strongly contribute to the source. However, for grains larger than a micron or so, sputtering yields are much larger than the yields of neutralized solar wind (Wimmer-Schweingruber and Bochsler, 2003). This suggests that the grains that give rise to the inner source are extremely small (hundreds of angstroms). A small grain population generated through catastrophic collisions of larger interplanetary grains would also yield a very large filling factor[1] which is consistent with observations of the inner source that require a net geometric cross-section typically 100 times larger than that inferred from zodiacal light observations (Schwadron et al., 2000). The very small grains act effectively as ultrathin foils for neutralizing solar wind (e.g., Funsten et al., 1993), but are not effective for scattering light owing to their very small size. The implications of this suggestion are striking: the inner source may come from a large population of very small grains near the Sun generated through catastrophic collisions of larger interplanetary grains (Wimmer-Schweingruber and Bochsler, 2003).

3.   Distant Cometary Tails

Direct sampling (von Rosenvinge et al., 1986; Nature, 1986) and remote sensing observations (Crovisier and Bockelée-Morvan, 1999; Huebner and Benkhoff, 1999) of material from a few comets have established the characteristic composition of cometary gas. Direct detection of cometary matter requires spacecraft on trajectories that take them close to targeted comets (von Rosenvinge et al., 1986; Nature, 1986). Two unplanned crossings of cometary tails by Ulysses (Gloeckler et al., 2000; Gloeckler et al., 2004) have shown that it is also possible to directly sample cometary ions with spacecraft that are not only far away but also at relatively large angular separation from the comet.

As a comet approaches the Sun it emits at an increasing rate volatile material consisting mainly of water group molecules. This neutral gas, evaporated from the comet's surface, moves out in roughly all directions at typical speeds of ~ 1 km/s and is then quickly dissociated and ionized by solar radiation and the solar wind. Immediately after being ionized the predominantly singly-charged cometary pickup ions are picked up

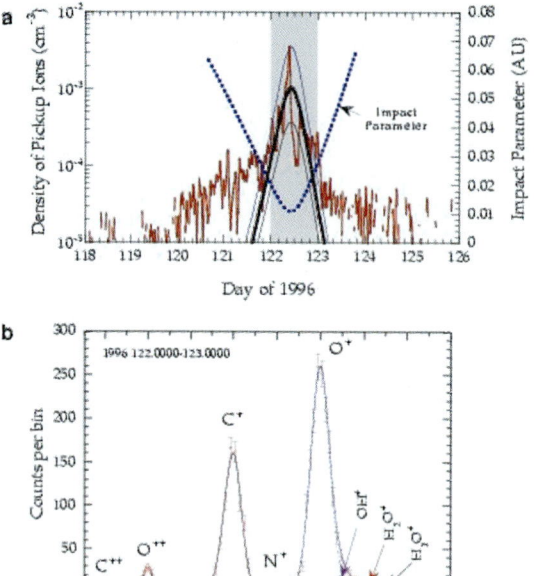

Figure 7.3 (a) Pickup O^+ density observed during the Hyakutake event shown in red. The blue dashed curve represents the distance of Ulysses from the Sun-comet line. The three black curves represent pickup ion distributions from a simple propagation model. (b) Abundances of various pickup ion species are derived from fitting counts binned according to the pickup ion mass-per-charge. These figures are reproduced from Gloeckler et al., 2000.

and swept away from the Sun by the solar wind, forming a thin, long ion tail that extends radially outward to distances of at least several AU. Pickup ions have a characteristic velocity spectrum with a sharp drop in density at a well defined cutoff speed which is related to the radial distance from where the gas was first ionized to the location where the pickup ions were observed.

The first unplanned comet tail crossing occurred in 1996 when comet Hyakutake was at 0.35 AU and Ulysses was at 3.7 AU (Gloeckler et al., 2000). In this case there was a nearly precise radial alignment between the comet and Ulysses. Figure 7.3 shows measurements from the SWICS instrument on Ulysses. The top panel shows the density of pickup O^+ as a function of the distance between Ulysses and the Sun-comet line. The second panel shows abundances inferred from the pickup ion observations.

The second unexpected comet tail crossing by Ulysses (Gloeckler et al., 2004) involved detection of ions from comet C/1999 T1, named McNaught-Hartley (IAU Circ., 1999, 7273), and perhaps comet C/2000 S5 (SOHO). Unlike the case of comet Hyakutake (Gloeckler et al., 2000; Jones et al., 2000), these comets were at large angular separations from the Ulysses spacecraft. The positions of Ulysses and comet McNaught-Hartley relative to the Sun and orbit of Earth are shown in Figure 10.6. While both Ulysses and McNaught-Hartley had almost the same lati-

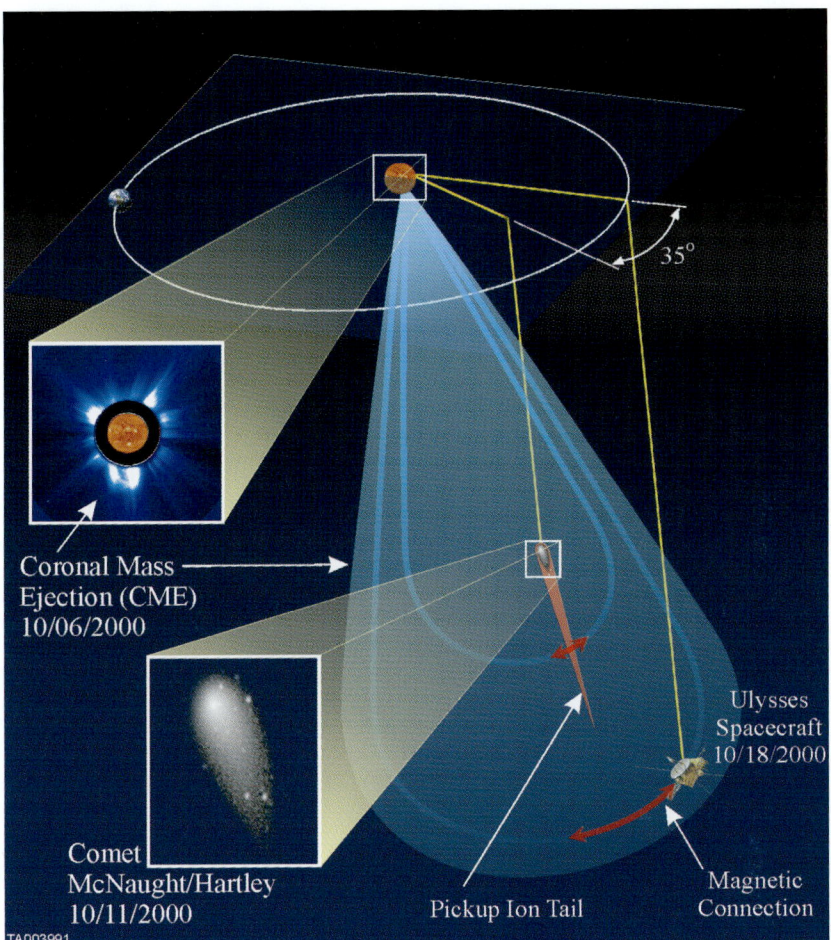

Figure 7.4.   Relative positions of the Sun, Ulysses spacecraft, and comet McNaught-Hartley. The orbit of Earth is shown for reference. A Coronal Mass Ejection erupted from the Sun on 2000 October 6, (SOHO LASCO C2 and 195A EIT images of release). The coronal mass ejection, represented by the light-blue, drop-shaped region, crossed the pickup ion tail of comet McNaught-Hartley shown in red on 2000 October 14. The CME reached the Ulysses spacecraft on 2000 October 18. The red arrowed arcs indicate the diffusive spread of the pickup ions along a magnetic flux tube (blue loop) as it is dragged by the solar wind first intercepting the comet close to its nucleus and then, about 4.5 days later, the Ulysses spacecraft. Photograph of comet McNaught-Hartley is courtesy of Maurice Clark. This figure is reproduced from Gloeckler et al., 2004

tude ($\sim 70°$ south) they were separated in longitude by $\sim 35°$ as shown in the figure. The minimum distance of McNaught-Hartley (at 1.492 AU from the Sun) to a line connecting the Sun and Ulysses' position (at 2.55

AU) ~ 4.5 days later (the impact parameter) reached its lowest value of ~ 0.318 AU or 4.77×10^7 km on day 293, when the maximum rate of O^+ was observed. The detection of these cometary ions was possible due to a coronal mass ejection (CME) that mixed and distorted the heliospheric magnetic field sufficiently to guide pickup ions produced near ($\lesssim 1$ million km) the comets to the Ulysses spacecraft located more than 150 million km from the comets. The ability of CMEs to carry cometary ions far from their radial paths significantly increases the probability of detecting these ions.

In many respects, the first unplanned comet-tail crossing by Ulysses was seen as an anomaly. However, the second crossing shows that chance detection of comet tails is more likely than we thought. Clearly, the presence of a CME increases the odds of an event. More importantly, however, these events suggest that a mission equipped with improved pickup ion instrumentation (e.g., with factor ~ 100 times the observing power of Ulysses/SWICS) would be capable of observing many cometary tails, and may significantly enhance our knowledge of cometary composition. It would also enhance the prospects of detecting pickup ions from a sungrazing comet destroyed near the Sun. The observation of such an event may be very valuable since the observed pickup ions would provide a measure of the net compositional inventory of the destroyed comet.

4.   Outer Source of Pickup Ions and Anomalous Cosmic Rays

Recent observations from the Voyager and Wind spacecraft have resolved ACR components comprised of easily ionized elements (such as Si, C, Mg, S, and Fe; Reames, 1999; Mazur et al., 2000; Cummings et al., 2002). An interstellar source for these "additional" ACRs, other than a possible interstellar contribution to C, is not possible (Cummings et al., 2002). Thus, the source for these ACRs must reside within the heliosphere. There are a number of potential ACRs sources within the heliosphere (see Schwadron et al., 2002). The solar wind particles are easily ruled out as a source since they are highly ionized, whereas ACRs are predominantly singly charged. Discrete sources such as planets are also easily ruled out since their source rate is not sufficient to generate the needed amounts of pickup ions. Another potential source is from comets. The net cometary source rate is sufficiently large only inside 1.5 AU, a location so close to the Sun that the generated pickup ions are likely to be strongly cooled by the time they reach the termination shock, making injection into diffusive shock acceleration extremely difficult. Moreover, a cometary source would naturally be rich in C, which is not consistent with the compositional observations of easily ionized ACRs.

The inner source (discussed in the preceding section) is another possibility, but again, adiabatic cooling poses a significant problem since these particles are picked up so close to the Sun. It has been suggested that the inner source may be substantially pre-accelerated in the inner heliosphere, thereby overcoming the effects of adiabatic cooling. Contradicting this suggestion, however, inner source observations in slow solar wind show clearly the pronounced effects of adiabatic cooling (Schwadron et al., 1999) and charge-state observations of energetic particles near 1 AU indicate no evidence for acceleration of the inner source (Mazur et al., 2002). *It appears that the additional population of ACRs requires a large source of pickup ions inside the heliosphere that is produced beyond 1 AU.*

Schwadron et al., 2002 suggest that there exists a strong outer heliospheric source of pickup ions that explains the presence of easily ionized ACRs. The source is material extracted from the Kuiper belt through a series of processes (shown schematically by the blue lines in Figure 7.1): First, micron-sized grains are produced due to collisions of objects within the Kuiper Belt; grains spiral in toward the Sun due to the Poynting-Robertson effect; neutral atoms are produced by sputtering and are converted into pickup ions when they become ionized; the pickup ions are transported by the solar wind to the termination shock and, as they are convected, are pre-accelerated due to interaction with shocks and due to wave-particle interactions; finally, they are injected into an acceleration process at the termination shock to achieve ACR energies. The predicted abundances are all within a factor of two of observed values, providing strong validation of this scenario. See Schwadron et al., 2002 for a detailed presentation of these points.

5. Ubiquitous Statistical Acceleration

Particles with energies of order 1 MeV per nucleon can be accelerated at the forward and reverse shocks which surround Co-rotating Interaction Regions (CIR's), where high and low speed streams interact in the solar wind (e.g., Barnes and Simpson, 1976; McDonald et al., 1976; Fisk and Lee, 1980). Gloeckler et al., 1994 and Schwadron et al., 1996 have suggested a two step acceleration process is called for. In the first step, pickup ions born with random speeds up to twice that of the solar wind are accelerated by transit-time damping of fluctuations in the magnetic field magnitude within CIRs, achieving random speeds between four and six times that of the solar wind. In the second step, these already

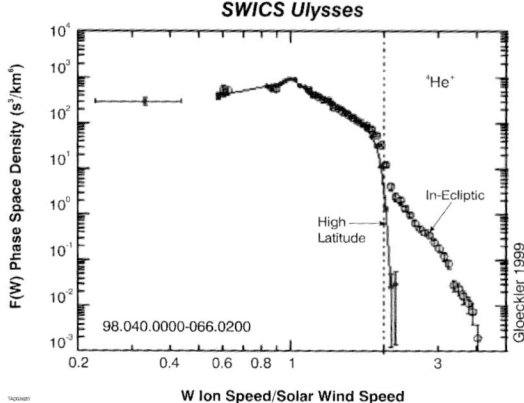

Figure 7.5 Distributions of interstellar pickup He$^+$ in high-latitude fast solar wind (black) and in-ecliptic slower solar wind (grey) observed by Ulysses/SWICS (Gloeckler, 1999).

pre-accelerated ions may efficiently move along magnetic field lines and are thereby injected into diffusive shock acceleration at the shocks that bound CIRs. Schwadron et al., 1996 found strong observational support for this two step acceleration process: a strong correlation was found between the presence of tails of accelerated pickup ions and the presence of fluctuations in the magnetic field magnitude that may be transit-time damped by pickup ions. Less than 10% of the pickup ion data with strong accelerated tails were within one day of a shock.

Gloeckler and Geiss, 1998 and Gloeckler, 2003 show the remarkable ubiquity (in slow solar wind) of suprathermal ion tails caused by statistical acceleration of pickup ions and solar wind ions (see Figure 7.5). Observations also show that these tails are much stronger in slow than in fast solar wind (Gloeckler et al., 1994; Collier et al., 1996; Gloeckler and Geiss, 1998; Chotoo et al., 2000; Gloeckler, 2003). These observations beg the question: *what are the properties intrinsic to slow solar wind that cause the ubiquitous statistical acceleration of solar wind ions and pickup ions?*

The answer may well be transit time damping. As already discussed, Schwadron et al., 1996 found a strong correlation between pickup ion tails and fluctuations in the magnetic field intensity that are the cause of transit time damping. Statistical acceleration through transit time damping was overviewed recently by Fisk et al., 2000. The close parallel was made to a magnetic pump which can be treated in a very straightforward manner.

Consider an ion that propagates along a field line in which the field magnitude varies. In the frame of reference of a compressive fluctuation (moving along the mean magnetic field) the ion's first adiabatic invariant is preserved. A fluctuation of field magnitude, with amplitude $(\delta|B|)/|B|$, results in a change of the perpendicular ion speed, $v_\perp$, given by $(\delta v_\perp)/v_\perp = (\delta|B|)/(2|B|)$. Since we are dealing here with compressive magnetosonic waves which propagate at an angle with respect to the mean magnetic field, the component of wave propagation parallel to the mean magnetic field with speed is $v_{\mathrm{ph}}\lambda_\parallel/\lambda_\perp$, where v_{ph} is the phase speed of the wave and $\lambda_\parallel$ and $\lambda_\perp$ are the parallel and perpendicular correlation lengths, respectively. Energy is conserved in the frame moving along the mean field with the fluctuation so that $v_\perp \delta v_\perp = -(v_\parallel - v_{\mathrm{ph}}\lambda_\parallel/\lambda_\perp)\delta v_\parallel$, where $v_\parallel$ is the ion speed parallel to the mean magnetic field in the plasma frame. This also yields an energy change in the plasma frame given by $\delta E = p\delta p/m = m\delta v_\parallel v_{\mathrm{ph}}\lambda_\parallel/\lambda_\perp$. Given also that the interaction time is $\delta t = \lambda_\parallel/|v_\parallel - v_{\mathrm{ph}}\lambda_\parallel/\lambda_\perp|$, the diffusion coefficient in momentum (or equivalently, speed) is

$$
\frac{D_{pp}}{p^2} \approx \left\langle \frac{(\delta p)^2}{p^2 \delta t} \right\rangle = \left(\frac{v_\perp}{v}\right)^4 \frac{\lambda_\parallel}{\lambda_\perp^2} \frac{v_{\mathrm{ph}}^2}{|v_\parallel - v_{\mathrm{ph}}\lambda_\parallel/\lambda_\perp|} \frac{\langle \eta^2 \rangle}{4} \tag{7.1}
$$

Detailed quasi-linear calculations of this effect were performed by, e.g., Fisk, 1976 and Schwadron et al., 1996, and Schwadron, 1996. The form for the coefficient of momentum diffusion depends critically on the ion gyroradius and on the nature of the compressive turbulence. Schwadron et al., 1996 found that they could account for observed pickup ion tails in slow solar wind if they assumed the turbulence was highly elongated (with parallel correlation lengths 30 times those of perpendicular correlation lengths). Correlation lengths perpendicular to the mean magnetic field of 0.01 AU were used and are typical of interplanetary turbulence.

6. Magnetic Footpoint Motions Through Speed Transitions and Resulting Particle Acceleration

Shearing of the magnetic field by the solar wind is inherent to the boundaries between faster and slower streams if there is an organized motion of open magnetic field footpoints on the Sun through these boundaries. More generally, shearing of the magnetic field may be an inherent property of slow solar wind, particularly on small spatial scales. The effect continuously transfers energy from the solar wind flow into compression and rarefaction regions. On small spatial scales, the effect helps to explain the general presence of compressive waves in slow solar wind (Schwadron, 2003). On large spatial scales, it suggests a new and different structure of the boundaries between fast and slow solar wind;

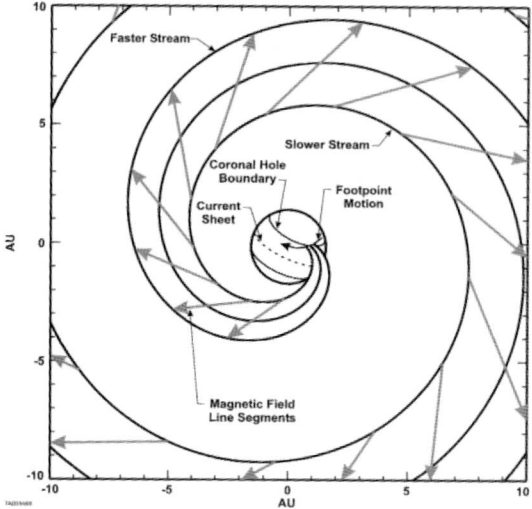

Figure 7.6 The orientation of the magnetic field in rarefaction regions under the influence of shearing and footpoint motion (Schwadron, 2002). Streamlines are shown by the solid spiral lines (black), and segments (grey) show the field orientation in the rarefaction regions. The speed gradient is given by a 300 km/s speed change over a 5 degree transition region.

structures that are quite favorable for particle acceleration, especially in the distant heliosphere (Schwadron and McComas, 2003).

Schwadron, 2002 considered the shearing effect in co-rotating rarefaction regions (CRRs) which are mapped out from the trailing edges of coronal holes. In CRRs, the heliospheric magnetic field is strongly underwound if drifts of open magnetic field footpoints through the coronal hole boundary are present. The underwinding is caused by solar wind shearing since the faster portions of the stream draw out the magnetic field more quickly than the slower portions (see Figure 7.6). The effect has been confirmed by observations of the heliospheric magnetic field (Murphy et al., 2002).

In the case of compression (rarefaction) regions, the shearing effect of solar wind speed gradients causes an increased (reduced) field magnitude. These magnitude changes are due to more than the mere expansion or compression of the plasma. This is exemplified by considering the magnetic field generated in heliospheric space when footpoints move across a *latitudinal* transition in the wind speed. Consider that the wind speed, V, is given as a function of co-latitude, θ: $V = V(\theta)$. In this case, the magnetic field can be solved exactly (Schwadron and McComas, 2003)

$$
\begin{aligned}
\mathbf{B} = B_{SS}\frac{R_{SS}^2}{r^2}\Bigg[&\left(1 - \frac{r\omega_\theta}{V^2}\frac{\partial V}{\partial \theta}\right)\hat{\mathbf{e}}_r \\
&- \frac{r\omega_\theta}{V}\hat{\mathbf{e}}_\theta - \frac{(\Omega_\odot + \omega_\phi)r\sin\theta}{V}\hat{\mathbf{e}}_\phi \Bigg]
\end{aligned}
$$

(7.2)

where ω_θ and ω_ϕ denote the angular rates (in a frame rotating with the Sun at the equatorial rotation rate) that footpoints move in co-latitude and longitude, and $\Omega_\odot$ is the equatorial rotation rate of the Sun. Here $\hat{\mathbf{e}}_r$, $\hat{\mathbf{e}}_\theta$, and $\hat{\mathbf{e}}_\phi$ are the unit vectors in the radial, co-latitude and azimuthal directions, respectively, and the heliocentric distance is r. The field strength B_{SS} is on the source surface at radial distance R_{SS} where the solar wind expansion first becomes radial. Note also the requirement that the footpoint motions are divergence free:

$$\frac{1}{\sin\theta}\frac{\partial}{\partial\theta}(\omega_\theta\sin\theta) + \frac{1}{\sin\theta}\frac{\partial\omega_\phi}{\partial\phi} = 0 \qquad (7.3)$$

The term in equation (7.2) caused by shearing, $-[B_{SS}R_{SS}^2\omega_\theta/(rV^2)]$ $(\partial V/\partial\theta)\hat{\mathbf{e}}_r$, leads to an additional component of the field in the radial direction which may either add to or subtract from the nominal radial component (depending on the direction of the speed gradient relative to the direction of footpoint motions). In this case, the plasma itself experiences no expansion or compression since the speed transitions occur over latitude. The fluctuations in magnitude arise purely from shearing.

The effects of shearing have been applied above to large-scale regions, but they apply on small spatial scales as well. As an example, we may consider a solar wind speed that varies on small spatial scales as a function of latitude in the presence of a large-scale footpoint drift in the latitudinal direction. Equation (7.2) may be used to solve for mean energy in compressive magnetic field fluctuations. The results show that modest 1% variations in solar wind speed lead to compressions of the field intensity that increase with distance from the Sun. By several AU, these intensity variations become larger than 10% of the mean field strength (Schwadron, 2003).

Small-scale speed fluctuations are observed to be larger in slow than in fast solar wind. This may be a natural consequence of forming slow solar wind from material stored and subsequently released by large coronal loops with highly variable properties (e.g., Schwadron et al., 1999; Fisk et al., 1999). This source for slow solar wind follows from the sporadic reconnection between open field lines and closed loops which is inherently tied to the dynamic organization and continuous reconfiguration of open magnetic flux on the Sun. Hence, the formation of slow solar wind from large loops and the global drift of open field footpoints naturally causes a state of slow solar wind in which shearing of the magnetic field (on small spatial scales) continuously feeds energy into fluctuations in the field strength – the required source of energy for the ubiquitous statistical acceleration of pickup ions observed in slow solar wind (e.g., Schwadron et al., 1996; Gloeckler and Geiss, 1998; Fisk et al., 2000; Gloeckler, 2003).

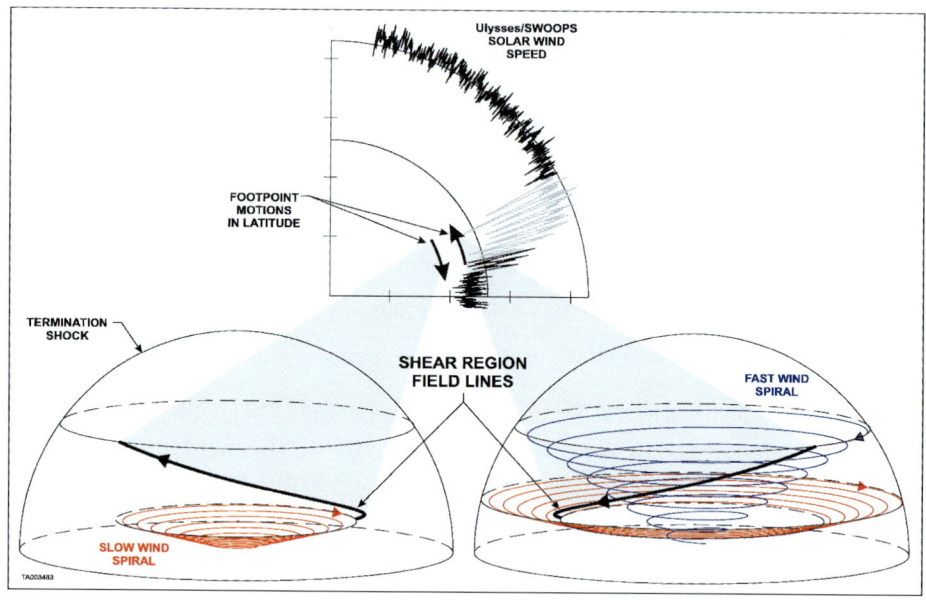

Figure 7.7. The three dimensional configurations of the magnetic fields in the distant interaction bands are indicated in the lower panels by black lines. The left (right) lower panel applies for footpoint motions from fast (slow) wind into the slow (fast) solar wind. The streamline in slow wind is indicated by the red curve, and the streamline in fast wind is indicated by the blue curve. The upper panel indicates the sense of footpoint motions and solar wind speeds measured by Ulysses as a function of latitude.

7.   FALTS

In order for pickup ions to undergo diffusive acceleration at the termination shock, they must propagate upstream against the solar wind flow, allowing them to travel back across the TS after being convected through it by the solar wind. This second crossing can be quite difficult since wrapping up of the interplanetary magnetic field spiral is thought to produce a quasi-perpendicular shock (the magnetic field is essentially perpendicular to the normal of the TS). There are regions where the field configuration is drastically non-spiral due to large-scale variations of the solar wind speed combined with magnetic footpoint motions back at the Sun (Schwadron, 2002; Schwadron and McComas, 2003). These regions naturally produce Favored Acceleration Locations at the Termination Shock - FALTS (Schwadron and McComas, 2003).

FALTS are generated by the coupled effects of large-scale latitudinal speed gradients of the solar wind coupled with motions of footpoints

back at the Sun between regions of fast and slow solar wind. Large-scale
latitudinal speed gradients of the solar wind are caused by the tilt of
the heliomagnetic axis with respect to the Sun's rotation axis, which
creates latitudinal bands in which fast and slow solar wind interact (
Burlaga, 1974; Hundhausen and Burlaga, 1975; Siscoe, 1976; Gosling
et al., 1978; Pizzo, 1989). Slow solar wind is emitted at low heliomagnetic latitudes and fast solar wind is emitted at higher heliomagnetic
latitudes. When the Sun's magnetic axis is tilted, interaction regions
form over the range of heliolatitudes where both fast and slow wind are
emitted. This structure rotates with the Sun, and at a fixed location the
solar wind varies between low and high speeds with each rotation. As
these streams propagate outward, the fast wind overtakes the slow wind,
forming compression regions and, typically within 2-3 AU of the Sun,
co-rotating shocks. These structures are called co-rotating interaction
regions or CIRs. Ulysses/SWOOPS observations compared the latitudinal bands of CIRs observed over Ulysses first orbit with the irregularly
structured wind observed approaching solar maximum (McComas et al.,
2000). During 1996, for example, these interaction regions extended
from about $10°$ to $30°$ latitude. As the CIRs move into the outer heliosphere, the fast and slow streams continue to interact until their speed
differences wear down, resulting in the formation of a large-scale bands
of intermediate speeds, referred to here as interaction bands. A large
distances, the speed in these interaction bands should change relatively
smoothly from a slow solar wind speed at the low latitude boundary to
a fast solar wind speed at the high latitude boundary. This monotonic
transition in speed is expected since fast solar wind is emitted during
a larger fraction of the rotation at high latitudes than at low latitudes
within the interaction band.

The second ingredient of FALTS is the motion of open magnetic footpoints between regions of fast and slow solar wind, for which there is
strong theoretical and growing observational support, as already discussed. As expressed by equation (7.2), the coupled effects of latitudinal
speed gradients and footpoint motions lead to a field configuration in
which a strong radial field component is generated by solar wind shearing. Figure 7.7 shows the three dimensional configuration of magnetic
field lines in the distant interaction bands (Schwadron and McComas,
2003). The field lines were drawn in this figure for an average value[2] of
$|\omega_\theta| = 0.15\Omega_s$, for an interaction region between a latitude $\lambda_s = 15°$ and
$\lambda_f = 35°$, and for solar wind speeds between 450 km/s and 750 km/s.
The upper panel in the figure shows a polar plot of solar wind speed as
a function of latitude observed by Ulysses. The bottom panels show the

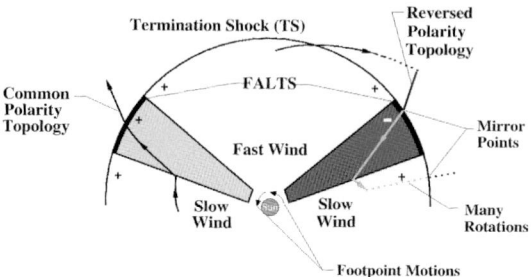

Figure 7.8. On the left (grey) is the FALTS topology when a footpoint moves from fast solar wind into slow solar wind leading to stretched fields in the interaction band with a polarity that agrees with the dominant polarity of the hemisphere. The line shows only the topology of this configuration, it does not represent a true 3D field line. On the right (dark grey) is the FALTS topology when footpoints move from the slow into fast solar wind. On the grey portion of this field line, the intersections with the TS act as converging mirror points, providing additional first-order Fermi acceleration.

3D field configurations in the interaction bands for two simple cases of footpoint motions: up and down in latitude.

The field configurations in the interaction bands have significant radial components. For the simplified case of a spherical TS, the normal is also directed radially, thus leading to significant field components parallel to the TS. Schwadron and McComas, 2003 show that these strong parallel components of the field significantly lower the energies required for particles to be injected into acceleration at the termination shock. These authors showed that the injection energy is lowered by almost two orders of magnitude at FALTS, and the flux of injected ions is increased by more than two orders of magnitude.

In addition to easier injection into diffusive shock acceleration along more radial fields, footpoint motions from slow into fast solar wind produce field configurations with local polarity inversions and introduce new large-scale current sheets in the outer heliosphere (Figure 7.8). This leads to severe modification of cosmic ray drifts in these region. This configuration also provides field lines that are connected to the termination shock at multiple locations. Particles on these field lines experience the termination shock intercepts as converging mirrors, and therefore experience additional first-order Fermi acceleration that may be stronger than diffusive acceleration (such topologies may also be associated on small-scales with turbulence, and may be important for accelerating low-energy particles).

8.   Summary

In this paper we have reviewed several important emerging areas in heliospheric physics:

- **The Inner Source** of pickup ions has been observed and appears to be caused by interactions between a very small grain population near the Sun and the solar wind (Geiss et al., 1995; Gloeckler and Geiss, 1998; Schwadron et al., 2000; Wimmer-Schweingruber and Bochsler, 2003). The discovery of the inner source suggests the important role played by grains in the production of heliospheric pickup ion populations.

- **Distant comet tails** have now been observed at least two times by the Ulysses: the first tail extended from comet Hyakutake (Gloeckler et al., 2000), and the second from comet McNaught-Hartley (Gloeckler et al., 2004). The odds of detecting distant comet tails are clearly much higher than once thought, suggesting that the properly designed pickup ion instrument may be able to sample and study cometary material through detection of tails from a variety of different comets. It also suggests the exciting possibility of detecting pickup ions from a sun-grazing comet destroyed during its approach close to the Sun; such pickup ions would provide a measure of the net compositional inventory of the destroyed sun-grazing comet.

- **The Outer Source** of pickup ions is generated due to sputtering of grains produced in the Kuiper Belt (Schwadron et al., 2002). The pickup ions are generated far from the Sun in regions where pre-acceleration within the heliosphere is natural (i.e., at corotating shocks and in the slow solar wind). Hence, outer source pickup ions are likely to have high energy tails at the termination and therefore may be efficiently injected into diffusive acceleration. The abundance and source location of outer source pickup ions suggest that they may contribute substantially to the anomalous cosmic rays, thereby explaining an additional population of easily ionized ACRs that cannot be accounted for with traditional heliospheric populations.

- **The dynamic nature of the Sun's open magnetic field** has far reaching effects in the heliosphere (Fisk, 1996; Fisk et al., 1999; Schwadron et al., 1999; Fisk and Schwadron, 2001; Schwadron, 2002; Schwadron and McComas, 2003). Footpoint motions across coronal hole boundaries produce nearly radial magnetic fields in co-rotating rarefaction regions which are related to energetic particle particle dwells (Schwadron, 2002). The motion of footpoints across random speed gradients in slow solar wind wind cause the development of strong magnetic field intensity variations and subsequent statistical acceleration (Schwadron, 2003).

- **Favored Acceleration Locations at the Termination Shock (FALTS)** are created by footpoint motions on the Sun coupled with large-scale solar wind speed variations as a function of heliolatitude (Schwadron and McComas, 2003). These coupled effects lead to the formation of highly distorted field structures with strong radial field components, providing relatively low speed ions with the ability to move back upstream in the solar wind. FALTS therefore allow for relatively low energy injection, and thus higher efficiency of injecting ions into diffusive shock acceleration at the TS. In addition, footpoint motions from slow into fast solar wind produce field configurations with local polarity inversions and introduce new large-scale current sheets in the outer heliosphere. This leads to severe modification of cosmic ray drifts in these regions. This configuration also provides field lines that are connected to the termination shock at multiple locations. Particles on these field lines experience the termination shock intercepts as converging mirrors, and therefore experience additional first-order Fermi acceleration.

The emergence of these new areas in heliospheric physics make the prospects of humankind's investigations of the inner and outer heliosphere richer than ever. Our Sun is capable of dynamism on even its largest scales, and this dynamism is essential to the particles and fields that comprise the system. There are important sources of energetic particles and anomalous cosmic rays both within and beyond our heliosphere. The heliosphere is a system composed of many interacting components. Comets, dust, pickup ions, energetic particles, the heliospheric magnetic field, and the evolution of magnetic fields at the Sun are intrinsically linked components of the heliospheric complex. As we resolve these links more clearly, the fabric of our heliosphere unfolds, revealing complexity and dynamism on all scales, but also an organization manifest in all its components.

Acknowledgments

This work was supported as a part of the SWOOPS effort by NASA's Ulysses program. NAS was also supported by a Southwest Research Institute Internal Research Grant and NSF (ATM-0244307).

Notes

1. the filling factor is the net area of the sky filled in by grains divided by the net area of the sky over which they are distributed; so, for example, a large solid object would have a filling factor of 1, whereas a finite set of points would have a filling factor of 0.

2. to satisfy the divergence-free condition for footpoint motions there is some small latitude dependence across the interaction band, $\omega_\theta \propto 1/\sin\theta$

References

Banks, P. M. (1971). *J. Geophys. Res.*, 76:4341.

Barnes, C. W. and Simpson, J. A. (1976). *Astrophys. J.*, 210:L91.

Burlaga, L. F. (1974). *J. Geophys. Res.*, 79:3717.

Chotoo, K., Schwadron, N. A., Mason, G. M., Zurbuchen, T. H., Gloeckler, G., Posner, A., Fisk, L. A., Galvin, A. B., Hamilton, D. C., and Collier, M. R. (2000). *J. Geophys. Res.*, 105:23107–23122.

Collier, M. R., Hamilton, D. C., Gloeckler, G., Bochsler, P., and Sheldon, R. B. (1996). *Geophys. Res. Lett.*, 23(10):1191–1194.

Crovisier, J. and Bockelée-Morvan, D. (1999). *Space Science Reviews*, 90:19–32.

Cummings, A., Stone, E., and Steenberg, C. (2002). *Astrophys. J.*, 578:194–210.

Fisk, L. A. (1976). *J. Geophys. Res.*, 81:4633–4640.

Fisk, L. A. (1996). *J. Geophys. Res.*, 101(A7):15547–15553.

Fisk, L. A., Gloeckler, G., Zurbuchen, T. H., and Schwadron, N. A. (2000). In Mewaldt et al., R. A., editor, *Acceleration and Transport of Energetic Particles Observed in the Heliosphere: ACE 2000 Symposium*, pages 229 – 233. AIP.

Fisk, L. A., Kozlovsky, B., and Ramaty, R. (1974). *Astrophys. J.*, 190:L35–L38.

Fisk, L. A. and Lee, M. (1980). *Astrophys. J.*, 237:620.

Fisk, L. A. and Schwadron, N. A. (2001). *Astrophys. J.*, 560:425.

Fisk, L. A., Zurbuchen, T. H., and Schwadron, N. A. (1999). *Astrophys. J.*, 521:868.

Funsten, H. O., Barraclough, B. L., and McComas, D. J. (1993). *Nuclear Instruments and Methods in Physics Research B*, 80/81:49–52.

Geiss, J., Gloeckler, G., Fisk, L. A., and von Steiger, R. (1995). *J. Geophys. Res.*, 100:23,373.

Gloeckler, G. (1999). *Space Science Reviews*, 89:91–104.

Gloeckler, G. (2003). In Velli, M., Bruno, R., and Malara, F., editors, *Solar Wind Ten: Proceedings of the Tenth International Solar Wind Conference*, page 583. AIP.

Gloeckler, G., Allegrini, F., Elliott, H. A., McComas, D. J., Schwadron, N. A., Geiss, J., von Steiger, R., and Jones, G. H. (2004). *Astrophys. J. Lett.*, 604:L121–L124.

Gloeckler, G., Fisk, L. A., Geiss, J., Schwadron, N. A., and Zurbuchen, T. H. (2000). *J. Geophys. Res.*, 105(A4):7459.

Gloeckler, G. and Geiss, J. (1998). *Space Sci. Rev.*, 86:127.

Gloeckler, G., Geiss, J., Roelof, E. C., Fisk, L. A., Ipavich, F. M., Ogilvie, K. W., Lanzerotti, L. J., von Steiger, R., and Wilken, B. (1994). *J. Geophys. Res.*, 99:17637.

Gloeckler, G., Geiss, J., Schwadron, N. A., Fisk, L. A., Zurbuchen, T. H., Ipavich, F. M., von Steiger, R., Balsiger, H., and Wilken, B. (2000). *Nature*, 404:576–578.

Gosling, J. T., Asbridge, J. R., Bame, S. J., and Feldman, W. C. (1978). *J. Geophys. Res.*, 83:1401–1412.

Huebner, W. F. and Benkhoff, J. (1999). *Space Science Reviews*, 90:117–130.

Hundhausen, A. J. and Burlaga, L. F. (1975). *J. Geophys. Res.*, 80:1845–1848.

Jokipii, J. R. and McDonald, F. B. (1995). *Scientific American*, 272:58.

Jones, G. H., Balogh, A., and Horbury, T. S. (2000). *Nature*, 404:574–576.

Mazur, J., Mason, G., Blake, J., Lkecker, B., Leske, R., Looper, M., and Mewaldt, R. (2000). *J. Geophys. Res.*, 105:21015–21023.

Mazur, J., Mason, G., and Mewaldt, R. (2002). *Astrophys. J.*, 566:555–561.

McComas, D. J., Gosling, J. T., and Skoug, R. M. (2000). *Geophys. Res. Lett.*, 27:2437.

McDonald, F. B., Teegarten, B. J., Trainor, J. H., von Rosenvinge, T. T., and Webber, W. R. (1976). *Astrophys. J. Lett.*, 203:L149.

Murphy, N., Smith, E. J., and Schwadron, N. A. (2002). *Geophys. Res. Lett.*, 29:23–1.

Nature (1986). "*in Nature (Suppl.)*", 321:259.

Pesses, M. E., Jokipii, J. R., and Eichler, D. (1981). *Astrophys. J.*, 246:L85.

Pizzo, V. J. (1989). *J. Geophys. Res.*, 94(A7):8673–8684.

Reames, D. V. (1999). *Astrophys. J.*, 518(1):473–479.

Schwadron, N. A. (1996). PhD thesis, U. Michigan.

Schwadron, N. A. (1998). *J. Geophys. Res.*, 103:20643.

Schwadron, N. A. (2002). *Geophys. Res. Lett.*, 29(14):doi:10.1029/2002GL015028.

Schwadron, N. A. (2003). In Velli, M., Bruno, R., and Malara, F., editors, *Solar Wind Ten: Proceedings of the Tenth International Solar Wind Conference*, volume CP679, pages 593–596. AIP.

Schwadron, N. A., Combi, M., Huebner, W., and McComas, D. J. (2002). *Geophys. Res. Lett.*, 29(20):`doi:10.1029/2002GL015829`.

Schwadron, N. A., Fisk, L. A., and Gloeckler, G. (1996). *Geophys. Res. Lett.*, 23:2871–2874.

Schwadron, N. A., Fisk, L. A., and Zurbuchen, T. H. (1999). *Astrophys. J.*, 521:859.

Schwadron, N. A., Geiss, J., Fisk, L. A., Gloeckler, G., Zurbuchen, T. H., and von Steiger, R. (2000). *J. Geophys. Res.*, 105(A4):7465.

Schwadron, N. A., Gloeckler, G., Fisk, L. A., Geiss, J., and Zurbuchen, T. H. (1999). In *AIP Conf. Proc. 471: Solar Wind Nine*, page 487.

Schwadron, N. A. and McComas, D. J. (2003). *Geophys. Res. Lett.*, 30: `doi:10.1029/2002GL016499`.

Siscoe, G. L. (1976). *J. Geophys. Res.*, 81:6235–6241.

von Rosenvinge, T. T., Brandt, J. C., and Farquhar, R. W. (1986). *Science*, 232:353–356.

Wimmer-Schweingruber, R. F. and Bochsler, P. (2003). *Geophys. Res. Lett.*, 30:49–1.

Chapter 8

A GLOBAL PICTURE OF CMES IN THE INNER HELIOSPHERE

N. Gopalswamy

Laboratory for Extraterrestrial Physics, NASA/GSFC, Greenbelt, MD 20771, USA

Abstract   This is an overview of Coronal mass ejections (CMEs) in the heliosphere with an observational bias towards remote sensing by coronagraphs. Particular emphasis will be placed on the results from the Solar and Heliospheric Observatory (SOHO) mission which has produced high quality CME data uniform and continuos over the longest stretch ever. After summarizing the morphological, physical, and statistical properties of CMEs, a discussion on the phenomena associated with them is presented. These are the various manifestations of CMEs observed at different wavelengths and the accompanying phenomena such as shocks and solar energetic particles that provide information to build a complete picture of CMEs. Implications of CMEs for the evolution of the global solar magnetic field are presented. CMEs in the heliosphere are then discussed including out-of-the-ecliptic observations from Ulysses and the possibility of a 22-year cycle of cosmic ray modulation by CMEs. After outlining some of the outstanding questions, a summary of the chapter is provided.

1.   Introduction

The white-light coronagraph on board NASA's seventh Orbiting Solar Observatory (OSO-7) detected the first "modern" coronal mass ejection (CME) on December 14, 1971 (Tousey, 1973). Just over an year before this detection, Hansen et al. (1971) observed the "rapid decay of the transient coronal condensation" using the Mauna Loa Coronal Activity Monitor during 1970 August 11-12, which is essentially a CME detection. They had also found temporal and spatial association of fast (1000 km s^{-1}) radio sources with the white-light transient feature. In fact, the concept of mass ejections existed as prominence eruptions (ac-

G. Poletto and S.T. Suess (eds.), The Sun and The Heliosphere as an Integrated System, 201–251.

tive and eruptive) since the first scientific observations of Secchi and de la Rue in the late 1800's (see, e.g., Tandberg-Hanssen, 1995): We now know that eruptive prominences form the inner core of many CMEs (see, e. g., House et al., 1981). Mass motions with speeds in the range 500-840 km s^{-1} were inferred from type II radio bursts (Payne-Scott et al., 1947). Moving type IV bursts, indicative of moving magnetized plasma structures in the corona with speeds of several hundred km s^{-1}, were discovered long ago (Boischot, 1957). Slow (< 10 km s^{-1}) and fast (($>$ 100 km s^{-1}) coronal green line transients were also known before the discovery of CMEs (DeMastus et al., 1973). At least two CMEs have been identified in eclipse pictures: during the Spanish eclipse on 1860 July 18 (see Eddy, 1974) and during the Indian eclipse on 1980 February 16 (Rusin et al., 1983). The concept of mass ejection from the Sun was very much in use for explaining geomagnetic storms (Lindemann, 1919). The idea that these plasma ejections might drive shocks (Gold, 1955) was soon confirmed by in situ observations (Sonett, 1964; Gosling, et al., 1968). Interplanetary disturbances were estimated to have a mass of 10^{16} g and an energy of 10^{32} erg (Hundhausen et al., 1970), which we now know are typical of CMEs.

Given the rapid explosion of knowledge on CMEs over the past four decades, it is impossible to review all the published material here. However, complementary reviews include Wagner (1984); Schwenn (1986), Hundhausen (1987), Kahler (1987), Gosling (1997), Howard et al. (1997), Low (1997), Hundhausen (1999), Webb (2002), St. Cyr et al. (2000), Gopalswamy et al. (2003b). In this chapter, we provide an overview of the new developments in CME research, drawing heavily on the results from the Solar and Heliospheric Observatory (SOHO) mission, which has made a significant impact on our current understanding of CMEs. Some of the results to be discussed in this chapter are: (i) Basic statistical properties of CMEs and their solar cycle variation, (ii) special populations such as halo and fast and wide CMEs, (iii) acceleration and deceleration CMEs in the inner heliosphere, (iv) CME-associated eruptive activities, (v) CME-CME interaction, (vi) CMEs in the heliosphere, (vii) role of high-latitude CMEs in solar polar magnetic reversals, (viii) the role of CMEs in modulating the galactic cosmic rays, and (ix) outstanding questions.

2. Solar Source of CMEs

From the early days of CME studies, it is known that CMEs are associated with flares and prominence eruptions (see, e.g. Munro et al., 1979). This means CMEs originate wherever flares and prominences occur. Flares occur in active regions, which contain high magnetic field with or without sunspots. Active regions consisting of sunspots of opposite polarity seem to produce the most energetic CMEs. Regions on the solar surface where cool prominences are suspended in the corona also contain closed magnetic field structures and they produce spectacular CMEs that carry the prominences out into the interplanetary (IP) medium. Prominences also reside along neutral lines in active regions. Even tiny bipoles observed as bright points in X-rays contain closed field structure producing small jet-like ejections (Shibata et al., 1992), although these are not typically counted as CMEs. CMEs observed at 1 AU by multiple spacecraft have revealed that the "legs of the CME" are probably connected to the Sun, with their feet anchored on either side of the magnetic neutral lines (Burlaga et al., 1981). There was an alternative suggestion that CMEs originated from low-latitude coronal holes (Hewish et al., 1985), but now it is fully established that CMEs originate from closed magnetic field regions on the Sun (see, e.g. Harrison, 1990). However, filaments near coronal holes seem to have a proclivity for eruption (Webb et al., 1978; Bhatnagar, 1996), which suggests that such eruptions can be mistakenly associated with coronal holes. Closed magnetic structure, thus, seems to be the basic characteristic of CME-producing regions on the Sun, which means the energy needed to carry billions of tons of ionized plasma in to the heliosphere must ultimately come from the magnetic field itself. How this energy is stored in the coronal magnetic fields and what triggers the energy release are topics of current research and debate.

3.   CME Morphology

The general appearance of a CME is shown in Fig. 1. The earliest activity observed on the Sun was a prominence eruption observed in microwaves from the southeast quadrant of the Sun. The prominence eruption was also observed by the Extreme-ultraviolet imaging telescope (EIT,) on board SOHO. In running difference images, a faint depletion can be seen surrounding the prominence. There are two dimming regions (D), one on each side of the neutral line, that mark the pre-eruption location of the prominence. After the eruption, a post eruption arcade forms (denoted by AF) with its individual loops roughly perpendicular to the neutral line. The dimming regions are located just outside the arcade, but at the opposite ends of the arcade axis. "Coronal dimming" represents the reduction in brightness in a certain region of the corona as compared to an earlier period, typically on either side of the polarity inversion line underlying the CME (see Sterling, 2003 for a review).

Dimming is a change in the physical conditions (density and temperature) of the emitting plasma, typically observed in X-rays (Hudson, 1999), EUV (Gopalswamy and Thompson 2000) and occasionally in microwaves (Gopalswamy, 2003b).

The white-light CME first appears an hour later above the occulting disk of the Large Angle and Spectrometric Coronagraph (LASCO) in the same position angle as the eruptive prominence. The bright frontal structure is loop-shaped, inside of which there is a bright core. From the morphological, position angle, and temporal coincidences, it is clear that the core seen in white light is nothing but the prominence. The EUV and microwave data alone give a speed of ~ 97 km s^{-1} that became higher by the time the CME entered the LASCO field of view. The legs of the frontal structure are thought to extend below the occulting disk with the feet located on either side of AF. There is a conspicuous void that separates the prominence core and the frontal structure, commonly referred to as cavity containing less coronal material and strong magnetic field. The cavity is also thought to have a flux-rope magnetic structure with the legs of the rope anchored on either side of the neutral line. The core and the frontal structure was about 5 R$_\odot$ by the time the CME left LASCO FOV. The average speed of the CME was 770 km s^{-1}. This CME could be thought of as a typical three-part structure CME. The classical three-part structure (Hundhausen et al., 1988) is well observed only in CMEs that are associated with prominences erupting from quiet regions. When prominences erupt from active regions, it is often difficult to discern the three-part structure. Prominences in active regions are thin and low-lying and may be heated and ionized before arriving in the coronagraph field of view.

Figure 2 shows another CME, in which the three-part structure is not very clear. This CME originated from an active region slightly behind the southwest limb. The white light CME was highly structured, but not similar to the one in Fig. 1. The CME was very dense with a compact internal structure that moved behind the frontal structure. The frontal structure was also flat-topped. The front moved with speed of ~ 2500 km s^{-1}, while the inner core had a speed of 1500 km s^{-1}. The core was much smaller within the overall volume of the CME. The main body of the CME is seen distinct from the two streamer displacements on either side of the CME. These disturbances are also likely to be present away from the plane of the sky.

From the above examples one can infer that a white-light CME is highly structured and is three-dimensional. Stereoscopic observation of a few CMEs by the Helios photometer and the Solwind coronagraph essentially demonstrated the 3D nature of CMEs (Jackson, 1985), and

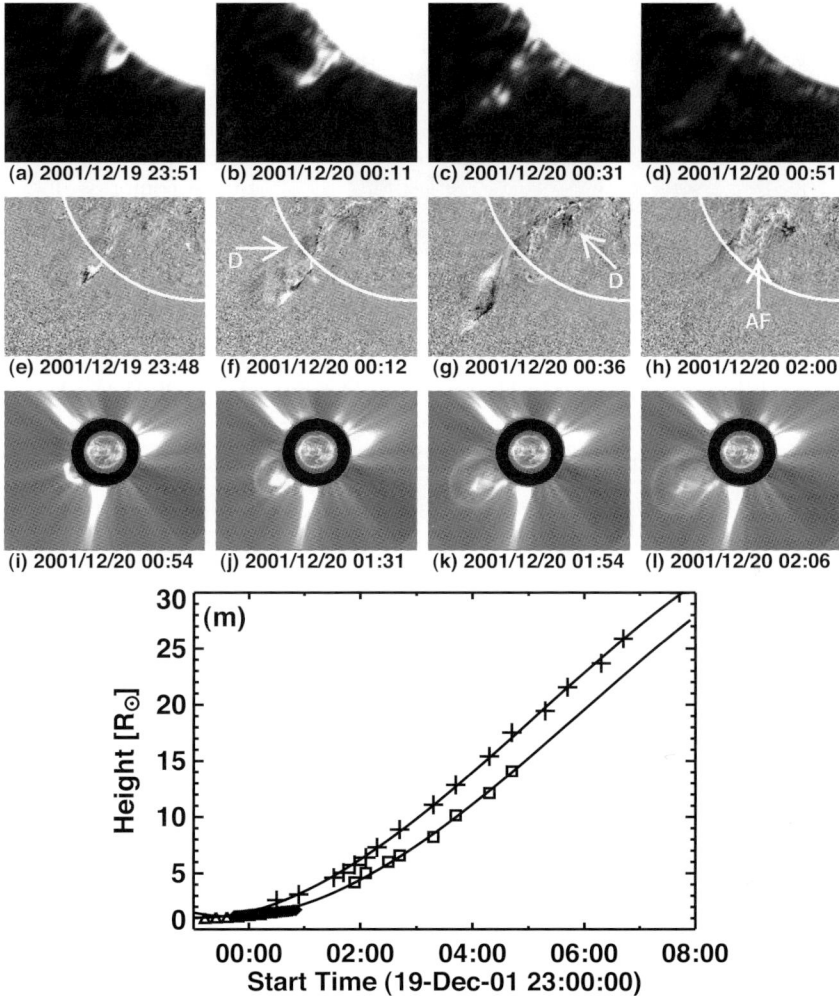

Figure 8.1. Morphology of a three-part CME and the associated solar surface activities: (a-d) prominence eruption in microwaves, (e-h) SOHO/EIT difference images showing the prominence eruption in EUV with dimming (D) and arcade formation (AF), (i-l): SOHO/LASCO images showing the core, void and frontal structure of the CME, and (m) height-time plots of the frontal structure ('plus' symbols, white-light) and the prominence from various sources (EUV -triangles; microwave - diamonds and white light -squares).

this was confirmed by numerical simulations (e. g., Crifo et al., 1983). LASCO has observed a number of different morphological types, which are yet to be surveyed and classified. Some CMEs are interpreted as flux ropes (Chen et al., 2000; Plunkett et al., 2000). Some CMEs have

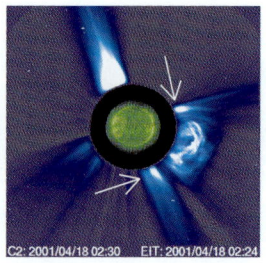

Figure 8.2 LASCO images of (left) the pre-event corona and (right) the 2001 April 18 CME. The pre-event corona can be seen on the left. Arrows point to the compressed regions of the streamers on either side of the CME.

voids with no prominence in them (Gopalswamy et al., 2001d). Jets and narrow CMEs with no resemblance to the three-part structure have also been observed (Wang and Sheeley, 2002; Yashiro et al., 2003).

4. Physical Properties

Since the material in CMEs is already present in the corona before ejection, we expect the CME to be at the coronal temperature. However, the core of the CME is prominence material and hence can be quite cool (4000 - 8,000 K). Not much is known about cavity, but is also thought to be at coronal temperatures. White light coronagraphs detect just the mass irrespective of the temperature. Non-coronagraphic observations are needed to infer temperatures. The magnetic field of the CMEs near the Sun is also unknown. Radio observations indicate a magnetic field strength of ≤ 1 G in the corona at a heliocentric distance of 1.5 $R_\odot$ (see, e.g., Dulk and McLean, 1978). Gyroresonance emission from active regions indicate that coronal magnetic fields above sunspots can be as high as 1800 G (White et al., 1991). When an eruption occurs in a strong field region, one might expect a strong field in the resulting CME. The field strength in the prominences are better known (Tandberg-Hanssen, 1995): 3-30 G in quiescent prominences and 20-70 G in active prominences, occasionally exceeding 100 G (Kim and Alexeyeva, 1994). The magnetic field in the cavity is virtually unknown. The idea that the cavity is a magnetic flux rope may have some support from the numerous dark threads observed in high resolution eclipse images (Engvold, 1997). The density in the inner corona is typically 10^{8-9} cm^{-3} and is expected to be present in the frontal structure of CMEs close to the Sun. Density estimates from white light observations (see e.g., Vourlidas et al., 2002) radio (Gopalswamy et al., 1993) and ultraviolet observations (Ciaravella et al., 2003) are consistent with such densities. The prominences are much denser (10^{10-11} cm^{-3}). The cavity is certainly of lower density compared to the frontal structure and prominence core.

Table 8.1. Summary of Space borne coronagraph observations of CMEs from OSO-7 (Tousey, 1973), Skylab (MacQueen et al., 1974), Solwind (Michels et al., 1980), SMM (MacQueen et al., 1980), and SOHO (Brueckner et al., 1995).

Coronagraph	OSO-7	Skylab	Solwind	SMM	LASCO
Epoch	1971	1973-74	1979-85	1980,84-89	1996-2003
FOV ($R_\odot$)	2.5-10	1.5 - 6	3 - 10	1.6 -6	1.2-32
# CMEs recorded	27	115	1607	1206	8008
Mean Speed (km/s)	-	470	460	350	489
Mean Width (deg.)	-	42	43	47	47
Mass (10^{15} g)	-	6.2	4.1	3.3	1.6

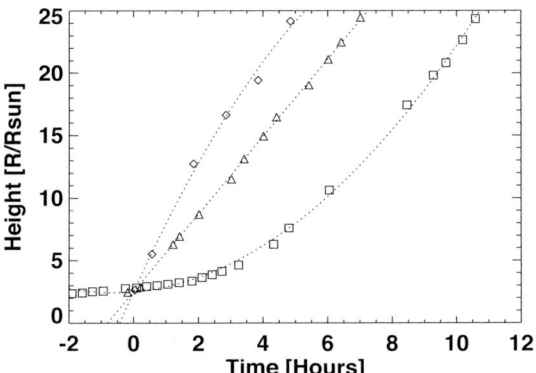

Figure 8.3 Height-time measurements of three representative CMEs observed by SOHO/LASCO: the accelerating CME of June 21, 1998 (squares), the constant speed CME of February 17, 2000 (triangles), and the decelerating CME of May 11, 1998 (diamonds). The curves are best-fit polynomials (linear for the constant speed case and quadratic for the other two). The plots are normalized to the time the CMEs reach 2.5 $R_\odot$. (See Gopalswamy et al., 2001e for more details).

5. Statistical Properties

The OSO-7 coronagraph detected only 27 CMEs over a period of 19.5 months. The Skylab ATM coronagraph recorded 110 CMEs during its 227 days of operation. The number shot up by an order of magnitude when the Solwind coronagraph on board P78-1 and the Coronagraph/Polarimeter on board the Solar Maximum Mission (SMM/CP) became operational. SOHO/LASCO has detected more than 8000 over a period of 8 years (1996-2003), confirming that CMEs are a common phenomenon. Table 1 summarizes these observations and updates a previous compilation by Hundhausen (1997).

5.1   CME Speed

Mass motion is the basic characteristic of CMEs, quantified by the speed. Coronagraphs obtain images with a certain time cadence, so when a CME occurs, the leading edge progressively appears at a greater heliocentric distance. By tracking a CME feature in successive frames, one can derive the speed of the feature. It must be pointed out that the height-time measurements are made in the sky plane so all the derived parameters such as speed are lower limits to the actual values. Figure 3 shows three examples of height-time (h-t) plots. A straight-line fit to the h-t measurements gives the average speed within the coronagraph field of view, but it may not be suitable for all CMEs. For studying the variation of speed, one has to use higher order fits. For SOHO/LASCO CMEs, the sky plane speed from linear fit ranges from tens of km s^{-1} to >2500 km s^{-1}, with an average value of 489 km s^{-1} (see Table 1 and Fig. 4). Skylab and P78-1 CMEs had similar average speeds, but the SMM value was relatively low (Hundhausen, 1997). The discrepancy may be due to poor data coverage and the inability to measure the speeds of many of the observed CMEs (Gopalswamy et al., 2003b). For similar reasons, the SMM data did not show a significant difference in the average speed of CMEs between solar activity minimum and maximum (Hundhausen, 1999), although other measurements did indicate a definite increase (Howard et al., 1985). SOHO data confirmed the increase beyond any doubt (Gopalswamy et al., 2003b) as demonstrated in Figure 5.

SOHO detected a number of CMEs with speeds exceeding 2000 km s^{-1} (Gopalswamy et al., 2003c). The largest speed (2657 km s^{-1}) observed was for the 2003 November 04 CME during the largest flare of cycle 23. These ultrafast CMEs constitute only a tiny fraction (25/8008) of the total number of CMEs, which suggests a possible upper limit to the energy that goes into mass motion in CMEs.

5.2   CME Acceleration

All CMEs have positive acceleration in the beginning as they lift off from rest (the propelling force (F_p) exceeds gravity (F_g) and other restraining forces). The moment a CME lifts off, it is subject to an additional retarding force - the drag, given by $F_d = CA\rho$—V_{cme}-V_{sw}—(V_{cme}-V_{sw}), where C is the drag coefficient (Chen, 1989; Cargill et al., 1996), A is the surface area of the CME, ρ is the plasma density, V_{cme} is the CME speed and V_{sw} is the solar wind speed (negligible close to the Sun). The three types of h-t profiles shown in Fig. 3 reflect various combinations of propelling and retarding forces: the accelerating profile

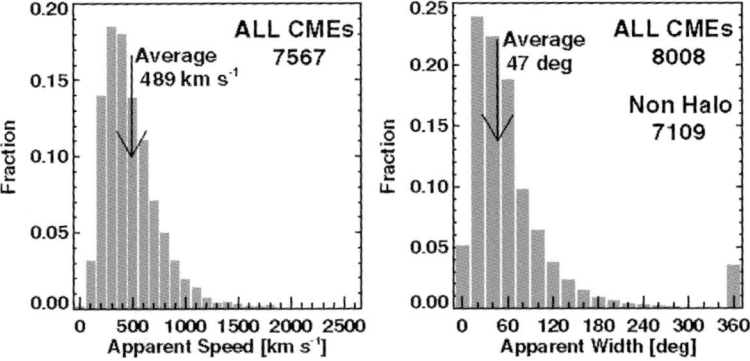

Figure 8.4. The speed (left) and width (right) distributions of all CMEs from 1996 to 2003. The width of a CME is measured as the angle subtended by the outer edges of the CME at the Sun center. The speed is obtained by straight-line fit to the height-time measurements. Even though 8008 CMEs were detected, the speed could be measured only for 7567 CMEs, giving an average speed of 489 km s^{-1}. The average width of 47° corresponds to the 7109 non-halo (width ≤120°) CMEs. Inclusion of all CMEs yields a width of 67°. The last bin in the width distribution contains the full halo CMEs, which constitute only ∼3.5% of all CMEs. The fraction of CMEs with width ≥120° is ∼11%. The speed and width are sky-plane projections and no attempt was made to correct for projection effects.

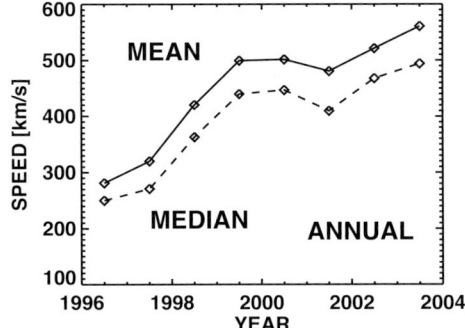

Figure 8.5 Annual mean and median speeds of SOHO/LASCO CMEs from 1996 to 2003 showing the clear increase towards solar activity maximum. Higher speeds prevailed even after the solar activity maximum.

indicates that the propelling force is still active in pushing the CME outward. The constant-speed and decelerating profiles suggest that the retarding forces either balance or exceed the propelling force. The average acceleration obtained from MLSO K-coronameter (FOV = 1.2 - 2.7 R$_\odot$) data is generally positive and high compared to those obtained from SMM (FOV = 1.8 - 5 R$_\odot$) and LASCO (FOV = 2 - 32 R$_\odot$) coronagraphs (Burkepile et al., 2002). Furthermore, combining data below the occulting disk with those from above clearly indicate that the ac-

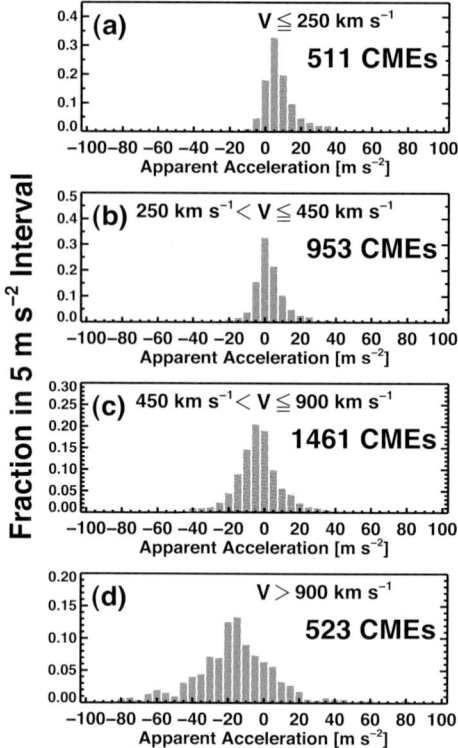

Figure 8.6 The average acceleration of CMEs (1996-2003) within the LASCO FOV for various speed ranges. Note the tendency for deceleration for faster CMEs.

celeration is variable (St. Cyr et al., 1999; Gopalswamy and Thompson, 2000; Wood et al., 1999; Zhang et al., 2001). Measurements of individual events give accelerations generally below a few km s^{-2}. Gopalswamy et al. (2001b) found that fast ($V > 900$ km s^{-1}) CMEs predominantly decelerated within LASCO FOV, suggesting that the deceleration is very general and must be due to drag. A number of recent studies suggest that the propelling forces fade out at heights below ~ 4 R$_\odot$ (Chen and Krall, 2003), so drag must play a significant role within LASCO FOV. Statistical analyses of the observed acceleration support this interpretation (Yashiro et al., 2004). Figure 6 shows the distribution of CME accelerations (a) for various speed ranges: (i) slow CMEs ($V_{cme} \leq 250$ km s^{-1}) are accelerated (median $a = 6$ m^{-2}), (ii) CMEs with speeds in the vicinity of solar wind speed (250 km s$^{-1} < V_{cme} \leq 450$ km s^{-1}) show little acceleration (median $a = 1.6$ m^{-2}), (iii) CMEs with speeds above the solar wind speed (450 km s$^{-1} < V_{cme} \leq 900$ km s^{-1}) show predominant deceleration (median $a = $ -4 m^{-2}), and the fast CMEs ($V > 900$ km s^{-1}) show clear deceleration (a = -16 m^{-2}). This behavior is also found when CME propagation is considered over the inner heliosphere (Gopalswamy et al., 2000a; Lyndsay et al., 1999).

5.3   CME Width

CME angular span (also referred to as CME width) is measured as the position angle extent in the sky plane. For CMEs originating from close to the limb, the measured width is likely to be the true width. For CMEs away from the limb, the measured width is likely to be an overestimate. Many CMEs show increase in width as they move out, so measurements are made when the width appears to approach a constant value. The average of the width distribution of SOHO/LASCO CMEs shown in Fig. 4 is 47° when we exclude CMEs with width > 120° (because they are unlikely to be actual widths). Annual averages of non-halo CME widths range from 47° to 61° (Yashiro et al., 2004); the average width is the smallest during solar minimum, peaks just before the maximum and then declines through the maximum. The average widths obtained from Skylab (42°), SMM (47°) and Solwind (43°) are remarkably similar and in good agreement with LASCO results (see Table 1). This is true only when we exclude CMEs with widths exceeding 120°, a population not present in significant numbers in pre-SOHO data. The average width is 67° when we include all CMEs (similarly to St. Cyr et al., 2000, who found a value of 72° during the rise phase of cycle 23).

5.4   CME Latitude

The latitude distribution of CMEs depends on how closed field regions are distributed on the solar surface. CME latitude is obtained from the central position angle of the CME, assuming that CMEs propagate radially away from the solar source region (Howard et al., 1986; Hundhausen 1993; Gopalswamy et al., 2003a). This assumption may not be always valid especially during the solar minimum periods when the CME trajectory is likely to be controlled by the global dipolar field of the Sun (Gopalswamy et al., 2000c). Figure 7 shows a plot of the CME latitude as a function of time along with the maximum excursions of the heliospheric current sheet (a good indicator of the presence of closed field structures at high latitudes) for CMEs associated with prominence eruptions. During the rising phase of cycle 23 (1997-1998), the CME latitudes were generally close to the equator and subsequently spread to all latitudes. During the maximum phase, there are many polar CMEs and the number of such CMEs was larger in the southern hemisphere and occurred over a longer time period than in the north. This behavior of CME latitudes with the solar activity cycle is consistent with previous measurements from Skylab/ATM (Hildner, 1977; Munro et al., 1979), P78-1/Solwind (Howard et al., 1985; 1986) and SMM/CP (Hundhausen, 1993).

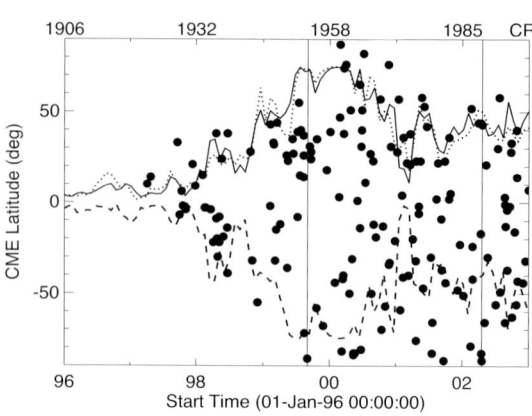

Figure 8.7 Latitudes of CMEs (filled circles) with known solar sources (identified from microwave prominence eruptions), plotted as a function of time. The Carrington Rotation numbers are marked at the top (CR). The dotted and dashed curves represent the tilt angle of the heliospheric current sheet in the northern and southern hemispheres, respectively; the solid curve is the average of the two. The two vertical lines indicate the start and end of the high-latitude CME activities.

5.5 CME Occurrence Rate

A CME rate of 0.5 CMEs/day was derived from the OSO-7 coronagraph data (Tousey et al., 1974). Skylab data indicated an average rate of ~ 1/day with a good correlation between sunspot number (SSN) and CME rate (Hildner et al., 1976). Combining Skylab, SMM, Helios (Photometer), and Solwind observations, Webb and Howard (1994) found a rate of 0.31 to 0.77 CMEs/day for the solar minimum years and 1.75 to 3.11 CMEs/day for the solar maximum years. The correlation between CME rate and SSN was also found to hold when the data were averaged over Carrington Rotation periods (Cliver et al., 1994). The early indication from SOHO was that the solar-minimum rate (0.8/day) was much higher than the uncorrected rate during previous minima (Howard et al., 1997); when more data came in, St. Cyr et al. (2000) concluded that the rate corresponding to the rise phase of cycle 23 was not significantly different from pre-SOHO observations. It finally turned out that the SOHO CME rate averaged over Carrington Rotation periods increased from less than 1 during solar minimum (1996) to slightly more than 6 during maximum (2002) (see Fig. 8). The solar-maximum rate of SOHO CMEs was nearly twice the highest corrected rate (3.11 per day) reported for previous cycles (Webb and Howard, 1994). We attribute this primarily to the better sensitivity and the enormous dynamic range (16000:1) of the LASCO coronagraphs. Additional factors include larger field of view and more uniform coverage over long periods of time (Howard et

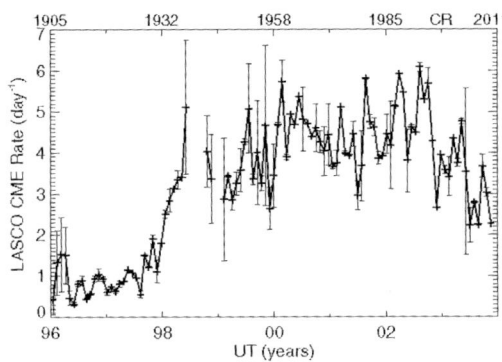

Figure 8.8 The CME occurrence rate (day^{-1}) averaged over Carrington Rotation (CR) periods as a function of time for the interval 1996-2003. There was a large data gap due to SOHO mission interruption during June to October 1998 and a smaller gap during January-February, 1999. The CR numbers are marked at the top. The error bars are based on the amount of SOHO downtime during each CR.

al., 1997). Note that LASCO CME rate is not corrected for duty-cycle, but an analysis by St. Cyr et al. (2000) suggested that such a correction may not be necessary for the LASCO data.

While SOHO data also confirmed the high correlation (r=0.86) between SSN and CME rate, the slope of the regression line was significantly different from pre-SOHO values (see Cliver et al., 1994) because of the higher maximum rate (Gopalswamy et al., 2003b). Furthermore, the CME rate peaked in CR 1993 (August 13-September 9, 2002), well after the maximum of the sunspot cycle (CR 1965, July 10-August 6, 2000). Figure 9 compares the CME rate with SSN averaged over longer periods of time (13 CRs). Clearly both have double peaks, but they are shifted with respect to each other. The difference between the two rates seems to be due the fact that CMEs originate not only from the Sunspot regions, but also from non-sunspot (quiescent filament) regions.

5.6   CME Mass and Energy

Skylab data indicated that a single CME could account for a mass of $\sim 4 \times 10^{15}$ g (Gosling et al 1974), which was soon confirmed (Hildner, 1977; Poland et al, 1981; Jackson and Howard, 1993; Howard et al, 1984). The mass in a CME is estimated by determining the CME volume and the number of electrons in the CME with the assumption that the CME is a fully ionized hydrogen plasma with 10% helium. Mass estimates have also been made using radio (Gopalswamy and Kundu 1992; 1993; Ramesh et al., 2003) and X-ray observations (Rust and Hildner, 1976; Hudson et al., 1996; Gopalswamy et al., 1996; 1997a; Hudson and Webb,

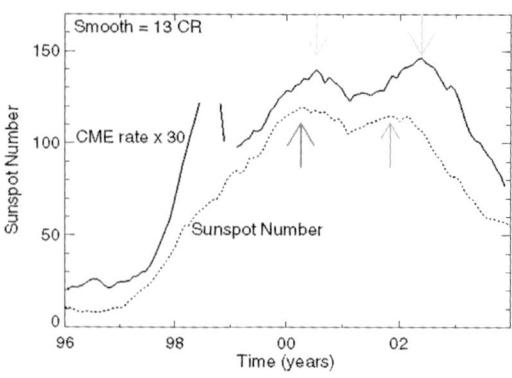

Figure 8.9 Time evolution of Sunspot number (SSN) and CME occurrence rate averaged over 13 Carrington Rotation periods. No smoothing was done for CME rates during the interval 1998 June to 1999 February, when there were large data gaps. The CME rate was multiplied by a factor of 30 to fit the scale. The arrows point to the two largest peaks in SSN and CME rate.

1997; Sterling and Hudson 1997; Gopalswamy and Hanaoka 1998). The radio and X-ray estimates (10^{14} - 10^{15} g) are generally lower than, but well within the range of, the white-light mass values. It must be pointed out that the X-ray and radio mass estimates of CMEs correspond to regions close to the Sun whereas the white light estimates correspond to larger heights (a few $R_\odot$). The X-ray and radio techniques are based on the thermal emission properties of the CME plasma (as opposed to Thomson scattering in white light), and hence provide an independent cross-check for mass estimates. However, routine estimates are done only in white light. Figure 10 shows a summary of mass and energy properties of 4297 LASCO CMEs for the period 1996-2002 (see also Table 1). The average mass (1.6×10^{15} g) of LASCO CMEs is somewhat lower than those of Solwind and SMM/CP CMEs (Vourlidas et al., 2002). This may be due to the fact that LASCO was able to measure CMEs of mass as low as 10^{13} g: $\sim$15% of CMEs had masses less than 10^{14} g. From the energy distribution shown in Fig. 10, it is found that the average (median) kinetic energy of the 4297 CMEs is 2.4×10^{30} erg (5×10^{30} erg), while the average (median) potential energy is 2.5×10^{30} erg (9.6×10^{30} erg). Figure 10 also shows the mass density (amount of mass in grams that corresponds to each pixel of the CME in LASCO images) as a function of height. The mass density increases rapidly to about 8 $R_\odot$ and then levels off. The fractional number of CMEs in each height bin (shown by the dashed-line histogram in the lower left panel of Fig. 10), suggests that those CMEs that reach greater heights have the largest mass density. We can see that $\sim$20% of CMEs reach their maximum mass at a height of $\sim$5 $R_\odot$, while almost half of the CMEs reach it within the LASCO/C2 FOV. In an earlier study, increases in mass by

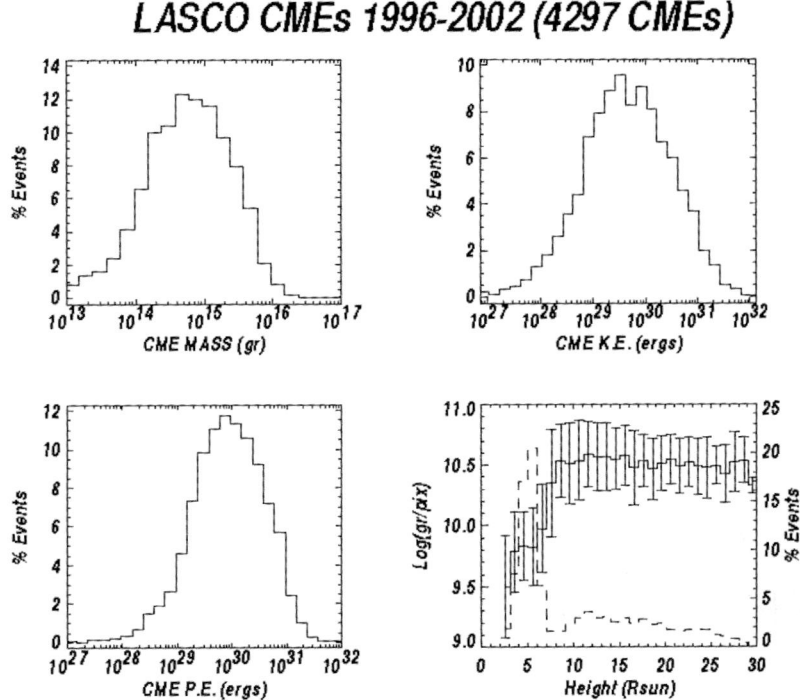

Figure 8.10. CME mass and energy (kinetic - K.E., and potential - P. E.) distributions and the evolution of mass density (grams/pixel) as a function of heliocentric distance. In the bottom right panel, the histogram (dashed line) shows that most of the CMEs were detected within the height range of increasing mass density. Not all detected CMEs have been included because mass measurements require (i) a good background image, (ii) three consecutive frames with CMEs, and (iii) CMEs well separated from preceding CMEs. Courtesy: A. Vourlidas.

a factor of up to 3 were found from the corona to the interplanetary medium (Jackson and Howard, 1993). Large mass increases (by a factor of 5-10) were also found from Yohokoh/SXT (Gopalswamy et al., 1996, 1997a) and SOHO/LASCO (Howard et al., 1997) observations. It is important to point out that LASCO movies show continued outflow of mass in the aftermath of CMEs for a day or so. A systematic study is needed to identify the origin and the magnitude of this mass compared to the CME mass obtained from snapshot images.

5.7 Halo CMEs

Halo CMEs are so named because of their appearance as approximately circular brightness enhancements surrounding the occulting disk.

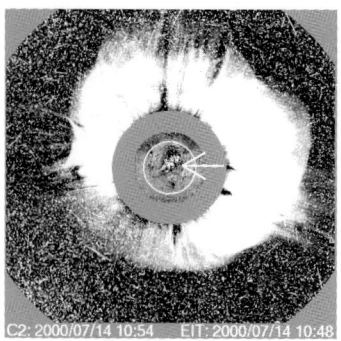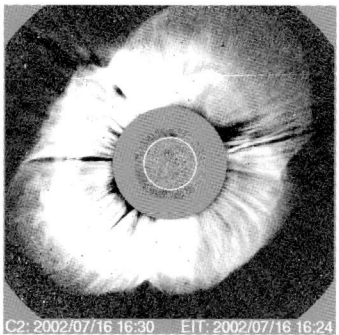

Figure 8.11. Front-side (left) and backside (right) full halo CMEs from SOHO/LASCO. The arrow points to the EUV eruption seen in the SOHO/EIT difference image superposed on the LASCO difference image; no EUV activity was observed for the backside halo because the solar source was occulted.

Although halo CMEs are known from pre-SOHO observations (Howard et al., 1982), their prevalence became clear in the SOHO data (Webb et al., 2000; St. Cyr et al., 2000; Webb, 2002; Gopalswamy et al., 2003b; Michalek et al., 2003; Yashiro et al., 2004). CMEs heading towards and away from the observer can appear as halos. Figure 11 shows two halo CMEs, one originating from the visible disk of the Sun and the other from the backside. From coronagraph images alone it is impossible to tell which way the halos are heading, so we need coronal images (such as the SOHO/EIT difference images in Fig. 11) to check if there is disk activity. It must be noted that the circular appearance of halos is due to projection on the sky plane. Figure 12 shows two CMEs originating from the same active region (AR 10486) when it was close to the disk center on 2003 October 28 and near the west limb on 2003 November 4. To an observer located above the west limb the October 28 event would appear as an east limb event, while the November 4 event would appear as a halo. Coronagraphs on the two STEREO spacecraft should be able to provide such a multiview for single CMEs. CMEs originating from close to the limb appear as asymmetric or partial halos (Gopalswamy et al. 2003b). Limb CMEs sometimes appear as halos because of faint enhancements seen above the opposite limb. These extensions may be shocks or magnetosonic waves propagating perpendicular to the direction of ejection (Sheeley et al., 2000).

The annual totals of halo CMEs are compared with those of the general population in Fig. 13. The number of halo CMEs had a broad peak during the solar maximum phase (2000-2002). However, the fraction of halo CMEs is always less than 5% (see also Fig. 4). The largest fraction resulted in 1997, during the rising phase of solar cycle 23. For the solar

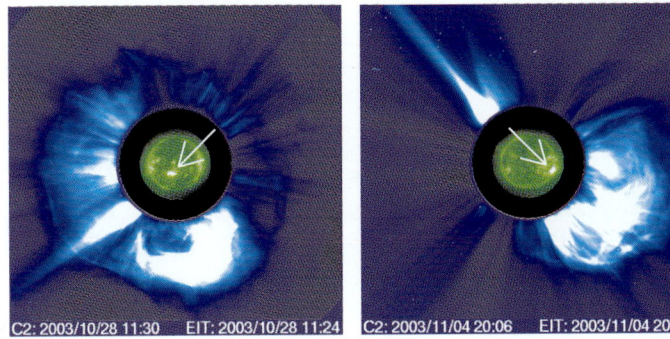

Figure 8.12. Two CMEs from the same active region (AR 10486) and similar speeds: (left) halo CME on 2003 October 28 (2459 km s^{-1}), and (right) limb CME on 2003 November 4 (2657 km s^{-1}). The arrows point to the EUV brightenings in the active region as observed by SOHO/EIT.

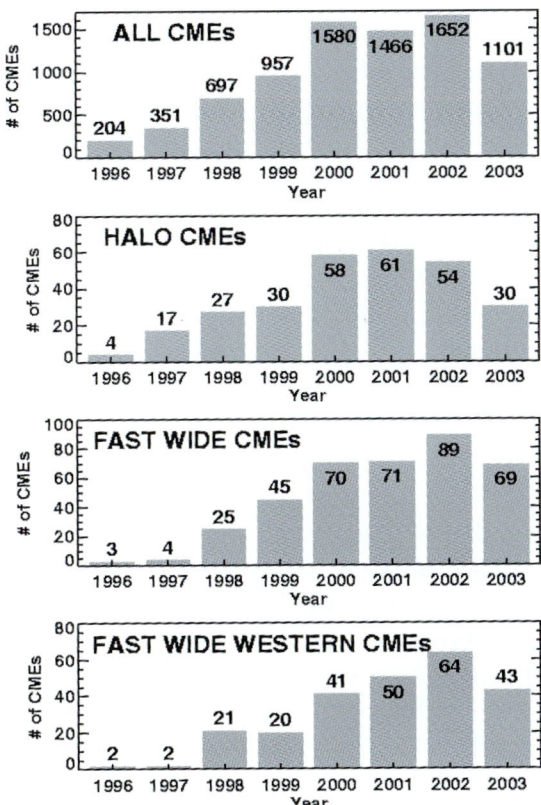

Figure 8.13 Annual numbers of the general population of CMEs compared with those of the special populations: halo, fast and wide, and fast-and-wide western CMEs. Fast and wide CMEs have speed > 900 km s^{-1} and width > 60°. Fast and wide western CMEs are the same as fast and wide CMEs, but their span includes position angle 270°. The numbers in each bin are marked. The special populations are similar in number but constitute a small fraction of the general population.

maximum phase (years 2000-2002), the number of halo CMEs exceeded 50 per year (100 per year if CMEs with width > 180° are considered).

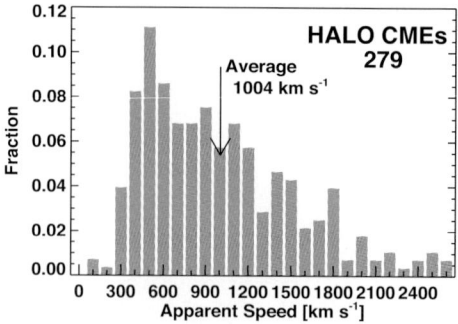

Figure 8.14 Speed distribution of the 279 halo CMEs for the period 1996-2003. Note that there are very few halo CMEs with speeds less than 300 km s^{-1} (most of these are from the solar minimum period).

What is special about halo CMEs? In principle, even narrow CMEs originating on the disk or backside should eventually become halo CMEs. These CMEs have to move far enough for their flanks to be "visible". Because of their large angle with respect to the sky plane and the distance from the Sun, they may not be detected by coronagraphs. Studying all the halo CMEs detected by LASCO, Yashiro et al. (2004) found that the average speed of the halo CMEs is roughly twice that of the general population of CMEs. Figure 14 shows the speed distribution of the 279 halo CMEs from 1996 to 2003. The average speed of the halo CME population shown is 1004 km s^{-1}, compared to 489 km s^{-1} for the general population (see Fig. 4). Thus, most of the halo CMEs seem to belong to a population known as fast-and-wide CMEs (speed (> 900 km s^{-1} and width $> 60°$), which are known for driving shocks and producing solar energetic particles and long-wavelength radio emission (Gopalswamy et al., 2003c). While it is not uncommon for CMEs from the eastern hemisphere to be associated with SEP events at Earth, western hemispheric fast and wide CMEs result in prompt increase of SEP intensity at 1 AU.

6.   Associated Activities

CMEs are associated with a number of phenomena starting all the way from the chromosphere (H-alpha flare ribbons, Moreton waves), and the corona (dimming, arcade formation, X-ray flares, prominence eruptions, X-ray and EUV ejecta, EUV wave transients, metric radio bursts) to the heliosphere (magnetic clouds, interplanetary radio bursts, shocks and energetic particles), that are observed as mass motion, waves and electromagnetic radiation. H-alpha and soft X-ray flares, prominence eruptions, and soft X-ray and EUV ejecta provide vivid pictures of the eruption during its early stages, generally not accessible to coronagraphs. Radio bursts produced by shocks (type II) and moving magnetic structures (type IV), are closely related to CMEs. Phenomena such as CME-related dimming (Hudson, 1999; Gopalswamy, 1999; Gopalswamy and Thompson, 2000; Klassen et al., 2000), EUV wave transients (Thompson et al., 1999; Gopalswamy and Thompson, 2000; Mann et al., 1999; Biesecker et al., 2002), and arcade formation (Hanaoka, 1994; Gopalswamy et al., 1999) have become benchmark signatures that are commonly used in identifying the solar sources of CMEs, in addition to the traditional H-alpha flare locations. SOHO's Ultraviolet Coronagraph Spectrometer (UVCS, Kohl et al., 1995) has turned out to be a useful source to estimate the true speed of CMEs (as opposed to sky plane speeds) and a number of physical parameters such as density and temperature (see, e.g., Ciaravella et al., 2003).

6.1   Flares and CMEs

Early statistical studies (see, e.g., Munro et al., 1979; Kahler, 1992) showed that $\sim 40\%$ of CMEs were associated with H-alpha flares and almost all flares (90%) with H-alpha ejecta were associated with CMEs. Thus the "mass motion" aspect of flares seems to be critical for a flare to be associated with CME. Flares have been classified (see, e.g. Pallavicini et al., 1977; Moore et al., 1999) as impulsive (short-duration (< 1h), compact (10^{26}-10^{27} cm^3), and low-lying (10^4 km)) and gradual (long duration (hours), large volumes (10^{28}-10^{29} cm^3), and great heights (10^5 km)). The probability of CME-flare association increases with flare duration (Sheeley et al., 1983): 26% for duration < 1h and 100% for duration > 6 h. It must be pointed out that some major flares associated with large-scale CMEs are not long-duration events (Nitta and Hudson, 2001; Chertok et al., 2004). Currently, there are three ideas about the flare-CME relationship: 1. Flares produce CMEs (see, e.g., Dryer, 1996), 2. Flares are byproducts of CMEs (Hundhausen, 1999), and 3. Flares and CMEs are part of the same magnetic eruption process (Harrison 1995; Zhang et al., 2001). Studies on temporal correspondence between CMEs and flares have concluded that CME onset typically precedes the associated X-ray flare onset by several minutes (e.g. Harrison 1991). This observational fact is considered to be a serious difficulty for flares to produce CMEs (Hundhausen, 1999). The flare process - reconnection that forms post flare loops - can be thought of as the force that propels overlying loops as CMEs (Anzer and Pnueman, 1982). Kahler et al. (1989) argued against such a model because they could not find evidence for a flare impulsive phase affecting the height-time history of CMEs. Zhang et al. (2001) investigated four CMEs and compared their time evolution with GOES X-ray flares. They found that the CMEs started accelerating impulsively until the peak of the soft X-ray flare, consistent with an earlier result that flare-associated CMEs are in general faster than other CMEs (MacQueen and Fisher, 1983). There is also weak correlation (r = 0.53) between soft X-ray flare intensities and associated CME energies (Hundhausen, 1999; Moon et al., 2002). The fact that flares with H-alpha ejecta are closely related to CMEs suggests that we need to understand how the free energy in the eruptive region is partitioned between heating (soft X-ray flares) and mass motion (CMEs). The connection between flares and CMEs needs to be revisited especially because of the availability of high quality multiwavelength data on flares and CMEs.

6.2 Prominence Eruptions

Prominence eruptions (PEs) are the near-surface activity most frequently associated with CMEs (Webb et al., 1976; Munro et al., 1979; Webb and Hundhausen, 1987; St. Cyr and Webb, 1991): 70% of CMEs are associated with PEs (Munro et al. 1979). Reverse studies indicate that the majority of PEs are associated with CMEs (Hori and Culhane, 2002; Gopalswamy et al., 2003a). Using microwave PEs, Gopalswamy et al. (2003a) found that (i) 73% of PEs had CMEs, while 16% had no CMEs at all, and the remaining PEs were associated streamer changes; (ii) the PE trajectories could be broadly classified as radial (R) and Transverse (T); (iii) most of the R events were associated with CMEs and the eruptive prominences attained larger heights, while most of the T events were not associated with CMEs; (iv) almost all of the PEs without CMEs were found to be T events (in which material does not leave the Sun). These results are consistent with those of Munro et al. (1979) who found that virtually all prominences that attained a height of at least $1.2 \, R_\odot$ were associated with *Skylab* CMEs. The source locations of CMEs and prominences spread to all latitudes towards the solar maximum in a similar fashion. During solar minimum, the central position angles of CMEs tend to cluster around the equator, while those of PEs were confined to the latitudes of active region belt, reflecting the stronger influence of the solar dipolar field on CMEs during solar minimum.

What is the physical connection between prominences and CMEs? Case studies have shown that eruptive prominences can be traced into the inner parts of the bright core (House et al. 1981; Illing and Athay, 1986; Gopalswamy et al., 1998), and this has been confirmed by statistical studies. There is also a close correspondence between the projected onset times of CMEs and PEs (Gopalswamy et al., 2003a). These results indicate that PEs form an integral part of CMEs. However, PEs are considered as a secondary phenomenon to the CME process because PEs may not have enough energy to drive CMEs (Hundhausen, 1999). Filippov (1998) has a different result: CMEs can be caused by the eruption of inverse-polarity prominences. Runaway reconnection in the magnetic field of the prominence is also thought to be fundamental for the onset of CMEs (Moore et al., 2001).

6.3   Are There Two types of CMEs?

On the basis of speed-height profiles of a dozen CMEs observed by the MLSO K-coronameter, MacQueen and Fisher (1983) suggested that different acceleration mechanisms may be operating in CMEs associated with prominence eruptions and flares. The flare-related CMEs were faster and characterized by constant speed, while the prominence-related CMEs were slower and accelerating within the coronameter FOV (see also St. Cyr et al., 1999). Tappin and Simnett (1997) used 149 LASCO CMEs and found that the constant speed CMEs were generally faster. Examples of constant speed and accelerating h-t profiles were also reported by others (Sheeley et al., 1999; Andrews and Howard, 2001; Gopalswamy et al., 2001b). The travel time of flare-related solar disturbances has also been found to be generally shorter than that of prominence-related ones (Park et al., 2002). Studying a much larger sample of LASCO CMEs, Moon et al. (2002) found a clear difference in speeds of flare-related (759 km s^{-1}) and prominence-related (513 km s^{-1}) CMEs. The flare-related CMEs also showed a tendency for deceleration, but this probably reflects the fact that they are faster (see Gopalswamy et al., 2001b). The question is whether the speed difference is qualitative or quantitative given that CMEs of both types involve closed magnetic regions with filaments. Studying the acceleration of CMEs, Chen and Krall (2003) conclude that one mechanism is sufficient to explain flare-related and prominence-related CMEs.

6.4   X-ray Ejecta

Klimchuk et al. (1994) found that the properties of 29 X-ray eruptions from Yohkoh/SXT were similar to those of white-light CMEs. Although they did not compare their data with white light observations, it is likely that they correspond to the frontal structure. X-ray ejecta were also frequently seen by SXT (see, e.g., Shibata et al., 1995), but their white-light counterpart was not checked. Gopalswamy et al. (1997b) reported an X-ray eruption followed by a disconnected X-ray plasmoid. Checking white-light data from MLSO, they concluded that the eruption was associated with a CME. The plasmoid was also associated with a moving type IV burst, which suggested that the X-ray plasmoid must have

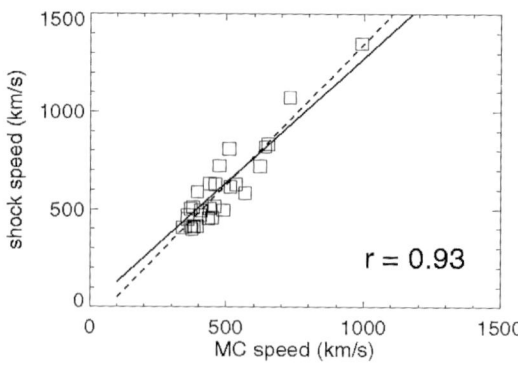

Figure 8.15 Scatter plot of shock speed versus magnetic cloud (MC) speed for a number of events detected by Wind at 1 AU. The correlation coefficient (r) is 0.93. The solid line is the best-fit to the data points. The dashed line is the gas dynamic piston-shock relationship.

also carried nonthermal particles, consistent with the scenario that the plasmoid is the heated prominence material (see, e.g., Wagner, 1984). Presence of nonthermal electrons can also be inferred occasionally from hard X-rays (Hudson et al., 2001). Recently, Nitta and Akiyama (1999) looked for X-ray ejecta in 17 limb flares and compared them with LASCO data. They found that (i) flares not associated with CMEs also lacked X-ray ejections, and (ii) the X-ray ejecta were inner structures of CMEs. These results are consistent with the dense prominence material present in the core of CMEs. To be "visible" in X-rays, they must have been heated. However, frontal structure of CMEs can also be occasionally seen in X-rays, as was reported by Gopalswamy et al. (1996). Spectroscopic observations also confirm that the prominence core can be hot (Ciaravella et al., 2003).

6.5   CMEs and Radio Bursts

Moving type IV bursts indicate magnetized plasma ejection; type II bursts indicate superAlfvenic mass motion. Therefore, these two bursts are expected to be closely related to CMEs. Moving type IVs come in three varieties: advancing fronts, expanding arches and isolated plasmoids (see Stewart, 1985 for a review). The isolated sources originate from the heated prominence material, also detected in X-rays and EUV. The advancing fronts and expanding arches must be structures associated with the CME itself (Gopalswamy and Kundu, 1989; Bastian et al., 2001), "visible" because of the nonthermal electrons trapped in them. The nonthermal electrons may be accelerated at the reconnection site beneath the CMEs or by the shock ahead of the CME.

Coronal and interplanetary shocks are inferred from metric and longer wavelength type II radio bursts, respectively (Wild et al, 1950; Malitson et a., 1973). Gosling et al. (1976) found that $\sim 85\%$ of CMEs with speed > 500 km s^{-1} were associated with type II and/or type IV bursts. In a reverse study, Munro et al. (1979) found that almost all type II or type IV bursts originating from within 45° of the limb were associated with CMEs. The speed distribution of coronal shocks was found to be similar to that of CMEs associated with type II bursts (Robinson, 1985). These observations clearly were consistent with the idea that CMEs moving faster than the local Alfven speed can drive an MHD shock. Later observations indicated metric type II bursts without CMEs and fast CMEs without metric type II bursts (Sheeley et al., 1984; Kahler et al., 1984, 1985). From these results it was inferred that some of the coronal shocks may be flare blast waves, consistent with the type II source location behind the leading edge of CMEs (Wagner and MacQueen, 1983; Gary et al., 1984; Robinson and Stewart, 1985; Gopalswamy et al., 1992).

Using Solwind (coronagraph) and Helios (in situ) data, Sheeley et al. (1985) found a near one-to-one correspondence between CMEs and IP shocks. All kilometric type II bursts observed by ISEE-3 are known to be associated with fast (> 500 km s^{-1}) and energetic CMEs and IP shocks (Cane et al., 1987). Recent data from Wind/WAVES (Bougeret et al., 1995) indicate that all decameter-hectometric (DH) type II bursts (1-14 MHz) are also associated with fast and wide CMEs capable of driving shocks (Gopalswamy et al., 2001b). Can we extend this CME-type II connection to metric type II bursts also? There are several arguments in favor of the idea that even metric type II bursts are due to CME-driven shocks: 1. Type II bursts without associated CMEs have been revisited by Cliver et al. (1999) to show that the CMEs might have been missed due to observational constraints. Further evidence came from a comparison of metric type II bursts with LASCO/EIT data: while no white-light CMEs were observed for some metric type II bursts, there were EUV eruptions from close to the disk center (Gopalswamy et al., 2001a), suggesting that these CMEs may have been masked by the occulting disk. 2. Lara et al. (2003) studied the CME properties of (i) metric type II bursts with no IP counterparts and (ii) IP type II bursts at frequencies ≤ 14 MHz. They found that the speed, width and deceleration of CMEs progressively increased for the general population of CMEs, CMEs associated with metric type II bursts and CMEs associated with IP type II bursts, in that order. This is clear evidence that the energy of a CME is an important factor in deciding whether it will be associated with a type II burst, consistent with an earlier conclusion by Robinson (1985) when comparing speeds of CMEs associated with metric type II bursts and those of IP shocks associated with km type II bursts. 3. The type II burst association with low-speed (200 km s^{-1}) CMEs and the lack of it for a large number of fast (speed > 900 km s^{-1}) and wide ($> 60°$) CMEs can be explained as a direct consequence of the Alfven speed profile in the ambient medium (Gopalswamy et al., 2001a). 4. The difference in drift rates of type II bursts below and above 1 MHz (Cane, 1983) and the lack of correlation between speeds derived from metric type II bursts and associated CMEs (Reiner et al., 2001) can be explained if we note that the CME speed changes rapidly in the inner corona and the CME is propagating through the region of highly variable Alfven speed. 5. The positional mismatch between CME leading edge and type II bursts can be explained by the preferential electron acceleration in the quasiperpendicular region of the CME bow shock (Holman and Pesses, 1983). 6. A blast wave is expected to be without a driver, but there is no evidence from in situ data for a shock without a driving ejecta. Almost all IP shocks followed by ICMEs seem to have a piston-shock relationship (see Fig. 15), and the corresponding white-light CMEs can be identified. Shocks detected "without drivers" can be attributed to limb CMEs so they are also driven, but only the flanks arrive at Earth (Schwenn, 1996; Gopalswamy et al., 2001a). It appears that all type II bursts can be associated with CMEs if we consider the combination of CME characteristics (speed, width) and the Alfven speed profile in the ambient medium.

The advent of EIT waves (Thompson et al., 1999) has provided some additional input to the problem of coronal shock source. Based on the good correspondence between EIT waves and metric type II bursts, Mann et al. (1999) suggested that EIT waves are of flare origin and might be the pre-shock stage of coronal shocks inferred from metric type II bursts. However, Gopalswamy and Kaiser (2002) pointed out that a CME-driven shock can form low in the corona where the Alfven speed is < 300 km s^{-1} and hence explain the metric type II burst and even the subsequent IP type II burst. One class of EIT waves known as "brow waves" (owing to their arc-like appearance in EIT images, - see Gopalswamy, 2000; Gopalswamy and Thompson, 2000) are spatially and temporally coincident with metric type II bursts (Gopalswamy et al., 2000d). Biesecker et al. (2002) later classified the brow waves as "events with sharp brightenings" and found them to be associated with metric type II bursts, flares, and CMEs. They also found an unambiguous correlation between EIT waves and CMEs, but a significantly weaker correlation between EIT waves and flares. There were also attempts to interpret wavelike features in soft X-ray images to be blast waves and associate them with Moreton waves (Moreton, 1960) and metric type II bursts (Narukage et al., 2002; Hudson et al., 2003). However, both of these reports did not take the presence of CMEs into consideration. For example, in the 1998 May 06 event at 08:03 UT studied by Hudson et al. (2003), there was also a 1100 km s^{-1} CME (see Table 1 Gopalswamy,

2003a), whose onset preceded the type II burst and hence cannot be ruled out as the source of the metric type II burst. It is quite likely that the EIT waves (at least the brow type) are coronal counterparts of Moreton waves, but are not inconsistent with a CME source. While we cannot completely rule out the possibility of flare blast waves causing metric type II bursts, the available and new evidence seem to favor CME-driven shocks (see also Mancuso and Raymond, 2004). Unfortunately, there is no reliable way of directly detecting shocks in the corona, except for possible shock signatures observed by UVCS (Raymond et al., 2000) and the white-light shock signatures (Sheeley et al., 2000; Vourlidas et al., 2003), which are not without CMEs.

6.6   CME Interaction and Radio Emission

Given the high rate ($\sim$ 6/day) of CME occurrence during solar maximum and the observed range of speeds, one would expect frequent interaction between CMEs. Although interactions among shocks and ejecta are known to happen in the heliosphere (Burlaga et al., 1987), SOHO images combined with the Wind/WAVES dynamic spectra provided direct evidence for CME interactions very close to the Sun (Gopalswamy et al., 2001c; 2002a). These interactions resulted in broadband nonthermal radio enhancements in the decameter-hectometric (DH) wavelength domain. Strengthening of shocks when propagating through the dense parts of preceding CMEs and trapping of particles in the closed loops of preceding CMEs were suggested as possible mechanisms that increase the efficiency of particle acceleration (Gopalswamy et al., 2002b). Shock strengthening can be seen from the fact the change in local Alfven speed (Va) is related to density (n) and magnetic field (B) changes: $dVa/Va = dB/B - (1/2)dn/n$. Recent numerical simulations support such shock strengthening (Wu et al., 2002). A shock traveling through a denser medium would be locally stronger and would accelerate more electrons resulting in enhanced radio emission, provided the magnetic field does not change significantly.

Figure 16 illustrates a recent CME interaction event: A sudden radio enhancement occurred over an existing Wind/WAVES type II radio burst on 2003 November 4. The radio enhancement is brighter than the associated type II burst and hence is nonthermal in nature. A very fast, shock-driving CME (CME2, 2657 km s^{-1}) approached a slower CME (CME1, $\sim$1000 km s^{-1}) and its dense core (CORE1, $\sim$700 km s^{-1}). The radio enhancement occurred when CME2 reached a heliocentric distance of 18 $R_\odot$, close to the core of CME1. The 21:18 UT SOHO image shows that the CME2 and CORE1 are very close when the ra-

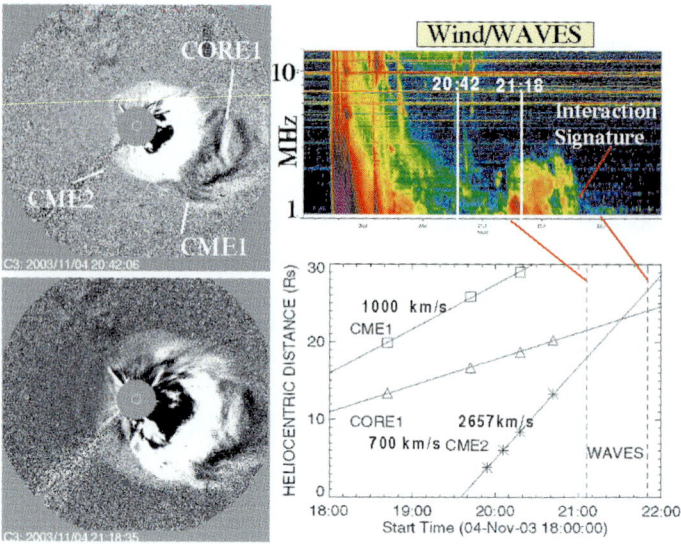

Figure 8.16. Wind/WAVES dynamic spectrum (top right) showing the interaction signature and the height-time diagram (bottom right) of the 2003 November 04 CMEs. The first CME (CME1), its core (CORE1) and the second CME (CME2) are marked in the SOHO image at 20:42 UT (top left). The SOHO image at 21:18 (bottom left) was taken when the Wind/WAVES interaction signature in radio was in progress. The times of the two SOHO images are marked on the WAVES dynamic spectrum. The duration of the interaction signature is denoted by the two vertical dashed lines on the height-time plot. The speeds of CME1, CME2, and CORE1 are also shown.

dio enhancement started. The radio emission lasted for about 40 min, roughly the time taken by the CME-driven shock to traverse CORE1 (size $\sim 7~R_{\odot}$). The high frequency edge of the type II burst was at ~ 1 MHz when the interaction signature started with a high-frequency edge of 3 MHz. A jump of 2 MHz in frequency would correspond to a density jump of 4 with respect to the ambient corona. This is also consistent with the relatively high white-light brightness of CORE1. The same interaction signature was observed by radio receivers on board Ulysses and CASSINI, which were at distances of 5 and 8.7 AU, respectively. The signatures arrived at CASSINI and Ulysses with a delay corresponding to the light travel times. Wind, Ulysses, and CASSINI were widely separated in heliocentric distance as well as angular separation, suggesting that the interaction signature is not narrowly beamed.

7. CMEs and Solar Energetic Particles

Kahler et al. (1978) found CMEs to be necessary requirements for the production of SEPs and hence suggested that SEPs may be accelerated

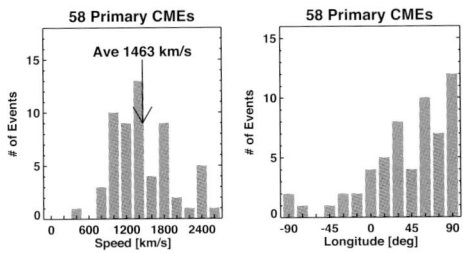

Figure 8.17 Distributions of speed (left) and source longitudes of CMEs associated with 58 major (proton intensity exceeding 10 particles per $(cm^2 \ s \ sr)$) SEP events from the period 1996-2002. The $90°$ bins also contain events from behind the limb.

by the shocks ahead of CMEs. The current paradigm is that impulsive, short-lived SEP events are due to flares and the large, gradual, long-lived events are accelerated in CME-driven shocks (see, e.g. Lin, 1987, Reames, 1999). Recent data also indicate that large SEP events are invariably associated with fast and wide CMEs (Fig. 17). CMEs from the western hemisphere typically result in high SEP intensity at Earth due to better connectivity (see Fig. 17), although it is not uncommon for CMEs from the eastern hemisphere to result in SEP events at Earth. Despite the general acceptance of CME-driven shocks as the source of large SEP events (Lee, 1997; Reames, 1999; Tylka, 2001), there is still no widely accepted theory that explains all the observed properties of SEPs. For example, the CME speed and SEP intensity are reasonably correlated, yet the scatter is very large (see Fig. 18): for a given CME speed, the SEP intensity has been found to vary over four orders of magnitude (Kahler, 2001; Gopalswamy et al., 2003c) with no satisfactory explanation. However, the SEP intensity is better correlated with the CME speed than with the flare size (Fig. 18).

A Type II burst is the primary indicator of shock near the Sun, where the SEPs are released (a few $R_\odot$ from the Sun - see, e.g., Kahler, 1994). The DH type II bursts also originate from this region and are known to have a 100% association with SEP events (Gopalswamy, 2003a). The occurrence rates (per Carrington Rotation) of large SEP events (>10 MeV protons from GOES), fast and wide CMEs from the frontside western hemisphere, IP shocks (detected in situ), DH type II bursts and major (GOES M and X-class) flares are quite similar, except for major flares, of which there were too many (Gopalswamy et al., 2003b,c). The close correlation among all these phenomena suggests that CME-driven shocks accelerate electrons (to produce type II bursts) and protons (detected as SEP events).

The simple classification of impulsive and gradual SEP events has recently been brought into question. Most of the CMEs associated with large SEP events are also associated with intense flares, so it is often dif-

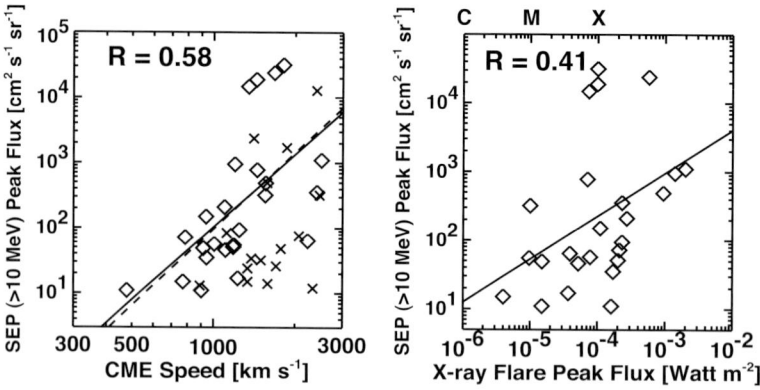

Figure 8.18. Scatter plot of the SEP intensities of > 10 MeV proton events with (left) CME speeds and (right) X-ray flare size. All events are plotted in the left panel, but only the 25 events with $0° <$ longitude $< 90°$ (diamonds) are included in the correlation. The solid lines are best fits to the diamonds. The correlation coefficients are r=0.58 for CME speeds (confidence level 99.9%) and 0.41 for X-ray flux (confidence level 98%). Excluding the outlier CME with a speed of 478 km s^{-1} results in r=0.54 (confidence level 99.75%) and the dashed line. See Gopalswamy et al. (2003c for details.

ficult to untangle the contributions from flare and shock sources (Cliver, 1996; Kocharov and Torsti, 2002). Flare particles (Mason et al., 1999) or SEPs from preceding CMEs (Kahler, 2001) may form seed particles for CME-driven shocks near the Sun as well as at 1 AU (Desai et al., 2003). Long rise times of some SEP events seem to be due to successive SEP injections (Kahler, 1993). SEP-producing shocks seem to propagate through the corona with preceding CMEs (Gopalswamy et al., 2002a). Large SEP events with preceding wide CMEs within a day from the same active region tend to have higher intensity (Gopalswamy et al., 2003c). Multiple shocks and CMEs can form configurations that can enhance the SEP intensity significantly (Kallenrode and Cliver, 2001; Bieber et al., 2002). Thus, the presence of preceding CMEs means disturbed conditions in the coronal and IP medium through which later CMEs propagate: density, flow velocity, magnetic field strength, magnetic field geometry, and solar wind composition may be different compared to normal solar wind conditions. Accelerated particles propagating through a medium denser than the normal solar wind (due to a preceding CME) may affect the observed charge states of the ions if the product of the density and the residence time is large enough to allow for additional electron stripping (Reames et al., 1999; Barghouty and Mewaldt, 2000).

8. CMEs in the Heliosphere

While the existence of magnetized plasma clouds was contemplated in the 1950s, their detection became possible with space borne measurements (Burlaga et al., 1981, Lepping et al., 1990). Helios 1 detected a magnetic loop behind an IP shock, which Burlaga et al. (1981) defined as a magnetic cloud (MC). The connection between CMEs and MCs was recognized when a Helios 1 MC was related to a white-light CME that left the Sun two days before (Burlaga et al., 1982). Analyzing the helium abundance enhancements (HAEs - Hirshberg et al., 1972) in the high speed plasmas behind IP shocks, Borrini et al. (1982) concluded that the HAEs must be the IP signatures of CMEs. At present a large number of IP signatures are used to identify the CME-related plasmas in the solar wind (see, e.g. Gosling et al., 1990): bidirectional streaming of superthermal electrons and ions, unusual abundances and charge states, low electron and proton temperatures, strong magnetic fields with flux rope structures, and Forbush decreases. It must be noted that not all of the signatures are present in all events (see Neugebauer and Goldstein, 1997). CMEs in the solar wind are commonly referred to as 'ejecta' or interplanetary CMEs (ICMEs). In situ observations of CMEs can be used to infer the magnetic field topology of the ICMEs and the physical conditions of their birthplace near the Sun (see, e.g., Henke, 1998; Lepri et al., 2001). When a CME moves past a spacecraft in the solar wind, the following sequence of structures would be detected: IP shock, sheath, and ejecta. On rare occasions, one observes cool dense material towards the end of the ejecta that resemble the prominence resting at the bottom of the coronal cavity in the pre-eruption phase of CMEs (Burlaga et al., 1998; Gopalswamy et al., 1998). As a working hypothesis, one can relate CMEs and ICMEs as follows: CME shock $\rightarrow$ IP shock, CME front $\rightarrow$ sheath, CME void $\rightarrow$ ICME (or ejecta), and CME core $\rightarrow$ density pulse (Gopalswamy, 2003b). Cliver et al. (2003) estimated that Earth is embedded within CME-related flows (shocks, sheaths and ejecta) for $\sim 35\%$ of the time during solar activity maximum and $\sim 10\%$ of the time during solar minimum. Only those CMEs, which originate close to the Sun center (within $30°$) are intercepted by Earth as ICMEs (Gopalswamy, 2002). Using bidirectional electron signatures, Gosling et al. (1992) found ~ 72 (8) ICMEs/year during solar activity maximum (minimum), similar to the variation in CMEs discussed in section 4. Klein and Burlaga (1982) found that $\sim 33\%$ of ICMEs were MCs. Recent studies show that the fraction of ICMEs that are MCs ranges anywhere from 11% to 100% (Cane and Richardson, 2003; see also Table 1 of Gopalswamy et al., 2000a).

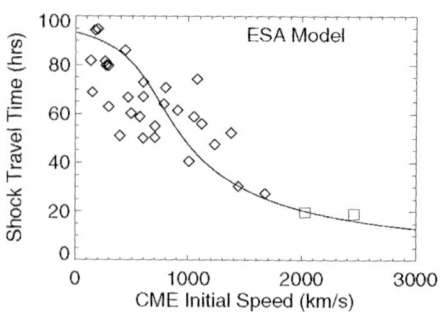

Figure 8.19 The empirical shock arrival (ESA) model, which predicts the shock travel time based on the initial speed of CMEs in the sky plane. The diamonds are for shocks driven by magnetic clouds. The squares represent the two fastest shocks of cycle 23, which originated from the ultrafast CMEs on 2003 October 28 (11:06 UT, 2459 km s^{-1}) and 2003 October 29 (20:41 UT, 2029 km s^{-1}).

ICMEs are responsible for the severest of geomagnetic storms and can be directly related to front-side halo CMEs (Gosling, 1993; St. Cyr et al., 2000). Webb (2002) finds that the fraction of halos associated with geomagnetic storms considerably decreased towards solar maximum. For example, 92% of the halos were associated with geomagnetic storms in the year 1997, while the fraction dropped to 35% in the year 2000. Detailed information on the internal structure (e.g., whether it contains southward magnetic field component) of halo CMEs is needed to understand why only certain halo CMEs result in geomagnetic storms. Thus the travel time of CMEs to 1 AU and their geoeffectiveness (magnitude and duration of geomagnetic storms) are of practical importance for space weather applications. Availability of simultaneous data on CMEs and ICMEs has made it possible to establish a relationship between their speeds (Lindsay et al., 1999). Influence of the solar wind on CMEs as they propagate away from the Sun can be postulated as an average IP acceleration (Gopalswamy et al., 2000a), which can be used to predict the travel time of CMEs (Gopalswamy et al., 2001e; Gopalswamy 2002) and shocks (Gopalswamy et al., 2003d) to various points in the heliosphere. Figure 19 shows the empirical shock arrival (ESA) model curve with observed travel times of MCs of cycle 23. The empirical model helps us understand the gross propagation of Earth-directed CMEs originating close to the disk center (within ±30°) and propagating through quiet solar wind. Drastically different conditions such as high speed wind, preceding CMEs (Manoharan et al., 2004) and significant projection effects (Gopalswamy et al., 2000b; Michalek et al., 2003) may also affect the predicted shock arrival times. As for geoeffectiveness, the ICME has to have southward magnetic field component, which is a difficult problem.

There have been several attempts to relate the magnetic field structure of the ejecta to that of filaments (e.g., Bothmer and Schwenn, 1994; Marubashi, 1997; Bothmer and Rust, 1997), arcades overlying filaments (Martin and McAllister, 1997), and the overlying global dipolar field of the Sun (Crooker 2000; Mulligan et al., 1998). However, there is no systematic scheme to predict the internal structure of an ICME based on magnetograms of the eruption regions.

8.1   High Latitude CMEs

Motion of magnetic clouds can continue in the heliosphere (Yeh, 1995) and has been observed beyond 11 AU (Burlaga et al., 1985; Funsten, et al., 1999), which means these objects must be commonplace in the heliosphere. While Voyager observations provided information on the heliospheric CMEs (HCMEs) in the ecliptic plane, high-latitude CMEs were first observed in situ by Ulysses (Gosling et al, 1994; Gosling and Forsyth, 2001). Ulysses CMEs observed during minimum conditions were fast compared to the ones observed during maximum conditions. A new class of CMEs known as "over-expanding" CMEs were discovered by Ulysses (Gosling et al., 1994). These CMEs have high internal pressure drive shocks due to the expansion into the heliosphere, rather than from the motion away from the Sun. Based on Ulysses observations (heliocentric distance 5.3 - 3.0 AU) over a 16-month interval at latitudes S30-S75, Gosling and Forsyth (2001) found that the HL HCMEs have an average duration of ∼67 h and a radial size of ∼0.7 AU (compared to the 0.25 AU at 1 AU); the occurrence rate of 15 per year is about 5 times smaller than the ecliptic rate at 1 AU. The ratio of HL to LL HCMEs seems to be very similar to the overall ratio (∼ 20%) with a slightly different definition of HL (latitude ≥ 60°) and LL (latitude < 60°) CMEs (Gopalswamy et al., 2003e).

Since the latitudes of CMEs are derived from the central position angle of CMEs in the sky plane, Gopalswamy et al. (2003e) considered HL CMEs as those with latitude ≥ 60°, and the LL CMEs with latitudes ≤40°. With this definition, they obtained a HL-to-LL CME ratio of 25%. To be consistent with this definition, we compared the Ulysses and SOHO/LASCO CMEs for the period July 10, 2000 to February 5, 2001, when Ulysses was poleward of S60. During these seven months, Ulysses detected 8 HCMEs, giving rate of 13.7 per year, very similar to Gosling and Forsyth's rate. Over the same interval, SOHO/LASCO observed 101 CMEs poleward of S60 and 602 LL CMEs, giving an HL-to-LL ratio of ∼17%. Interestingly, the ratio of HL CMEs at the Sun and HCMEs at Ulysses is ∼8%. The ratio of LL CMEs (602) at the Sun

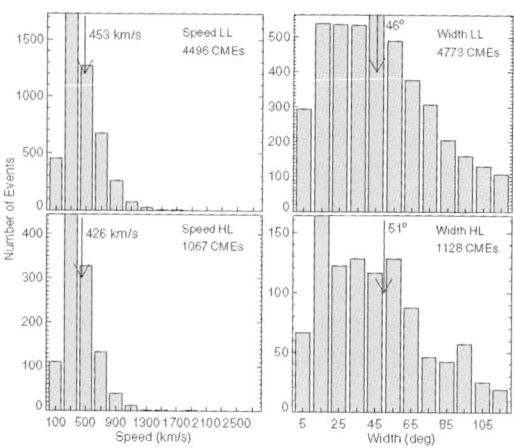

Figure 8.20 The speed and width distributions of high-latitude ($\geq 60°$, HL) and low-latitude ($\leq 40°$, LL) CMEs, with average values marked. CMEs with latitudes between 40 and $60°$ are not included for a clear separation between the two populations. CMEs with widths $\geq 120°$ are excluded because it is difficult to obtain their latitudes. The distributions of HL and LL CMEs are not very different.

to the ICMEs at 1 AU (25, see Cane and Richardson, 2003) is $\sim 4\%$, which becomes $\sim 8\%$ if we assume that half of the SOHO LL CMEs are backsided. Figure 20 shows the overall comparison between HL and LL CMEs for the entire cycle 23 until August 2003. The HL-to-LL ratio ($\sim 14\%$) is not too different from the numbers above. Although these comparisons indicate that the latitudinal distribution of CMEs at the Sun and in the heliosphere may be similar, Reisenfeld et al. (2003) reported that two of the equatorial CMEs observed by SOHO/LASCO were observed at Earth as well as by Ulysses at high latitudes (above N75). The large separation between Earth and Ulysses in latitude ($73°$) and longitude ($64°$) for one of the events is consistent with the large width of some white-light CMEs ($>120°$). However, these huge events are rare at the Sun.

9.   CMEs and Solar Polarity Reversal

Two magnetic cycles have been completed since Horace Babcock first noted the reversal of polarities of solar polar magnetic fields in 1959 (cycle 19). The Sun faithfully reversed the sign of its polar magnetic fields during all the sunspot cycles (20-23) since then. Common signatures of magnetic polar reversals on the Sun are the disappearance and reformation of polar coronal holes (Webb, et al., 1984; Bilenko, 2002; Harvey and Recely, 2002) and the disappearance of the polar crown filaments (PCFs) following a sustained march to the poles (Waldmeier, 1960; Cliver et al., 1994; Makarov, Tlatov, and Sivaraman, 2001). Studying the polarity reversals of cycle 23, Gopalswamy et al. (2003e) found that the epochs of solar polar reversal are closely related to the cessation of HL CME activities, including the non-simultaneous reversal in the north and south poles (see Fig. 21). Before complete reversal, several temporary reversals take place with corresponding spikes in the HL CME rates. The high-latitude CMEs also provide a natural explanation for the disappearance of closed field structures that approach the poles, which need to be removed before the reversal could be accomplished. The polarity reversal seems to be a violent process involving CMEs of mass a few times 10^{15} g and a velocity of hundreds of km s^{-1}. The kinetic energy of each of these CMEs is typically a few times 10^{30} erg. Figure 20 shows that there were $\sim 10^3$ HL CMEs over a period of $\sim 10^3$ days, during which the reversal was completed. This amounts to an energy dissipation rate of $\sim 10^{30}$ erg/day. The results presented here also support the hypothesis that CMEs may represent the process by which the old magnetic flux and helicity are removed and replaced by the those of the new magnetic cycle (Low 1997; Zhang and Low, 2001). Inclusion of CMEs along with the photospheric and subphotospheric processes completes the full set phenomena that need to be explained by any successful theory of the solar dynamo.

10.   CMEs and Cosmic Ray Modulation

Newkirk et al. (1981) identified CMEs as the solar origin of the low-frequency power in the interplanetary magnetic field fluctuations and suggested that the solar cycle dependent modulation of galactic cosmic rays (GCRs) can be explained by the presence of CME-related magnetic inhomogeneities in the heliosphere. Although they explored CMEs at latitudes below 60°, we now know that CMEs are present at all latitudes at least during solar maximum as isolated (Gosling and Forsyth, 2001; Balogh, 2002) or as merged interaction regions (Burlaga et al., 1993). The relationship recently found between HL CMEs and the reversal of global solar magnetic field (Gopalswamy et al., 2003e), and the relationship of the latter with the drift of GCRs into the solar system (Jokipii et al., 1977), suggest that HL CMEs may play an important role in long-term GCR modulation. SOHO results have conclusively shown the existence of a higher and more cycle-dependent CME occurrence rate (varying by factors up to 10) than pre-SOHO data indicated (Wagner, 1984). It was recently found that the inverse of the GCR intensity was correlated well with the HL CME rate during the rise phase of cycle 23 (Lara et al., 2004). This provides a clue to the 22-year pattern of GCR modulation, which does not directly follow solar activity indices such as SSN (see, e.g., Potgeiter et al., 2001). As we pointed out before, CME rate need not follow SSN, owing to the PCF-related CMEs,

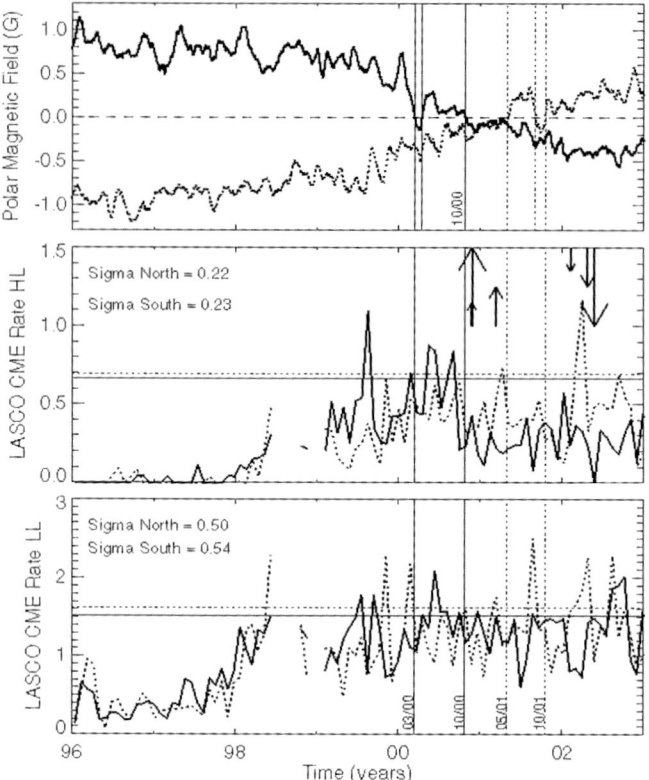

Figure 8.21. (top) The polar field strength averaged over regions poleward of 70°
(from NSO/Kitt Peak). Times of polarity reversal are marked by the vertical lines
(solid -north; dashed - south). CME rates from high (middle) and low (bottom)
latitudes are distinguished by the hemispheres (solid - north and dotted - south).
Times when the PCF branch disappeared are marked by small (Lorenc et al., 2003)
and medium (Harvey and Recely 2002) arrows. Large arrows mark the times of
cessation of high latitude prominence eruptions from Fig. 20. The direction of the
arrows indicates the hemisphere (up - north; down - south). The horizontal lines
in the middle and bottom panels show the 3-sigma levels of the CME rates (solid -
north; dotted - south). The standard deviation (sigma) of the rates in the north and
south are marked in the respective panels.

so treating the HL and LL CMEs separately may help understand the
GCR modulation pattern.

The drift of positively-charged GCRs in the solar system is known to
be poleward when the polarity at the solar north pole is positive (known
as A>0 cycle) and equatorward when the solar polarity is negative (A<0
cycle). The poleward and equatorward approaches are switched for the
negatively charged GCRs such as electrons and antiprotons (Bieber et

al., 1999). The A>0 and A<0 epochs commence when polarity reversal completes during even and odd numbered solar activity cycles, respectively. For example, there was a switch from A>0 to A<0 in the first half of 2002. This preferential direction of approach of GCRs immediately suggests that HL and LL CMEs should be alternately important for blocking GCRs during the A>0 and A<0 cycles. During A>0 cycles, GCR ions enter the heliosphere from the polar direction, so the HL CMEs must be effective in blocking them. For A<0 cycles, the approach of GCR ions is equatorward, so LL CMEs effectively block them. In order to see this effect, we compared the GCR intensity (from Climax neutron monitor) and HL and LL CME rates for the declining phase of cycle 21 (after the start of the A<0 epoch) and the rising phase of cycle 23 (before the end of the A>0 epoch). For these two phases, complete CME observations exist and the CME rates can be easily separated into LL and HL parts (Gopalswamy et al., 2003e). Figure 22 shows two cross-correlation plots comparing GCR intensity and HL and LL CME rates averaged over Carrington Rotation (CRot) periods (27.34 days). The cycle 21 data are somewhat noisy because the CME data were obtained with lower sensitivity compared to the SOHO/LASCO data for cycle 23. Nevertheless one can clearly see that the roles of HL and LL CMEs are reversed in the two epochs: higher anticorrelation was obtained between GCR intensity and LL CMEs for the A<0 epoch. On the other hand, the GCR intensity showed a better anticorrelation with the HL CME rate for the A>0 epoch. The tandem influence of HL and LL CMEs is thus consistent with the 22 year modulation cycle of GCRs.

On the basis of the above discussion, we can provide a tentative explanation as to how the GCRs and CMEs are coupled, as follows. An A>0 cycle begins right after the completion of the polarity reversal during the maximum of an even-numbered cycle. GCRs start entering the heliosphere from the poles. Since HL CMEs have subsided around this time (Gopalswamy et al., 2003e), the GCR intensity recovers quickly, reaching peak intensity at the solar minimum, when the new odd-numbered cycle begins. As the solar activity builds up, LL CMEs become more abundant, but there are no HL CMEs during the rise phase so GCR intensity is still relatively high. When HL CMEs start appearing in the pre-maximum phase of the odd cycle the GCR intensity drops precipitously until the solar polarity reverses at the odd-cycle maximum to begin the A<0 epoch. In the A<0 epoch, GCRs start entering equatorward. Since the LL activity during the declining phase is relatively high, GCR intensity continues to be affected by the LL CMEs until the activity approaches the activity minimum. Then comes the rise phase of the next even cycle, with continued blocking of GCRs solely by LL

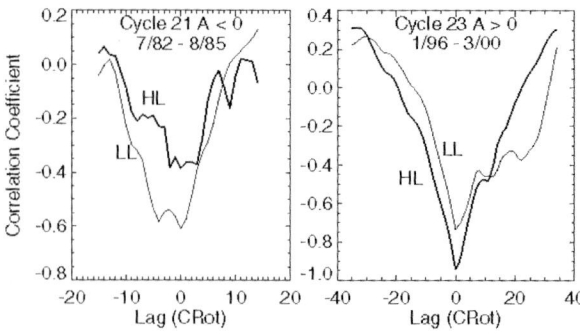

Figure 8.22 Cross-correlation between GCR intensity and HL and LL CME rates for (left) the A<0 epoch of cycle 21 and (right) the A>0 epoch of cycle 23.

CMEs. The appearance of HL CMEs before the maximum of the even cycles is of no consequence because the GCRs still approach equatorward and hence severely affected by the LL CMEs of the even-numbered cycle. This completes the 22-year cycle consisting of flat-topped and pointy components. The pointy component is tightly correlated with SSN (because of modulation by LL CMEs); the flat-top component is correlated with HL CMEs (PCF-related) owing to their appearance just before the solar maximum. The flat-topped response naturally explains the lag between solar activity and GCR recovery (Cliver and Ling, 2001).

11. Some Outstanding Questions

11.1 CME Initiation

Even after three decades of CME observations, we do not fully understand how CMEs are initiated. We do understand the details of the pre-CME structure: a set of one or more closed flux systems that eventually erupt. This could be a simple bipole with a core-envelope structure (Moore et al., 2001; Magara and Longcope, 2001), a flux rope with overlying restraining field (Low and Zhang, 2002; Forbes et al., 1994; Linker et al., 2001; Wu et al., 2000), a combination of bipoles (Machado et al., 1988) or multipolar structure (Antiochus et al., 1999; Chen and Shibata, 2000; Feynman and Martin, 1995). A successful CME model should account for the observed range of speed, mass, acceleration of CMEs, and the distribution of energy into heating, particle acceleration and mass motion. The current level of sophistication of CME models is less than adequate to account for all the observed characteristics (see e.g., Forbes, 2000; Klimchuk, 2000). Initial models based on the assumption of flare-produced CMEs (e.g., Dryer, 1982) have largely been abandoned because CME onset precedes flare onset (e.g., Wagner et al., 1981). After this, the emphasis shifted to loss of equilibrium (Low, 1996), primarily motivated by the three-part structure (frontal structure, cavity, and core) of CMEs and the coronal helmet streamers well observed in eclipse pictures (Saito and Tandberg-Hanssen, 1973). The cavity is identified as a flux rope of low plasma density and high magnetic field strength. In the pre-eruption state, the flux rope is held down by the prominence mass, the mass of the plasma contained in the overlying fields, and the magnetic pressure of these overlying fields. A CME is produced when the confinement of the flux ropes breaks down for a variety of reasons, such as loss of prominence mass (Low and Zhang, 2002). The interaction between the current in the flux rope and in the current sheets in the overall configuration decides the eruption and dynamics of the flux rope. This way, it is even possible to account for the accelerating CMEs from inverse polarity prominences and the constant speed CMEs from normal polarity prominences.

It is currently believed that the energy required to propel the CME has to come from the magnetic fields of the solar source region (see, e.g., Forbes, 2000). To illustrate the maximum energy that may be needed in CMEs, let us consider the 2003 November 04 CME, the fastest ($\sim$2700 km s^{-1}) event of cycle 23 (see Fig. 12): The CME had a mass of $\sim$ 2×10^{16}g, so that we can estimate the kinetic energy to be $\sim 7 \times 10^{32}$ erg. There is probably no other CME with an energy larger than this, so we can take that the largest energy released from an eruption is $\sim 10^{33}$ erg, and might represent the maximum free energy in the magnetic fields of the source. Considering a large active region (photospheric diameter $\sim$5 arcmin), we can estimate its coronal volume of 10^{30} cm^3. An average coronal field of 200 G over this volume implies a magnetic potential energy of $\sim 10^{33}$ erg. Microwave observations of the corona above sunspots have shown magnetic fields exceeding 1800 G (White et al., 1991), so an average of 200 G is not unreasonable. The highest value of potential magnetic energy in active regions surveyed by Venkatakrishnan and Ravindra (2003) is also $\sim 10^{33}$ erg. Since the potential magnetic energy is probably smaller than the total magnetic energy by only a factor $<$2 (Forbes, 2000), we infer that occasionally a substantial fraction of the energy contained in an active region may be released in the form of a CME. How this much free energy builds up in active regions is not fully understood.

11.2   How do CMEs Evolve?

In section 8, we saw that only a small fraction of CMEs originating at the Sun seem to reach 1 AU and beyond. This means a large number of CMEs may not survive as distinct entities for too long. Figure 23 shows the evolution of a white-light CME, which faded within the LASCO/C2 FOV above the northeast limb. The final height up to which the CME

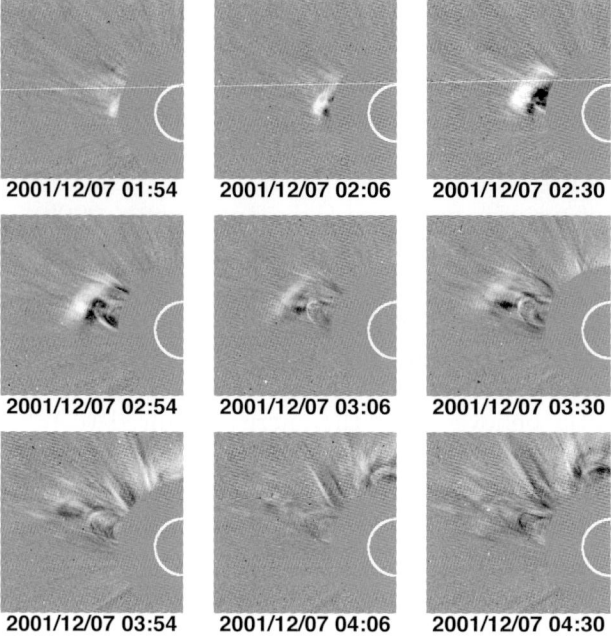

2001/12/07 01:54	2001/12/07 02:06	2001/12/07 02:30
2001/12/07 02:54	2001/12/07 03:06	2001/12/07 03:30
2001/12/07 03:54	2001/12/07 04:06	2001/12/07 04:30

Figure 8.23. Evolution of the 2001 December 07 CME at 01:54 UT as observed by SOHO/LASCO from the above the northeast limb. The CME is well defined but within an hour falls apart and fades away.

could be tracked was ~ 5.5 $R_\odot$. The CME was slow (273 km s^{-1}) and did not show deceleration. Figure 24 shows the distribution of the final heights of all the CMEs within the $32R_\odot$ FOV of LASCO. Clearly, many CMEs could not be tracked beyond $\sim 10R_\odot$. It is not clear if these CMEs faded because their density became too small to be detected or they ceased to exist as an entity different from the solar wind. The distribution also shows a second peak close to the edge of the field of view. Preliminary investigation shows that these are indeed the fast and wide CMEs (including halo CMEs). What causes the rapid dissipation of the smaller CMEs? Speculations such as the presence of enhanced turbulence in the 10-20 $R_\odot$ region (Mullan, 1997) need to be explored to understand them.

The shock-driving CMEs constitute a small fraction (a few percent) of all CMEs (Gopalswamy et al., 2003c), much smaller than the 20% estimated by Hundhausen (1999). The majority of CMEs are likely to be subAlfvenic and supersonic. These CMEs must be driving slow and intermediate shocks, as suggested by simulation studies (Whang, 1987; Steinolfson, 1992). Flat-top and concave upward morphology observed

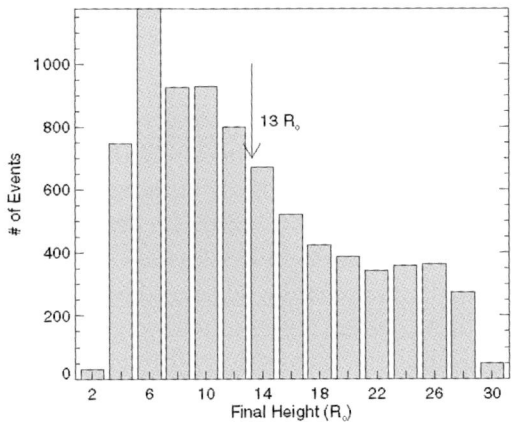

Figure 8.24 Distribution of the final heights of all the LASCO CMEs from 1996 to 2003. The average value is only 13 $R_\odot$, compared to the LASCO FOV of $32 R_\odot$. Note the second peak in the distribution close to the edge of LASCO FOV. The second bin from the left corresponds to the edge of the LASCO/C2 FOV. The large number in this bin is likely due to the fact that many CMEs could not be tracked from C2 to C3 FOV.

in some SMM CMEs are thought to indicate the presence of slow and intermediate shocks (Hundhausen 1999). The extensive SOHO database needs to be exploited to fully understand the slow and fast mode shocks. Such studies will be helpful in understanding the lateral structure of CMEs.

CMEs are observed as density enhancements within the coronagraph field of view. On the other hand there are many in situ signatures of CMEs as discussed in section 7. Most models dealing with CME initiation assume that the CME is a flux rope coming out of an eruption region to be either preexisting (Low and Zhang, 2002) or formed during eruption (Gosling et al., 1995). The current paradigm is that the flux of the envelope field is transferred to the flux rope during the eruption, so at 1 AU only the flux rope is observed. If the envelope field continues to be present, one should be able to observe counterstreaming electrons in the sheath of shock-driving CMEs. There seems to be no evidence for such counterstreaming (Gosling 2004, private communication). This may be a good test to understand the interplanetary evolution of CMEs. Another kind of evolution is the change in topology of the CME field lines from closed to open via interchange reconnection process (Larson et al., 1997; Crooker et al., 2002), which might explain the magnetic flux balance in the heliosphere. While important, such reconnection between background solar wind and the ICME field lines does not appear to be able to destroy the overall ICME structure over distances of 1 AU (Schmidt and Cargill, 2003). More studies are needed to assess how common such evolution is, given the fact that magnetic clouds are observed throughout the heliosphere (Burlaga et al., 1993).

12.   Summary

CMEs are multithermal structures in general, carrying coronal ($\sim$2 MK) material in the front followed by cool prominence ($\sim$8000 K) material in some cases and hot flare-material ($\sim$10 MK) in others. In some CMEs, there is a void between the frontal structure and the prominence core, with coronal temperature and a magnetic field stronger than in the ambient corona. The prominence core can be observed in X-rays, microwaves, H-alpha, and EUV as hot ejecta (thermal emission) or as a moving type IV burst due to nonthermal electrons trapped in the ejecta. After the eruption, hot post-eruption arcades or flare loops form, marking the location of eruption on the Sun. CME speeds vary over three orders of magnitude, from $\sim$ 20 km s^{-1} to more than 2500 km s^{-1} and the average speed shows a clear increase towards the solar activity maximum. Typical CMEs are $\sim$ 47° wide, but a small fraction of fast and wide CMEs have far-reaching heliospheric consequences. Fast CMEs drive powerful fast mode shocks, which in turn accelerate electrons and ions over extended periods of time. All the CME substructures are likely to propagate into the heliosphere, producing various observational signatures. The shock-accelerated electrons produce type II radio bursts in the IP medium. While the protons and heavier ions accelerated by the shock near the Sun reach 1 AU in a few tens of minutes, locally accelerated energetic particles arrive with the shocks on the time scale of days. CMEs undergo varying acceleration due to a combination of propelling and retarding forces. Far from the Sun, most CMEs tend towards the speed of the solar wind and the magnitude of acceleration is several m s^{-2}. CMEs arriving at Earth can also cause major geomagnetic storms if they possess southward magnetic field component that can reconnect with Earth's magnetic field. These storms are associated with a number of phenomena in other layers of Earth's atmosphere and on the ground. A fraction of the CMEs observed near the Sun can propagate to the far reaches of the heliosphere and become part of the merged interaction regions of various scales. These CMEs also can scatter off galactic cosmic rays, probably contributing to the 22-year modulation cycle.

The superior capability of the SOHO/LASCO coronagraphs enabled us to observe CMEs with unprecedented continuity and spatial coverage and hence we have a better picture of the whole phenomenon over a significant fraction of solar cycle 23, and we hope that SOHO will acquire data over the remaining part of the solar cycle. The solar maximum CME rate was found to be much higher than previously thought. Even though there is good correlation between CME rate and Sunspot number, their peaks were nearly two years apart. To fully understand this relationship, we need to consider both sunspot (active region) and non-spot (filament region) sources of CMEs. High-latitude CMEs associated with polar crown filaments and low-latitude CMEs from the active region belt naturally form into two groups, which have wider implications than just the initiation issue. The cessation of high-latitude activity seems to clearly mark the completion of polarity reversals at least for cycles 23 and 21. This leads to an important conclusion that the polarity reversal is an energetic process involving the release of large amounts of energy. The rate of high latitude CMEs is clearly related to the migration of closed field structures to the poles (one indication is the rush to the poles of the polar crown filaments as signified by the high tilt angles.) Occasionally, the high-latitude rate can be as high as the low-latitude rate, but overall the low-latitude activity dominates.

Halo CMEs and fast-and-wide CMEs are important from the space weather point of view. These CMEs constitute a small subset of all CMEs and can be studied independent of the thousands of ordinary CMEs. We need to focus on front-sided halo CMEs for assessing their geoeffectiveness. The shock-driving capability of the fast and wide CMEs is an important aspect consistent with the current paradigm that energetic particles (in large events) are accelerated by these shocks. The structural and magnetic connection between CMEs near the Sun and in the heliosphere is clear in a crude sense, but the details are still missing. The vast amount of CME data put out by SOHO and the availability of a wealth of complementary data from space and ground is likely to lead to appreciable progress in CME research. The birth, life, and death of CMEs involve an intriguing chain of physical processes on a grand scale, fully observable using ground and space borne instruments. Thus the study of CMEs is of enormous interest in uncovering the underlying physics of interaction between plasma and magnetic field. Studying CMEs is also crucial in understanding the space environment, into which humans often venture, because they cause intense geomagnetic storms and drive shocks that rise the particle radiations to hazardous levels.

Acknowledgments: I thank L. Burlaga, H. Hudson, S. Kahler, A. Ciaravella, S. T. Suess, and O. C. St. Cyr for critical comments on the manuscript. I thank S. Yashiro and A. Lara for help with some of the figures, and S. Nunes for cross-checking the references. I acknowledge the LASCO team headed by R. A. Howard and the Wind/WAVES team headed by J.-L. Bougeret and M. L. Kaiser for making their data available on line. SOHO is a project of international cooperation between ESA and NASA.

References

Andrews, M. & Howard, R. A., 2001, Space Sci. Rev., 95, 127

Antiochos, S. K., Dahlburg, R. B., & Klimchuk, J. A., 1999, Astrophys. J., 510, 485

Anzer, U. & Pnueman, G. W., 1982, Solar Phys., 17, 129.

Balogh, A., 2002, The evolving Sun and its influence on Planetary Environments, ed. B. Montosinos, A. Gimenez. & E.F. Guinan, ASP Conf. Ser., 269, 37

Barghouty, A. F. & Mewaldt, R. A., 2000, AIP Conf. Proc. 528: Acceleration and Transport of Energetic Particles Observed in the Heliosphere, 528, p. 71

Bastian, T. S., Pick, M., Kerdraon, A., Maia, D., & Vourlidas, A., 2001, Astrophys. J., 558, L65

Bhatnagar, A., 1996, Astrophys. Space Sci., 243, 105

Bieber, J. et al., 1999, Phys. Rev. Lett., 83, 674

Bieber, J. et al., 2002, Astrophys. J., 567, 622

Bilenko, I. A., 2002, Astron. Astrophys, 396, 657

Boischot, A., 1957, Comptes Rendus Acad. Sci., Paris, 244, 1326

Borrini, G., Gosling, J. T., Bame, S. J., & Feldman, W. C., 1982, J. Geophys. Res., 87, 4365

Bothmer, V., & Rust, R. M., 1997, Coronal Mass Ejections, ed. N. Crooker, J. Joselyn, & J. Feynman (Washington DC: Amer. Geophys. Union), p. 139

Bothmer, V. & Schwenn, R., 1994, Space Sci. Rev., 70, 215

Bougeret, J.-L., et al., 1995, Space Sci. Rev., 71, 231

Brueckner, G. E. et al., 1995, Solar Phys., 162, 357

Burkepile, J. et al., 2002, Fall AGU Meeting 2002, abstract #SH21A-0

Burlaga, L., Sittler, E., Mariani, F., & Schwenn, R., 1981, J. Geophys. Res., 86, 6673

Burlaga, L. et al ., 1982, Geophys. Res. Lett., 9, 1317

Burlaga, L. F., Goldstein, M. L., McDonald, F. B., & Lazarus, A. J., 1985, J. Geophys. Res., 90, 12027

Burlaga, L., Behannon, K. W. & Klein, L. W, 1987, J. Geophys. Res., 92, 5725

Burlaga, L. F., McDonald, F. B., & Ness, N. F., 1993, J. Geophys. Res., 98, 1

Burlaga, L. et al., 1998, J. Geophys. Res., 103, 277

Cane, H., 1983, JPL Solar Wind Five, p. 703.

Cane, H., Sheeley, N. R., & Howard, R. A., 1987, J. Geophys. Res., 92, 9869

Cane, H. & Richardson, I. G., 2003, J. Geophys. Res., 108(4), SSH 6-1

Cargill, P. J., Chen, J., Spicer, D. S., & Zalesak, S. T., 1996, J. Geophys. Res., 101, 4855

Chen, J., 1989, Astrophys. J., 338, 453

Chen, J. et al., 2000, Astrophys. J., 533, 481

Chen, J. & Krall, J., 2003, J. Geophys. Res., 108, A11, 1410

Chen, P. F. & Shibata, K., 2000, Astrophys. J., 545, 524

Chertok, I. M., Grechnev, V. V., Hudson, H. S., & Nitta, N. V., 2004, J. Geophysical Res., 109, 2112

Ciaravella, A. et al., 2003, Astrophys. J., 597, 1118

Cliver, E. W., 1996, High Energy Solar Physics, ed. R. Ramaty, N. Mandzhavidze, & X.-M. Hua, AIP Conf. Proc., Vol. 374. Woodbury, NY, p.45

Cliver, E. W., & Ling, A. G., 2001a, Astrophys. J., 551, L189

Cliver, E. W., & Ling, A. G., 2001b, Astrophys. J., 556, 432

Cliver, E. W., St. Cyr, O. C., Howard, R. A., & McIntosh, P. S., 1994, in Solar coronal structures, ed. V. Rusin, P. Heinzel & J.-C. Vial, VEDA Publishing House of the Slovak Academy of Sciences, p.83

Cliver, E. W., Webb, D.F. & Howard, R. A., 1999, Solar Phys., 187, 89

Cliver, E. W., Ling, A. G. & Richardson, I. G., 2003, Astrophys. J., 592, 574

Crifo, F., Picat, J. P., & Cailloux, M., 1983, Solar Phys., 83, 143

Crooker, N. U., 2000, JASTP, 62, 1071

Crooker, N. U., Gosling, J. T., & Kahler, S. W., 2002, J. Geophys. Res., 107(A2), SSH 3-1

DeMastus, H. L., Wagner, W. J., & Robinson, R. D., 1973, Solar Phys., 31, 449

Delaboudiniere, J.-P. et al., 1995, Solar Phys., 162, 291

Desai, M. et al., 2003, Astrophys. J., 588, 1149

Dulk, G. A., & McLean, D. J., 1978, Solar Phys., 57, 279

Dryer, M., 1996, Solar Phys., 169, 421

Eddy, J. A., 1974, Astron. Astrophys., 34, 235

Engvold, O., 1997, in New Perspectives on Solar Prominences (IAU Colloquium 167), ed. D. Rust, D. F. Webb, & B. Schmieder, Vol. 150, p. 23

Feynman, J. & S. F. Martin, 1995, J. Geophys. Res., 100, 3355

Filippov, B. P. 1998, in New Perspectives on Solar Prominences (IAU Colloquium 167), ed. D. Rust, D. F. Webb, & B. Schmieder, Vol. 150, p. 342

Forbes, T., 2000, J. Geophys. Res., 105, 23153

Forbes, T. G., P. A. Isenberg, & E. R. Priest, 1994, Solar Phys., 150, 245

Funsten, H. O. et al., 1999, J. Geophys. Res., 104, 6679

Gary, D. et al., 1984, Astron. Astrophys., 134, 222

Gold, T, 1955, J. Geophys. Res., 64, 1665

Gopalswamy, N. 1999, Solar Physics from Radio Observations, Eds.: T. S. Bastian, N. Gopalswamy & K.Shibasaki, NRO Report No. 479., p.141

Gopalswamy, N., 2002, Solar-Terrestrial Magnetic Activity and Space Enviroment, COSPAR Colloquia Series, Vol. 14, edited by H. N. Wang & R. L. Xu, p. 157

Gopalswamy, N., 2003a, Geophys. Res. Lett., 30(12), SEP 1-1

Gopalswamy, N., 2003b, Adv. Space Research, 31(4), 869

Gopalswamy, N. & Kundu, M. R., 1989, Solar Phys., 122, 145

Gopalswamy, N. & Kundu, M. R., 1992, Particle acceleration in cosmic plasmas, AIPC 264, p. 257

Gopalswamy, N. & Kundu, M. R., 1993, Solar Phys., 143, 327

Gopalswamy, N., Kundu, M. R., Hanaoka, Y., Enome, S., & Lemen, J. R. 1996, New Astronomy, 1, 207

Gopalswamy, N. et al., 1997a, Astrophy. J., 475, 348

Gopalswamy, N. et al., 1997b, Astrophy. J., 486, 1036

Gopalswamy, N., & Hanaoka, Y, 1998, Astrophy. J., 498, L179.

Gopalswamy, N. et al., 1998, Geophys. Res. Lett., 25, 2485

Gopalswamy, N., Yashiro, S., Kaiser, M. L., Thompson, B. J., & Plunkett, S., 1999, Solar Physics with Radio Observations, Eds.: T. S. Bastian, N. Gopalswamy & K. Shibasaki, NRO Report No. 479., p.207

Gopalswamy, N. & Thompson, B. J., 2000, JASTP, 62, 1457

Gopalswamy, N. et al., 2000a, Geophys. Res. Lett., 27, 145

Gopalswamy, N. et al., 2000b, Geophys. Res. Lett., 27, 1427

Gopalswamy, N. Hanaoka, Y. & Hudson, , 2000c Adv. Space Res., 25(9), 1851

Gopalswamy, N., Kaiser, M. L., Sato, J., & Pick, M., 2000d, in High Energy Solar Physics, Ed. R. Ramaty & N. Mandzhavidze, PASP Conf Ser., vol. 206, p. 355

Gopalswamy, N., Lara, A., Kaiser, M. L., & Bougeret, J.-L., 2001a, J. Geophys. Res., 106, 25261

Gopalswamy, N., Lara, A., Yashiro, S., Kaiser, M. L. & Howard, R. A., 2001b, J. Geophys. Res., 106, 29107

Gopalswamy, N., Yashiro, S., Kaiser, M. L., Howard, R. A., & Bougeret, J.-L., 2001c, Astrophys. J., 548, L91

Gopalswamy, N., St. Cyr, O. C., Kaiser, M. L., Yashiro, S., 2001d, Solar Phys., 203, 149

Gopalswamy, N., Yashiro, S., Kaiser, M. L., Howard, R. A. & Bougeret, J.-L., 2001e, J. Geophys. Res., 106, 29,219

Gopalswamy, N. & Kaiser, M. L., 2002, Adv. Space Res., 29(3), 307

Gopalswamy, N., Yashiro, S., Kaiser, M. L., Howard, R. A., & Bougeret, J.-L., 2002a, Geophys. Res. Lett., 29(8), DOI:10.1029/2001GL013606

Gopalswamy, N., et al., 2002b, Astrophys. J., 572, L103

Gopalswamy, N. Shimojo, M., Lu, W., Yashiro, S., Shibasaki, K. & Howard, R. A., 2003a, Astrophys. J., 586, 562

Gopalswamy, N., Lara, A., Yashiro, S., Nunes, S., & Howard, R. A., 2003b, Solar variability as an input to the Earth's environment, Ed.: A. Wilson. ESA SP-535, Noordwijk: ESA Publications Division, p. 403

Gopalswamy, N. et al., 2003c, Geophys. Res. Lett., 30(12), SEP 3-1

Gopalswamy, N., Manoharan, P. K., & Yashiro, S., 2003d, Geophys. Res. Lett., 30(24), SSC 1-1

Gopalswamy, N., Lara, A., Yashiro, S., & Howard, R. A., 2003e, Astrophys. J., 598, L63

Gosling, J. T., 1993, J. Geophys. Res., 98, 18937

Gosling, J. T., 1997, Coronal Mass Ejections, ed. N. Crooker, J. Joselyn, & J. Feynman (Washington DC: Amer. Geophys. Union), p. 9

Gosling, J. T., Asbridge, J. R., Bame, S. J., Hundhausen, A. J., & Strong, I. B., 1968, J. Geophys. Res., 73, 43

Gosling J. T. et al., 1974, J. Geophys. Res., 79, 4581

Gosling, J. T. et al., 1976, Solar Phys., 48, 389

Gosling, J. T., Bame, S. J., McComas, D. J., & Phillips, J. L., 1990, Geophys. Res. Lett., 17, 901

Gosling, J. T., McComas, D. J., Phillips, J. L., & Bame, S. J., 1992, J. Geophys. Res., 97, 6531

Gosling, J. T. et al., 1994, Geophys. Res. Lett., 21, 2271

Gosling, J. T., Birn, J., & Hesse, M., 1995, Geophys. Res. Lett., 22, 869

Gosling, J. T. & Forsyth, R. J., 2001, Space Sci. Rev., 97, 87

Hanaoka et al., 1994, PASJ, 46, 205

Hansen, R. T., Garcia, C. J., Grognard, R. J.-M., & Sheridan, K. V., 1971, Proc. ASA, 2, 57

Harrison, R. A., 1990, Solar Phys., 126, 185

Harrison, R. A., 1991, Adv. Space res., 11(1), 25

Harrison, R. A., 1995, Astron. Astrophys., 304, 585

Harvey, K. & Recely, F., 2002, Solar Phys., 211, 31

Henke, T., 1998, Geophys. Res. Lett., 25, 3465

Hewish, A., Tappin, S. J. & Gapper, G. R., 1985, Nature, 314, 137

Hildner, E. et al., 1976, Solar Phys., 48, 127

Hildner, E., 1977, Study of Travelling Interplanetary Phenomena, ed. M.A. Shea, D.F. Smart, & S.T. Wu., ASSL, Vol. 71, p.3

Hirshberg, J., Bame, S. J., & Robbins, D. E., 1972, Solar Phys., 23, 467.

Holman, G. D. & Pesses, M. E., 1983, Astrophys. J., 267, 837

Hori, K. & Culhane, J. L., 2002, Astron. Astrophys., 382, 666.

House, L. L., Wagner, W. J., Hildner, E., Sawyer, C., & Schmidt, H. U. 1981, Astrophys. J., 244, L117

Howard, R. A., Michels, D. J., Sheeley, N. R., Jr., & Koomen, M. J., 1982, Astrophys. J., 263, L101

Howard, R. A., Sheeley, N. R., Jr., Michels, D. J., & Koomen, M. J., 1984, Adv. Space Res., 4(7), 307

Howard, R. A., Michels, D., Sheeley, N. R., & Koomen, M. J., 1985, J. Geophys. Res., 90, 8173

Howard, R. A., Michels, D., Sheeley, N. R., & Koomen, M. J., 1986, ASSL Vol. 123: The Sun and the Heliosphere in Three Dimensions, ed. R. Marsden, D. Reidel, Norwell, Mass., p. 107

Howard, R. A. et al., 1997, Coronal Mass Ejections, ed. N. Crooker, J. Joselyn, & J. Feynman (Washington DC: Amer. Geophys. Union), 17

Hudson, H., 1999, Eds.: T. S. Bastian, N. Gopalswamy & K. Shibasaki, NRO Report No. 479, p.15

Hudson, H. & D. F. Webb, 1997, in Coronal Mass Ejections, ed. N. Crooker, J. A. Joselyn, & J. Feynman, AGU Monograph 99, p. 27

Hudson, H., Acton, L. W. & Freeland, S. L., 1996, Astrophys. J., 470, 629

Hudson, H. S. & Cliver, E. W., 2001, J. Geophys. Res., 106, 25199

Hudson, H. S., Kosugi, T., Nitta, N. V., & Shimojo, M., 2001, Astrophys. J., 561, L211

Hudson, H. S., Khan, J. I., Lemen, J. R., Nitta, N. V., & Uchida, Y., 2003, Solar phys., 212, 121

Hundhausen, A. J. 1987, Proc. Sixth international Solar Wind Conference, Vol. 1, ed. V. J. Pizzo, T. E. Holzer, & D. G. Sime, High Altitude Observatory, NCAR, Boulder, Colorado, p. 181

Hundhausen, A. J., 1993, J. Geophys. Res., 98, 13,177

Hundhausen, A. J., 1997, in Coronal Mass Ejections, ed. N. Crooker, J. A. Joselyn, & J. Feynman, AGU Monograph 99, p. 1

Hundhausen, A. J., 1999, Many Faces of the Sun, ed. K. T. Strong, J. L. R. Saba, & B. M. Haisch, Springer-Verlag, New York, p. 143

Hundhausen, A., Bame, S. J., & Montgomery, M. D., 1970, J. Geophys. Res., 75, 4631

Illing, R. M. E. & Athay, G., 1986, Solar Phys., 105, 173

Jackson, B., 1985, Solar Phys., 100, 563

Jackson, B. & Howard, R. A., 1993, Solar Phys., 148, 359

Jokipii, J. R., Levy, E. H., & Hubbard, W. B., 1977, Astrophys. J., 213, 861

Kahler, S. W., 1992, ARA&A, 30, 113

Kahler, S. W., 1993, J. Geophys. Res., 98, 5607

Kahler, S. W., 1994, Astrophys. J., 428, 837

Kahler, S. W., 2001, J. Geophys. Res., 106, 20947

Kahler, S. W.,Hildner, E. & van Hollebeke, M. A. I., 1978, Solar Phys., 57, 429

Kahler, S. W., Sheeley, N. R., Howard, R. A., Michels, D. J., & Koomen, M. J., 1984, Solar Phys., 93, 133

Kahler, S. W., et al., 1985, J. Geophys. Res., 90, 177

Kahler, S. W., 1987, Sixth International Solar Wind Conference, Ed. V.J. Pizzo, T. Holzer, & D.G. Sime, NCAR Technical Note NCAR/TN-306+Proc, Volume 2, 1987., p.215

Kahler, S. W., Sheeley, N. R., Jr. & Liggett, M., 1989, Astrophys. J., 344, 1026

Kallenrode, M.-B., & Cliver, E. W., 2001, Proc. ICRC 2001, 3319

Kim, I. & Alexeyeva, V., 1994, Solar Active region Evolution: Comparing Models with Observations, ed. K. S. Balasubramaniam & G. Simon, ASP Conf. ser. 68, p. 403

Klassen, A., H. Aurass, G. Mann & B. J. Thompson, 2000, Astron. Astrophys. (Sup.), 141, 357.

Klein, L. W. & Burlaga, L. F., 1982, J. Geophys. Res., 87, 613

Klimchuk, J. A., 2000, Space Weather, ed. P. Song, H. J. Singer, & G. L. Sisco, AGU Monograph 125, p. 143

Klimchuk, J. A., et al., 1994, in X-Ray Solar Physics from Yohkoh, ed. Y. Uchida (Tokyo: Universal Academy Press), 181

Kocharov, L. & Torsti, J., 2002, Solar Phys., 207, 149

Kohl, J. L. et al., 1995, Solar Phys., 162, 313

Lara, A., Gopalswamy, N., Nunes, S., Muñoz, G., & Yashiro, S., 2003, Geophys. Res. Lett., 30, No.12, SEP 4-1

Lara, A., Gopalswamy, N., Caballero-Lopez, R. A., Yashiro, S. & Valdes-Galicia, J. F., 2004, Astrophys. J. (submitted)

Larson, D. E. et al., 1997, Geophys. Res. Lett., 24, 1911

Lee, M., 1997, in Coronal Mass Ejections, ed. N. Crooker, J. A. Joselyn, & J. Feynman, AGU Monograph 99, p. 227

Lepping, R. L., J. A. Jones, & L. F. Burlaga, 1990, J. Geophys. Res., 95, 11957

Lepri, S. et al., 2001, J. Geophys. Res., 106, 29231

Lin, R. P., 1987, Solar Phys., 113, 217

Lindemann, F. A., 1919, Phil. Mag., 38, 669

Lindsay, G. M., Luhmann, J. G., Russell, C. T., & Gosling, J. T., 1999, J. Geophys. Res., 104, 12515

Linker, J. A., Lionello, R., Mikić, Z., & Amari, T., 2001, J. Geophys. Res., 106, 25165

Lorenc, M., Pasorek, L., & Rybanský, M., 2003, in Solar Variability as an input to the Earth's Environment, ESA-SP, 535, Noordwijk: ESA Publications Division, p. 129

Low, B. C., 1996, Solar Phys., 167, 217

Low, B. C. 1997, in Coronal Mass Ejections, ed. N. Crooker, J. A. Joselyn, & J. Feynman, AGU Monograph 99, p. 39

Low, B. C., 2001, J. Geophys. Res., , 106, 25141

Low, B. C. & Zhang, M., 2002, Astrophys. J., 564, L53

Machado, M. et al., Astrophys. J., 326, 425

MacQueen, R. M. & R. Fisher, 1983, Solar Phys., 89, 89

MacQueen, R. M. et al., 1974, Astrophys. J., 187, L85

MacQueen, R. M. et al., 1980, Solar Phys., 65, 91

Magara, T. & Longcope, D. W., 2001, Astrophys. J., 559, L55

Makarov, V. I., Tlatov, A. G., & Sivaraman, K. R., 2001, Solar Phys., 202, 11

Malitson, H. H., Fainberg, J., & Stone, R. G., 1973, Astrophys. Lett., 14, 111

Mancuso, S. & Raymond, J. C., 2004, Astron. Astrophys., 413, 363

Mann, G., Klassen, A., Estel, C., & Thompson, B. J., 1999, in Proc. of 8th SOHO Workshop, Edited by J.-C. Vial & B. Kaldeich-Schmann., p.477

Manoharan, P. K., Gopalswamy, N., Lara, A. & Yashiro, S., 2004, J Geophys. Res., in press.

Martin, S. F. & McAllister, A. H., 1997, Coronal Mass Ejections, ed. N. Crooker, J. Joselyn, & J. Feynman (Washington DC: Amer. Geophys. Union), p. 127

Marubashi, K., 1997, Coronal Mass Ejections, ed. N. Crooker, J. Joselyn, & J. Feynman (Washington DC: Amer. Geophys. Union), p. 147

Mason, G. M., Mazur, J. E., & Dwyer, J. R., 1999, Astrophys. J., 525, L133

Michalek, G., Gopalswamy, N. & Yashiro, S., 2003, Astrophys. J., 584, 472

Michels, D. J., Howard, R. A., Koomen, M. J., & Sheeley, N. R., Jr., 1980, Radio physics of the sun, ed. M.R. Kundu & T. E. Gergely, Dordrecht, D. Reidel, p.439

Moon, Y.-J. et al., 2002, Astrophys. J., 581, 94

Moore, R. L., Falconer, D. A., Porter, J. G., & Suess, S. T. 1999, Astrophys. J., 526, 505

Moore, R. L., Sterling, A. C., Hudson, H. S., & Lemen, J. R. 2001, Astrophys. J., 552, 833

Moreton, G. E., 1960, Astron. J., 65, 494

Mullan, D. J., 1997, AIP Conf. Proc. 385: Robotic Exploration Close to the Sun: Scientific Basis, 385, 235

Mulligan, T., Russell, R. T., & Luhmann, J., 1998, Geophys. Res. Lett., 25, 2959

Munro, R. H. et al. 1979, Solar Phys. 61, 201

Newkirk, G., Hundhausen, A. J. & Pizzo, V., 1981, J. Geophys. Res., 86, 5387

Narukage, N., Hudson, H. S., Morimoto, T., Akiyama, S., Kitai, R., Kurokawa, H., & Shibata, K., 2002, Astrophys. J., 572, L109

Nuegebauer, M. & Goldstein, B. E., 1997, in Coronal Mass Ejections, ed. N. Crooker, J. A. Joselyn, & J. Feynman, AGU Monograph 99, p.

Nitta, N. & Akiyama, S. 1999, Astrophys. J., 525, L57

Nitta, N. V. & Hudson, H. S., 2001, Geophys. Res. Lett., 28, 3801

Obayashi, T., 1962, J. Geophys. Res., 67, 1717

Pallavicini, R.; Serio, S.; Vaiana, G. S., 1977, Astrophys. J.,216, 108

Park, Y. D., Moon, Y.-J., Kim, I., & Yun, H. S. 2002, Ap&SS, 279, 343

Payne-Scott, R., Yabsley D. E., & Bolton, J. G., 1947, Nature, 160, 256

Plunkett, S. P. et al., 2000, Solar Phys., 61, 201

Potgeiter, M. S., Burger, R. A., & Ferreira, S. E. S., 2001, Space Sci. Rev., 97, 295

Poland, A., Howard, R. A., Koomen, M. J., Michels, D. J., Sheeley, N. R., Jr., 1981, Solar phys., 69, 169

Ramesh, R., Kathiravan, C., & Sastry, C. V., 2003, Astrophys. J., 591, L163

Raymond, J. et al., 2000, Geophys. Res. Lett., 27, 1439

Reames, D. V., 1999, Space Sci. Rev., 90, 413

Reames, D. V., 2001, Geophysical Monograph 125, AGU, Washington DC, 101

Reames, D. V., Ng, C. K., & Tylka, A. J., 1999, Geophys. Res. Lett., 26, 3585

Reiner, M. J. & Kaiser, M. L., 1999, JGR, 104, 16979

Reiner, M. J. et al., 2001, JGR, 106, 25279

Reisenfeld, D. B., Gosling, J. T., Forsyth, R. J., Riley, P., & St. Cyr, O. C., 2003, Geophys. Res. Lett., 30, 5

Robinson, R. D., 1985, Solar Phys., 95, 343

Robinson, R. D. & Stewart, R. T., 1985, Solar Phys., 97, 145

Rusin, V. & Rybansky, M., 1983, BAICz, 34, 25

Rust, D. M. & Hildner, E., 1976, Solar Phys., 48, 381

Saito, K. & Tandberg-Hanssen, E., 1973, Solar Phys., 31, 105

Schmidt, J. M. & Cargill, P. J. 2003, J. Geophysical Res., 108, SSH 5-1

Schwenn, R., 1986, Space Sci. Rev., 44, 139

Schwenn, R., 1996, Astrophys. Space Sci., 243, 187

Sheeley, N., Howard, R. A., Koomen, M. J., & Michels, D. J., 1983, Astrophys. J., 272, 349

Sheeley, N. R. et al., 1984, Astrophys. J., 279, 839

Sheeley, N. R., et al., 1985, J. Geophys. Res., 90, 163

Sheeley, N. R., J. H. Walters, Y.-M. Wang, & R. A. Howard, 1999, J. Geophys. Res., 104, 24739

Sheeley, N. R., Hakala, W. N., Wang, Y.-M., 2000, J. Geophys. Res., 105, 5081

Shibata, K., et al., 1992, PASJ, 44, 173

Shibata, K., et al., 1995, ApJ, 451, L83

Sonett, C. P., Colburn, C. D. S., Davis, L., Smith, E. J., & Coleman, P. J., 1964, Phys. Rev. Lett., 13, 153

St. Cyr, O. C. & Webb, D. F., 1991, Solar Phys., 136, 379

St. Cyr, O. C., Burkepile, J. T., Hundhausen, A. J., & Lecinski, A. R., 1999, J. Geophys. Res. , 104, 12493

St. Cyr, O. C. et al., 2000, J. Geophys. Res., 105, 18169

Steinolfson, R. S., 1992, Proc. Of the 26th ESLAB Symposium on the Study of the Solar-terrestrial physics

Sterling, A., 2003, Solar variability as an input to the Earth's environment, Ed.: A. Wilson, ESA SP-535, Noordwijk: ESA Publications Division, p. 415

Sterling, A., & Hudson, H., 1997, Astrophys. J., 491, L55

Stewart, R. T., 1985, Solar Radiophysics, Cambridge and New York, Cambridge, University Press, p. 361

Tandberg-Hanssen, E. 1995, The Nature of Solar Prominences, Kluwer, Dordrecht

Tappin, S. J. & Simnett, G. M. 1997, Correlated Phenomena at the Sun, in the Heliosphere and in Geospace, Edited by A. Wilson. European Space Agency, ESA SP-415, p.117

Thompson, B. J. et al., 1999, Astrophys. J., 517, L151

Tokumaru, M., Kojima, M., Fujiki, K., & Yokobe, A., 2000, J. Geophys. Res., 105, 10,435

Tousey, R., 1973, The solar corona, Space Res., 13, 713

Tousey, R., Howard, R. A., & Koomen, M. J., 1974, Bull. American Astron. Soc., 6, 295

Tylka, A., 2001, J. Geophys. Res., 106, 25,233

Uchida, Y., 1960, PASJ, 12, 376

Venkatakrishnan, P. V. & Ravindra, B., 2003, Geophys. Res. Lett., 30(23), SSC 2-1

Vourlidas, A., Buzasi, D., Howard, R. A., & Esfandiari, E., 2002, Solar variability: from core to outer frontiers, Ed. A. Wilson. ESA SP-506, Vol. 1. Noordwijk: ESA Publications Division, p. 91

Vourlidas, A., Wu, S. T., Wang, A. H., Subramanian, P., & Howard, R. A, 2003, Astrophys. J., 598, 1392

Wagner, W. J., 1984, ARA&A, 22, 267

Wagner, W. J. et al., 1981, Astrophys. J., 244, L123

Wagner, W. J. & MacQueen, R. M., 1983, Astron. Astrophys., 120, 136

Wang, Y.-M., Sheeley, N. R., & Andrews, M. D., 2002, J. Geophys. Res., 10

Webb, D. F., 2002, Half a Solar Cycle with SOHO, ed. A. Wilson, ESA SP-508, Noordwijk: ESA Publications, p. 409

Webb, D. F., Krieger, A. S. & Rust, D. M., 1976, Solar phys., 48,159.

Webb, D. F., E. W. Cliver, N. U. Crooker, O. C. St. Cyr, & B. J. Thompson, J. Geophys. Res., 105, 7491, 2000.

Webb, D. F., Davis, J. M. & McIntosh P. S., 1984, Solar Phys., 92, 109

Webb, D. F., Nolte, J. T., Solodyna, C. V. & McIntosh, P., 1978, Solar Phys., 58, 389

Webb, D. F. & Hundhausen, A. J. 1987, Solar Phys., 108, 383.

Webb, D. F., & Howard, R. A., 1994, J. Geophys. Res., 99, 4201

Webb, D. F., Cliver, E. W., Crooker, N. U., St. Cyr, O. C., & Thompson, B. J., 2000, J. Geophys. Res., 105, 7491

Whang, Y.-C., 1987, J. Geophys. Res., 92, 4349

White, S. M., Kundu, M. R. & Gopalswamy, N., 1991, Astrophys. J., 366, L43

Wild, J. P., 1950, Aust. J. Sci. Ser. A, 3, 541

Wilson, R. M., & E. Hildner, 1984, Sol. phys., 91, 169

Wood, B. E. et al., 1999, Astrophys. J., 512, 484

Wu, S. T., Guo, W. P., Plunkett, S. P., Schmieder, B., & Simnett, G. M., 2000, JATP, 62, 1489

Wu, S. T., Wang, A. H., & Gopalswamy, N, 2002, in SOLMAG 2002, Ed. H. Sawaya-Lacoste, ESA SP-505. Noordwijk, Netherlands, p. 227

Yashiro, S., Gopalswamy, N., Michalek, G., & Howard, R. A., 2003, Adv. Space Res., 32, 2631

Yashiro, S. et al., 2004, J. Geophys. Res., in press

Yeh, T, A dynamical model of magnetic clouds, Asrophys. J., 438, 975, 1995

Zhang, M. & Low, B. C., 2001, Asrophys. J., 561, 406

Zhang, J., Dere, K. P., Howard, R. A., Kundu, M. R., White, S. M., 2001, Asrophys. J., 559, 452

Chapter 9

MHD TURBULENCE IN THE HELIOSPHERE

Evolution and Intermittency

Bruno Bavassano
Istituto di Fisica dello Spazio Interplanetario (CNR)
Roma, Italy
bavassano@ifsi.rm.cnr.it

Roberto Bruno
Istituto di Fisica dello Spazio Interplanetario (CNR)
Roma, Italy
bruno@ifsi.rm.cnr.it

Vincenzo Carbone
Dipartimento di Fisica, Università della Calabria
Rende (CS), Italy
carbone@fis.unical.it

Abstract   The solar wind is an excellent laboratory to study the behaviour of MHD turbulence in a collisionless plasma. This is a fundamental topic in both plasma physics and astrophysics. The impressive amount of observations at different solar distances and latitudes collected in last three decades has allowed us to reach a good understanding on many aspects of the complex phenomenon of solar wind turbulence. The present review focuses on two major topics. One is the type of evolution followed by the turbulence as the solar wind expands into the interplanetary space. A comparison is performed between variations observed in low- and high-latitude solar wind, as derived from old measurements in the ecliptic and from recent data by Ulysses. Implications about processes of local generation of turbulence are discussed. The second topic is related to the fact that we are dealing with fluctuating fields which are neither isotropic nor single scale–invariant. These features cause intermittency in the solar wind MHD turbulence. In this review

G. Poletto and S.T. Suess (eds.), The Sun and The Heliosphere as an Integrated System, 253–282.
© 2004 *Kluwer Academic Publishers. Printed in the Netherlands.*

we will address both interplanetary observations and theoretical models which allow us to shed some light on the nature of the intermittency as observed in interplanetary space.

Keywords: Solar Wind, Turbulence, Intermittency

1.   Introduction

The heliosphere is a region of space influenced by the Sun and by its expanding corona, the solar wind. This is a high Reynolds' number plasma flow that offers one of the best opportunities to study collisionless plasma phenomena by in-situ measurements. Here we will focus on a topic of primary importance for both plasma physics and astrophysics, the magnetohydrodynamic (MHD) turbulence. It should be stressed that the use of the term MHD only refers to the fact that we are looking at phenomena falling in the MHD regime (i.e., frequency well below the proton gyrofrequency). The name does not imply that these phenomena may necessarily be described by the MHD theory.

The MHD turbulence strongly affects several aspects of heliospheric behaviour, such as plasma heating, solar wind generation, high-energy particles acceleration, and cosmic rays propagation. In the seventies and the eighties impressive advances have been made in the knowledge of turbulent phenomena in solar wind. In those years, however, with spacecraft observations confined within a small latitudinal belt around the solar equator, only a limited fraction of the heliosphere was accessible. In the nineties, with the launch of the Ulysses spacecraft, investigations have been extended to the high-latitude regions of the heliosphere. This has allowed us to study how MHD turbulence evolves in polar solar wind, a plasma flow in which the effects of large-scale inhomogeneities are considerably less important than in low-latitude wind. With this new laboratory, relevant advances have been made. In the following, the Ulysses observations of turbulence evolution in polar wind will be discussed and compared to those typical of near-equatorial solar wind. For an exhaustive review on turbulence in near-equatorial wind, with a historical perspective and a discussion of proposed theories, reference can be made to Tu and Marsch (1995). An overview of the Ulysses results on polar turbulence can be also found in Horbury and Tsurutani (2001).

The other topic of the present review is the intermittency of solar wind fluctuations. Intermittency has been the object of several studies in the past few years. This phenomenon is due to the fact that solar wind fluctuations are not isotropic and poorly single scale–invariant, two of

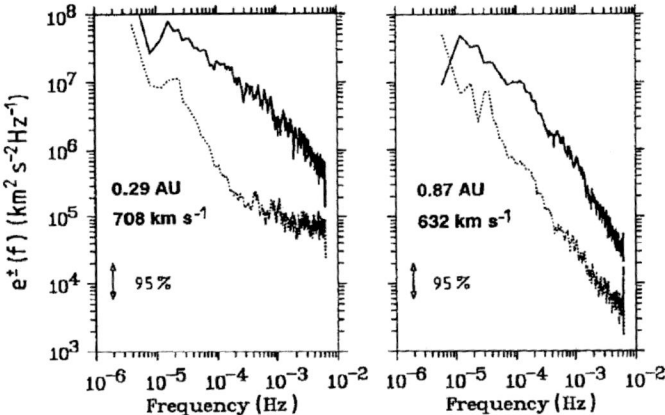

Figure 9.1. Power spectra of e_+ and e_- (solid and dotted line, respectively) in the trailing edge of fast streams at 0.29 and 0.87 AU (adapted from Marsch and Tu, 1990a, copyright 1990 American Geophysical Union, modified by permission of American Geophysical Union).

the fundamental hypotheses at the basis of Kolmogorov's (1941) theory called K41. The first direct consequence is that smaller and smaller turbulent eddies are less and less space filling if turbulence is looked at in the framework of the classical Richardson's cascade. This phenomenon causes solar wind turbulence to be intermittently distributed in space. In other words, the global scale invariance required in the Kolmogorov theory would evolve towards a local scale invariance where different fractal sets characterized by different scaling exponents can be found. In this work we will address both interplanetary observations and the theoretical models which allowed us to shed some light on the nature of this phenomenon as observed in interplanetary MHD turbulence.

2.   MHD Turbulence Evolution

As clearly indicated by spacecraft observations, MHD turbulence in solar wind has a strongly Alfvénic character. This is because Alfvénic modes have a longer lifetime than other MHD modes (Barnes, 1979). Thus, it seems useful to briefly recall some of the parameters that are generally used to describe Alfvénic fluctuations. Basic quantities are the Elsässer's variables ($z_\pm$). Tu and Marsch (1995) exhaustively discussed the use of these variables in solar wind turbulence studies. The Elsässer's variables are defined as $z_\pm = v \pm b$, where v and b are the velocity and magnetic field vectors, respectively. Note that in the above definition the magnetic field b has to be given in Alfvén units (i.e., the magnetic values have to be divided by $\sqrt{4\pi\rho}$, with ρ the mass density). Taking into account how the sign of the Alfvénic correlation depends on the propagation direction with respect to the background magnetic field, it has become common to use the above definition in the case of a background magnetic field pointing to the Sun, while the equation $\mathbf{z}_\pm = \mathbf{v}\mp\mathbf{b}$ is taken for the opposite polarity. With this choice we have that, whatever the polarity is, $\mathbf{z}_+$ ($\mathbf{z}_-$) fluctuations always correspond to modes with an outward (inward) direction of propagation, with respect to Sun, in the plasma frame. The relative weight of the energies (per unit mass) e_+ and e_- associated to $\mathbf{z}_+$ and $\mathbf{z}_-$ fluctuations, respectively, is measured by the Elsässer ratio $r_E = e_-/e_+$. An analogous measure for the energies e_V and e_B of the $\mathbf{v}$ and $\mathbf{b}$ fluctuations is given by the Alfvén ratio $r_A = e_V/e_B$. Other related parameters are the normalized cross-helicity $\sigma_C = (e_+-e_-)/(e_++e_-) = (1-r_E)/(1+r_E)$ and the normalized residual energy $\sigma_R = (e_V-e_B)/(e_V+e_B) = (r_A-1)/(r_A+1)$.

Both outward and inward propagating fluctuations are observed in the solar wind. Outward fluctuations mainly have a solar origin (or, more precisely, inside the Alfvén critical point). Conversely, inward propagating fluctuations can only be generated outside such a critical distance. As well known, the presence of both kinds of fluctuation leads to the development of nonlinear interactions.

The main features of the Alfvénic turbulence evolution versus radial distance (i.e., versus transit time to the observer) will be now discussed for ecliptic and polar wind. Ecliptic results are mainly based on Helios and Voyagers observations, while polar results obviously rely on Ulysses measurements.

2.1 Ecliptic Turbulence

Since the first studies on Alfvénic fluctuations in ecliptic wind (e.g., Belcher and Davis, 1971) it was clear that fast streams (or, more precisely, their trailing edge) are the best places to observe them. Helios spacecraft, with a systematic covering of the inner heliosphere from 0.3 to 1 AU, gave the first opportunity to study the turbulence evolution with solar distance (Bavassano et al., 1982, Denskat and Neubauer, 1982).

Figure 1 (from Marsch and Tu, 1990a) shows how e_+ and e_- power spectra (solid and dotted line, respectively) vary when solar distance increases from 0.3 to 0.9 AU. The e_+ spectrum declines faster than that of e_-, with the result that the two spectra approach each other. At the same time the spectral slopes evolve in such a way that an extended inertial regime expands to low frequencies.

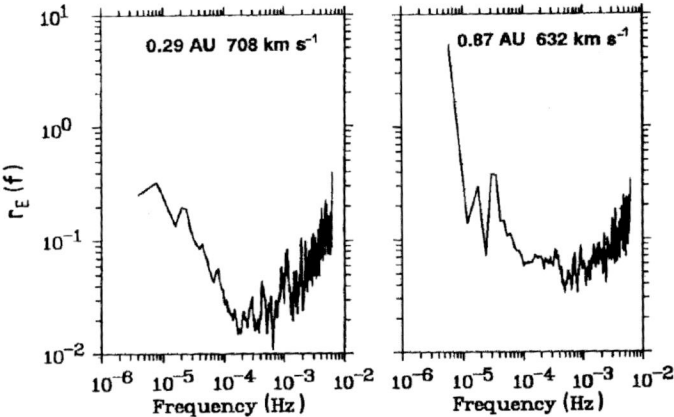

Figure 9.2. Power spectrum of the Elsässer ratio r_E ($= e_-/e_+$) in the trailing edge of fast streams at 0.29 and 0.87 AU (adapted from Marsch and Tu, 1990a, copyright 1990 American Geophysical Union, modified by permission of American Geophysical Union).

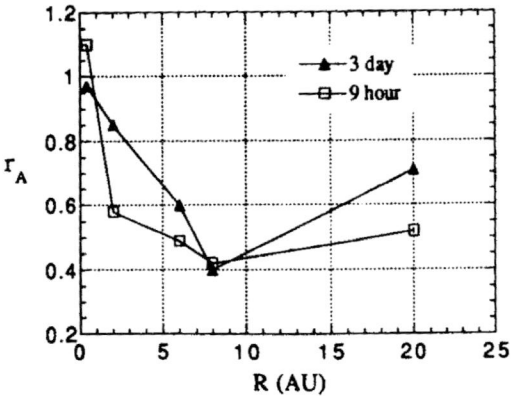

Figure 9.3 Radial variation of the Alfvén ratio r_A ($=e_V/e_B$) as seen by Helios and Voyagers between 0.3 and 20 AU (from Roberts et al. [1990], copyright 1990 American Geophysical Union). The 9-hr curve (see squares) shows that r_A, after the fast decrease at small distance, remains nearly unchanged.

The corresponding variation in the relative weight of outward and inward fluctuation energies is shown by Figure 2 (Marsch and Tu, 1990a). The small values of r_E observed at 0.3 AU in the core of the Alfvénic regime (frequencies around $10^{-4} - 10^{-3}$ Hz) have disappeared at 0.9 AU. However, in spite of this r_E increase, the predominance of $\mathbf{z}_+$ fluctuations remains a clear feature of the Alfvénic turbulence observed by Helios inside 1 AU. Bavassano et al. (2001), using Ulysses data from the ecliptic phase of the mission, have shown that this situation persists at least up to 5 AU.

Another important feature of the solar wind Alfvénic fluctuations is that the magnetic fluctuation energy tends to become dominant as the solar distance increases. This decreasing trend for the Alfvén ratio

r_A ($=e_V/e_B$) clearly appears in Helios data (Bruno et al., 1985, Marsch and Tu, 1990a). It has been shown by Bavassano and Bruno (2000) that this change in the relative amplitude of **v** and **b** fluctuations occurs without a remarkable variation of their correlation. The decreasing trend of r_A, however, is not without a limit, in other words a seemingly final stage is reached in which the value of r_A remains nearly unchanged. This is seen in Figure 3 (from Roberts et al., 1990), combining Helios and Voyagers observations to give an overall view of the radial variation of r_A between 0.3 and 20 AU. The 9-hr curve, the one of the two reported curves that best describes the typical MHD scales, shows that r_A, after a fast and pronounced decrease, remains nearly unchanged.

2.2 Polar Turbulence

Observations by Ulysses during its first out-of-ecliptic orbit have shown that, at low solar activity, the solar wind at high latitudes is a fast and relatively steady flow (e.g., McComas et al., 1998). A remarkable feature of this so-called polar wind is the ubiquitous presence of an intense flow of Alfvénic fluctuations (e.g., Goldstein et al., 1995, Horbury et al., 1995b, Smith et al., 1995, Bavassano et al., 1998). Similar to previous ecliptic observations in fast streams, a largely dominant fraction of these fluctuations is outward propagating, with respect to the Sun, in the solar wind frame. It should be underlined that the Ulysses observations, besides their exploratory character, are important in view of the fact that the relative absence of structure in the polar flow offers the opportunity of studying the evolution of Alfvénic turbulence under almost undisturbed conditions.

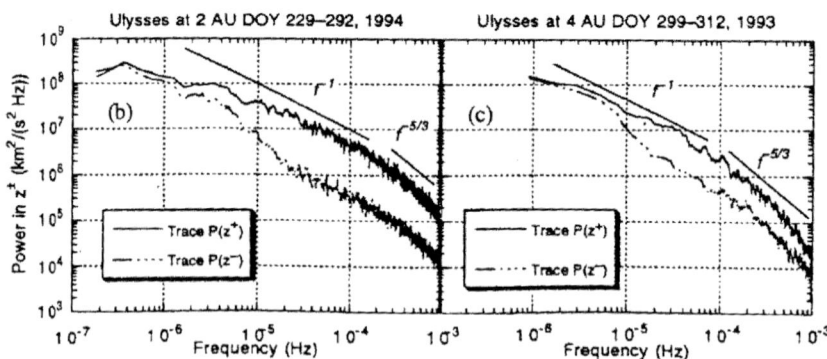

Figure 9.4. Power spectra of **z$_+$** and **z$_-$** (upper and lower curve, respectively) in polar wind at (left panel) ~ 2 AU and (right panel) ~ 4 AU (adapted from Goldstein et al., 1995, copyright 1995 American Geophysical Union, modified by permission of American Geophysical Union).

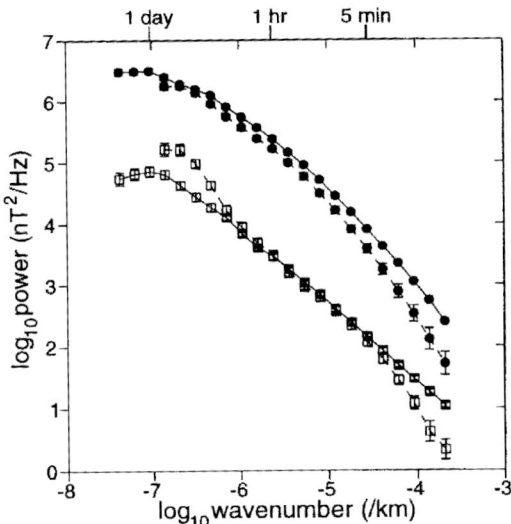

Figure 9.5 Power spectra of magnetic field components (solid circles) and magnitude (squares) as estimated at 1 AU from power scalings at Ulysses (solid lines) in polar wind between 1.4 and 4.1 AU and Helios (dashed lines) in ecliptic fast wind inside 1 AU (from Horbury and Balogh, 2001, copyright 2001 American Geophysical Union).

In Figure 4 power spectra of z_+ and z_- at about 2 and 4 AU in polar wind as obtained by Goldstein et al. (1995) are shown. The spectral evolution appears qualitatively similar to that observed in ecliptic wind, with the development of a turbulent cascade with increasing distance that moves to lower frequencies the breakpoint between the f^{-1} and $f^{-5/3}$ regimes. However, in polar wind the breakpoint is at higher frequency than at similar distances in ecliptic wind (see Horbury et al., 1996a). Thus, the spectral evolution in polar wind is slower than in ecliptic wind.

Further evidence on this is given in Figure 5. The plot (from Horbury and Balogh, 2001) shows the power spectra of magnetic field components (solid circles) and magnitude (squares) as estimated at 1 AU from power scalings measured by Ulysses in polar wind between 1.4 and 4.1 AU (solid lines) and by Helios in ecliptic fast wind inside 1 AU (dashed lines). At time scales smaller than few hours (see scale on top) the general agreement appears good, in both shape and magnitude. However, it is clearly seen that the spectrum of magnetic components is steeper in ecliptic wind, as expected for a faster evolution.

A general agreement exists on the fact that the slower evolution for polar turbulence has to be ascribed to the lack of a large-scale stream structure. The role of such a structure in accelerating turbulence evolution has been stressed by Bavassano et al. (1998) using Ulysses data at mid-latitudes (in the first out-of-ecliptic orbit), where strong gradients were dominant in the velocity pattern. It should be also mentioned that the turbulence appears younger in polar flow, since fluctuations at

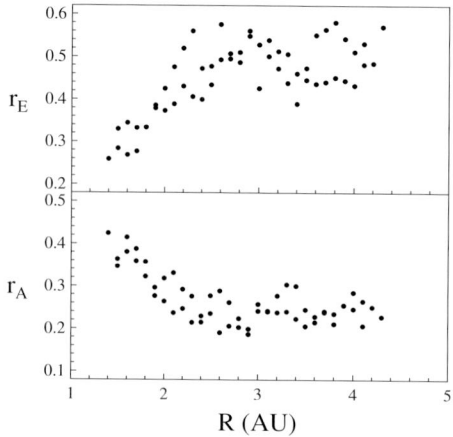

Figure 9.6 Radial variation of (top) Elsässer ratio r_E ($=e_-/e_+$) and (bottom) Alfvén ratio r_A ($=e_V/e_B$) in polar wind, as obtained from hourly total variances of the fluctuating vectors (from Bavassano, 2003, by courtesy of the American Institute of Physics).

a given distance have had less time to evolve due to the higher wind speed (e.g., see Matthaeus et al., 1999).

The radial variation of the Elsässer ratio r_E and the Alfvén ratio r_A in polar wind is shown in Figure 6 (see also Bavassano et al., 2000a). Both the decline of the e_+ predominance and the increase of the e_B predominance do not go beyond some limits, in agreement with ecliptic observations.

Agreement between polar and ecliptic observations is found also for the radial variation of magnetic field fluctuations. Figure 7 shows the decline with solar distance for hourly variances of magnetic field components and magnitude in polar wind (Forsyth et al., 1996). The radial gradient for data in the range from 1.5 to 3 AU (data at larger distances have not been included to avoid effects related to compressive features) is in a good agreement with that found in ecliptic wind, at the same scale and for a similar range of distances, by Bavassano and Smith (1986) with Pioneer 10 and 11 data.

Alfvénic turbulence evolution, however, is better described by the variation of $\mathbf{z}_+$ and $\mathbf{z}_-$, rather than by magnetic field. Figure 8 is a composite plot combining the Ulysses observations in polar wind with those by Helios in the trailing edge of fast streams on the ecliptic. As regards polar wind, the e_+ values (hourly variances) exhibit the same radial gradient over all the investigated range of distances. In contrast, for e_- a change of slope around 2.5 AU is clearly apparent. Another remarkable feature is the good agreement of the Ulysses gradients with Helios data.

The e_- behaviour highlighted by Figure 8 has probably to be related to local generation effects that disappear outside ∼2.5 AU. Turbulence generation in polar wind is a very relevant point. Velocity shear probably is an important factor in generating turbulence and in driv-

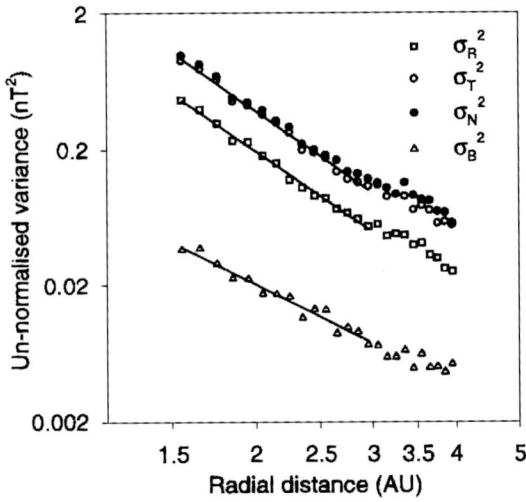

Figure 9.7 Radial variation of hourly variances of magnetic field components, in RTN Solar-Heliospheric coordinates, and (bottom curve) magnetic field magnitude (from Forsyth et al., 1996, copyright 1996 American Geophysical Union).

ing turbulence evolution for ecliptic solar wind, but certainly is not so for polar wind, where large-scale velocity gradients are almost absent. It has been proposed (e.g., see Malara et al., 2000, and Del Zanna, 2001) that in polar wind a role might be played by the parametric decay. Simulations of the non-linear development of the parametric decay for large-amplitude non-monochromatic Alfvénic fluctuations (Malara et al., 2000) have shown that the final state strongly depends on the value of β (thermal to magnetic pressure ratio). For $\beta < 1$ the normalized cross-helicity σ_C decreases, from an initial value of 1, to values close to 0. Thus, the instability appears able to completely destroy the initial Alfvénic correlation. In sharp contrast, for $\beta = 1$ (a value closer to real solar wind conditions) σ_C remains different from 0 in the final state. In this case the parametric instability is not able to go beyond some limit in the disruption of the initial correlation between velocity and magnetic field fluctuations. This solution surely is qualitatively reminiscent of the Ulysses data behaviour shown in Figure 8 (see also the r_E variation in Figure 6). Analogous results have been reached by Del Zanna (2001) for arc-polarized Alfvén waves. It should be stressed, however, that these models have many limitations (e.g., see discussion by Malara et al., 2001). Moreover, very recently it has been suggested (Matthaeus, 2004) that an almost constant (and different from zero) cross-helicity could be the result of a balancing between weak velocity shear, which reduces σ_C, and dynamic alignment (Dobrowolny et al., 1980, Grappin et al., 1982, Matthaeus et al., 1983), a process that increases σ_C. In conclusion, turbulence generation for a flow like the polar solar wind still is an open issue.

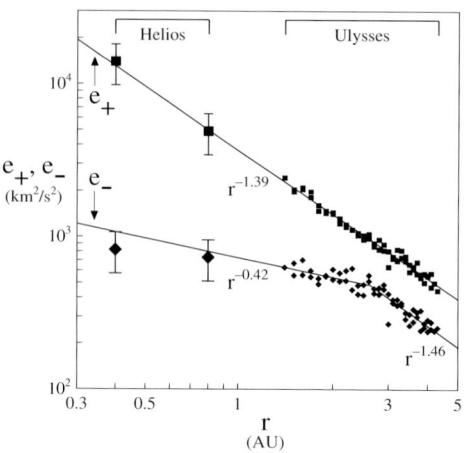

Figure 9.8 A composite plot combining Ulysses observations in polar wind with those by Helios 1 and 2 inside 1 AU on the ecliptic plane (adapted from Bavassano et al., 2000b, copyright 2000 American Geophysical Union, modified by permission of American Geophysical Union). The values of e_+ (e_-) are shown as squares (diamonds), small for Ulysses and large for Helios. Best fit lines and radial power laws for Ulysses data are given.

A last comment is about the nature of the Alfvénic turbulence variations observed in polar wind (i.e., radial, or latitudinal, or both). Both distance and latitude concurrently vary along the Ulysses trajectory, thus this point needs to be carefully examined. Several analyses of Ulysses data have indicated that the variation of turbulence properties in high-latitude wind is essentially radial, rather than latitudinal, in nature (e.g., Goldstein et al., 1995, Horbury et al., 1995b, Forsyth et al., 1996). A further robust argument in favor of the radial character of the turbulence variation comes from the agreement between gradients observed in high-latitude wind and in fast streams on the ecliptic. This has been seen to hold both for magnetic field fluctuations (see above) and for outward and inward Alfvénic fluctuations (as shown by Bavassano et et al., 2000b, 2001). However, Horbury and Balogh (2001) claimed that fluctuations in magnetic field components at time scales shorter than about 1 day (in the spacecraft frame) have a non negligible dependence upon latitude. Bavassano et al. (2002) have reexamined this issue by analysing, with the same method of Horbury and Balogh (2001), the behaviour of the Elsässer variables $\mathbf{z}_\pm$, instead of the magnetic field. This is not a minor point, since magnetic fluctuations, though obviously related to the Alfvénic turbulence, also include non-negligible contributions from other kinds of fluctuation and from structures convected by the plasma flow. Moreover, spurious effects related to compressive features in polar wind (as those seen in Figure 7) were removed. The conclusion of Bavassano et al. (2002) is that, at least for the core of the Alfvénic regime, a robust evidence exists in favour of a radial nature for the turbulence evolution in polar wind.

2.3 Conclusions on Turbulence Evolution

Ulysses observations, combined with previous results in the ecliptic, allow us to get a quite complete view of the Alfvénic turbulence evolution in the heliosphere from 1 to 5 AU.

Polar wind observations, when compared to results in the ecliptic plane, do not appear as a dramatic break. Polar evolution is similar to that in the ecliptic, although slower. This agrees well with the view proposed by Bruno (1992), a middle course between a non-relaxing turbulence (lack of velocity shear) and a quickly evolving turbulence (due, for instance, to the large amplitude of fluctuations relative to the mean magnetic field).

The identification of the processes driving the polar turbulence evolution is not firm yet. The parametric decay could be a good candidate, however the proposed models need improvements. Very recently it has been suggested that velocity shear, though weak, and dynamic alignment could account for the observed behaviours.

It has been seen that in all the examined region the Alfvénic fluctuations appear characterized by 1) a predominance of outward fluctuations (i.e., a positive cross-helicity) and 2) a predominance of magnetic fluctuations (i.e., a negative residual energy).

As regards outward fluctuations, in high-latitude wind their dominant character probably extends well away from the Sun. At low solar activity the polar wind fills a large fraction of the heliospheric cavity. For such conditions the outward fluctuations may play a leading role in acceleration and diffusion of high-energy particles. In other words, models based on the assumption of a negligible cross-helicity should not be able to give a good description of these phenomena.

As regards the imbalance in favour of magnetic energy, it does not appear to go beyond some limit. Several ways to get a non-zero residual energy have been proposed, for instance, 2-D processes (e.g., Oughton et al., 1994, Matthaeus et al., 1996) or propagation in a non uniform medium (e.g., Hollweg and Lee, 1989). However, convincing arguments to account for the existence of a limit value have not yet been given.

In conclusion, though non negligible advances have still to be made, a satisfactory understanding of the physical processes driving the evolution of the solar wind Alfvénic turbulence does not seem too far away.

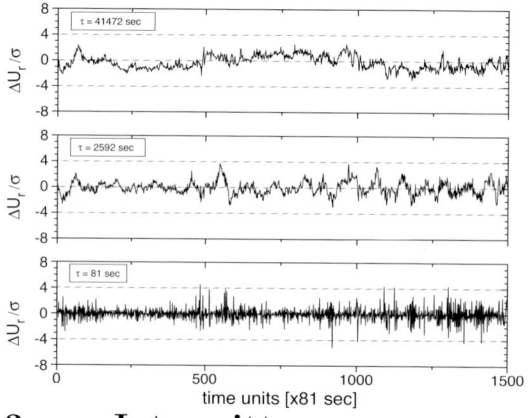

Figure 9.9 Differences of the radial component of the solar wind velocity at three different time scales r, normalized to their respective standard deviation σ. From top to bottom: $r = 41472 sec$, $r = 2592 sec$ and $r = 81 sec$, respectively. The whole time interval lasts about 1.4 days

3.   Intermittency

Turbulence is a phenomenon where chaotic dynamics and power law statistics coexist, and is characterized by randomness in both spatial and temporal domain (e.g. Frisch (1995); Bohr et al. (1998)). Power law statistics is visible for example in the behavior of spectral energy (see fig. 1.1). However, for a stochastic process monitored by a time serie $u(t)$, the spectral energy at frequency f is related to statistical behavior of the 2–th order moment of differences $\delta u_r = u(t + r) - u(t)$ (where $r \sim 1/f$). The 2–th order moment fully characterizes the stochastic process only if the probability density function (PDF) $P(\delta u_r)$ are gaussian. On the contrary we have to investigate the whole set of moments defined as $S_r^{(p)} = \langle [u(t + r) - u(t)]^p \rangle$ and called structure functions (see section 8 of Tu and Marsch, 1995). In a turbulent field the stochastic variables δu_r represent characteristic fluctuations across a turbulent eddy at the scale r. Phenomenology of the turbulent cascade, under the hypothesis of statistical homogeneity, isotropy and assuming that the energy transfer rate in the cascade is constant (see e.g. Frisch (1995)), yields in the inertial range a relation $S_r^{(p)} \sim r^{\zeta_p}$ where $\zeta_p = p/m$. In the fluid–like case $m = 3$ (Kolmogorov (1941)), while $m = 4$ in the magnetically dominated case (Kraichnan (1965);Carbone (1993b);Carbone (1993a)). Experiments on real fluid flows showed in all cases a departure from the linear scaling (Frisch (1995)), attributed to intermittency in fully developed turbulence. To look at what intermittency is in a turbulent field, we report in figure 9.9 the time behavior of δu_r for three different scales r and normalized to the relative standard deviation σ. The time series $u(t)$ is the radial component of the velocity field as observed by Helios 2 spacecraft at 0.9 AU, within a time interval chosen between day 49 and 51 of 1976. It appears evident that, as r becomes smaller, "intermittent" intense fluctuations become more and more enhanced, and

Table 9.1. Normalized scaling exponents ζ_p/ζ_3 for velocity and magnetic variables for slow wind calculated through ESS (Benzi et al. (1993)). Errors represent the standard deviations of the linear fitting. As a reference we reported the scaling exponents of structure functions for velocity and temperature, as calculated in a wind tunnel(Ruiz Chavarria et al. (1995)).

p	$u(t)$ (solar wind)	$B(t)$ (solar wind)	$v(t)$ (fluid)	$T(t)$ (fluid)
1	0.36 ± 0.06	0.56 ± 0.06	0.37 ± 0.01	0.46 ± 0.02
2	0.70 ± 0.05	0.83 ± 0.05	0.70 ± 0.01	0.78 ± 0.01
3	1.00	1.00	1.00	1.00
4	1.28 ± 0.02	1.14 ± 0.02	1.28 ± 0.02	1.18 ± 0.02
5	1.53 ± 0.03	1.25 ± 0.03	1.54 ± 0.03	1.31 ± 0.03
6	1.79 ± 0.05	1.35 ± 0.05	1.78 ± 0.05	1.40 ± 0.03

they eventually will dominate the statistics. On the contrary at large scale fluctuations appear to be smoother. At the smallest scales the time behavior of δu_r is dominated by regions where fluctuations are low, in between regions where fluctuations are intense and turbulent activity is very high. This behavior suggests that statistics can be changed by these intense fluctuations and scaling laws can be "anomalous". Anomalous scaling characterizes intermittency in fully developed turbulence (Frisch, 1995). Anomalous scaling laws can be investigated through the departure from a linear scaling by the exponents ζ_p. In table 9.1 we report the normalized scaling exponents ζ_p/ζ_3 for the radial component of both the velocity and the magnetic field $B(t)$. Here, scaling exponents have been obtained by using $S_r^{(3)}$ as generalized scale rather than r (this is the so called ESS technique, see Benzi et al. (1993)). In the same Table we report also the normalized scaling exponents of both the velocity and the temperature fields in ordinary fluid flows. It can be seen that there is, in all cases, a significant departure from the Kolmogorov $p/3$ linear scaling. Scaling exponents for the velocity field are similar to that obtained in turbulent flows on earth, thus showing a kind of universality in the intermittency. If we measure the degree of intermittency through the distance between the curve ζ_p/ζ_3 and the linear scaling $p/3$, it can be seen that magnetic field is more intermittent than velocity field. The same difference is observed between the velocity field and a passive scalar (in this case the temperature) in ordinary fluid flows. That is the magnetic field, as long as intermittency properties are concerned, behaves like a passive field convected by the flow.

The starting point for the investigation of intermittency in the solar wind dates back to 1991 when Burlaga (1991a) started to study fluctuations of the bulk velocity field at 8.5 AU using data coming from

Voyager 2 satellite. He found that anomalous scaling laws for structure functions can be recovered in the range $0.85 \leq r \leq 13.6$ hours. This range of scales has been arbitrarily identified as a kind of "inertial range", a region where a linear scaling exists between $\log S_r^{(p)}$ vs. $\log r$, and the scaling exponents have been calculated as the slope of these curves. Moreover, data sets consist of about 4500 data points, too low to determine structure functions of order $p \leq 20$. Nevertheless the scaling was found to be quite in agreement with that found in ordinary fluid flows. Even if the data can be in agreement with the random–β model (Benzi et al. (1984)), from a theoretical point of view Carbone (1993a) and Carbone (1994) showed that normalized scaling exponents ζ_p/ζ_4 calculated by Burlaga (1991a) can be better fitted by using a p–model derived from the Kraichnan phenomenology (Kraichnan (1965); Carbone (1993a)) in the framework of general fragmentation models, which read

$$\zeta_p = 1 - \log_2 \left[\mu^{p/m} + (1 - \mu)^{p/m} \right] \tag{9.1}$$

The parameter $1/2 \leq \mu < 1$ characterizes the "strength" of intermittency, while $m = 3$ or $m = 4$ describe respectively the fluid–like or the magnetically–dominated phenomenology. Data are compatible with the value $\mu \simeq 0.77$ for the free parameter and $m = 4$. The same author (Burlaga (1991b)) investigated the multifractal structure of the interplanetary magnetic field near 25 AU, and analyzed positive defined fields such as magnetic field strength, temperature and density using the multifractal machinery of dissipation fields (Meneveau and Sreenivasan (1991); Paladin and Vulpiani (1987)). Burlaga (1992) showed that intermittent events observed in corotating streams at 1 AU should be described by a multifractal geometry. Even in that case the number of points used was very low to assure the reliability of high–order moments.

Marsch and Liu (1993) investigated the structure of intermittency of turbulence observed in the inner heliosphere by using Helios 2 data. This analysis, performed on both bulk velocity and Alfén speed, calculated structure functions in the whole range 40.5 sec. (the instrumental time resolution) up to 24 hours, and in this region p–th order scaling exponents were obtained. Note that the number of data points used in this paper is very small to assure some reliability for order $p = 20$ structure functions, as instead reported Marsch and Liu (1993). Similarly to findings reported by Burlaga (1991a), these authors found evidence for the presence of anomalous scaling laws. A comparison between fast and slow streams at two heliocentric distances, namely 0.3 AU and 1 AU, allowed these authors to conjecture a scenario for high speed streams were Alfvénic turbulence, originally self–similar (or poorly intermittent) near the Sun, "... loses its self–similarity and becomes more multifractal in nature" (Marsch and Liu (1993)), which means that intermittent corrections slightly increase when the wind expands from 0.3 AU to 1 AU. However, no such behaviour seemed to occur in the slow solar wind. From a phenomenological point of view, Marsch and Liu (1993) found that data can be fit by a piece–wise linear function for the scaling exponents ζ_p, namely a β–model $\zeta_p = 3 - D + p(D - 2)/3$, where $D \simeq 3$ for $p \leq 6$ and $D \simeq 2.6$ for $p > 6$. Moreover, the same authors said: "We believe that we see similar indications in the data by Burlaga, who still prefers to fit his whole ζ_p data set with a single fit according to the non–linear random β–model". This conclusion was certainly due to the fact that the number of data points used was very small. Only structure functions of order $p \leq 4$ are reliably described by the number of points used by Burlaga (1991a).

In the above quoted data analysis, which in some sense presents quite contradictory results, two main criticisms must be raised: 1) too poor data sets have been used in order to have high–order statistics, and 2) the range of scales where scaling laws have been recovered is arbitrary. To overcome these difficulties, Carbone et al. (1996a) investigated the behavior of the normalized ratios ζ_p/ζ_3 through the Extended Self–similarity (ESS) procedure (Benzi et al. (1993)), using data coming from low–speed stream measurements of Helios 2. Using ESS the whole range covered by measurements is linear, and scaling exponent ratios can be reliably calculated. Moreover, to have a data set with a high number of points, authors mixed into the same statistics data coming from different heliocentric distances (from 0.3 AU up to 1 AU). This is not correct as far as fast wind fluctuations are concerned, because (Marsch and Liu (1993), Bruno et al. (2003), Pagel and Balogh (2003)) there is a radial evolution of intermittency. Results showed that intermittency is a real characteristic of turbulence in the solar wind, and that the curve ζ_p/ζ_3 is a nonlinear function of p, as soon as $p \leq 6$ is considered.

Looking at the papers in literature, it can be seen that often scaling exponents ζ_p as observed mainly in high–speed streams within the inner heliosphere, cannot be properly explained by any cascade model for turbulence. This feature has been attributed to the fact that this kind of turbulence is not in a fully–developed state with a well defined spectral index. Models developed by Tu et al. (1984) and Tu (1988) were successful in describing the observed radial evolution of the power spectra. Using the same idea, Tu et al. (1996) investigated the behavior of an extended cascade model developed on the basis of the p–model (Meneveau and Sreenivasan (1987); Carbone (1993a)). Authors conjectured that:

i) the scaling laws for fluctuations are still valid even when turbulence is not fully developed; ii) the energy cascade rate is not consistent, rather the moments depend not only on the generalized dimensions D_p but also on the spectral index α of the power spectrum. The model reads

$$\zeta_p = 1 + \left(\frac{p}{m} - 1\right) D_{p/m} + \left[\alpha\frac{m}{2} - \left(1 + \frac{m}{2}\right)\right]\frac{p}{m} \qquad (9.2)$$

where the generalized dimensions are recovered from the usual p–model

$$D_p = \frac{\log_2\left[\mu^p + (1 - \mu)^p\right]}{(1 - p)}$$

In the limit of "fully developed turbulence", say when the spectral slope is $\alpha = 2/m + 1$, the usual expression (9.1) is recovered. Helios 2 data are consistent with this model as long as the parameters are $\mu \simeq 0.77$ and $\alpha \simeq 1.45$, and the fit is relatively good (Tu et al. (1996)). Recently, Horbury and Balogh (1997) and Horbury et al. (1997) studied the magnetic field fluctuations of the polar high–speed turbulence from Ulysses measurements at 3.1 AU and at 63° heliolatitude. They showed that the scaling exponents of structure functions of order $p \leq 6$, in the scaling range $20 \leq r \leq 300$ sec. can be reliably described by the Kolmogorov–like model. On the contrary, (Horbury et al. (1996)) a p–model model where the parameter p is allowed to change at every step of the cascade, does not describe the scaling of the structure functions.

Analysis of scaling exponents of p–th order structure functions have been performed using different portions of Ulysses data. Horbury et al. (1995a) and Horbury et al. (1995b) investigated the structure functions of magnetic field as obtained by observations at distances between 1.7 and 4 AU, and covering a heliographic latitude from 40° to 80°. Investigating the spectral index of the 2–th order structure function, they found a decrease with heliocentric distance attributed to the radial evolution of the fluctuations. Further investigations (Ruzmaikin et al. (1995)) used structure functions to study the Ulysses magnetic field data in the range of scales $1 \leq r \leq 32$ min. These authors showed that intermittency is at work and developed a bi–fractal model to describe Alfvénic turbulence. They found that intermittency may change the spectral index of the 2–th order structure function and this (Carbone (1993b)) modifies the calculation of the spectral index. Ruzmaikin et al. (1995) found that polar Alfvénic turbulence should be described by a Kraichnan phenomenology (Kraichnan (1965)). However, the same data can be fit also by a fluid–like scaling law (Tu et al. (1996)), and it is difficult to decide with the analysis of 2–th order structure function which scaling relation describes more appropriately intermittency in the solar wind.

In a further paper, Carbone et al. (1995) investigated differences in the ESS scaling laws between ordinary fluid flows and solar wind turbulence. Through the analysis of different data sets collected in the solar wind and in ordinary fluid flows, it was shown that normalized scaling exponents ζ_p/ζ_3 are the same as far as $p \leq 8$ is considered. This indicates a kind of universality in the scaling exponents for velocity structure functions. Differences between scaling exponents calculated in ordinary fluid flows and solar wind turbulence are confined to high–order moments. Nevertheless the differences found in the data sets have been related to different kinds of singular structures in the She and Leveque model (She and Leveque (1994)). This model, which coincides with the p–model, reads

$$\zeta_p = \frac{p}{m}(1 - x) + C\left[1 - \left(1 - \frac{x}{C}\right)^{p/m}\right] \qquad (9.3)$$

The parameter $C = x/(1 - \beta)$ is identified as the codimension of the most singular structures. In the case of fluid flows $C = 2$ and the most intermittent structures are identified as filaments. Solar wind data can be fit by that model when the most intermittent structures are assumed to be planar sheets $C = 1$ and $m = 4$, which suggests the presence of the Kraichnan scaling. On the contrary ordinary fluid flows can be fit only when $C = 2$ and $m = 3$, which suggests that the structures are filaments and the Kolmogorov scaling is present. However it is worthwhile to remark that differences have been found for high–order structure functions, just where measurements are unreliable.

3.1   Probability Distribution Functions of Fluctuations and Self–similarity

The presence of scaling laws for fluctuations is a signature for the presence of self–similarity in the phenomenon. In fact the observable δu_r (see the beginning of section 2 for definition), which depends on a scaling variable r, is invariant with respect to the scaling relation $r \to \lambda r$, when there exists a parameter $\mu(\lambda)$ such that $\delta u_r = \mu(\lambda)\delta u_{\lambda r}$. The solution of this last relation is a power law $\delta u_r \sim r^h$ where the scaling exponent is $h = -\log_\lambda \mu$. Then the ratio of fluctuations at two scales $\delta u_{\lambda \ell}/\delta u_r \sim \lambda^h$ depends only on the value of h (no characteristic scales are present). This means that PDFs (see the beginning of section 2 for definition) are related such that $P(\delta u_{\lambda r}) = P(\lambda^h \delta u_r)$. Let us consider the standardized variables $y_r = \delta u_r/\langle(\delta u_r)^2\rangle^{1/2}$. It can be easily shown that when h is unique, say in a pure self–similar situation, PDFs are such that $P(y_r) = P(y_{\lambda r})$, say by changing scale PDFs coincide. As far as PDFs are concerned, in Figure 9.10 we report the PDFs for the

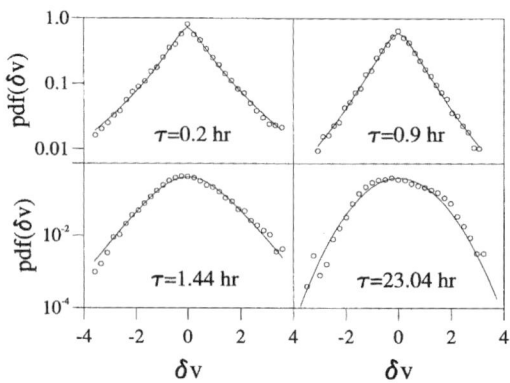

Figure 9.10 PDFs of solar wind velocity fluctuations at different scales. Solid lines represent the results of the fit obtained using Castaing's model. (from Sorriso et al. (1999), copyright 1999 American Geophysical Union).

standardized velocity $\Delta u_r = \delta u_r / \langle \delta u_r^2 \rangle^2$ at three different scales r (for sake of brevity, we omit similar graphs obtained for magnetic fluctuations $\Delta b_r = \delta b_r / \langle \delta b_r^2 \rangle^2$). It appears evident that the global self–similarity in real turbulence is broken. PDFs at different scales do not coincide, their shape seems to depend on r (Marsch and Tu (1997)). At large scales PDFs look almost gaussian, but they become more stretched as r decreases. At the smallest scale PDFs are stretched exponential. This scaling dependence of PDFs is a different way of saying that the scaling exponents of fluctuations are anomalous, or it can also be adopted as another definition of intermittency. Note that the wings of the PDFs are higher than they would be if these PDFs were gaussian. This implies that intense fluctuations have a probability of occurrence grater than that they should have if they were gaussianly distributed. Said differently, intense stochastic fluctuations are less rare than we should expect from the point of view of a gaussian approach to the statistics of turbulence. These fluctuations play a key role in the statistics of turbulence.

Marsch and Tu (1994) started to investigate the behavior of PDFs of fluctuations against scales and they found that PDFs are rather spiky at small scales and quite gaussian at large scales. The same behavior have been obtained by Sorriso et al. (1999) who investigated Helios 2 data for both velocity and magnetic field. Besides the idea of self–similarity underlying the process of energy cascade in turbulence, a different point of view can be introduced. That is a model that tries to characterize the behavior of the PDFs through the scaling laws of the parameters which describes how the shape of the PDFs changes going towards small scales (Castaing and Gagne, (1990)). In its simplest form the model can be introduced by saying that PDFs of increments δu_r at a given scale are built up by a convolution of gaussian distributions of width

$\sigma = \langle (\delta u_r)^2 \rangle^{1/2}$, whose distribution is given by $G_\lambda(\sigma)$, namely

$$P(\delta u_r) = \frac{1}{2\pi} \int_0^\infty G_\lambda(\sigma) \exp\left(-\frac{\delta u_r^2}{2\sigma^2}\right) \frac{d\sigma}{\sigma} \qquad (9.4)$$

In a purely self–similar situation, where the energy cascade generates only a trivial variation of σ with scales, the width of the distribution $G_\lambda(\sigma)$ is zero, and invariably we recover a gaussian distribution for $P(\delta u_r)$. On the contrary when the cascade is not strictly self–similar, the width of $G_\lambda(\sigma)$ is different from zero, and the scaling behavior of the width λ^2 of $G_\lambda(\sigma)$ can be used to characterize intermittency.

In order to make a quantitative analysis of the energy cascade leading to the scaling dependence of PDFs described above, the distributions obtained in the solar wind have been fit (Sorriso et al. (1999)) by using the log–normal ansatz

$$G_\lambda(\sigma) = \frac{1}{\sqrt{2\pi}\lambda} \exp\left(-\frac{\ln^2 \sigma/\sigma_0}{2\lambda^2}\right) \qquad (9.5)$$

The width of the log–normal distribution of σ is given by $\lambda^2(r) = \sqrt{\langle (\Delta\sigma)^2 \rangle}$, while σ_0 is the most probable value of σ.

The expression (9.4) has been fit to the experimental PDFs of both velocity and magnetic intensity, and the corresponding values for the parameter λ have been recovered. In Figure 9.10 we plotted, as full lines, the curves relative to the fit. It can be seen that the scaling behavior of PDFs, in all cases, is very well described by (9.4). At every scale r, we get a single value for the width $\lambda^2(r)$, which can be approximated by a power law $\lambda^2(r) = \mu r^{-\gamma}$ for $r < 1$ hour. The values of parameters μ and γ obtained in the fit are $\mu = 0.75 \pm 0.03$ and $\gamma = 0.18 \pm 0.03$ for the magnetic field ($\sigma_0 = 0.90 \pm 0.05$), while $\mu = 0.38 \pm 0.02$ and $\gamma = 0.20 \pm 0.04$ for the velocity field ($\sigma_0 = 0.95 \pm 0.05$), in the range $r \leq 0.72$ hours, for slow–speed streams. For high–speed streams we found respectively $\mu = 0.90 \pm 0.03$ and $\gamma = 0.19 \pm 0.02$ for the magnetic field ($\sigma_0 = 0.85 \pm 0.05$), while $\mu = 0.54 \pm 0.03$ and $\gamma = 0.44 \pm 0.05$ for the velocity field ($\sigma_0 = 0.90 \pm 0.05$). This means that magnetic field is more intermittent than velocity field.

3.2 Radial Evolution of Intermittency

Bruno et al. (2003) studied the radial dependence of solar wind intermittency looking at magnetic field and velocity fluctuations within the inner heliosphere, between 0.3 and 1 AU. They analyzed compressive (intensity differences between consecutive vectors) and directional (vector differences between consecutive vectors) fluctuations for both fast and slow wind, focusing their attention on the behavior of the flatness of the PDF at different scales. The behaviour of this parameter was then used to estimate the degree of intermittency of different time series. Their main results can be summarized in the following points: a) compressive fluctuations, defined as $\delta|\underline{\xi}(t)|_\tau = |\underline{\xi}(t+\tau)| - |\underline{\xi}(t)|$, are more intermittent than directional fluctuations, defined as $|\delta\underline{\xi}(t)|_\tau = (\sum_{i=x,y,z}(\xi_i(t+\tau) - \xi_i(t))^2)^{1/2}$, $\underline{\xi}(t)$ being either velocity or magnetic field and τ the generic small scale close to the end of the inertial range; b) magnetic fluctuations are more intermittent than velocity fluctuations; c) large scale fluctuations are always rather Gaussian, regardless of type of wind or heliocentric distance; d) slow wind is more intermittent than fast wind but it does not depend on heliocentric distance; e) intermittency for both compressive and directional fluctuations, within fast wind, increases with distance. Because of the respective definition of compressive and directional fluctuations, they concluded that intermittency of directional fluctuations is influenced by the contribution of both compressive phenomena and uncompressive fluctuations like Alfvén waves. These considerations helped to understand why directional fluctuations are always less intermittent than compressive fluctuations and why only fast wind shows radial evolution. These authors assumed that the two major ingredients of interplanetary MHD fluctuations are compressive fluctuations due to a sort of underlying, coherent structure convected by the wind and stochastic Alfvénic fluctuations propagating in the wind. The coherent nature of the convected structures increases intermittency while the stochastic nature of Alfvénic modes makes the PDFs of the fluctuations more Gaussian, decreasing intermittency. In particular, directional fluctuations will have their intermittency more or less reduced depending on the amplitude of Alfvénic modes with respect to that of compressive fluctuations. Thus, directional fluctuations would always be less intermittent than the compressive counterpart. Moreover, the same authors provided an explanation for the radial dependence of intermittency within fast wind. During the radial expansion, slow wind MHD turbulence, characterized by a remarkable low level of Alfvénicity, is found in a sort of frozen state which is convected by the wind into the interplanetary space without major changes. On the contrary, fast wind turbulence strongly evolves in the inner heliosphere. Alfvénic fluctuations, which, close to the sun, largely dominate the other fluctuations, become weaker and weaker during the wind expansion to the extent that at 1 AU their amplitude is much reduced (see Tu and Marsch(1995) and references therein). Then, intermittency would radially evolve within fast rather than slow wind. On the other hand, the reason why compressive fluctuations also become more intermittent within fast wind has to be ascribed to the fact that fast wind becomes more and more compressive with radial distance while the compressive level of slow wind remains the same, as shown by Marsch and Tu (1990).

Other important results were reached by similar studies performed on the components Bruno et al. (2003). Because of the presence of the large scale spiral magnetic field, the spatial symmetry is broken and a preferential direction, parallel to the mean field, has to be taken unto account. It was shown that rotating magnetic data into the mean field reference system, characterized by one component along the mean field and the other two components lying on a plane perpendicular to this direction, intermittency of the parallel component is higher than that of the two perpendicular components which, in addition, have quite the same level. These results, which were found in fast but not in slow wind, corroborated the idea that Alfvénic fluctuations play an important role in this problem. As a matter of fact, perpendicular components are more directly influenced by Alfvénic fluctuations, whose $\delta \underline{B}$ is perpendicular to the ambient field $\underline{B}$, and as a consequence fluctuations of these components are more stochastic and less intermittent. On the contrary, Similar studies performed on velocity fluctuations highlighted the more intermittent character of the radial component, suggesting that, for this parameter, the usual RTN reference system is more appropriate. The same authors concluded that the radial dependence of intermittency of interplanetary fluctuations is strongly related to the radial evolution of their turbulent spectrum, as also shown by Pagel and Balogh (2003).

3.3   Identifying Intermittent Events

Intermittent events within turbulence can be identified by using techniques based on wavelets. One of these techniques is called Local Intermittency Measure (LIM) and was introduced by Farge (1992). Identifying these events can be used to eliminate them and to make conditioned statistics, that is to find structure functions separately by using the full data set or by using only the gaussian field. Veltri and Mangeney (1999), using ISEE data, found that when conditioned statistics is used, a linear scaling is recovered for scaling exponents ζ_p. Interestingly enough authors found that the radial velocity displays the characteristic Kolmogorov scaling $p/3$, while the other components of the velocity displays a Kraichnan slope $p/4$. Moreover, the same authors concluded that the most intermittent structures were shocks or current sheets.

Stimulated by this study, Bruno et al. (2001) focused on the nature of interplanetary events contributing to intermittency and, adopting the method introduced by Farge (1992), were able to locate and analyze

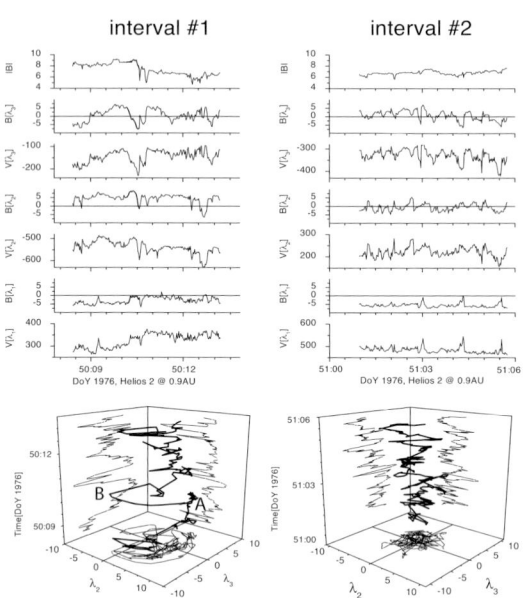

Figure 9.11 Left–hand–side, from top to bottom: magnetic field intensity, magnetic field and velocity fluctuations along maximum λ_3, intermediate λ_2 and minimum λ_1 variance directions of magnetic field fluctuations, and the magnetic field hodogram on the maximum variance plane $\lambda_3 - \lambda_2$, as a function of time (vertical axis). The arc–like fluctuation from A to B visible on the $\lambda_3 - \lambda_2$ plane, corresponds to the thick line in the upper panels. The same parameters are shown for interval #2, which doesn't have intermittent events, on the right column. (copyright 2001 Elsevier Science Ltd., from Bruno et al., 2001).

a typical intermittent event within fast solar wind. They showed that these kinds of events were characterized by a large coherent rotation of the magnetic field direction, across which the magnetic field intensity and total plasma pressure were rather discontinuous. The top left panel of Figure 9.11 shows magnetic field intensity and the three components of magnetic field and wind velocity rotated into the minimum variance reference system of the magnetic field fluctuations. The panel below shows the hodogram on the maximum variance plane $\lambda_3 - \lambda_2$, as a function of time, showed on the vertical axis. The same analysis, performed on another interval not affected by intermittency, is shown on the right column of the same Figure and in the same format. Clearly, only within the first interval the magnetic field vector does describe an arc–like structure larger than 90° on the maximum variance plane (see rotation from A to B on the 3–D graph at the bottom of the left column) in correspondence with the time interval marked on the top panel by the solid heavy line, right where the intermittent event was located. Moreover, the behavior of magnetic field and velocity components across the

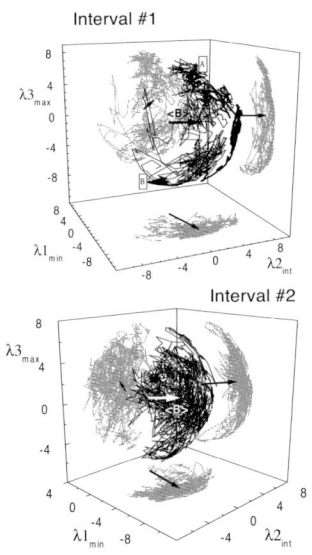

Figure 9.12 Top: path followed by the tip of the magnetic vector in the minimum variance reference system for interval #1. The thick line extending from label A to label B refers to the arc–like discontinuity shown in Figure 9.11. Bottom: path observed during interval #2. Projections on the three coordinate planes and the average magnetic field vector with its projections on the same planes are also shown. (copyright 2001 Elsevier Science Ltd., from Bruno et al., 2001).

rotation suggested that this was a 2–D structure co–moving with the wind. Further insights could be obtained studying the path followed by the tip of the vector within the minimum variance reference system, during these two time intervals. It was found that the interval characterized by intermittency showed a rather patchy distribution while the one with no intermittency showed a more uniform distribution as shown in fig.9.12. In other words, intermittency was reflected in the fact that several small directional jumps around a given average orientation were interrupted by a larger jump which forced the vector to be oriented and fluctuate around another average direction. In particular, the tip of the vector would follow a trajectory similar to a random walk. More in detail, the path followed by the tip of the vector was reminiscent of a Lévy–flight process, which differs from a Brownian motion for the probability distribution of the jumps which is not longer Gaussian but rather a Lévy–stable. This distribution, which highlights the presence of long range correlations, asymptotically behaves like a power-law (Pareto (1906)) $f(\xi) \propto \xi^{-(1+\alpha)}$ with the exponent in the range $0 < \alpha < 2$. However, since the second and higher moments of this distribution are infinite and, as such, in contrast with experimental data which show finite variance, a Truncated Lévy Flight (TLF) distribution has been proposed (Mantegna and Stanley (1994)) introducing a cutoff in the standard Lévy. A TLF statistics with a smooth cutoff is generated by a random variable ξ which is characterized by the following distribution: $f(\xi) = Ae^{-\lambda|\xi|} \mid \xi \mid^{-1-\alpha}$, where λ represents the cutoff and $0 < \alpha < 2$ is the characteristic exponent. A very good example of TLF statistics is represented by the dispersion of a passive scalar in a turbulent flow (Klafter et al. (1996)) and the magnetic field, convected by the solar wind into the interplanetary space, might act in the same way. Bruno et al. (2004) showed that interplanetary observations can be reasonably fitted by a TLF distribution. Moreover, they found a clear radial dependence for the PDFs of these fluctuations to evolve from Gaussian like to possible TLF only within fast wind. They suggested that the observed behavior was due to a competing action between quasi–stochastic, propagating fluctuations and convected structures, both contributing to solar wind turbulent fluctuations.

All these observations gave new insights about the topology which characterizes velocity and magnetic field vector orientations within solar wind MHD turbulence and contributed to re–formulating the idea of the presence of a main structure convected by the wind that is formed by flux tubes tangled up in space. Each hypothetical flux tube is characterized by local values of magnetic and plasma parameters. Moreover, the presence of Alfvénic fluctuations would make the magnetic field vector randomly wander about the local average direction. The idea of a solar wind turbulence made of propagating Alfvén modes and convected structures is not new and recalls the model suggested by Tu and Marsch(1993) following which the solar wind fluctuations are mainly due to the coexistence of pressure balance structure (PBS) type flux tubes and Alfvén waves. In the inner heliosphere these PBS–type flux tubes are imbedded in the large structure of fast solar wind streams and would form a kind of spaghetti–like sub–structure (McCracken and Ness(1966)), which probably has its origin at the base of the solar atmosphere. A tangential discontinuity could mark the border between two contiguous flux tubes. This seems to be the case of interval #1 in Figure 9.11 although the total pressure is not balanced across the discontinuity. Each flux tube would be characterized by an average magnetic field vector whose fluctuations would mainly happen on a plane perpendicular to this direction. Passing to the next tube, the average field would more ore less rapidly change its intensity and direction and, its fluctuations would cluster around this new direction. It follows that moving across a series of these tubes would increase the intermittent character of the recorded fluctuations since we would preferentially sample coherent structures. On the contrary, moving along one of these tubes would decrease the intermittency of our sample since we would be sampling preferentially stochastic fluctuations. This geometric effect should be added to the spectral evolution of the fluctuations since, due to the spiral configuration of interplanetary magnetic field, it contributes to the observed radial dependence of intermittency.

Finally, the possibility of removing intermittent events offered by the LIM technique (Farge (1992)) allowed to solve one of the long lasting problems concerning the anisotropy character of interplanetary magnetic field fluctuations as observed in the inner heliosphere. As a matter of fact, solar wind magnetic field fluctuations are strongly anisotropic, being the power perpendicular to the minimum variance direction is much larger than that along it (Bavassano et al., (1982)). Moreover, anisotropy of field and velocity fluctuations has a different radial behavior. While velocity anisotropy doesn't evolve with distance, magnetic field anisotropy clearly increases with increasing heliocentric distance (Bavassano et al., (1982)). It has recently been shown (Bruno et al. (1999a); Bruno et al. (1999b)) that magnetic field intermittency, mainly caused by events generated by the dynamical stream–stream interaction as the wind expands, plays a relevant role in the radial trend of magnetic anisotropy. As a matter of fact, these authors showed that removing intermittency events from the original magnetic field would completely eliminate the observed radial trend and magnetic field anisotropy would then closely resemble velocity anisotropy. These results finally ended the long withstanding problem of reconciling the radial increase of anisotropy observed by Bavassano et al., (1982) with the observed $\delta\underline{B}$ and $\delta\underline{V}$ decoupling which would suggest an increase of isotropy (Klein et al.(1993)).

3.4   Conclusions on Intermittency

Since the first investigations by Burlaga (1991a), several other investigators have focused on intermittency and the role that this phenomenon plays in solar wind turbulence. In Section 3 we gave just a short and, unavoidably, not complete summary of the results which have been obtained and the studies which have been conducted on this important topic. Today we know much about the phenomenological behavior of solar wind intermittency, for example within different types of solar wind, as a function of scale or heliocentric distance. Moreover, the efforts of several investigators and the use of new techniques introduced in data analysis allowed us to single out those events causing intermittency. It was found that most of them were almost 2–D Tangential Disontinuities, across which the magnetic field experienced a large directional rotation, separating regions characterized by different values of the main plasma parameters. These coherent structures, possibly the border of adjacent flux tubes, are imbedded in the solar wind and convected into the interplanetary space. A simultaneous presence of these structures with propagating, stochastic Alfvénic fluctuations is to be expected especially within fast wind. As a matter of fact, it has been found that, within fast wind, the level of intermittency is due to a competing action between Alfvénic fluctuations and convected structures. The predominance of one with respect to the other would result in a different level of intermittency. In particular, data samples characterized by a strong Alfvénicity are much less intermittent than others. However, so far, it has been impossible to establish whether these coherent magnetic and velocity structures are convected directly from the source regions of the solar wind or they are locally generated by stream–stream dynamic interaction or, as an alternative view would suggest (Primavera et al. (2002)), they are locally created by parametric decay instability of large amplitude Alfvèn waves or by sporadic and localized interactions of coherent structures that emerge naturally from plasma resonances (Wu and Chang(2000)). Probably all these origins coexist at the same time and only future space missions designed to disentangle temporal from spatial effects could give an answer to this question.

References

Barnes, A.: 1979, in *Solar System Plasma Physics*, vol. 1, E. N. Parker, C. F. Kennel, and L. J. Lanzerotti (eds.), North-Holland Publishing Company, Amsterdam, p. 249.

Bavassano, B.: 2003, in *Solar Wind Ten*, M. Velli, R. Bruno, and F. Malara (eds.), AIP Conference Proceedings **679**, p. 377.

Bavassano, B., and Bruno R.: 2000, J. Geophys. Res. **105**, 5113.

Bavassano, B., and Smith, E. J.: 1986, J. Geophys. Res. **91**, 1706.

Bavassano, B., Dobrowolny, M., Mariani, F., and Ness, N. F.: 1982, J. Geophys. Res. **87**, 3617.

Bavassano, B., Pietropaolo, E., and Bruno, R.: 1998, J. Geophys. Res. **103**, 6521.

Bavassano, B., Pietropaolo, and Bruno, R.: 2000a, J. Geophys. Res. **105**, 12,697.

Bavassano, B., Pietropaolo, E., and Bruno, R.: 2000b, J. Geophys. Res. **105**, 15,959.

Bavassano, B., Pietropaolo, E., and Bruno, R.: 2001, J. Geophys. Res. **106**, 10,659.

Bavassano, B., Pietropaolo, E., and Bruno, R.: 2002, J. Geophys. Res. **107**, 1452, doi 10.1029/2002JA009267.

Belcher, J. W., and Davis, L., Jr.: 1971, J. Geophys. Res. **76**, 3534.

Benzi, R., G. Paladin, G. Parisi, A. Vulpiani: J. Phys. A **17**, 3521 (1984)

Benzi, R., Ciliberto, S., Tripiccione, R., Baudet, C., Massaioli, F., and Succi, S.: 1993, Phys. Rev. E **48**, R29.

Bohr,T., M.H. Jensen, G. Paladin, A. Vulpiani: *Dynamical system approach to turbulence*, (Cambridge University Press, Cambridge, U.K. 1998)

Bruno, R.: 1992, in *Solar Wind Seven*, E. Marsch and R. Schwenn (eds.), COSPAR Colloquia Series vol. 3, Pergamon Press, Tarrytown, N.Y., p. 423.

Bruno, R., Bavassano, B., and Villante, U.: 1985, J. Geophys. Res. **90**, 4373.

Bruno, R., Bavassano, B., Pietropaolo, E., Carbone, V., and Veltri, P.: 1999a, Geophys. Res. Lett. **26**, 3185.

Bruno, R., Bavassano, B., Bianchini, L., Pietropaolo, E., Villante, U., Carbone, V., and Veltri, P.: 1999b, *Proceedings of the 9th European Meeting on Solar Physics, Magnetic Fields and Solar Processes*, ESA SP–**448**, 1147.

Bruno, R., Carbone, V., Veltri, P., Pietropaolo, E., and Bavassano, B.: 2001, Planetary Space Sci. **49**, 1201.

Bruno, R., Carbone, V., Sorriso-Valvo, L., and Bavassano, B.: 2003, J. Geophys. Res. **108**, 1130.

Bruno, R., Sorriso-Valvo, L., Carbone, V., and Bavassano, B.: 2004, Europhys. Lett., in press.

Burlaga, L. F.: 1991a, J. Geophys. Res. **96**, 5847.

Burlaga, L. F.: 1991b, Geophys. Res. Lett. **18**, 69.

Burlaga, L. F.: 1992, J. Geophys. Res. **97**, 4283.

Burlaga, L. F. and Klein, L. W.: 1986, J. Geophys. Res. **91**, 347.

Carbone, V.: 1993b, Ann. Geophys. **11**, 866.

Carbone, V.: 1993a, Phys. Rev. Lett. **71**, 1546.

Carbone, V.: 1994, Ann. Geophys. **12**, 585.

Carbone, V., Veltri, P., and Bruno, R.: 1995, Phys. Rev. Lett. **75**, 3110.

Carbone V., Bruno, R., and Veltri, P.: 1996a, Geophys. Res. Lett. **23**, 121.

Carbone, V., Veltri, P., and Bruno, R.: 1996b, Nonlin. Proc. in Geophys. **3**, 247.

Castaing B., Gagne Y., Physica D, 1990, **46**, 177.

Del Zanna, L.: 2001, Geophys. Res. Lett. **28**, 2585.

Denskat, K. U., and Neubauer, F. M.: 1982, J. Geophys. Res. **87**, 2215.

Dobrowolny, M., Mangeney, A., and Veltri, P.: 1980, Phys. Rev. Lett. **45**, 144.

Farge, M.: 1992, Ann. Rev. Fluid Mech. **24**, 395.

Frisch, U.: *Turbulence: the legacy of A.N. Kolmogorov*, (Cambridge University Press, Cambridge, U.K. 1995)

Forsyth, R. J., Horbury, T. S., Balogh, A., and Smith, E. J.: 1996, Geophys. Res. Lett. **23**, 595.

Goldstein, B. E., Smith, E. J., Balogh, A., Horbury, T. S., Goldstein, M. L., and Roberts, D. A.: 1995, Geophys. Res. Lett. **22**, 3393.

Grappin, R., Frisch, U., Léorat, J., and Pouquet, A.: 1982, Astron. Astrophys. **105**, 6.

Hollweg, J. V. and Lee, M. A.: 1989, Geophys. Res. Lett. **16**, 919.

Horbury, T. S., Balogh, A., Forsyth, R. J., and Smith, E. J.: 1995a, Ann. Geophys. **13**, 105.

Horbury, T. S., Balogh, A., Forsyth, R. J., and Smith, E. J.: 1995b, Geophys. Res. Lett. **22**, 3401.

Horbury, T. S., Balogh, A., Forsyth, R. J., and Smith, E. J.: 1996a, Astron. Astrophys. **316**, 333.

Horbury, T. S., Balogh, A., Forsyth, R. J., and Smith, E. J.: 1996b, J. Geophys. Res. **101**, 405.

Horbury, T. S., Balogh, A.: 1997, Nonlin. Process in Geophys. **4**, 185.

Horbury, T. S., Balogh, A., Forsyth, R. J., and Smith, E. J.: 1997, Adv. Space Res. **19**, 847.

Horbury, T. S. and Balogh, A.: 2001a, J. Geophys. Res. **106**, 15,929.

Horbury, T. S. and Tsurutani, B. T.: 2001b, in *The heliosphere near solar minimum: The Ulysses perspective*, A. Balogh, R. G. Marsden, and E. J. Smith (eds.), Springer-Verlag, Berlin, p. 167.

Klafter, J., Shlesinger, M. F., and Zumofen, G.: 1996, Physics Today **49**, 33.

Klein, L., Bruno, R., Bavassano, B., and Rosenbauer, H.: 1993, J. Geophys. Res. **98**, 17461.

Kolmogorov, A. N.: 1941, Dokl. Akad. Nauk. SSSR **30**, 9, in Russian, translated: 1995, *Kolmogorov ideas 50 years on*, J. C. R. Hunt and O. M. Phillips (eds.), Proc. R. Soc. Lond. A **434**, 9.

Kraichnan, R. H.: 1965, Phys. Fluids **8**, 1385.

Kraichnan, R. H.: 1974, J. Fluid Mech. **62**, 305.

Malara, F., Primavera, L., and Veltri, P.: 2000, Phys. Plasmas **7**, 2866.

Malara, F., Primavera, L., and Veltri, P.: 2001, Nonlin. Proc. in Geophys. **8**, 159.

Mantegna, R. and Stanley, H. E.: 1994, Phys. Rev. Lett. **73**, 2946.

Marsch, E. and Tu, C.-Y.: 1990a, J. Geophys. Res. **95**, 8211.

Marsch, E. and Tu, C.-Y.: 1990b, J. Geophys. Res. **95**, 11945.

Marsch, E. and Liu, S.: 1993, Ann. Geophys. **11**, 227.

Marsch, E. and Tu, C.-Y.: 1994, Ann. Geophys. **12**, 1127.

Marsch, E. and Tu, C.-Y.: 1997, Nonlin. Proc. in Geophys. **4**, 101.

Matthaeus, W. H.: 2004, private communication.

Matthaeus, W. H., Ghosh, S., Oughton, S., and Roberts, D. A.: 1996, J. Geophys. Res. **101**, 7619.

Matthaeus, W. H., Goldstein, M. L., and Montgomery, D. C.: 1983, Phys. Rev. Lett. **51**, 1484.

Matthaeus, W. H., Zank, G. P., Smith, C. W., and Oughton, S.: 1999, Phys. Rev. Lett. **82**, 3444.

McComas, D. J., Bame, S. J., Barraclough, B. L., Feldman, W. C., Funsten, H. O., Gosling, J. T., Riley, P., and Skoug, R.: 1998, Geophys. Res. Lett. **25**, 1.

McCracken, K. G. and Ness, N. F.: 1966, J. Geophys. Res. **71**, 3315.

Meneveau, C. and Sreenivasan, K. R.: 1987, Phys. Rev. Lett. **59**, 1424.

Meneveau, C. and Sreenivasan, K. R.: 1991, J. Fluid Mech. **224**, 429.

Oughton, S., Priest, E. R., and Matthaeus, W. H.: 1994, J. Fluid Mech. **280**, 95.

Pagel, C. and Balogh, A.: 2003, J. Geophys. Res. **108**, 1012, doi 10.1029/2002JA009498.

Paladin, G., A. Vulpiani: Phys. Rep. **4**, 147 (1987)

Pareto, V.: *Manuale di economia politica con una introduzione alla scienza sociale.* (Milan: Società Editrice Libraria 1906)

Primavera, L., Malara, F., and Veltri, P.: 2003, in *Solar Wind Ten*, M. Velli, R. Bruno, and F. Malara (eds.), AIP Conference Proceedings **679**, p. 505.

Ruiz Chavarria, G., Baudet, C. , Ciliberto, S., Europhys. Lett. **32**, 319 (1995)

Roberts, D. A., Goldstein, M. L., and Klein, L. W.: 1990, J. Geophys. Res. **95**, 4203.

Ruzmaikin, A., Feynman, J., Goldstein, B. E., Smith, E. J., and Balogh, A.: 1995, J. Geophys. Res. **100**, 3395.

She, Z.-S. and Leveque, E.: 1994, Phys. Rev. Lett. **72**, 336.

Smith, E. J., Balogh, A., Neugebauer, M., and McComas, D.: 1995, Geophys. Res. Lett. **22**, 3381.

Sorriso-Valvo, L., Carbone, V., Veltri, P., Consolini, G., and Bruno, R.: 1999, Geophys. Res. Lett. **26**, 1801.

Tu, C.-Y.: 1988, J. Geophys. Res. **93**, 7.

Tu, C.-Y and Marsch, E.: 1993, J. Geophys. Res. **98**, 1257.

Tu, C.-Y and Marsch, E.: 1995, Space Sci. Rev. **73**, 1.

Tu, C.-Y., Pu, Z.-Y., and Wei, F.-S.: 1984, J. Geophys. Res. **89**, 9695.

Tu, C.-Y., Marsch, E., and Rosenbauer, H.: 1996, Ann. Geophys. **14**, 270.

Veltri, P. and Mangeney, A.: 1999, in *Solar Wind Nine*, S. R. Habbal, J. V. Hollweg, and P. A. Isenberg (eds.), AIP Conference Proceedings **471**, p. 543.

Wu, C.-C and Chang, T.: 2000, Geophys. Res. Lett. **27**, 863.

Chapter 10

WAVES AND TURBULENCE IN THE SOLAR CORONA

Eckart Marsch

Max Planck Institute for Solar System Research

Max-Planck-Institut für Sonnensystemforschung

(Former: Max-Planck-Institut für Aeronomie)

marsch@linmpi.mpg.de

Abstract   In the solar corona (and solar wind) waves and turbulence occur at all scales ranging from the particles' gyroradii to the size of a solar radius (or even to an astronomical unit). A concise review of some new observations and theories of waves in the sun's atmosphere and corona is given, with the focus being on coronal waves that are magnetically confined to loops, as well as on waves in the open coronal funnels and holes. In the corona all kinds of kinetic (plasma) waves and fluid (magnetohydrodynamic) waves, at wavelengths that may range from the size of a loop (about several Mm) down to the inertial lengths of a coronal ion (about a km), are believed to play a key role in the transport of mechanical energy from the chromosphere to the Sun's outer corona and wind, and through the dissipation of wave energy in heating and sustaining the solar corona. Recent evidence obtained from spectroscopy of lines emitted by coronal ions points to cyclotron resonance absorption as a possible cause of the observed emission-line broadenings. Novel remote-sensing solar observations reveal low-frequency loop oscillations of the type expected from MHD theory. They appear to be excited by magnetic activity and are strongly damped. Kinetic models of the corona indicate the importance of wave-particle interactions, which may hold the key to understand coronal particle acceleration and heating by high-frequency waves.

Keywords: Solar corona and solar wind, waves and turbulence, coronal oscillations, network activity, turbulence generation and dissipation, ion velocity distributions

G. Poletto and S.T. Suess (eds.), The Sun and The Heliosphere as an Integrated System, 283–317.
© 2004 *Kluwer Academic Publishers. Printed in the Netherlands.*

1.   Introduction

This concise review addresses selected current issues of kinetic (plasma) waves and fluid (magnetohydrodynamic, MHD) waves and turbulence in the solar corona. We start with briefly describing the basic coronal magnetic field structures, in which the waves and flows occur. Coronal loops and funnels are the magnetic building blocks of the lower corona, and their fields guide the distribution and propagation of wave-mechanical energy which is generated by steady photospheric magnetoconvection and transient activity of the magnetic network. Ultimately, the main energy source for waves and fluctuations in the heliosphere is solar magnetoconvection and coronal magnetic activity. Magnetohydrodynamic waves and kinetic plasma waves, as well as the eigen-oscillations of coronal flux tubes are addressed. Finally, some ideas about the generation, transfer and dissipation of energy associated with waves and turbulence in the corona are presented. The emphasis in this chapter will be on the corona. Other reviews in this book discuss waves and turbulence in the interplanetary solar wind.

Waves in the solar corona, solar wind and heliosphere is a rather wide research field which we can here, because of the lack of space and size of the subject, deal with only in a superficial and selective way. Therefore, at the outset we refer to some broader reviews of this field for further details and in depth studies. MHD structures, waves and turbulence in the solar wind, including observations and models, have been reviewed extensively by Marsch (1991a) and Tu and Marsch (1995), with emphasis on the Helios observations in the ecliptic plane and inner heliosphere. The Ulysses observations at high latitudes and radial distances between 1 AU and about 5 AU are described by Horbury and Tsurutani (2001), and observations made in the outer heliosphere mostly by the Voyagers are contained in the book of Burlaga (1995) and the reviews of Goldstein et al. (1997) and Goldstein and Roberts (1999), which also include some numerical simulation results.

2.   Coronal Magnetic Field Structures

Waves in the solar corona is a difficult subject, because the corona is known to be highly structured and nonuniform. Furthermore, the direct measurement of the magnetic field vector in the corona is still not possible. For the corona we have to rely on models of the magnetic field, usually obtained by extrapolation through potential-field, force-free, or MHD numerical methods (see e.g. the review by Neugebauer (1999)), starting from the bottom of the corona. In *Figure* 10.1 we show an example of the large-scale coronal magnetic field obtained this way.

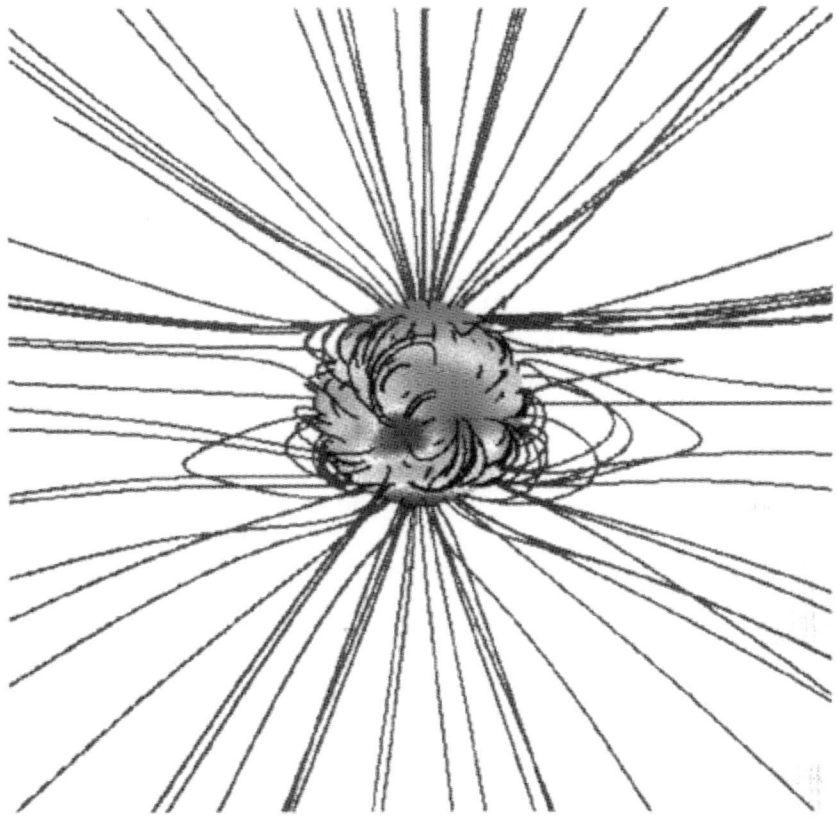

Figure 10.1. Model magnetic field of the solar corona, in particular showing the mapping of the open coronal flux from the photosphere to the source surface at about $2.5\,R_s$, from which the entire heliospheric magnetic field originates (after Neugebauer, 1999).

The coronal field determines the magnetic field of the entire heliosphere, which is dealt with in other chapters of this book.

Whereas at the photosphere/chromosphere interface the magnetic field vector can routinely be measured by means of full Stokes polarimetry, or its single components by the simple Zeeman effect, it is only very recently that the magnetic vector at the coronal base was measured directly by solanki03, which led to the detection of a current sheet and illustrated the topological change of the field with height in the transition region. Modern X-ray imagers on Yohkoh, or the extreme ultraviolet (EUV) imagers on SOHO (Solar and Heliospheric Observatory) and TRACE (Transition Region and Coronal Explorer) provide "images" of the coro-

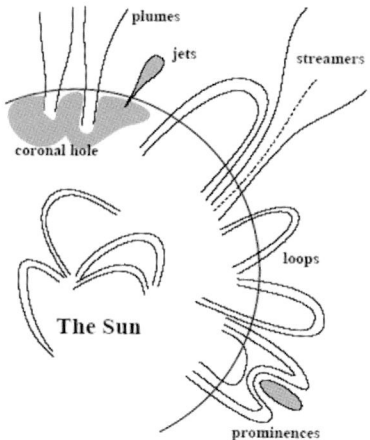

Figure 10.2 Schematic presentation of magnetic structures in the solar corona: Coronal holes, funnels, plumes, jets, streamers, prominences, and loops (after Roberts and Nakariakov, 2003).

nal field, which is made visible through the emission of plasma confined therein.

These space observatories yielded a wealth of coronal images and provided new diagnostic information on the plasma state, and thus enabled many new insights be made into the physical processes and the morphology of the solar corona. The new picture arising from all these observations is that the corona is a restless, intricate, and magnetically coupled system. This "new solar corona" is reviewed by Aschwanden et al. (2001). The complex dynamics and broad range in spatial and temporal scales constitute the main challenge in the contemporary physics of the solar corona. Typically, coronal images reveal a multitude of bright closed loops (on all scales down to 1 arcsec and perhaps below) and conspicuous streamers, and in striking contrast dark voids, corresponding to coronal funnels and coronal holes (CHs). The upper-chromosphere and lower-transition-region emissions (at temperatures below about 10^5 K) exhibit the magnetic network with typical cell sizes of 20-30 Mm, and seem to originate from a myriad of small loops.

All these magnetic structures, of which some are shown schematically in *Figure* 10.2, together constitute a complex background field, in which the waves (either kinetic or fluid-type and linear or nonlinear) are excited (e.g., by reconnection or foot-point motion in the photosphere), and through which they have to propagate. While the waves may escape to the heliosphere on open field lines, they will remain trapped in the corona when bouncing in closed loops.

From space missions as Helios, Ulysses, Yohkoh, SOHO or TRACE it has become abundantly clear that three characteristic types of solar corona and associated solar wind exist, variably prevailing at different heliographic latitudes and longitudes over the solar cycle. Solar wind comes in the form of (1) steady fast, (2) variable slow and (3) transient (sometimes very fast) flows, mostly as coronal mass ejections. For a recent short review of the three-dimensional structure of the solar wind over the solar cycle see McComas (2003). The global coronal magnetic field, as shown in *Figure* 10.1, and the related heliospheric flow pattern is most conspicuous near solar minimum, when fast streams emanate from open polar CHs, and slow streams originate from the cusps and boundary layers of the equatorial streamer belt. The closed corona (active loops and prominences) intermittently opens and abruptly produces huge coronal mass (and magnetic flux) ejections (CMEs).

The wave and turbulence pattern coming with these types of flows reflects the coronal conditions as reviewed by Tu and Marsch (1995). Fast streams are permeated by large-amplitude Alfvén waves (with flat spectra, index -1) accompanied by low-level sonic noise, whereas slow streams often carry sizable compressive fluctuations (with Kolmogorov $-5/3$-spectra), show hardly any Alfvén waves but convect embedded pressure-balanced structures such as discontinuities. In contrast, CMEs are usually characterized by strong and quiet magnetic fields with neither magnetosonic nor Alfvénic disturbances. The coronal reasons for these distinct differences in the occurrence and nature of waves and turbulence in the solar wind are largely unknown.

The basic theory of MHD waves in the solar corona was developed a quarter of a century ago (see the early work of Roberts et al. (1984) and Roberts (1985) on MHD waves), however its verification lacked in the old days adequate coronal observations, which today are available and some of which are described in the next sections. Facing these recent observations the theory has been revisited and reviewed extensively by Roberts (2002), Roberts (2003) and Roberts and Nakariakov (2003). Selected results from these reviews are cited below. Based upon the modern observations coming mainly from the TRACE and SOHO missions, coronal oscillations were in the past years observed and theoretically analysed in much more detail than previously possible. A new field of solar physics developed, *coronal seismology*, to which an introduction is given by Nakariakov (2003).

3. Magnetic Network Activity and Coronal Heating

The corona is hot even when there is no obvious magnetic activity. This general notion seems to indicate uniformity of the heating process. SOHO has shown that the coronal holes are, in terms of proton and heavy ion kinetic temperatures (see Kohl et al. (1997), Wilhelm et al. (1998), and Tu et al. (1998)), the hottest places on the Sun. Observationally, the magnetic (chromospheric) network when seen in cool lines at altitudes of several Mm looks the same beneath the closed (quiet) corona in loops and open corona in holes, but it differs substantially when seen in hot transition region lines as reported by Hassler et al. (1999). The energy flux required to power the fast solar wind and the energy flux needed to produce the quiet magnetically confined corona (mainly compensation for radiative losses) are both about $5 \ 10^5$ erg cm^{-2}s^{-1}. Because of the uniformity of the network over the quiet sun, is seems natural to postulate the energy source for coronal heating to be about the same everywhere.

SOHO indicates that the solar wind coming from the polar coronal holes emanates directly from the chromospheric magnetic network as observed by Hassler et al. (1999) and Wilhelm et al. (2000), whereby the whole coronal base is involved, with relatively high upward initial speeds up to 10 km/s. Below the base, the coronal field (of about 10 G) is anchored mainly in the supergranular network (see *Figure* 10.3), which occupies merely 10% of the base area in holes. In the network lanes the magnetic pressure dominates the thermal pressure. The network field (of about 10-100 G) is rooted in the photosphere in small, kG-field flux tubes (about 100 km in size), expands rapidly with height in the transition region and then fills the entire overlying corona.

Solar magnetograms (as shown below in *Figure* 10.4) clearly indicate that the magnetic network field exists side-by-side in two characteristic components, i.e. in uncancelled unipolar flux tubes (funnels) and in closed, multiple flux tubes (loops). The only conceivable energy source for the corona and wind is the dynamic network field itself. An open-funnel potential field according to Gabriel (1976) has no free energy, a complex static field assumed by Dowdy et al. (1986) does not envisage magnetic energy release, but only time-varying fields undergoing reconnection as suggested by Axford and McKenzie (1997) allow flaring and can provide energy to the corona. The small loops (at a scale of 1 Mm and smaller) will emerge or submerge and collide, and thus constitute a permanent source of energy. The small-scale activity of this magnetic carpet, the dynamics of ephemeral regions and their consequences for coronal heating are discussed in the paper by schrijver98. The related scenario of transition region dynamics is illustrated in *Figure* 10.3. We assume that the very small-scale reconnection events, here called "pico-flares", are similar to other intra-network flares in the sense that remnant closed loops containing hot EUV emitting plasma are produced.

parker88 first developed a coronal heating scenario involving so-called "nano-flares", in which the dissipation is conceived to take place in small

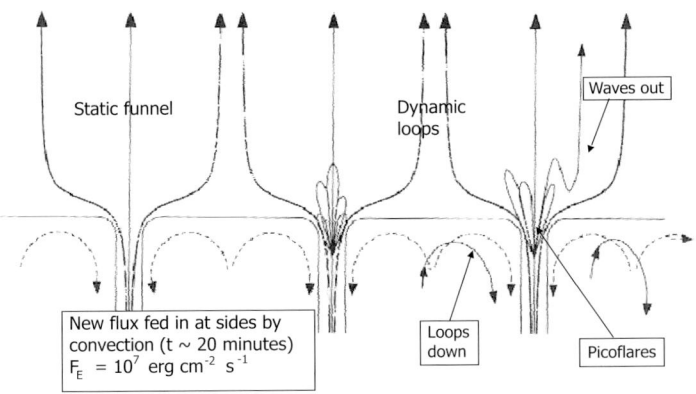

Figure 10.3. Schematic of the supergranular magnetic network consisting of coronal funnels and embedded small-scale loops, which may intermittently reconnect while being driven by magnetoconvection (after Axford and McKenzie, 1997).

current sheets arising from a relaxation towards equilibrium of the highly stressed and distorted coronal fields, when being shuffled around by foot-point motions. This dynamical process may lead to coronal heating, and via reconnection to topology changes of the magnetic field. The distribution and statistics of the nanoflares and the coronal turbulence and dissipation associated with them were discussed in detail by Einaudi and Velli (1999), who also carried out numerical simulations of the possible small-scale dynamics of a magnetically forced model corona. The resulting bursts of energy release may manifest themselves in various explosive plasma processes and radiation events, such as blinkers or bright points.

In the lower transition region, the network "pico-flares" must occur frequently and should have a dimension down to the presently unresolved 10-100 km scale. The energy burned in the associated "magnetic furnace" must be replaced by magnetoconvection in the photosphere, which ensures adequate supply of randomly-looped magnetic flux. If the energy thus generated by magnetic reconnection is released in frequent "pico-flares" introduced by Axford and McKenzie (1997), with typical energies of about 10^{21} erg, then such small-scale magnetic activity is also expected to continually produce a sizable amount of waves at low but also relatively high frequencies (larger than 1 Hz), as schematically shown in *Figure* 10.3. Wave dissipation would rapidly occur within a

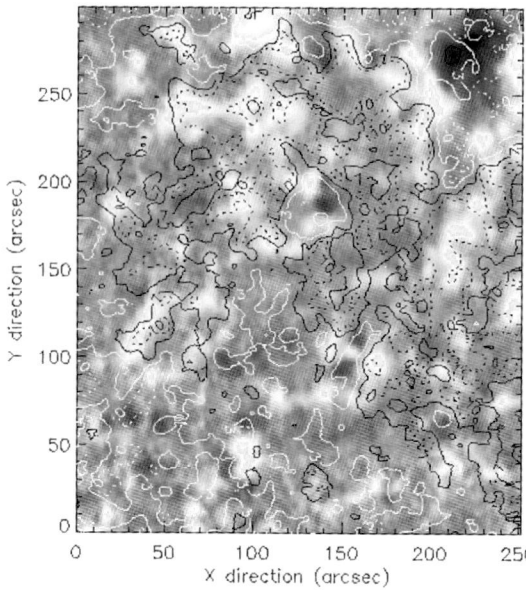

Figure 10.4 Photospheric magnetogram and neon-ion outflow inferred from SUMER observations of Doppler blue-shifts of Ne $\sigma_C$viii (λ 77 nm), relative to C σ_Ci (λ 154.2 nm) in an equatorial coronal hole. Black contours: dotted 10 km/s, solid, 7 km/s; white: solid 3 km/s. Kitt Peak magnetogram: white, positive, black, negative field polarity. Note the close correlation between blueshift and positive field, indicating significant upflows in open unipolar magnetic regions (after Xia et al., 2003).

fraction of a solar radius and could involve an essentially linear, kinetic mechanism such as cyclotron and Landau wave damping.

Thus the magnetic network not only is a continuous site of wave and turbulence generation by reconnection and shaking of flux tubes anchored therein, but it also is the source region of the solar wind and its mass and magnetic flux filling the whole heliosphere. The outflow of the nascent solar wind and the photospheric source magnetic field are shown together in *Figure* 10.4, which presents a magnetogram of an equatorial coronal hole analysed by Xia et al. (2003) and the overlaid contours of the sizable Doppler-shifts (as measured by SUMER, Solar Ultraviolet Measurements of Emitted Radiation) of Neon ions, indicating blue-shifts, i.e. outflows, at the cell boundaries and lane junctions in the network below the coronal hole, and red-shifts, i.e. downflows, in the regions underlying the magnetically closed parts of the corona.

4.   Waves and Flows in Loops and Funnels

It is difficult to observe coronal waves, because of the uncomplete and insufficient plasma diagnostics possible in the solar atmosphere. In-situ measurements, such as made in other solar-system plasmas (e.g. in the solar wind or Earth's magnetosphere), are in the corona impossible. Only remote-sensing is feasible, directly by using photons (in X-rays, ultraviolet, visible, and infrared light), or electromagnetic waves (radio, plasma waves), and somewhat more indirectly by means of corpuscular radiation (solar wind and energetic particles). Waves manifest themselves in spectral lines through intensity (i.e., electron density) modulation and the Doppler effect which leads to line shifts being either resolved as flows or unresolved as thermal and turbulent broadenings. For a comprehensive review of EUV spectroscopy and coronal diagnostics see Wilhelm et al. (2004). However, spectroscopic diagnostics suffers from the line-of-sight (LOS) problem, which requires even for optically thin lines to disentangle the integrated signal along the LOS, and from the radiative source and transfer problem, which implies to consider for the corona conditions far from local collisional (thermal) equilibrium.

Concerning the largest resolvable wave scales, it is required that the wavelength $\lambda \ll L$ and period $P \ll T$, where L is the extent of the field of view and T the duration of an observational sequence. This requirement is readily fulfilled by modern instruments. Also, their spectral resolutions are sufficient to measure Doppler shifts and broadenings. The integrated effects of the resolved low-frequency waves or the unresolved high-frequency turbulence are observationally considered in the LOS turbulent amplitude ξ, which corresponds to an effective temperature of an ion (with kinetic temperature T_i and mass m_i) according to the formula (k_B is the Boltzmann constant):

$$T_{i,eff} = T_i + \frac{m_i}{2k_B}\xi^2 . \tag{10.1}$$

Ample evidence was in the recent past provided by the TRACE and SOHO missions for the existence of waves trapped in loops or propagating in flux tubes (see, e.g., Aschwanden et al. (2002) or Wang (2003) for a review) and plasma flowing out in open funnels. For this see again Xia et al. (2003) and the results shown in *Figure* 10.4. Coronal funnels and their associated ion flows have been modelled by Marsch and Tu (1997) and Hackenberg et al. (1999). In their two-fluid model they considered ion heating which is assumed to be caused by cyclotron-wave sweeping. A steep inward temperature gradient resulting from electron heat conduction was also found. The sonic point was located close to the sun at $2R_s$ ($1R_s = 700$ Mm).

Theoretically, coronal waves confined in loops are, when being excited by flaring at the loop apex, expected to travel from loop top to bottom and thus will produce mainly redshifts after Hansteen (1993). On the contrary, the plasma streaming out on open field lines is supposed to produce mainly blueshifts. Concerning the observational evidence, both expected types of Doppler shift have indeed been observed by many authors: Hassler et al. (1999), Wilhelm et al. (2000), Xia et al. (2003),

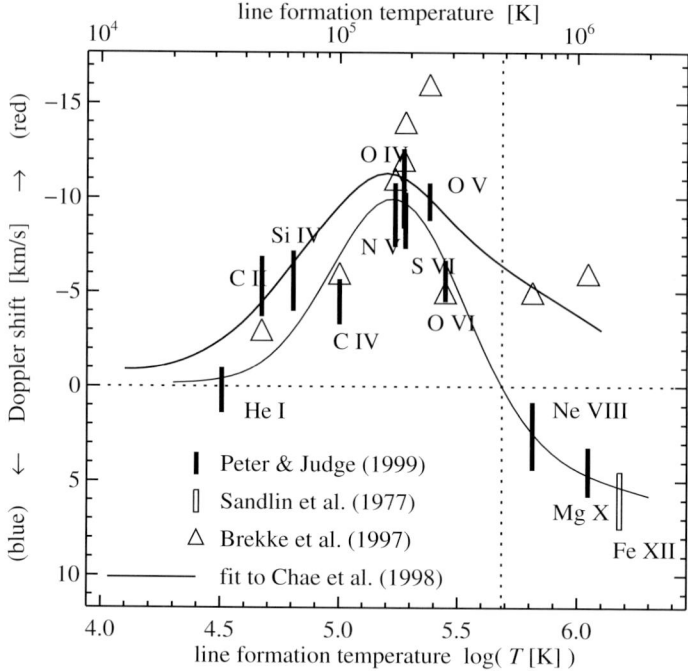

Figure 10.5. Doppler shifts of EUV emission lines according to Peter and Judge (1999).

Seely et al. (1997), Marsch et al. (1997), Tu et al. (1998), Tu et al. (1999), Peter and Judge (1999), and Peter (2001).

A survey of the average observational situation is given in *Figure* 10.5 and *Figure* 10.6, which show the line shifts and widths (separately for the core and wing parts), respectively versus the line-formation temperature, i.e. essentially the electron temperature in the range from $4\,10^4$ to $6.3\,10^5$ K, values corresponding to lower (below $1.3R_s$) heights in the solar corona. Note the distinct change from red to blue shifts in *Figure* 10.5 with increasing temperature. Considering *Figure* 10.6 taken from Peter (2001), you will note a clear separation in line width between the core and tail components for all the lines shown, with the width typically ranging between 20 and 30 km/s in the core (lower points) but steadily growing with height in the corona from about 40 to 90 km/s in the line tail (upper points). The major contributions, to perhaps the core but certainly the tail component, of the line broadening come from ξ, which is an empirical wave amplitude that is conveniently also used to estimate the amount of turbulent wave energy existing in the corona; see formula (10.1) again.

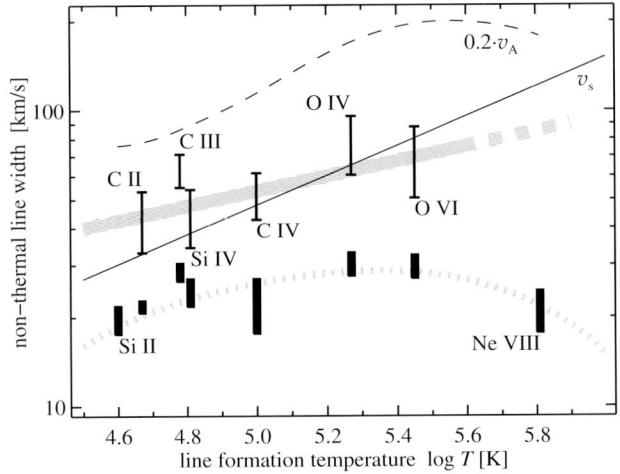

Figure 10.6. Emission line widths corresponding to wave amplitudes as a function of height in the lower corona. The widths of the core (indicated by grey dots) and wing parts (indicated by the grey-shaded bar) of each line are plotted separately. Alfvén speed (dashed line) and sound speed (continuous line) are given as well (after Peter, 2001).

5. Magnetohydrodynamic Waves and Flux Tube Oscillations

5.1 Observation and Theory

Let us now turn to magnetohydrodynamic waves in the corona and discuss some observations and their consequences. Detectability of coronal MHD waves requires that spatial (pixel size) and temporal (exposure/cadence) resolution be less than the wave lengths and periods considered. The spectral resolution must be sufficient to resolve Doppler shifts and broadenings (with SUMER this can be done with subpixel accuracy of a few km/s).

Nakariakov (2003) has put together an instructive table of some modern imagers / spectrometers on SOHO, Yohkoh and TRACE with the relevant parameters, such as spatial resolution and minimum pixel size (ranging from 0.5 arcsec to 4 arcsec), temporal resolution and maximal cadence (from 10 s to 30 s), and the spectral bands. TRACE first detected longitudinal waves in intensity (density) variations which were interpreted as slow magnetoacoustic waves. TRACE loop images in Fe $\sigma_C$ix (λ 17.1 nm) were taken at a 15 s cadence. De Moortel et al. (2000) analysed the properties of such longitudinal oscillations and provided a statistical overview of the physical properties of many cases of longitudinal oscillations detected at the base of large coronal loops, having lengths of a fraction of a solar radius.

Frequent Doppler oscillations were also measured by the SUMER spectrometer; see the review of Wang (2003). Some typical properties are: The oscillations are mainly seen in hot ($T_e > 7$ MK) flare lines, sometimes correlated with cool emissions. During solar activity they are correlated with the flux of soft X-rays and often appear to be triggered by mass ejecta and shock fronts. The periods range from 5 to 25 minutes, with typical damping times of several 10 minutes. Many events re-occur while the loops are still present.

When considering compressive MHD waves, an important dynamic time scale is the transit time, $t_s = L/c_s \approx 10L_{10}/T_6^{1/2}$ minutes ($L_{10} = L/10^{10}$cm, and $T_6 = T/10^6$K), where c_s is the sound speed. For other relevant time scales in loops see Walsh (2002), from whom we quote typical times (wave periods) for various physical processes occurring in loops: Alfvén and fast magnetosonic waves, 5 s; slow mode wave, 200 s; gravity induced motions, 40 s; thermal equilibration by electron heat conduction, 600 s; radiative cooling time, 3000 s; and convection, 300 s. All these parameters of course vary with height in the loop (or more generally the solar atmosphere).

There is general agreement that low-frequency coronal MHD waves are generated and released in the photosphere, housing a prolific energy source that is connected with magnetoconvection. The transport and propagation of wave energy in various forms and modes is not well understood, given the nonuniformity and variability of the coronal magnetic field, as discussed in the previous sections. Even less understood is the final conversion of ordered mechanical into random thermal energy of the particles. Understanding this irreversible energy conversion at collisional or kinetic scales is the main problem of coronal heating, which is addressed below.

The governing equations, describing waves in the corona, are the standard MHD equations for the magnetic field, **B**, the flow velocity, **V**, mass density ρ, and pressure p in the solar atmosphere and the gravitational acceleration **g**, see e.g. Roberts and Nakariakov (2003). Often an adiabatic equation of state with $p \sim \rho^\gamma$ and adiabatic index $\gamma = 5/3$ is employed. Applicability of MHD requires that the speeds involved are small against the speed of light c, which is always the case, that length scales are larger than the proton gyroradius or inertial length, which usually is true with the exception of current sheets, and also large in comparison with the collisional free path, which may become as large as 1 Mm in the corona. To describe the microphysics of wave dissipation and coronal heating, the MHD theory cannot be applied, but multi-fluid theory or kinetic physics is required.

5.2   Oscillations of Thin Flux Tubes

Oscillations of magnetic flux tubes have been theoretically studied for a long time, see e.g. the reviews by Roberts (1985) and Roberts (1991). The main restoring forces for perturbations stem from the magnetic and thermal pressure for compressive (transverse kink or longitudinal sausage) waves or the magnetic tension for incompressive (such as torsional Alfvén) waves.

In the light of the rather detailed modern wave observations, the theory was recently revisited by Roberts (2002), Roberts (2003) and Roberts and Nakariakov (2003). From these reviews we reiterate some of the basics of waves as described in the MHD picture. For linear wave motion and when disregarding stratification, the following coupled set of second-order equations for the perturbations is obtained:

$$\rho_0 \left(\frac{\partial^2}{\partial t^2} - c_A^2 \frac{\partial^2}{\partial z^2} \right) \mathbf{v}_\perp = -\nabla_\perp \left(\frac{\partial p_T}{\partial t} \right) , \tag{10.2}$$

$$\rho_0 \left(\frac{\partial^2}{\partial t^2} - c_t^2 \frac{\partial^2}{\partial z^2} \right) v_z = -\left(\frac{c_t}{c_A} \right)^2 \frac{\partial}{\partial z} \left(\frac{\partial p_T}{\partial t} \right) , \tag{10.3}$$

where the total pressure variation, p_T, is given by:

$$p_T = p + \frac{1}{4\pi} \mathbf{B}_0 \cdot \mathbf{B} , \tag{10.4}$$

which is coupled to the flow velocity components through the first-order equation:

$$\frac{\partial p_T}{\partial t} = \rho_0 \left(c_A^2 \frac{\partial v_z}{\partial z} - (c_s^2 + c_A^2) \nabla \cdot \mathbf{v} \right) . \tag{10.5}$$

The magnetic pressure of the non-uniform background field, $\mathbf{B}_0 = B_0 \hat{\mathbf{z}}$, in equilibrium balances the thermal pressure, p_0, through

$$\nabla \left(p_0 + \frac{B^2}{8\pi} \right) = 0 . \tag{10.6}$$

The flow is $\mathbf{v} = v_z \hat{\mathbf{z}} + \mathbf{v}_\perp$, with a component v_z along and $\mathbf{v}_\perp$ perpendicular to the field $\mathbf{B}_0$. These equations apply to Cartesian (magnetic slab) as well as cylindrical (magnetic flux tube) coordinates.

Here three relevant phase speeds were introduced. The sound speed is given by: $c_s = \sqrt{\gamma p_0/\rho_0}$. The slow magnetoacoustic tube speed is:

$$c_t^{-2} = c_s^{-2} + c_A^{-2}, \qquad c_t = c_s c_A / \sqrt{c_s^2 + c_A^2} . \tag{10.7}$$

The Alfvén speed is as usually defined, $c_A = B_0/\sqrt{(4\pi\rho_0)}$, with the height-dependent field strength inside the tube denoted by $B_0(z)$. Considering the typical numbers, one finds for the photosphere that $c_t \approx$ 7 km/s, which is the slowest possible phase speed and generally subsonic as well as sub-Alfvénic. The kink-mode phase speed is:

$$c_k = c_A \sqrt{\rho_0/(\rho_0 + \rho_e)} , \qquad (10.8)$$

where the density inside the tube is ρ_0, and outside in the field-free external environment it is denoted as ρ_e. The density ratio can by means of total pressure equilibrium be expressed by the plasma beta as: $\rho_0/\rho_e = \beta/(1+\beta)$, where $\beta = 2c_s^2/(\gamma c_A^2)$.

However, the waves in the solar atmosphere are according to Roberts (1991) strongly influenced by the gravitational stratification of the matter, which leads for many wave modes to a cutoff at a characteristic frequency, $\omega_\alpha = k_\alpha c_\alpha$ (e.g., with index $\alpha = k$ for the kink, t for the tube, and A for the Alfvén mode). The characteristic wave vector k_α relates to the inverse of the typical gradient scale of the restoring force involved, such as gravity or buoyancy. The wave generally will have a height-dependent phase speed, $c_\alpha(z)$, with the coordinate z for the altitude. The wave equation for the upper chromosphere and transition region often can be characterized by external forcing, F_α, and intrinsic dispersion, and be cast in a form resembling a driven Klein-Gordon equation. This may, by following e.g. Rae and Roberts (1982), Hasan and Kalkofen (1999), or Nakariakov (2003), be written as:

$$\frac{\partial^2 Q_\alpha}{\partial z^2} - \frac{1}{c_\alpha^2}\frac{\partial^2 Q_\alpha}{\partial t^2} - k_\alpha^2 Q_\alpha = F_\alpha . \qquad (10.9)$$

The related free-wave linear dispersion then shows a minimum frequency (the cut-off) and reads:

$$\omega(k) = \sqrt{(kc_\alpha)^2 + \omega_\alpha^2} . \qquad (10.10)$$

The amplitude when compensated for barometric stratification is given by $Q(z,t) = \xi(z,t)\exp(-z/(4H))$, where $\xi(z,t)$ denotes the original wave amplitude (for the sake of simplicity we here just consider a single component), and H is the scale height assumed to be constant.

The significance of the Klein-Gordon equation after Roberts (2003) is that through the cutoff frequency it introduces the natural oscillation period of the nonuniform background medium in MHD equilibrium, and further that it implies evanescence below that cutoff, which means no wave can propagate vertically into the atmosphere if $\omega < \omega_\alpha$. Another consequence is that an impulsive (e.g., flare produced) disturbance may create a wavefront at speed c_α in the medium, and thus sets up a wake oscillating at the natural frequency ω_α behind the front.

Generally, nonuniformity of the atmosphere may lead to very complex nonlinear wave mode-couplings; see e.g. the reviews by Roberts (1991) and Goossens (1991). The simpler stratification effects are most easily demonstrated for vertically propagating sound and Alfvén waves. To give examples for the height-dependence of the cutoff frequencies, we first quote after Roberts (2003) the one for sound waves:

$$\omega_s^2(z) = \frac{c_s^2}{4\Lambda_0^2}(1 + 2\Lambda_0'), \qquad (10.11)$$

where $\Lambda_0 = p_0(z)/(g\rho_0(z))$ is the pressure scale height of the nonuniform background atmosphere and g the gravitational acceleration (assumed to be constant since it only varies on the scale of 1 R_s). The prime in (10.11) indicates a spatial derivative with respect to z. In an isothermal atmosphere with $\Lambda_0' = 0$, we obtain the acoustic cutoff frequency: $\omega_s = c_s/(2\Lambda_0) = \gamma g/(2c_s)$. With $\gamma = 5/3$ and $g = 0.274$ km s^{-2} and a sound speed of $c_s = 7.5$ km/s, one obtains $\Lambda_0 = 125$ km, and thus $\omega_s = 0.03$ s^{-1}, or a cyclic frequency of 4.8 mHz and period of 3.5 minutes. In the hotter corona, stratification is less significant and the corresponding period would be much longer, i.e. 1-2 hours.

After Roberts (2003), the linear wave equation for sausage waves in a thin magnetic flux tube in the stratified solar atmosphere also corresponds to a Klein-Gordon equation, such as given in (10.9). Incompressible kink-mode oscillations yield transverse displacements of amplitude $\xi(z,t)$ for the slender flux tube and can, after a transformation such as above used for compensating the barometric stratification, also be written in terms of a Klein-Gordon wave equation with a cutoff at:

$$\omega_k^2(z) = \frac{c_k^2}{4\Lambda_0^2}(\frac{1}{4} + \Lambda_0'). \qquad (10.12)$$

Finally, the incompressible Alfvén mode may also satisfy a Klein-Gordon equation of the type (10.9), where the variable now is $Q = \rho_0^{1/4} v_\phi$, with the azimuthal torsional velocity displacement v_ϕ, and the corresponding cutoff frequency:

$$\omega_A = c_A/(4\Lambda_0). \qquad (10.13)$$

Note that for a bounded Q, the above density scaling leads to an exponential growth in the stratified (e.g., isothermal) atmosphere, and thus to a substantial amplification of the small (photospheric) wave amplitude with height in the chromosphere or transition region (as shown in the *Figure* 10.7 and *Figure* 10.8). Consequently, shocks will occur or

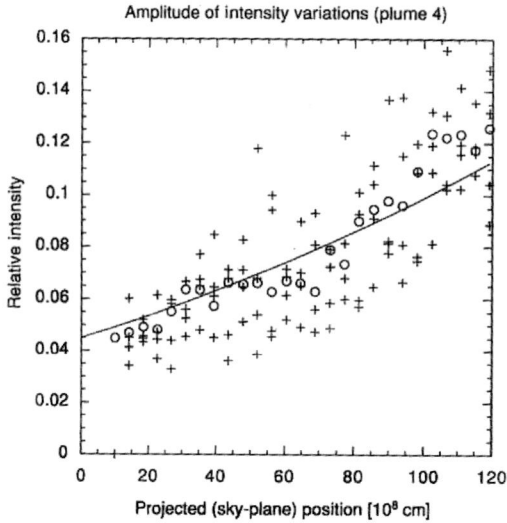

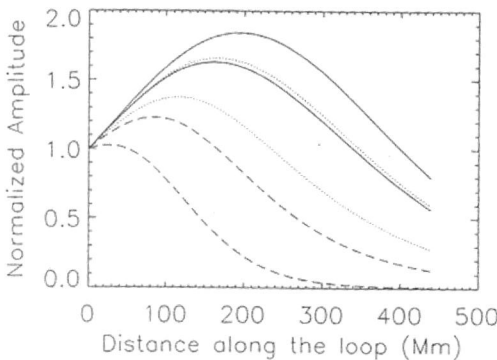

Figure 10.7 Compressive wave amplitude relative to the background density in a plume (with an open field) versus height. The observational data points from the measured amplitudes are plotted by circles. Estimates of their upper and lower limits are given by the crosses. The solid line shows the analytical model wave amplitude, reflecting with its exponential increase the barometric density stratification (after Ofman et al., 1999).

Figure 10.8 The slow-magnetosonic wave amplitude versus height, with an initial amplitude of $\delta V_0 = 0.02c_s$ in a coronal magnetic loop with a closed field. The wave periods are 900 s (solid curves), 600 s (dotted curves) and 300s (dashed curves). The lower curves respectively correspond to stronger dissipation with a coefficient being larger by a factor of 2.5 (after Nakariakov et al., 2000).

other nonlinear effects, causing sharp transitions, steepening and subsequent wave dissipation as found by Carlsson and Stein (1997). These effects will become weaker and stretched out with height in the corona, owing to its higher temperature and larger scale height.

5.3   Wave Amplitudes Versus Height from Numerical Models

Instead of a quasilinear analytic perturbation treatment of the MHD equations one may seek for fully nonlinear direct numerical solutions. This has in the recent past been done by various authors. Here we just give two examples for such an approach and refer to the references in the here cited literature for further work. For example, the evolution of the slow-magnetosonic wave amplitude, having an initial value of $\delta V_0 = 0.02c_s$, was calculated by Nakariakov et al. (2000) as a function of height in a coronal magnetic loop. Expectedly, such models reveal strong wave steepening with height in the solar atmosphere, as shown in *Figure* 10.7 which indicates the steepening and subsequent limitation via dissipation of the amplitude of a propagating sound wave.

Ofman et al. (1999) and coworkers have calculated numerically the turbulent outflow of the solar wind driven by MHD waves on open field lines. The nascent solar wind is strongly heated in their models by an effective viscosity and ohmic resistance, whereby the classical functional dependence of the dissipation terms is retained. However, the transport coefficients are drastically enhanced by orders of magnitude over their collisional values, so as to locate the dissipation region at low heights (fraction of 1 R_s). One result from their model for a polar plume in a CH is shown in *Figure* 10.8, showing an increasing wave amplitude versus height.

Using TRACE observations of transverse oscillations of a large coronal loop, which was excited by a solar flare, Nakariakov et al. (1999) studied the decay of the oscillations and the associated effective dissipation coefficient. They empirically estimated it to be eight to nine orders of magnitude larger than the theoretically predicted, classical value. The collisional time for wave damping, if interpreted as being due to resistive or viscous dissipation, is well known and given by Ofman et al. (1994). The observed dissipation has a much shorter time constant, which implies a (magnetic) Reynolds number that is orders of magnitude smaller than expected from Coulomb collisions. This evaluation should be confirmed by future observations, and certainly the issue of transport coefficients in the weakly collisional corona remains unsolved and direly needs further study. Also, observations with a much-higher resolution are required to detect the conjectured velocity shear layer within the loop at scales near 15 km.

5.4 A Standing Slow Magnetoacoustic Wave

Many clear cases of coronal loop oscillations could be associated with solar magnetic activity in connection with flares. Some of the post-flare loops as seen by Extreme Ultraviolet Imaging telescope (EIT) on SOHO were studied with the help of SUMER in much detail, and the results reviewed recently by Wang et al. (2003a). *Figure* 10.9 shows perhaps the clearest example of a standing slow-mode sound wave observed by Wang et al. (2003b), which was excited during such a flare event. The

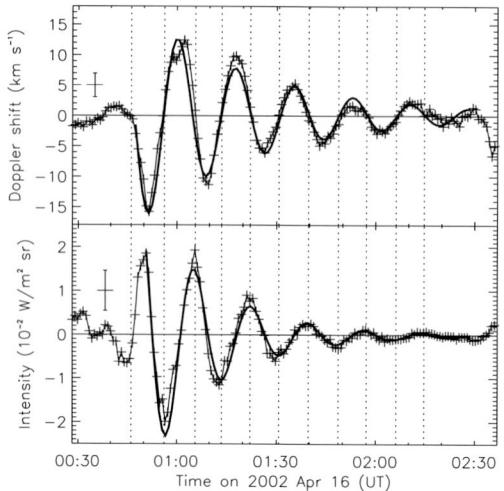

Figure 10.9 Slow-mode standing wave detected in a post-flare loop (after Wang et al., 2003b).

geometrical parameters of the related loop were determined as: Azimuth angle (in East-West direction) $\alpha = 19$ deg, inclination angle $\theta = 27$ deg, angle between magnetic field and LOS near the slit $\gamma = 15$ deg, loop length $L = 191$ Mm, length of segment from apex (A) to foot point (F) $L_{AF} = 64$ Mm, with $L_{AF} = 1/3\ L$.

Doppler shift, intensity and background continuum of the line Fe $\sigma_C$xix (λ 111.81 nm) were analysed together. After removal of the background, a damped sine function was used to describe the oscillation as follows:

$$V = V_0 + V_m \sin(\omega t + \phi) \exp(-t/T_d) , \qquad (10.14)$$

with the derived parameters: $V_m = 18 \pm 1.5$ km/s, period $P = 17.6$ min, and damping time $T_d = 36.8 \pm 2.6$ min ($T_d/P \sim 2.1$). Velocity and intensity oscillations were found to have a quarter-period phase difference, pointing to slow tube-mode standing waves. Also the background continuum intensity shows quasi-periodic fluctuations, roughly in phase with the oscillations in Fe $\sigma_C$xix. With the same functional form as in (10.14), one finds for the intensity: $I_m/I_0 = 0.19$, $P = 17.1$ min, and $T_d = 21 \pm 1.6$ min ($T_d/P \sim 1.2$).

The damping rates were empirically determined from SUMER data for some 49 cases in 27 events. The values found by Wang et al. (2003a) are in agreement with classical dissipation rates, due to a higher thermal conduction and viscosity when the temperature as seen in the post-flare loops is much higher than 10^6 K. The theoretical scaling relation, $T_d \sim P$, compares favourably with their empirical result: $T_d \sim 0.68 P^{1.06\pm0.18}$. The results obtained from TRACE for 11 cases considered by Ofman and Aschwanden (2002) are in agreement with dissipation by phase mixing for kink-mode oscillations, yielding a scaling relation $T_d \sim P^{4/3}$, which compares favourably with their empirical result: $T_d \sim 0.9 P^{1.30 \pm 0.21}$.

6. Plasma Waves and Heating of Particles

In what follows we will discuss kinetic plasma waves in the corona. Whereas with SOHO and TRACE the low-frequency fluid modes can easily be resolved, it is presently not possible to detect high-frequency MHD and plasma waves with frequencies above 1 Hz. To do this the spatial (pixel size) and temporal (exposure/cadence) resolution of the observations must be less than the wave lengths and periods, a requirement which is missed by the existing instrumentation by orders of magnitude.

The shortest relevant ion kinetic scales are: The proton inertial length or the wavelength of a cyclotron wave, $\lambda_p = 2\pi c_A/\Omega_p = 1434(n/\mathrm{cm}^3)^{-1/2}$ km (about 100 m in a CH), and its gyroperiod, $P_p = 2\pi/\Omega_p = 0.66(B /\mathrm{G})^{-1}$ ms. The resonant wave amplitude, δV_g, may be less than 1 km/s, i.e. $\delta V_g \leq 0.001 c_A$. Here $\Omega_j = e_j B/(m_j c)$ is the gyrofrequency of species j with charge e_j, B is the magnetic field, and c_A again is the Alfvén speed. $R_p = v_p/\Omega_p$ is the proton gyroradius, and $v_p = \sqrt{k_B T_p/m_p}$ its thermal speed.

Heavy ion temperatures were estimated from measured line widths by means of equation (10.1). The ion cyclotron resonance frequency is $f_i/\mathrm{kHz} = 1.5 Z_i/(A_i B/\mathrm{G})$, with the ionic charge number, Z_i, and atomic mass number, A_i. By resonance absorption of waves near the gyro frequency, the magnetic moment of an ion will increase, and consequently the perpendicular ion temperature as well, especially for an ion moving in the declining field (mirror configuration) of an expanding funnel or CH. This effect may lead to a scaling of the line widths with the wave period, i.e. $P_i \sim A_i/Z_i$, which was indeed found from SUMER observations obtained by Tu et al. (1998) and Tu et al. (1999). The resulting ion kinetic temperature is $T_i = (2-6)$ MK at $r = 1.15\,R_s$. By comparison with the small length λ_p, the collisional free path, $\lambda_c \approx 3\,(T_6^2/n_{10})$ km (with $T_6 = T/10^6$K and $n_{10} = n/10^{10}\mathrm{cm}^3$), for a proton is rather large in the corona. In a CH we find $\lambda_c \approx 1000$ km, which is about the pixel size of present EUV imagers and spectrometers. Therefore, the corona should largely behave like a collisionless and tightly magnetized plasma (because of the low plasma beta of only a few percent), and thus kinetic wave effects will dominate over collisional effects.

The empirical wave/turbulence amplitude ξ was shown versus line formation temperature in the summary *Figure* 10.6 (see also Wilhelm et al. (1998) for further results). For magnetohydrodynamic waves travelling on open field lines we expect, when they escape without damping

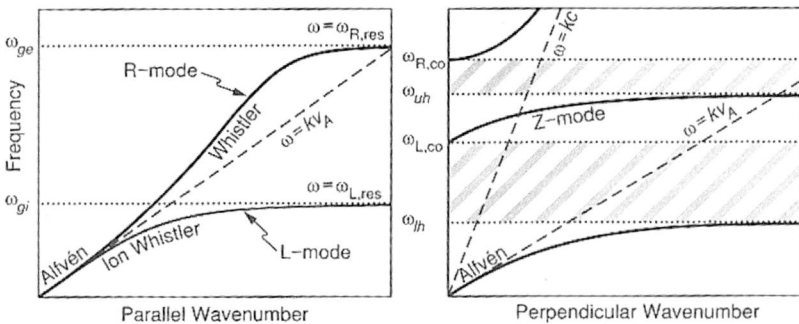

Figure 10.10. Left: Dispersion branches for parallel propagating low-frequency plasma waves. Right: Dispersion branches for perpendicular wave propagation, with the resonant frequencies indicated by dotted lines. The shaded areas refer to wave stop bands. The dashed lines refer to the Alfvén (left) and fast magnetosonic (right) MHD wave branches. The steeper dashed line on the right indicates the free-space electromagnetic wave (after Treumann and Baumjohann, 1996).

or suffer only weak dissipation, that their amplitudes follow a WKB-type evolution. According to Nakariakov (2003), this means a scaling like $\xi \sim \rho^{-1/4}$ for Alfvén waves, which seems consistent with the results presented in the upper curve of *Figure* 10.6. We consider that waves confined in loops are more prone to strong dissipation with consequent loop heating than the ones escaping in funnels. This might explain why there is a maximum in the lower curve.

High-frequency (above 1 Hz at the base) waves propagating to the outer corona may finally dissipate within a few R_s. The SOHO measurements, see Kohl et al. (1997) and Cranmer et al. (1999), indicate that oxygen and hydrogen mean thermal speeds are in excess of the canonical value of about 150 km/s corresponding to 1 MK. In coronal holes strong heating of hydrogen (protons) was found, with $T_p = 2 - 3$ MK. The oxygen ions showed even stronger (mostly perpendicular) heating, leading to $T_o \leq 100$ MK. Moreover, a large temperature anisotropy, with $T_{o\perp}/T_{o\parallel} > 10$, was found. As to the possible origin of this heating ion-cyclotron resonance is commonly assumed. For a comprehensive review of the theory see Hollweg and Isenberg (2002). There may be many other waves involved, though, which could be absorbed through Landau damping, an old idea going back to Barnes (1969).

There is a rich variety of kinetic plasma waves being relevant for the corona: Parallel ion-cyclotron waves, with $\omega \leq \Omega_i = e_i B/(m_i c)$, and kinetic Alfvén waves with finite gyroradius effects leading to dispersion; or whistler waves, with $\omega \leq \Omega_e = e_e B/(m_e c)$. In addition, for oblique propagation there are the upper and lower hybrid waves, the latter with

$\omega_{lh} = \sqrt{|\,\Omega_p \Omega_e\,|}$. The standard dispersion curves (at a given location in the nonuniform corona) for parallel and perpendicular propagating plasma waves are given in *Figure* 10.10. For further details see any space-plasma text book, e.g. of Treumann and Baumjohann (1996), or for the corona especially the book of Benz (1993).

The wave absorption coefficient (or growth rate in case of an instability), i.e. the wave opacity sensitively depends upon the ion and electron velocity distributions. The heating and acceleration of the ions depend further on the unknown wave spectrum in the corona. The corresponding rates based on quasilinear theory were calculated by Marsch and Tu (2001) and Marsch (2002) and are discussed below.

What processes could generate ion-cyclotron waves? Theoretically, an ion core-temperature anisotropy can drive the waves unstable, similarly hot ion beams (coronal ion jets), electric currents, e.g. due to an ion-electron drift, ion loss-cone distributions, the parametric decay of large-amplitude Alfvén waves, energy cascading from the MHD regime, or inhomogeneity leading to frequency sweeping for a spatially varying $\Omega_i(z)$. The ion velocity distribution functions (VDFs) observed by Helios in the fast solar wind reveal distinct nonthermal features indicating kinetic wave activity as described by Marsch (1991b). However, do such kinetic processes really operate in the corona? We presently do not know and therefore consider the origin and level of kinetic ion waves in the solar corona an open issue.

7. Generation, Transfer and Dissipation of Coronal Turbulence

7.1 Generation of Magnetohydrodynamic Waves

Coronal magnetic fields are rooted deeply in the turbulent photosphere. Magnetoconvection there will invariably lead to waves in the corona. Therefore, coronal MHD waves and turbulence are most probably driven by magnetoconvection or released by the relaxation of strained fields through reconnection. The gentle shaking of flux tubes through steady convection or the violent pushing of loops by emerging flux bumping into them, may lead to various kinds of coronal eigen-oscillations and propagating coronal waves. As a result, flux tubes (loops and funnels) will carry frequent magnetoacoustic waves and/or torsional Alfvén waves into the corona.

In a series of papers Musielak and Ulmschneider (2002), Musielak and Ulmschneider (2003a) and Musielak and Ulmschneider, (2003b) worked out a detailed model for the source and excitation of coronal turbulence, involving photospheric shaking and pushing of thin magnetic flux tubes.

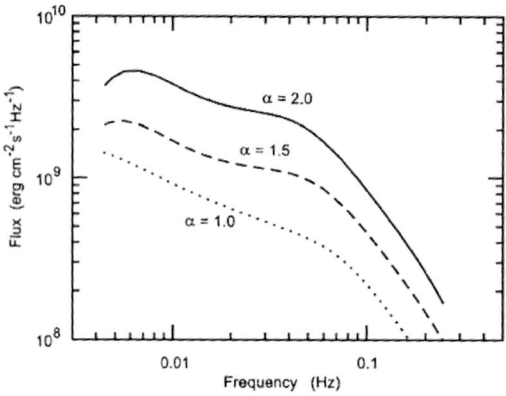

Figure 10.11 Spectral energy flux in dependence on the frequency of waves generated by gentle shaking of flux tubes. Here α is a measure of the mixing length, $\ell = \alpha H$, and the barometric scale height H is about 300 km (after Musielak and Ulmschneider, 2002).

They act as wave guides conducting the wave mechanical energy into the solar corona. At the photospheric boundary they assumed various forms of turbulence spectra, such as the famous Kolmogorov $-5/3$ law. In their papers these authors provide detailed calculations of the wave spectra and consider the energetics of chromospheric and coronal heating. Their model spectra are particularly relevant for a discussion of a turbulent cascade in the corona. A model spectrum is shown in *Figure* 10.11. Here α is a measure of the mixing length, $\ell = \alpha H$, and the barometric scale height H is about 300 km. For further references on coronal turbulence also see the review of Petrovay (2003).

7.2 Wave Energy Transfer and Turbulent Cascade

Whereas progress was made in the question of how and where waves and turbulence in the solar corona might be generated, the problem of how turbulent energy spreads in wave-vector and/or real space, i.e. how it propagates and is distributed between fluctuations at (vastly) different scales in the corona, remains unsolved. Matthaeus et al. (2003) have described a scenario, in which 2-D turbulence in the strongly magnetized open corona prevails, with an anisotropic magnetohydrodynamic cascade by which energy is transferred mostly perpendicular to the magnetic field to dissipation at the gyrokinetic scales. This may lead to quasi-2-D turbulence, which can be described by reduced MHD. A parallel cascade can be driven by Alfvén waves counterpropagating along the background field, and was shown by Tu (1987) to yield sufficient replenishing of wave energy in the cyclotron domain. This scenario was successfully applied to explain the spectral evolution of Alfvénic fluctuations in the interplanetary solar wind, which was reviewed by Tu and Marsch (1995).

Kinetic effects and particle heating by Alfvénic turbulence in the extended corona were recently studied by Cranmer and van Ballegooijen

(2003) in a comprehensive article containing most of the references relevant to this topic. A new spectral transfer model incorporating advection and diffusion in wave-number space was proposed, with dominant transfer perpendicular to **B**. The model equation for the $k_\parallel$-integrated spectrum, $W_\perp(k_\perp)$, contains wave energy advection, diffusion, injection and dissipation and reads:

$$\frac{\partial W_\perp}{\partial t} = \frac{\partial}{\partial x}\left[\frac{1}{\tau}(-\beta W_\perp + \gamma\frac{\partial W_\perp}{\partial x})\right] + S(k_\perp) - D(k_\perp)W_\perp , \quad (10.15)$$

where t is the time, $x = \ln k_\perp$, and $\tau(k_\perp)$ is a characteristic time of spectral transfer or eddy turn over. The parameters β and γ are model constants of order unity supposed to quantify the advection and diffusion in $k_\perp$-space. Furthermore, S is the spectral forcing localized at low $k_\perp$, and D the kinetic dissipation at high $k_\perp$. This transfer equation was tailored to describe coronal turbulence, similarly to theories e.g. derived in the book of Melrose and McPhedran (1991) on weakly nonuniform, turbulent dielectric media. Some numerical results obtained by integration of (10.15) are presented in *Figure* 10.12.

The theory of Cranmer and van Ballegooijen (2003) makes some ad-hoc (yet plausible) assumptions. It attempts to combine a turbulent cascade provided by random, scale-invariant convected fluctuations with the dispersive and dissipative properties of coherent propagating waves, a difficult approach that appears contradictory in itself. The theory still lacks rigourous derivation from kinetic or MHD equations, admittedly a difficult task that remains for future research. The invoked perpendicular cascade leads through Landau damping to preferential electron heating, which also is obtained in the work of Shukla et al. (1999) considering high-frequency kinetic Alfvén waves with $k_\parallel = \Omega_p/c_A$. However, preferred electron heating is in obvious contradiction to observations from SOHO, indicating cool electrons and hot ions in the extended open corona as shown by Marsch (1999) and Marsch et al. (2003b).

Theoretically, there are numerous alternative wave-turbulence generation scenarios, where the free energy may reside in non-thermal particle populations driving kinetic instabilities, or in which mode conversion or decay of large-amplitude waves may ensure the supply of energy to the dissipation domain (see e.g. the review of Goossens (1991) or recently the paper by Voitenko and Goossens (2002)). Kinetic Alfvén waves were anew discussed by Hollweg (1999) and Shukla et al. (1999) and received much attention. Because of the oblique propagation, their dispersion properties naturally involves the ion gyrokinetic scale, but they show a propensity to dissipate by electron Landau damping.

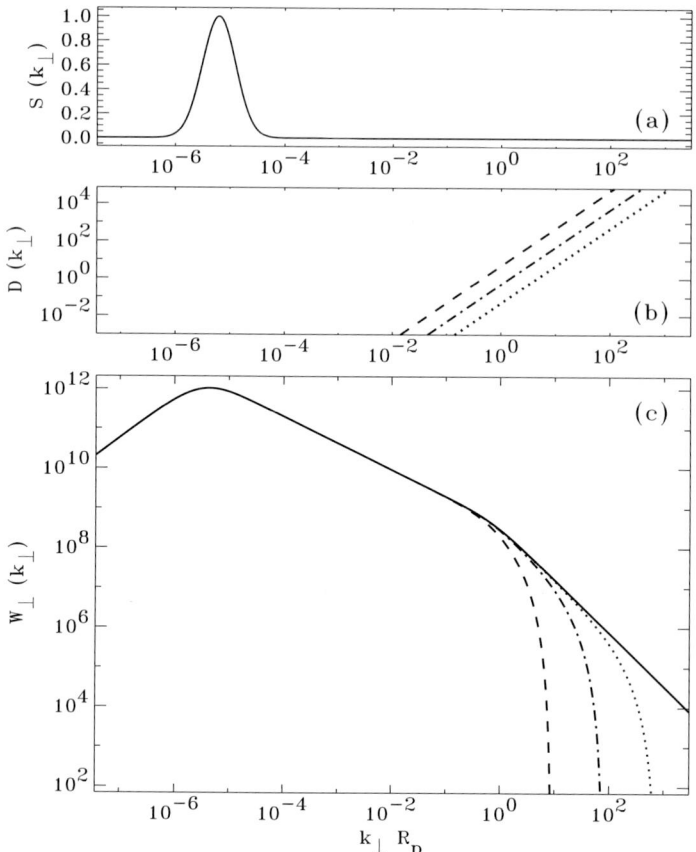

Figure 10.12. Representative turbulence spectrum at 2 R_s shown in dependence on the spatial frequency of coronal fluctuations. It was obtained from a numerical solution of the spectral transfer equation (10.15); (a) outer-scale localized source spectrum; (b) dissipation rate, $D(k_\perp) = D_0(k_\perp R_p)^2$, where $D_0 = 0.05\,\mathrm{s}^{-1}$ (dotted line), $D_0 = 0.5\,\mathrm{s}^{-1}$ (dot-dashed line), and $D_0 = 5\,\mathrm{s}^{-1}$ (dashed line); (c) time-steady reduced power spectrum, $W_\perp(k_\perp)$ for the three dissipation rates and without dissipation (solid line). The proton gyroradius is $R_p = 0.024$ km (after Cranmer and van Ballegooijen, 2003).

To describe all the work done on kinetic waves is beyond the scope of this article. But equally important for omission here than simple lack of space is that many wave-particle scenarios were not even verified empirically in laboratory experiments, and therefore their application to the corona is all the more questionable and speculative.

In summary, concerning the possible replenishment of high-frequency fluctuations from MHD waves by cascading, various assumptions were made in coronal turbulence models about the unknown energy spectral transfer mechanism. As discussed above, the basic problems of energy cascading - oblique as well as parallel - remain unsolved. Large-scale coronal MHD structures may preferentially excite perpendicular short-scale fluctuations, the dissipation of which will involve Landau damping coupled to kinetic processes acting on oblique wavevectors as suggested by Leamon et al. (2000). Whatever the coronal wave processes - they must be most effective for heavy ions (more than for protons and electrons), given their high kinetic temperatures inferred from SOHO observations.

7.3   Wave Dissipation in the Kinetic Domain

Let us briefly discuss the issue of kinetic wave dissipation. Assume that a nonlinear cascade or an instability have somehow delivered sufficient wave energy to the kinetic scales. Every linear kinetic wave mode (indicated by the subscript M) can fully be described by the Fourier components of its electric field vector, which may be written as $\tilde{\mathbf{E}}(\mathbf{k}, \omega) = 2\pi \sum_M \delta(\omega - \omega_M(\mathbf{k})) E_M(\mathbf{k}) \mathbf{e}_M(\mathbf{k})$, in terms of the wave frequency, $\omega_M(\mathbf{k})$, unimodular polarization vector, $\mathbf{e}_M(\mathbf{k})$, and amplitude, $E_M(\mathbf{k})$. A similar expression holds for the magnetic field, which is directly given through the induction equation. We define the spectral energy density of the electric field of mode M by the expression $\mathcal{E}_M(\mathbf{k}) = \mid E_M(\mathbf{k}) \mid^2 /(8\pi)$. Marsch and Tu (2001) have calculated the general resonant heating and acceleration rates for coronal ions and electrons by plasma waves. These rates can, for a quasi-linear superposition at random phases of all waves of type M, be cast in the compact form:

$$
\begin{bmatrix} \frac{\partial}{\partial t} U_j \\ \frac{\partial}{\partial t} V_{j\parallel}^2 \\ \frac{\partial}{\partial t} V_{j\perp}^2 \end{bmatrix} = \int_{-\infty}^{+\infty} \frac{d^3k}{(2\pi)^3} \frac{\omega_j^2 \, \mathcal{E}_M(\mathbf{k})}{\rho_j \, \omega_M^2(\mathbf{k})} \sum_{s=-\infty}^{+\infty} \mathcal{R}_j(\mathbf{k}, s) \begin{bmatrix} k_\parallel \\ 2k_\parallel w_j(\mathbf{k}, s) \\ s\Omega_j \end{bmatrix}
\tag{10.16}
$$

Here $\omega_j = ((4\pi e_j^2 n_j)/m_j)^{1/2}$ is the plasma frequency of species j. The resonance function, $\mathcal{R}_j$, or wave opacity is a functional of the particle distribution function, $f_j(w_\parallel, w_\perp, t)$, and essentially involves its pitch-angle derivative which is evaluated in the respective wave frame at the Landau ($s = 0$) or cyclotron (any integer Bessel function index, $s = \pm 1, \pm 2, ..$) resonance:

$$
\mathcal{R}_j(\mathbf{k}, s) = -(2\pi)^2 \frac{k_\parallel}{\mid k_\parallel \mid} \int_0^\infty dw_\perp \times
$$

$$
\left[\mid \mathbf{e}_M^*(\mathbf{k}) \cdot \mathbf{V}_j(\mathbf{k}, \mathbf{w}, s) \mid^2 (w_\perp \frac{\partial}{\partial w_\parallel} + \frac{s\Omega_j}{k_\parallel} \frac{\partial}{\partial w_\perp}) f_j(w_\perp, w_\parallel) \right]_{w_\parallel = w_j(\mathbf{k}, s)}
\tag{10.17}
$$

This resonance function is by definition dimensionless. It essentially contains the squared scalar product between the velocity vector $\mathbf{V}_j$ (related to the current carried by a gyrating particle of species j), and the polarization vector of mode M, as well as the pitch-angle gradient of the velocity distribution $f_j(\mathbf{w})$. The random particle velocity $\mathbf{w}$ refers to the rest frame of species j, and the resonant speed is $w_j(\mathbf{k}, s) = (\omega_M(\mathbf{k}) - k_\parallel U_j - s\Omega_j)/k_\parallel$, whereby the mean drift speed is denoted as U_j and directed along the magnetic field.

To calculate these rates explicitly, one must know $f_j(\mathbf{w})$ and $\mathcal{E}_M(\mathbf{k})$ and also self-consistently solve the wave dispersion relation to obtain the required wave frequency $\omega_M(\mathbf{k})$. Simply assuming those quantities as being done in many models (e.g. a Maxwellian $f_j(\mathbf{w})$ and power-law spectrum, $\mathcal{E}_M(\mathbf{k}) \sim k^{-5/3}$) is not satisfying, since spectra and velocity distributions are theoretically expected to evolve. In the solar wind they are indeed observed (see the reviews by Marsch (1991a) and Marsch (1991b)) to evolve substantially, quickly within a few wave damping times on small scales, and slowly owing to the inhomogeneity on large scales. In kinetic model calculations for the corona this was also found by Vocks and Marsch (2001).

Equations (10.16) and (10.17) are rather general and apply to the corona as a dissipative linear dielectric medium permeated by weak plasma-wave turbulence. Simpler versions or variants of them were applied in many solar wind models, see e.g. Marsch et al. (2003b) and Cranmer and van Ballegooijen (2003) and the references therein. But to fully exploit them in multi-fluid models, with wave-mode spectra including absorption features and realistic particle VDFs based on weakly-collisional ion kinetics, remains an issue for future work.

7.4   Origin and Generation of Coronal High-Frequency Waves

The origin and existence of high-frequency (cyclotron) waves in the corona remains to be shown. In the empirical model of Cranmer et al. (1999) for a polar coronal hole, spectroscopic constraints were placed on the cyclotron resonance heating. Additional empirical results, derived from solar remote-sensing and solar-wind in-situ measurements, that constrain coronal heating theories were reviewed by Marsch (1999). In many solar wind models, a simple power-law spectrum for the waves was assumed, the intensity of which was determined by an extrapolation of the in-situ measurements to the corona. High-frequency waves (most likely in the Alfvén/ion-cyclotron mode) may provide the necessary coronal heating through rapid dissipation within $1 \, R_s$. As mentioned already, these waves were proposed by Axford and McKenzie (1997) to originate through reconnection from the active magnetic network in the lower solar transition region.

A key feature of this model is that the damping of Alfvén waves at the cyclotron frequency in a rapidly declining magnetic field (frequency sweeping), provides strong heating close to the Sun. This idea was originally developed for the solar wind by Tu (1987) and then corroborated by Tu and Marsch (1997) in a two-fluid turbulence model, including parametric studies made by Marsch and Tu (1997) of the wind properties in dependence on the average wave amplitude, ξ, at the coronal base.

Presently, we do not have direct evidence for high-frequency waves that may provide the heating. The various models discussed by Marsch et al. (2003b) imply properties which still require observational verification. For example, the high-frequency waves should:

- be fairly linear with $\delta B \ll B$,

- have frequencies even up to 1 kHz,

- dissipate quickly within 1 R_s and disappear beyond $\sim 10\ R_s$,

- preferentially heat heavy ions, discriminating them according to Z_i/A_i,

- be mostly left-hand polarized to dissipate by ion-cyclotron absorption.

All these properties can only be verified by direct in-situ measurements, since remote-sensing or spectroscopic evidence cannot give the required complete plasma diagnostics. The waves therefore remain a problem from the observational point of view. On the other hand, given our knowledge from solar wind in-situ measurements and other accessible space plasmas, such as planetary magnetospheres, we have every reason to believe that high-frequency waves do exist in the corona, because they are, along with the particles, a natural component of any plasma, particularly if it is like the corona far from thermal equilibrium and contains ample free magnetic energy. This of course is common wisdom for the solar radio astronomy community, which has for decades looked into plasma waves and related processes generating the solar nonthermal radio emissions in flares. The book by Benz (1993) describes the related kinetic plasma processes in the solar corona.

7.5   Ion Velocity Distribution and Wave Absorption

In a series of papers written by Vocks and Marsch (2001), Vocks (2002) and Vocks and Marsch (2002), a semi-kinetic model involving a reduced Boltzmann equation which only depends on $v_{\parallel}$ as well as a diffusion model, studied by Cranmer (2001), have been constructed to describe the subtle kinetic effects of wave-particle interactions in coronal funnels and holes. For a short review of these efforts see Marsch et al. (2003a), and for a comprehensive review of the possible role of ion-cyclotron interaction in the generation of the fast solar wind see Hollweg and Isenberg (2002).

It was demonstrated in the semi-kinetic model of Vocks and Marsch (2001) that it is meaningful to reduce the 3-D ion VDF by an integration over $v_{\perp}$. This procedure yields two relevant reduced velocity distributions:

$$\begin{pmatrix} F_{j\parallel} \\ F_{j\perp} \end{pmatrix}(v_{\parallel}) = 2\pi \int_0^{\infty} dv_{\perp} v_{\perp} \begin{pmatrix} 1 \\ v_{\perp}^2/2 \end{pmatrix} f_j(v_{\perp}, v_{\parallel}), \qquad (10.18)$$

where a negative $v_{\parallel}$ points in the sunward direction. The evolution equations for these reduced VDFs are obtained by taking the corresponding moments of the Boltzmann equation after Vocks (2002). A gyrotropic 2-dimensional VDF, $f_j(v_{\parallel}, v_{\perp})$, of species j can be constructed from the two reduced VDFs as follows:

$$f_j(v_{\parallel}, v_{\perp}) = F_{j\parallel}(v_{\parallel}) \frac{F_{j\parallel}(v_{\parallel})}{2\pi F_{j\perp}(v_{\parallel})} \exp\left(-\frac{v_{\perp}^2 F_{j\parallel}(v_{\parallel})}{2F_{j\perp}(v_{\parallel})}\right). \qquad (10.19)$$

By solving the kinetic equations for (10.18) numerically, it is found that heavy ions are preferentially heated, and that sizable temperature anisotropies can thus form, results which are in accord with SOHO observations. The VDFs of the heavy ions deviate strongly from a Maxwellian, an effect which increases with height due to the decrease of the density, and thus to the decline in the efficiency of Coulomb collisions.

To give some exemplary results, we discuss a model plasma consisting of H^{1+}, He^{2+} and O^{5+}, the latter being considered as a typical minor ion representing the cumulative effect of all heavy ions. At the lower boundary, the relative densities are: 0.1 for helium and 0.001 for oxygen. Ionization dynamics are not accounted for in the model. The computational domain extends from the transition region by more than $0.6\,R_s$ into the lower corona. The model O^{5+} VDF obtained at $r = 1.44\,R_s$ is shown in *Figure* 10.13. The anisotropy can be recognized by the elliptic deformation of the contours. Furthermore, the preferred pitch-angle

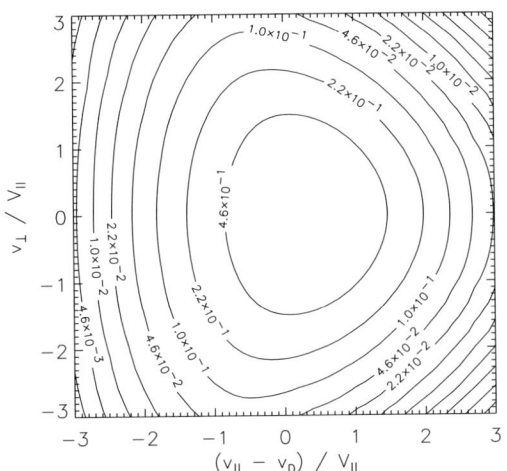

Figure 10.13 Two-dimensional gyrotropic model VDF of O^{5+} ions at $1.44\,R_s$, displaying a distinct temperature anisotropy, which is caused by resonant diffusion in the wave field, and sizable skewness (heat flux) away from the Sun, which reflects the magnetic mirror force acting on the weakly collisional ions. The ion bulk drift speed is denoted as v_D. Here the capitalized $V_\parallel$ is the ion thermal speed (parallel to the magnetic field) which is used for normalization (after Vocks and Marsch, 2002).

scattering of oxygen ions at negative speeds leads to a skewness of the VDF, which declines in velocity space much faster on the sunward than anti-sunward side. This result demonstrates the necessity of the kinetic approach in modelling coronal plasma processes.

The results from the model of Vocks and Marsch (2002) show that in the transition region the temperatures of protons and He^{2+} ions rapidly increase to coronal values, but then stay fairly constant. The proton temperature develops no anisotropy. It was found that the wave heating mechanism prefers ions with the lowest local resonance frequencies, i.e. O^{5+} ions with negative $v_\parallel$. The oxygen ions show a different kinetic behaviour than protons. The profile of $T_{O\parallel}$ is very similar to the proton temperature, but $T_{O\perp}$ is strongly enhanced. It rises continuously to very high values of the order 10^7 K in the extended corona. Thus, a strong anisotropy with $T_{O\perp} > T_{O\parallel}$ is formed. This preferred heating of the heavy ions and concurrent formation of a temperature anisotropy is consistent with the SOHO observations reported by Cranmer et al. (1999).

The corresponding wave spectra, as calculated numerically in the diffusion models of Vocks and Marsch (2001) or Cranmer (2001), show rather deep absorption edges that occur at and below the ion gyrofrequencies of the species involved. As the cross-section area of the magnetic flux tube or funnel under consideration increases with height, the magnetic field strength decreases, and so does the gyrofrequency. For example, the local proton gyrofrequency at $r = 1.44\,R_s$ has a value equivalent to $0.02\,\Omega_p$ at the lower bound of the computational domain. Therefore, the frequency of a wave entering the simulation box at the lower bound will, during upward propagation, increase relative to the local ion gyrofrequency. At a certain height, waves with frequencies above the lowest ion gyrofrequency will have suffered severe damping.

This is the essence of the "frequency sweeping" mechanism discussed by Tu and Marsch (1997) and applied to coronal funnels by Marsch and Tu (1997), and with a more realistic MHD magnetic field model of the funnel by Hackenberg et al. (1999) and Hackenberg et al. (2000). However, because of this strong absorption it was argued by Cranmer (2001) that the waves solely originating from the coronal base would not suffice to heat the coronal ions, but that local wave production in coronal holes was required.

The wave damping rate, γ, can be derived by using the full dispersion relation. For waves propagating parallel to the background magnetic field, and for a multi-component plasma, it may be written in a form involving only reduced VDFs (here $\hat{\rho}_j$ denotes the fractional mass density of species j):

$$\frac{\gamma}{\omega} = \frac{\pi}{2} \sum_j \hat{\rho}_j \left(\frac{\Omega_j}{\omega}\right)^2 \left(-\frac{\Omega_j}{k_\parallel} F_{j\parallel} + \frac{\partial F_{j\perp}}{\partial v_\parallel}\right)\Bigg|_{v_\parallel = \frac{\omega - \Omega_j}{k_\parallel}}. \qquad (10.20)$$

Apparently, the damping (or growth) rate γ depends primarily on the number of resonant ions that are able to absorb (or emit) the waves. Therefore, γ should not only be normalized to Ω_p, but also to the number of resonant ions given by $F_{j\parallel}(v_\parallel)$. This normalized damping rate is plotted in *Figure* 10.14 with a startling result. Over a wide range of negative speeds it is close to zero, meaning that the reduced VDFs reached marginal stability, at which wave absorption does not take place any more. For positive speeds, γ has also some negative values, indicating that here the waves are still being absorbed.

That the ion VDF will approach the limit of stability over a wide range of velocities such that wave absorption ceases, is what one expects from resonant pitch-angle diffusion leading to plateau formation. *Figure* 10.14 shows that a major fraction of the oxygen ions have a wave absorption coefficient close to zero, i.e. their opacity vanishes and the plasma becomes transparent for these waves. Ions with low cyclotron frequencies could absorb a significant amount of the spectral wave energy in spite of their low densities, an effect which has been investigated by Cranmer (2000), considering the summed absorption effect of practically all ions that might possibly exist in the corona. But when instead of being

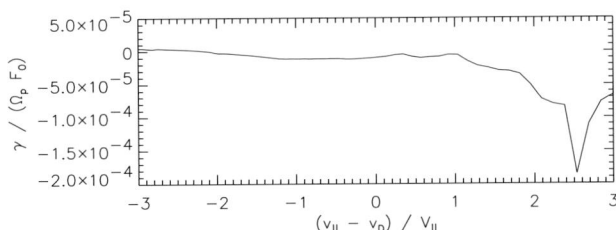

Figure 10.14. Normalized damping rate, per resonant particle of O^{5+}, evaluated from model results obtained at $2.46\,R_s$. The plot displays a distinct flatness of the curve over a wide range of speeds, corresponding to a quasilinear plateau in the VDF (after Vocks and Marsch, 2002).

Maxwellians their VDFs attained plateaus, and thus reached the limit of stability, this absorption would cease, whereupon direct wave heating of ions with higher resonance frequencies (such as alpha particles and protons) is enabled in the extended corona. This complicated issue certainly requires further study.

8.   Summary and Conclusion

Recent space solar observations showed that there exists a rich variety of waves on all scales, ranging from R_s down to R_p in the magnetically structured corona. Coronal loops were found to oscillate and flux tubes to support sound, kink and Alfvén waves, and also slow-mode standing waves. Spectroscopic evidence was obtained for propagating waves in plumes and small-scale flows in coronal funnels, and corresponding numerical fluid simulations were carried out. Ion heating by cyclotron waves was concluded to be a prominent mechanism to create the large EUV emission line broadenings, which were spectroscopically inferred to exist in the extended corona. Turbulence models involving oblique kinetic Alfvén waves and a transverse cascade were developed, which mainly led to electron heating by Landau damping. The ion kinetics was shown to determine the plasma-wave opacity in the corona and thus to regulate MHD-wave absorption. Coronal waves were argued to be most likely generated by magnetoconvection and the magnetic activity in the network. To better understand the transport properties and dissipation processes of waves and turbulence in the multi-scale and inhomogeneous corona remains a difficult task for future research.

References

Aschwanden, M.J., Poland, A.I., and Rabin, D.M.: 2001, *Annu. Rev. Astron. Astrophys.* **39**, 175

Aschwanden, M.J., De Pontieu, B., Shrijver, C.J., and Title, A: 2002, *Solar Phys.* **206**, 99

Axford, W.I., and McKenzie, J.F.: 1997, in *Cosmic Winds and the Heliosphere*, J.R. Jokipii, C.P. Sonett, and M.S. Giampapa (Eds.), Arizona University Press, Tucson, 31

Barnes, A.: 1969, *Astrophys. J.* **155**, 311

Burlaga, L.: 1995, *Interplanetary Magnetohydrodynamics*, Oxford University Press, Oxford, U.K.

Benz, A.O.: 1993, *Plasma Astrophysics*, Kluwer Academic Publishers, Dordrecht, The Netherlands

Carlsson, M., and Stein, R.F.: 1997, *Astrophys. J.* **481**, 500

Cranmer, S.R., G.B. Field, and J.L. Kohl: 1999, *Astrophys. J.* **518**, 937

Cranmer, S.R.: 2000, *Astrophys. J.* **532**, 1197

Cranmer, S.R.: 2001, *J. Geophys. Res.* **106**, 24937

Cranmer, S.R., and van Ballegooijen, A.A.: 2003, *Astrophys. J.* **594**, 573

De Moortel, I., Ireland, J., Hood, A.W., and Walsh, R.W.: 2002, *Solar Physics* **209**, 61

Dowdy J.F., Rabin D., Moore R.L.: 1986, *Solar Phys.* **105**, 35

Einaudi, G. and Velli, M.: 1999, *Physics of Plasmas* **6**, 4146

Gabriel A.H.: 1976, *Phil. Trans. Roy. Soc.* **A281**, 339

Goldstein, M.L., Roberts D.A., and W. H. Matthaeus: 1997, Magnetohydrodynamic Turbulence in Cosmic Winds, in *Cosmic Winds and the Heliosphere*, 521

Goldstein, M.L., and Roberts D.A.: 1999, *Physics of Plasmas* **6**, 4154

Goossens, M.: 1991, in *Advances in Solar System Magnetohydrodynamics*, ed. E. R. Priest and A. W. Hood, Cambridge University Press, 137

Hackenberg, A., Mann, G., and Marsch, E.: 1999, *Space Sci. Rev.* **87**, 207

Hackenberg, A., Marsch, E., and Mann, G.: 2000, *Astron. Astrophys.* **360**, 1139

Hansteen, V. H.: 1993, *Astrophys. J.* **402**, 741

Hasan, S.S., and Kalkofen, W.: 1999, *Astrophys. J.* **519**, 899

Hassler, D.M., Dammasch, I.E., Lemaire, P., Brekke, P., Curdt, W., Mason, H.E., Vial, J.C., and Wilhelm, K.: 1999, *Science* **283**, 810

Hollweg, J.V.: 1999, *J. Geophys. Res.* **104**, 14811

Hollweg, J.V., and Isenberg, P.A.: 2002, *J. Geophys. Res.* **107**, SSH 12-1, 10.1029/2001JA000270

Horbury, T., and Tsurutani, B.: 2002, Ulysses measurements of waves, turbulence and discontinuities, in *The Heliosphere near Solar Minimum, The Ulysses Perspective*, Eds. A. Balogh, R.G. Marsden, and E.J. Smith, Springer Praxis, Praxis Publishing, Chichester, U.K., 167

Kohl, J. L., Noci, G., Antonucci, E., et al.: 1997, *Solar Phys.* **175**, 613

Leamon, R.J., W.H. Matthaeus, C.W. Smith, G.P. Zank, and D.J. Mullan: 2000, *Astrophys. J.* **537**, 1054

Marsch, E.: 1991a, MHD Turbulence in the Solar Wind, in *Physics of the Inner Heliosphere*, Vol. II, ed. R. Schwenn and E. Marsch, 159

Marsch, E.: 1991b, Kinetic Physics of the Solar Wind Plasma, in *Physics of the Inner Heliosphere*, Vol. II, ed. R. Schwenn and E. Marsch, 45

Marsch, E. 1999, *Space Sci. Rev.* **87**, 1

Marsch, E. 2002, *Nonlinear Proc. Geophys.* **9**, 69

Marsch, E., and Tu, C.Y.: 1997, *Astron. Astrophys.* **319**, L17

Marsch, E., and Tu, C.Y.: 2001, *J. Geophys. Res.* **106**, 227

Marsch, E., Tu, C.Y., Wilhelm, K., Curdt, W., Schühle, U., and Dammasch, I.E.: 1997, *ESA-SP-* **404**, 555

Marsch, E., Vocks, C., and Tu, C.Y.: 2003a, *Nonlinear Proc. Geophys.* **10**, 110

Marsch, E., Axford, W.I., and McKenzie, J.M: 2003b, Solar Wind, *Dynamic Sun*, ed. B.N. Dwivedi, Cambridge University Press, 374

Matthaeus, W.H., Dmitruk, P., Oughton, S., and Mullan, D.: 2003, in *Solar Wind Ten*, Eds.: M. Velli, R. Bruno, F. Malara, AIP Conference Proceedings Volume 679, 427

McComas D.J.: 2003, in *Solar Wind Ten*, Eds.: M. Velli, R. Bruno, F. Malara, AIP Conference Proceedings Volume 679, 33

Melrose, D.B., and McPhedran, R.C.: 1991, *Electromagnetic Processes in Dispersive Media*, Cambridge University Press, Cambridge, U.K.

Musielak, Z.E., and Ulmschneider, P.: 2002, *Astron. Astrophys.* **386**, 606

Musielak, Z.E., and Ulmschneider, P.: 2003a, *Astron. Astrophys.* **400**, 1057

Musielak, Z.E., and Ulmschneider, P.: 2003b, *Astron. Astrophys.* **406**, 725

Nakariakov, V.M., Ofman, L., DeLuca, E.E., Roberts, B., and Davila, J.M.: 1999, *Science*, **285**, 862

Nakariakov, V.M., Ofman, L., and Arber, T.D.: 2000, *Astron. Astrophys.*, **353**, 741

Nakariakov, V.M.: 2003, in *Dynamic Sun*, ed. B.N. Dwivedi, Cambridge University Press, Cambridge, U.K., 314

Neugebauer, M.: 1999, *Rev. Geophys.*, **37**, 107

Ofman, L., Davila, J.M., and Steinolfson, R.S.: 1994, *Astrophys. J.* **421**, 360

Ofman, L., Nakariakov, V.M., and DeForest, C.E.: 1999, *Astron. Astrophys.* **514**, 441

Ofman, L., and Aschwanden, M.J.: 2002, *Astron. Astrophys.* **576**, L153

Parker, E.N.: 1988, *Astrophys. J.* **330**, 474

Petrovay, K.: 2003, in *Turbulence, Waves and Instabilities in the Solar Plasma*, NATO Advanced Workshop, Eds.: R. Erdélyi, K. Petrovay, E. Loránd, B. Roberts, and M. J. Aschwanden, Kluwer Academic Publishers, Dordrecht, The Netherlands

Peter, H.: 2001, *Astron. Astrophys.* **374**, 1108

Peter, H. and Judge, P.: 1999, *Astrophys. J.* **522**, 1148

Rae, I.C., and Roberts, B.: 1982, *Astrophys. J.*, **256**, 761

Roberts, B.: 1985, in *Solar System Magnetic Fields*, Ed. E. Priest, D. Reidel Publishing Company, 37

Roberts, B.: 1991, in *Advances in Solar System Magnetohydrodynamics*, Eds. E. R. Priest and A. W. Hood, Cambridge University Press, Cambridge, U.K., 105

Roberts, B.: 2002, in *Solar Variability: From Core to Outer Frontiers*, Ed. A. Wilson, *ESA SP-506*, 481

Roberts, B.: 2003, in *SOHO 13, Waves, Oscillations and Small Scale Transient Events in the Solar Atmosphere: A Joint View From SOHO and TRACE, ESA SP-547*, 1-14

Roberts, B., Edwin, P.M., and Benz, A.: 1984, *Astrophys. J.* **279**, 857

Roberts, B., and V.M. Nakariakov: 2003, in *Nato Advanced Research Workshop, NATO ASI Series, Turbulence, Waves, and Instabilities in the Solar Plasma, Kluwer Academic Publishers*, 165-188

Schrijver, C.J., Title, A.M., Harvey, K.L., et al.: 1999, *Nature* **394**, 152

Seely, J.F., Feldman, U., Schühle, U., Wilhelm, K., Curdt, W., and Lemaire, P.: 1997, *Astrophys. J.* **484**, L87

Shukla, P.K., Bingham, R., McKenzie, J.F., and Axford, W.I.: 1999, *Solar Phys.* **186**, 61

Solanki, S.K., Lagg, A., Woch, J., Krupp, N., and Collados, M.: 2003, *Nature* **425**, 692

Tu, C.-Y.: 1987, *Solar Phys.* **109**, 149

Tu, C. Y., and Marsch, E.: 1995, MHD Structures, Waves and Turbulence in the Solar Wind, *Space Sci. Rev.* **73**, 1-210

Tu, C.Y., and Marsch, E.: 1997, *Solar Phys.* **171**, 363

Tu, C. Y., Marsch, E., Wilhelm, K. and Curdt, W.: 1998, *Astrophys. J.* **503**, 475

Tu, C. Y., Marsch, E. and Wilhelm, K.: 1999, *Space Sci. Rev.* **87**, 331

Treumann, R. A., and Baumjohann, W.: 1996, *Basic Space Plasma Physics*, Imperial College Press, London, p.237

Vocks, C.: 2002, *Astrophys. J.*, **568**, 1017

Vocks, C., and Marsch, E.: 2001, *Geophys. Res. Lett.* **28**, 1917

Vocks, C., and Marsch, E.: 2002, *Astrophys. J.* **568**, 1030

Voitenko, Y., and Goossens, M.: 2002, *Solar Phys.* **209**, 37

Walsh, R.: 2002, *ESA SP-* **508**, 253

Wang, T.J., Solanki, S.K., Curdt, W., Innes, D.E., Dammasch, I.E., and Kliem, B.: 2003a, *Astron. Astrophys.* **406**, 1105

Wang, T.J., Solanki, S.K., Innes, D.E., Curdt. W., and Marsch, E.: 2003b, *Astron. Astrophys.* **402** 402, L17

Wang, T.J.: 2003, *ESA SP-* **547**, 1

Wilhelm, K., Marsch, E., Dviwedi, B.N., Hassler, D.M., Lemaire, P., Gabriel, A., and Huber, M.C.E.: 1998, *Astrophys. J.* **500**, 1023

Wilhelm, K., Dammasch, I.E., Marsch, E., and Hassler, D.: 2000, *Astron. Astrophys.* **353**, 749

Wilhelm, K., Dwivedi, B.H., Marsch, E., and Feldman, U.: 2004, *Space Science Reviews*, in press

Xia, L., Marsch, E. and Curdt, W.: 2003, *Astron. Astrophys.* **353**, L5

Chapter 11

THE INFLUENCE OF THE CHROMO-SPHERE-CORONA COUPLING ON SOLAR WIND AND HELIOSPHERIC PARAMETERS

Øystein Lie-Svendsen

Norwegian Defence Research Establishment

P.O. Box 25, NO-2027 Kjeller, Norway

Also at the University of Oslo, Norway

Oystein.Lie-Svendsen@ffi.no

Abstract   Based on fluid models of the solar wind that extend from the chromosphere out to 1 AU we discuss how the chromosphere-corona coupling affects the ma ss flux, flow speed, and helium abundance of the solar wind in the heliosphere. The downward transport of energy from the corona to the transition region and upper chromosphere can determine the location of the chromosphere-transition region interface and hence the pressure of the corona, and thus strongly influence the solar wind mass flux. In rapidly expanding magnetic field geometries, simulating coronal holes, very little heat is conducted down if only protons are heated in the corona, and in this geometry a reasonable solar wind mass flux is difficult to achieve without additional electron heating in the corona or transition region heating. In a radial geometry downward heat conduction is efficient, resulting in high coronal densities. The rapidly expanding geometry therefore favours a high-speed wind with a low mass flux, and conversely for a radially expanding geometry. In all cases most of the energy deposited in the corona ends up in the solar wind. In a rapidly expanding geometry the helium abundance tends to be low everywhere and the helium flux is limited by the supply from the chromosphere. The radial geometry favours high coronal helium abundances, which may even be close to unity, and the helium flux is then determined by the amount of energy available in the corona to heat helium.

G. Poletto and S.T. Suess (eds.), The Sun and The Heliosphere as an Integrated System, 319–352.

1.   Introduction

At first glance the coupling between the Sun's atmosphere and the solar wind seems almost trivial; after all, this atmosphere is the source of material for the wind. However, in this chapter we aim to demonstrate, through recent numerical modelling studies, that the nature of this coupling is highly complex and "nonlinear," and that this nontrivial coupling needs to be accounted for in order to understand such fundamental properties of the solar wind as its mass flux and elemental composition. Understanding this coupling for our nearest star may also have implications for other stars. For instance, to what extent does the matter ejected by a star, and hence the interstellar matter, reflect the elemental composition of the star itself?

The concept of treating the solar atmosphere and wind as one integrated system is not new. Already Hammer (Hammer, 1982a; Hammer, 1982b), modelling coronae of other stars, included processes such as downward heat conduction and radiation, processes that are important when, e.g., the transition region between the chromosphere and corona is included. In modelling the solar wind, the importance of including the transition region and corona, and hence the coronal heating, rather than imposing the heating as a boundary condition (i.e., setting the coronal temperature), has been emphasized, e.g., by Hollweg, 1986, and Hollweg and Johnson, 1988. Withbroe, 1988, studied the mass and energy flow of the inner solar wind, using a model similar to Hammer's. As emphasized by Withbroe, in these "radiative energy balance" models there are essentially only two sets of free parameters, namely those specifying the location and amount of coronal heating and those specifying the flow geometry.

These models only extended into the transition region, down to a temperature of a few times 10^5 K, and hence did not include the upper chromosphere. Moreover, with the exception of the study by Hollweg and Johnson, 1988, they were one-fluid models that could not account for different electron and proton temperatures. Extending a solar wind model into the chromosphere is numerically challenging because the model needs to accommodate the (potentially) very steep temperature gradient of the lower transition region. This gradient is a consequence of Coulomb collisions in a collision-dominated plasma, in which the "classical" heat flux vector (which is mainly carried by electrons) is

$$\mathbf{q} \approx -\kappa_e T^{5/2} \nabla T \qquad (11.1)$$

where κ_e is only weakly dependent on temperature. As long as radiation and other energy loss processes can be neglected, maintaining a constant heat flux $\mathbf{q}$ then requires that $\nabla T \propto T^{-5/2}$. As an example, maintaining a constant heat flux from the corona, where $T \sim 10^6$ K, down to the chromosphere, where $T \sim 10^4$ K, would then require that ∇T increases by a factor 10^5. Solving the partial differential transport equations using finite differencing techniques, the "standard" method, would then require very densely spaced grid points in the lower transition region. Adding to the difficulties, the transition region itself may be a highly dynamic region, as we shall discuss later, so that the location of this steep gradient will not be known in advance, and may even change in time. Using a fixed numerical grid, one may therefore need a huge number of grid points in order to "capture" the transition region, which in turn makes the code prohibitively expensive to run. Fortunately, numerical techniques have been developed that solve this problem (see e.g., Dorfi and Drury, 1987). These so-called adaptive grid techniques allow the numerical grid itself to move in time and concentrate grid points in regions where, e.g., the temperature gradient is large. Hence a dynamic transition region can be accommodated with a modest number of grid points.

A solar wind model based on these adaptive grid techniques has been developed by Hansteen and Leer, 1995 (see also Hansteen et al., 1997). In this model the chromosphere, transition region, corona, and solar wind are treated consistently as one coupled system. Moreover, a multifluid model is used that allows for different temperatures for electrons, protons, and (helium) ions. The discussion in this chapter will be based on an extension of this model.

Why do we need to include the chromosphere if we are primarily interested in the corona and solar wind? The simple answer is that the lower boundary must be located in a region that is unperturbed by the processes taking place in the corona and (inner) solar wind. Most importantly, the lower boundary must be located where the densities of the species considered are known. If the lower boundary is placed in a region affected by coronal processes, such as the upper transition region, the assumed densities at the lower boundary may turn out to be physically inconsistent with the (heating) processes taking place in the corona; as we shall argue below this density may be set by the downward conduction of heat from the corona. Hence parameters that depend on coronal densities, such as the solar wind mass flux, may turn out to be inconsistent with the processes assumed to take place in the corona. Additionally, by lowering the lower boundary to a region where densities are (presumably) known, the model solutions no longer depend on more or less arbitrary boundary conditions, which increases the predictive capability of the model. Finally, and most pertinent to the topic of this chapter, having a solar wind model with the lower boundary in the chromosphere allows us to study the processes taking place in the region from the upper chromosphere to the corona, and how these affect solar wind properties such as its mass flux and elemental composition.

Before we discuss the solar wind, in Sect. 2 we discuss the energy balance on closed coronal loops, since this offers the most simple demonstration of how the heating of the corona is related to its density. In Sect. 3 we then go on to present the transport equations that we use to study the chromosphere-solar wind coupling. In Section 4 we discuss a pure hydrogen-proton-electron solar wind, while in Section 5 we discuss helium in the corona and solar wind. The results are briefly summarized in Sect. 6.

2.   Closed Coronal Loops

The reason why the coronal density is so tightly coupled to the heating process is that any excess energy (that is, energy that is not lost to the solar wind) will be conducted down towards the transition region, where it will eventually be converted into radiation, and this radiative loss is proportional to density (squared). The tight interplay between the coronal density, or more precisely its pressure, and the downward heat flux is most clearly seen on magnetic field lines that are not open to the solar wind. In this case all the "mechanical" energy deposited along the field line must eventually be converted into radiation, either locally or the energy is transported downwards by heat conduction to regions that are sufficiently dense to radiate the energy.

We consider for simplicity a cylindrical magnetic flux tube with constant cross section A. Denoting the mechanical energy flux by $F_M(s)$, where s is the coordinate along the flux tube, and the radiative energy flux, which is the cumulative radiative energy loss integrated along the flux tube from the base up to the point s, by $F_R(s)$, energy conservation at any point along the field line in a steady state can simply be expressed as

$$F_M(s) + F_R(s) + F_q(s) = \text{constant}, \qquad (11.2)$$

where $F_q(s)$ is the heat flux along the flux tube. We furthermore assume that the mechanical energy flux is deposited in the upper part of the loop and restrict ourselves to the region below. Taking the divergence of (11.2) we then have

$$\frac{dF_R}{ds} + \frac{dF_q}{ds} = 0. \qquad (11.3)$$

The radiative cooling is mainly caused by collisional excitation of minor ions by electrons, and is hence proportional to the electron density n_e. If in addition we assume that the plasma is well mixed along the flux tube (an assumption that may be questioned), in other words that the minor ion abundances are constant, the minor ion densities are proportional to the electron density, in which case the cooling rate per unit length along the flux tube may be written

$$\frac{dF_R}{ds} = An_e(s)^2 L(T) \tag{11.4}$$

where the loss function $L(T)$ is only a function of the (electron) temperature. Since we restrict ourselves to a region of the flux tube below the temperature maximum, the temperature is monotonically increasing and we may use T instead of s as the independent variable. Using (11.1) we can therefore express the heat flux divergence as (Landini and Monsignori Fossi, 1975)

$$\frac{dF_q}{ds} = A\frac{dq}{ds} = A\frac{dq}{dT}\frac{dT}{ds} = -\frac{A}{2\kappa_e}T^{-5/2}\frac{dq^2}{dT}. \tag{11.5}$$

With a hot corona and a thin transition region the pressure will be nearly constant from the top of the chromosphere and well into the corona. Assuming then that the pressure $P = 2n_e kT$ (for equal electron and proton temperatures) is constant, and using (11.4) and (11.5), (11.3) may be rearranged as

$$\frac{1}{2}P^2\frac{\kappa_e}{k^2}L(T)T^{1/2} = \frac{dq^2}{dT} \tag{11.6}$$

where k is Boltzmann's constant. This equation may be formally integrated from a temperature T_1 to a maximum temperature T_2, yielding (Rosner et al., 1978)

$$\frac{1}{2}P^2\frac{\kappa_e}{k^2}K(T_1, T_2) = q_2^2 - q_1^2 \tag{11.7}$$

where q_1 and q_2 are the heat fluxes at T_1 and T_2, respectively, and $K(T_1, T_2) \equiv \int_{T_1}^{T_2} L(T)T^{1/2}\,dT$ is a function of T_1 and T_2 only. As shown, e.g., by Rosner et al., 1978, the loss rate L becomes very small at temperatures of order 10^4 K or less. For sufficiently low T_1 we may therefore write $K(T_1, T_2) \simeq K(T_2)$. Equation (11.7) says that the total radiative loss in the transition region is balanced by heat flux from the corona.

So far the pressure P has been treated as an arbitrary parameter. Require now that the given downward heat flux q_2 at temperature level T_2 be absorbed by the time we reach temperature T_1, so that $q_1 = 0$. In that case (11.7) can be rewritten

$$P = \sqrt{\frac{2}{\kappa_e K(T_2)}}\,k\,|q_2|. \tag{11.8}$$

Since $K(T_2)$ may be regarded as a constant (for a given upper level temperature T_2), (11.8) says that the transition region (and corona) pressure is directly proportional to the downward heat flux q_2. In other words, for the transition region to be able to radiate all of the energy flux q_2, and nothing more, the pressure has to be at exactly the value given by (11.8).

What happens if we prescribe a pressure that is different from (11.8)? In that case we do not have energy balance and the steady state assumption can no longer hold. If the prescribed pressure is lower than the value obtained from (11.8), the transition region receives more energy from the corona than it is able to radiate. The excess energy then reaches the upper chromosphere, which is heated and ionized. Hence the upper chromosphere is "eroded" and the chromosphere-transition region interface moves downwards to a higher chromospheric density level. The pressure of the transition region (and corona), which roughly equals the pressure at the top of the chromosphere, will therefore increase. The erosion continues until the pressure has become sufficiently high to satisfy (11.8), at which point the energy balance is reestablished. Conversely, if the prescribed pressure is too low, the transition region radiates more energy than it receives from the corona, causing cooling of the lower transition region. This causes in turn a decrease in the degree of ionization and hence the chromosphere-transition region interface moves upwards to a lower density level. This process continues until the pressure has become sufficiently low to satisfy (11.8).

The simple analytical result (11.8) demonstrates the intimate relation between energy transport in the corona and the density (pressure) of the corona, and why the coronal heating cannot be considered independently of the processes setting the coronal density. Moreover, (11.8) shows that the pressure must be expected to be a dynamical quantity: The downward heat flux will change in time and therefore the pressure must change, too. The chromosphere-transition region interface thus moves, illustrating the need for an adaptive grid in the numerical implementation of the transport equations.

On magnetic field lines open to the solar wind the energy balance is less simple since a large fraction of the energy deposited in the corona will not be conducted downwards, but lost to the solar wind. In addition, the flow through the transition region implies that some of the downward heat flux will be absorbed by the enthalpy flux (that is, heating of the plasma as it streams upwards) rather than being converted into radiation. Still there is a tight coupling between the coronal heating process and the coronal pressure, and in some of the solutions we present below the pressure is still to a good approximation given by (11.8), where q_2

is now the fraction of the deposited mechanical energy flux that ends up as downward heat conduction. However, we shall also demonstrate that the fraction of the deposited energy that is conducted downwards is itself a function of the coronal density, which means that the coupling between the chromosphere and corona is highly nonlinear.

The derivation above was based on the assumption that the radiative loss in the transition region is caused by heat conduction from the corona. It is of course also possible that the upper chromosphere and transition region are heated by processes unrelated to the coronal heating process, e.g., by damping of upward propagating sound waves (evidence for waves in the transition region are given e.g., by Doyle et al., 1998). If this deposited energy flux is much larger than the heat flux from the corona, the pressure of the transition region and corona will be set by this direct energy deposition, by requiring that the pressure has the value needed for the transition region radiative loss rate to equal the energy deposition rate. In this case the coronal density should not be very sensitive to the coronal heating process. In Section 4 we show how direct deposition of energy in the transition region can have a large impact on the solar wind mass flux, even when this energy flux is much smaller than the mechanical energy flux deposited in the corona in order to drive the solar wind.

3.   The Modelling Tools

In order to describe the chromosphere-solar wind system, we need transport equations that are able to describe the collision-dominated and mostly neutral chromosphere as well as the collisionless supersonic flow of the solar wind. Hammer, 1982a, and Withbroe, 1988, used one-fluid models in which electron and proton temperatures are equal. However, SOHO Ultraviolet Coronagraph Spectrometer (UVCS) observations (Kohl et al., 1999) indicate that protons (and heavier ions) are much hotter than electrons in the corona. This is corroborated by model studies (Hansteen and Leer, 1995), showing that because protons are much poorer conductors of heat than electrons and cannot easily escape from the corona unless they become hot, protons tend to become much hotter than electrons. A multifluid model, in which the electron and proton energy conservation equations are treated separately, is needed to handle such temperature differences. (Two-fluid models of the solar wind date back to the work by Hartle and Sturrock, 1968, who solved separate energy equations for electrons and protons, from the temperature maximum and outwards.) UVCS observations and empirical modelling of Doppler dimming also indicate large temperature anisotropies for oxygen ions and neutral hydrogen (Kohl et al., 1997; Cranmer et al., 1999; Antonucci et al., 2000; Zangrilli et al., 2002), with much larger thermal motion perpendicular to the magnetic field than along the field. To capture such anisotropies we need a fluid model that solves separate energy equations for the parallel and perpendicular thermal motion. Equally important, by allowing for temperature anisotropies, the magnetic mirror force can also be accounted for; when the perpendicular temperature is much higher than the parallel temperature the rapidly expanding magnetic field near the Sun creates a strong outwardly-directed force on the particles which is important for their dynamics. (Transport equations that can handle such anisotropies have been presented a long time ago e.g., by Holzer and Axford, 1970, and solutions have been obtained e.g., by Leer and Axford, 1972.)

Finally, the discussion in Sect. 2 showed how important heat conduction can be for the pressure in the transition region and corona. We therefore need equations that can give a reasonable description of the heat flow. The model used by Hansteen and Leer, 1995, did not solve explicitly heat flux equations, but rather assumed that both electrons and protons behaved "classically" in the corona (meaning that heat fluxes were given by expressions of the form (11.1)). However, for low coronal densities and high proton temperatures, which one gets naturally in multifluid models with proton heating (e.g., Hansteen and Leer, 1995), there is no reason to expect that protons behave classically even in the corona. Departures from classical heat conduction have previously been modelled empirically by requiring that the solar wind solution match observations (Cuperman et al., 1972). Hollweg, 1986, modelling the flow beyond the temperature maximum, invoked an *ad hoc* transition from classical to "collisionless" heat conduction at 10 R_S (R_S is the solar radius). Our goal is to have equations that can generate this transition selfconsistently. We therefore need to solve explicitly heat flux equations that attempt to capture the transition from classical (i.e., collision-dominated) to nonclassical heat flow.

For these reasons we shall base our discussion on solutions to so-called gyrotropic transport equations, more precisely the gyrotropic (strong magnetic field) approximation to the 16-moment equations developed by Demars and Schunk, 1979. These equations were first applied to the solar wind by Olsen and Leer, 1999, and later by Li, 1999. However, their models only extended down to the low corona, and hence could not account for the chromospheric coupling. In this formulation, for each particle species s we solve for the density n_s, for the flow velocity u_s along the radial magnetic field, for the temperature parallel ($T_{s\|}$) and perpendicular ($T_{s\perp}$) to the magnetic field, and for the heat flux moments along the magnetic field of parallel ($q_{s\parallel}$) and perpendicular ($q_{s\perp}$) thermal motion. For each species the six conservation equations can be written as follows: particle number conservation,

$$\frac{\partial n_s}{\partial t} = -\frac{1}{A}\frac{\partial}{\partial r}(n_s u_s A) + \frac{\delta n_s}{\delta t} \tag{11.9}$$

momentum,

$$\frac{\partial u_s}{\partial t} = -u_s\frac{\partial u_s}{\partial r} - \frac{k}{m_s}\frac{\partial T_{s\parallel}}{\partial r} - \frac{kT_{s\parallel}}{n_s m_s}\frac{\partial n_s}{\partial r} - \frac{1}{A}\frac{dA}{dr}\frac{k}{m_s}(T_{s\parallel} - T_{s\perp})$$
$$+ \frac{e_s}{m_s}E - \frac{GM_S}{r^2} + \frac{1}{n_s m_s}\frac{\delta M_s}{\delta t} \tag{11.10}$$

parallel and perpendicular thermal energy,

$$\frac{\partial T_{s\parallel}}{\partial t} = -u_s\frac{\partial T_{s\parallel}}{\partial r} - 2T_{s\parallel}\frac{\partial u_s}{\partial r} - \frac{1}{n_s k}\frac{\partial q_{s\parallel}}{\partial r} - \frac{1}{A}\frac{dA}{dr}\frac{q_{s\parallel}}{n_s k} + \frac{2}{A}\frac{dA}{dr}\frac{q_{s\perp}}{n_s k}$$
$$+ \frac{1}{n_s k}Q_{sm\parallel} + \frac{1}{n_s k}\frac{\delta E_{s\parallel}}{\delta t} \tag{11.11}$$

$$\frac{\partial T_{s\perp}}{\partial t} = -u_s\frac{\partial T_{s\perp}}{\partial r} - \frac{1}{A}\frac{dA}{dr}u_s T_{s\perp} - \frac{1}{n_s k}\frac{\partial q_{s\perp}}{\partial r} - \frac{2}{A}\frac{dA}{dr}\frac{q_{s\perp}}{n_s k} + \frac{1}{n_s k}Q_{sm\perp}$$
$$+ \frac{1}{n_s k}\frac{\delta E_{s\perp}}{\delta t} \tag{11.12}$$

and heat flux density of parallel and perpendicular thermal motion,

$$\frac{\partial q_{s\parallel}}{\partial t} = -u_s\frac{\partial q_{s\parallel}}{\partial r} - 4q_{s\parallel}\frac{\partial u_s}{\partial r} - u_s q_{s\parallel}\frac{1}{A}\frac{dA}{dr} - 3\frac{k^2 n_s T_{s\parallel}}{m_s}\frac{\partial T_{s\parallel}}{\partial r}$$
$$+ \frac{\delta q_{s\parallel}'}{\delta t} \tag{11.13}$$

$$\frac{\partial q_{s\perp}}{\partial t} = -u_s\frac{\partial q_{s\perp}}{\partial r} - 2q_{s\perp}\frac{\partial u_s}{\partial r} - 2u_s q_{s\perp}\frac{1}{A}\frac{dA}{dr} - \frac{k^2 n_s T_{s\parallel}}{m_s}\frac{\partial T_{s\perp}}{\partial r}$$
$$- \frac{1}{A}\frac{dA}{dr}\frac{k^2 n_s T_{s\perp}}{m_s}(T_{s\parallel} - T_{s\perp}) + \frac{\delta q_{s\perp}'}{\delta t}. \tag{11.14}$$

Here t denotes time, r is the distance from the centre of the Sun, G is Newton's gravitational constant, M_S is the mass of the Sun, and m_s and e_s are the atomic mass and charge of the particle, respectively. The average temperature and total heat flux are then given as

$$T_s = \frac{1}{3}(T_{s\parallel} + 2T_{s\perp}) \tag{11.15}$$

$$q_s = \frac{1}{2}(q_{s\parallel} + 2q_{s\perp}). \tag{11.16}$$

In (11.11) and (11.12) $Q_{sm\|}$ and $Q_{sm\perp}$ specify the "mechanical" heating terms — the energy that is supplied to the system — and will be specified shortly. We do not include momentum deposition (pushing) as an additional energy source in the models presented here.

For the flow tube area $A(r)$ ($\propto 1/B(r)$ where B is the strength of the magnetic field), we choose the analytical form given by Kopp and Holzer, 1976,

$$A(r) = A_0 \left(\frac{r}{R_S}\right)^2 \frac{f_{\max}e^{(r-r_g)/\sigma_g} + f_g}{e^{(r-r_g)/\sigma_g} + 1}, \qquad (11.17)$$

where $A_0 = 1$ m^2 is the flux tube area at the lower boundary of the model, $f_g = 1 - (f_{\max} - 1)e^{(R_S - r_g)/\sigma_g}$, and the expanding geometry is specified through the parameters $f_{\max}$, r_g and σ_g. In all models shown here we use $r_g = 1.3\,R_S$ and $\sigma_g = 0.5\,R_S$ (Munro and Jackson, 1977). A radially expanding solar wind is obtained by setting $f_{\max} = 1$, while $f_{\max} = 5$ leads to a rapidly expanding flow in which the flux tube area at infinity is 5 times larger than a radially expanding flow tube. Even for the rapidly expanding flow geometry we consider only a purely radial magnetic field; that is, the flow tube expands superradially, but it does not twist.

In (11.9) $\delta n_s/\delta t$ specifies the rate of production and loss of particles due to ionization and recombination; since the model is started in the chromosphere where the hydrogen (and helium) gas is mostly neutral, and the transition region, where ionization takes place, is allowed to move dynamically, the production and loss of particles must also be carried out dynamically. It is through this process that the transition region pressure is allowed to change in response to changes in the downward heat flux. In the model radiative recombination is taken from Allen, 1976, collisional ionization and three-body recombination from Arnaud and Rothenflug, 1985, while photoionization is roughly modelled assuming constant rates from Vernazza et al., 1981.

The collision terms $\delta M_s/\delta t$, $\delta E_{s\|(\perp)}/\delta t$, and $\delta q'_{s\|(\perp)}/\delta t$, as given by Demars and Schunk, 1979, are very complex. We therefore use simplified expressions for these, that agree with the 8-moment collision terms (Schunk, 1977) in the collision-dominated limit with no temperature anisotropies. The momentum equation collision term reads

$$\frac{\delta M_s}{\delta t} = -n_s m_s \sum_t \nu_{st}(u_s - u_t) + \frac{5}{6}\sum_t \nu_{st}\frac{z_{st}\mu_{st}}{kT_{st}}\left(q_{s\|} - \frac{m_s n_s}{m_t n_t}q_{t\|}\right)$$
$$+ m_s \sum_t (n_t u_t R_{ts} - n_s u_s R_{st}), \qquad (11.18)$$

where the sums extend over all other species t, ν_{st} is the collision frequency, z_{st} are pure numbers that depend on the type of collisions (e.g., $z_{st} = 3/5$ for Coulomb collisions), and $\mu_{st} = m_s m_t/(m_s + m_t)$ and $T_{st} = (m_t T_s + m_s T_t)/(m_s + m_t)$ are the reduced mass and reduced temperature, respectively. The last term in (11.18) accounts for addition and loss of momentum due to production and loss of particles, where R_{ts} is the production rate for species s due to ionization, recombination and charge exchanges reactions of species t.

The second term in (11.18) is the so-called *thermal force*; for minor ions and helium ions in the transition region collisions with protons and electrons lead to a very strong force, proportional to q_e and q_p, that pushes the heavier ions up. This force may even be stronger than gravity and can thus lead to a large accumulation of helium ions in the corona.

The radiative losses are accounted for in the electron energy equations assuming an optically thin atmosphere and using collisional coefficients supplied by Judge and Meisner, 1994.

The remainder of the collision terms are given and discussed by Lie-Svendsen et al., 2003.

When the transition region pressure is allowed to adjust itself in response to the energy supply, essentially the only adjustable parameters of the model are those specifying the geometry and the energy supply (see Withbroe, 1988). Our focus is not on how energy is delivered to the corona and solar wind (the coronal heating problem), but rather on how the corona and wind respond to energy deposition. In order to illustrate the effect of the chromospheric coupling on the solar wind we therefore choose simple, analytical expressions for the various heating terms included in the model, since this gives us precise control over where and how we heat the particles. (An example of a more physically "realistic" heating function, based on a turbulent cascade of Alfvén waves, is provided by Lie-Svendsen et al., 2001.)

The radiative cooling of the chromosphere is balanced by (isotropic) heating of electrons through heating terms

$$Q^{CH}_{em\|} = Q^{CH}_{em\perp} = \frac{2}{3} C_e n_e (n_p + n_H) \qquad (11.19)$$

in (11.11) and (11.12). Choosing $C_e = 5 \times 10^{-39}$ W m^3 leads to a chromospheric temperature of about 7-8000 K with the chosen radiative loss function.

In Sect. 4 we discuss the coupling of the chromosphere and solar wind through modifying the energy supply to the transition region. For this purpose we make use of an additional isotropic electron heating term, intended to add heat directly to the transition region, and modelled as a Gaussian,

$$Q^{TR}_{em\parallel} = Q^{TR}_{em\perp} = \frac{2}{3} \frac{F_t}{\sqrt{\pi}\sigma_t} \exp\left(-\left(\frac{r - r_t}{\sigma_t}\right)^2\right). \tag{11.20}$$

Here the electrons are heated merely out of convenience; since electrons and protons are strongly coupled in the lower transition region, the energy is quickly shared among all species.

The main part of the solar wind energy is supplied in the corona through a simple exponential damping of a prescribed mechanical energy flux F_{ms0} for each species s (protons, electrons, and α-particles), starting at a prescribed heliocentric distance r_{1s},

$$
\begin{aligned}
F_{ms}(r) &= F_{ms0} \quad \text{for } r < r_{1s} \\
F_{ms}(r) &= F_{ms0} \exp\left(-\frac{r - r_{1s}}{H_m}\right) \quad \text{for } r \geq r_{1s}, \tag{11.21}
\end{aligned}
$$

and where the "damping length" H_m specifies how close to the Sun the energy is deposited. Based on the observations indicating that hydrogen and oxygen ions are heated largely in the perpendicular direction (Kohl et al., 1997; Cranmer et al., 1999), protons and α-particles will be heated only in the perpendicular direction in the results presented below, so that the heating terms in (11.11) and (11.12) read

$$Q_{pm\parallel} = 0 \tag{11.22}$$

$$Q_{pm\perp} = -\frac{1}{A}\frac{dF_{mp}}{dr} \tag{11.23}$$

$$Q_{\alpha m\parallel} = 0 \tag{11.24}$$

$$Q_{\alpha m\perp} = -\frac{1}{A}\frac{dF_{m\alpha}}{dr}. \tag{11.25}$$

The transport equations (11.9)–(11.14) are integrated in time until a steady state is reached, using a (partially) implicit finite differencing scheme. As pointed out in the introduction, an adaptive grid is essential in order to capture the dynamic transition region. The grid point density, $\rho_z(z)$, is chosen to be proportional to the electron temperature gradient (Dorfi and Drury, 1987),

$$\rho_z(z) \propto \sqrt{1 + \left(\frac{z_{av}}{T_{e,av}}\frac{dT_e}{dz}\right)^2}, \tag{11.26}$$

where z is the distance from the lower boundary and "av" denotes an average of the subscripted variable. This equation is solved simultaneously with the transport equations, so that the grid density follows the electron temperature gradient at all times. The total number of grid points is fixed at 480 between the lower boundary in the chromosphere and 1 AU.

At the lower boundary we specify an unperturbed chromosphere with neutral hydrogen and helium densities set to $n_H = 10^{20}$ m^{-3} and $n_{He} = 10^{19}$ m^{-3}. The temperature at the lower boundary is allowed to vary and is effectively set by the choice for C_e above. The flow speeds at the lower boundaries are allowed to change, too, in order to maintain force balance. Hence the hydrogen and helium fluxes are not set in advance, but are model results that will change in response to changes in heating rates.

4.   The Electron-Proton Solar Wind

Applying the model we shall first consider a "pure" electron-proton-hydrogen solar wind, thus neglecting helium entirely. The purpose is to get some understanding of how changes in the flow geometry and energy flow in the transition region affect basic solar wind properties, particularly the solar wind mass flux. As always, we learn how a process influences the system by making changes to that process, specifically by changing the energy deposition rate in the transition region. The results of this section are presented in more detail by Lie-Svendsen et al., 2002.

First, to see how the flow geometry alone affects the system we run the model with a radial ($f_{max} = 1$) and a rapidly expanding ($f_{max} = 5$) flow geometry, but keeping the heating rates essentially unchanged. That means, if the wind emanating from the rapidly expanding geometry is to have roughly the same energy flux density (the sum of gravitational potential and kinetic energy fluxes) at 1 AU as the radially expanding wind, we need to apply a mechanical energy flux that is a factor 5 larger. For the $f_{max} = 1$ case we therefore apply $F_{mp0} = 80$ W and for $f_{max} = 5$ we use $F_{mp0} = 400$ W (in a flow tube with $A(r = R_S) = 1$ m^2 at the bottom), while the location of the coronal heating is the same; we set $r_{1p} = 1.01\ R_S$ and $H_m = 0.5\ R_S$ in (11.21). In addition, in both cases we heat electrons with $F_t = 13$ W, using $r_t = 1.006\ R_S$ and $\sigma_t = 0.002\ R_S$ in (11.20), ensuring that this energy is deposited in the lower transition region.

The main results are summarized in Figure 11.1. In the bottom panels only the total heat fluxes, given by (11.16), are shown. With essentially all the energy deposited in the protons, and most of it within one solar radius above the surface ($H_m = 0.5\ R_S$), the protons become very hot, reaching an average temperature of nearly 10^7 K in the rapidly expanding flow. The rapid deposition of the mechanical energy flux also

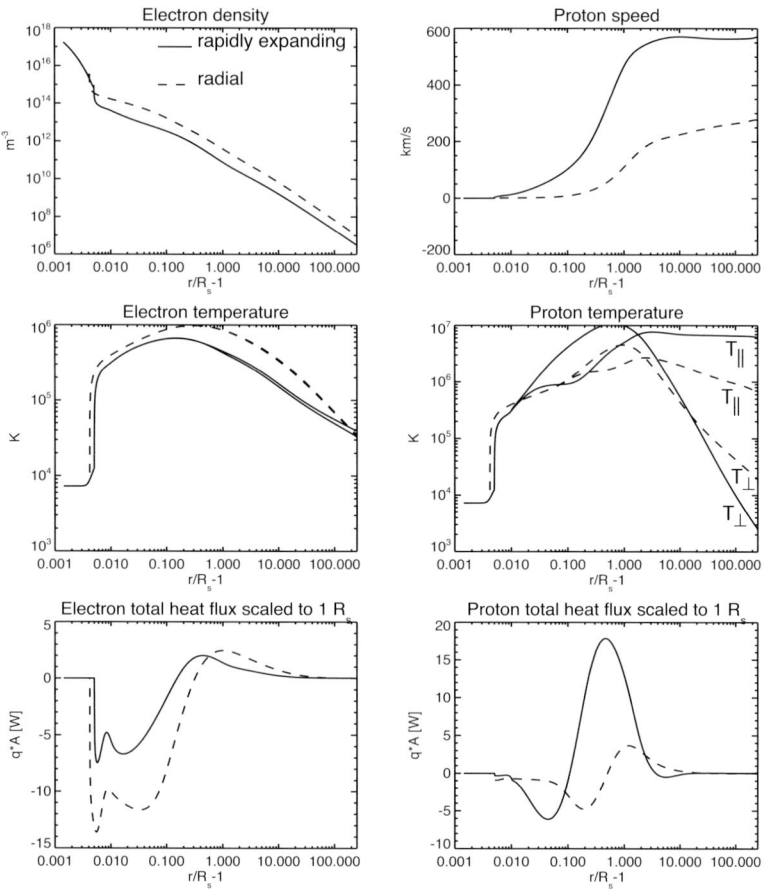

Figure 11.1. Summary of solar wind model solutions in rapidly expanding ($f_{max} = 5$) and radially expanding ($f_{max} = 1$) geometries, from Lie-Svendsen et al., 2002.

means that the acceleration must be rapid, and we note that the rapidly expanding solution reaches its terminal flow speed of nearly 600 km/s within $r = 4\ R_S$. (Higher flow speeds can easily be obtained by depositing more energy in the corona.) That heating close to the Sun leads to very high proton temperatures and rapid acceleration of the wind is also seen in the models by McKenzie et al., 1997. Empirical modelling based on UVCS/SOHO observations of the HI Lyα emission line indicates that neutral hydrogen reaches a tempeature of about $2 - 3 \times 10^6$ K in coronal holes, while the Doppler dimming indicates a more gradual acceleration of the wind, with neutral H having a flow speed of 200–300 km/s at $r = 3\ R_S$ (Antonucci et al., 2000; Zangrilli et al., 2002) and 300–400 km/s at $r \approx 4\ R_S$ (Cranmer et al., 1999). The wind from low-latitude streamer and coronal holes accelerate even more slowly, with $u_p \approx 100$ km/s at $r = 4 \; R_S$ (Poletto et al., 2002). In order to get a proton temperature in better accordance with the results obtained by Cranmer et al., 1999, while still obtaining a high-speed solar wind, we would have to deposit the energy more gradually (that is, increase H_m). A more gradual energy deposition would necessarily cause the wind to accelerate more gradually also. Alternatively, the proton temperature may be lowered by increasing the coronal density, and thus the collisional coupling between electrons and protons, which requires that more energy be supplied to the upper chromosphere and lower transition region, e.g., by explicit heating of electrons.

Because electrons are not heated explicitly in the corona in these models, only by thermal contact with protons, they are much colder, with a maximum temperature that does not exceed 10^6 K, and in the rapidly expanding model T_e does not exceed 7×10^5 K. Observed coronal line intensity ratios indicate a coronal hole electron temperature in the range $7.5 - 9 \times 10^5$ K (Habbal et al., 1993; David et al., 1998; Wilhelm et al., 1998). In the rapidly expanding model electron and proton temperatures decouple already at $r \approx 1.02 \; R_S$, illustrating the poor coupling between electrons and protons.

Because we heat protons only in the perpendicular direction, large anisotropies with $T_{p\perp} \gg T_{p\|}$ develop in the corona, particularly in the rapidly expanding solution where densities are low. However, as the protons move outwards $T_{p\perp}$ decreases rapidly within a few solar radii, while $T_{p\|}$ increases. In the asymptotic wind $T_{p\|}$ is nearly constant and much larger than $T_{p\perp}$. This behaviour is a consequence of conservation of the first adiabatic invariant (magnetic moment): As protons move outwards and the magnetic field weakens the Lorentz force converts perpendicular velocity into parallel velocity. In the 16-moment formulation this conversion is mediated by the heat flux (one reason why heat flux equations were included). Indeed it can be shown analytically that once the wind has reached its asymptotic flow speed, then $T_{p\perp} \propto r^{-2}$ while $T_{p\|} \approx$ constant. This result is in striking disagreement with *in situ* measurements in the solar wind. For instance, *Helios* found $T_{p\perp}/T_{p\|} \approx 1.7$ at 0.3 AU, and even at 1 AU the ratio was only slightly less than unity (Marsch et al., 1982). This observed behaviour cannot be modelled without some process other than Coulomb collisions to transfer energy from parallel to perpendicular motion (Li et al., 1999). A velocity distribution with $T_\| \gg T_\perp$ should be susceptible to the firehose instability which would reduce the anisotropy. However, this instability cannot account for $T_\| < T_\perp$, which would require other forms of wave particle interactions. Moreover, even if one could somehow create a velocity distribution far from the Sun with $T_{p\perp} > T_{p\parallel}$, as these protons move further from the Sun magnetic moment conservation would again cause a reduction in $T_{p\perp}$ and an increase in $T_{p\parallel}$. Maintaining $T_{p\perp} > T_{p\parallel}$ throughout the heliosphere therefore requires *continuous* wave-particle interactions as the protons move into the outer heliosphere. The shape of the proton velocity distribution, as measured, e.g., by *Helios*, therefore tells us a lot about *in situ* plasma processes in the solar wind, but little about the process that heated the protons in the corona.

In these models electrons were heated isotropically, and the electron temperature anisotropy is small everywhere (and barely visible in Figure 11.1), with $1 < T_{e\parallel}/T_{e\perp} < 1.2$ everywhere. This is in qualitative agreement with *Helios* observations, which generally found $T_{e\parallel} > T_{e\perp}$ in the solar wind between 0.3 and 1 AU (Pilipp et al., 1987). These observations show that electrons and protons behave very differently in the solar wind: While the protons are heavily influenced by local plasma processes in the solar wind, the observed electron velocity distribution is consistent with a simple adiabatic expansion of the electron gas from the corona to the observation point (Lie-Svendsen et al., 1997).

Perhaps the most remarkable feature of these solutions is the large difference that the change in geometry causes. Despite that we use the same heating, apart from scaling it to have the same energy flux at 1 AU, we get densities in the inner corona that differ by about a factor 10. The rapidly expanding model leads to a high-speed wind with a particle flux at 1 AU, $(n_p u_p)_E = 2.4 \times 10^{12}$ m^{-2} s^{-1}, typical of the observed high-speed wind, while the radially expanding flow takes on properties of a typical slow solar wind, with a particle flux that is nearly twice as large, $(n_p u_p)_E = 4 \times 10^{12}$ m^{-2} s^{-1}, and a flow speed of less than 300 km/s at 1 AU. In the rapidly expanding solution the proton temperature is significantly higher than in the radial solution, and conversely for electrons. The lower electron temperature is consistent with the lower density in the rapidly expanding solution leading to a lower energy transfer rate from protons to electrons by collisions.

The key to these large differences lies in the energy budget of the solutions, which we shall discuss later. However, we note from the lower left panel of Figure 11.1 that the downward electron heat flux in the transition region reaches a minimum of about -14 W in the radial model and only about -7 W in the rapidly expanding model, despite that we deposit five times as much energy in the corona in the latter model. (The minimum of q_e, located at $r = r_t = 1.006\ R_S$, coincides with the maximum of the transition region heating term (11.20).) Since this electron heat flux must have its origin in the protons (that are heated) this shows again that the coupling between electrons and protons is very different in the two models. In light of the discussion of Sect. 2, the higher down-ward heat flux in the radial model translates into a higher transition region pressure, and the chromosphere-transition region interface moves downwards to a higher chromospheric density, as is most clearly seen in the electron temperature profiles of the figure.

Figure 11.1 shows that the proton heat flux remains positive (meaning outwards) well below the proton temperature maximum, particularly for the rapidly expanding solution. This shows that protons become colli-sionless close to the Sun, and illustrates why it is so important to have solar wind models that can allow for nonclassical heat conduction. By contrast, the electron heat flux changes sign at the temperature maxi-mum, showing that the electrons behave like a collision-dominated gas in the corona, despite that we solve the same transport equations for electrons as for protons (and thus allow for nonclassical heat transport also in the electrons).

In order to understand why the two geometries give such different solutions we need to look at the energy budgets, shown in Figure 11.2. In this figure the gravitational potential energy flux is defined as

$$F_G \equiv AnuGM_S m_p \left(\frac{1}{R_S} - \frac{1}{r} \right), \qquad (11.27)$$

where n and u are the electron and proton density and flow speed and m_p the proton mass. The enthalpy flux is defined as

$$F_T \equiv Anuk \left(\frac{3}{2}(T_{e\|} + T_{p\|}) + T_{e\perp} + T_{p\perp} \right). \qquad (11.28)$$

The radiative loss energy flux has been obtained by integrating the radia-tive loss rate over the volume of the flow tube from the lower boundary. The figure shows only a small region, with the lowest value of $r/R_S - 1$ (the distance from the solar surface) corresponding to $T_e \approx 7300$ K (up-per chromosphere) and the highest r-value corresponds to $T_e \sim 6 \times 10^5$ K.

Figure 11.2 shows how the various energy inputs to the model are converted into radiation, gravitational potential energy (mass flux) and particle energy. The chromospheric heating merely balances radiative losses in the chromosphere and ceases to be important at the top of the chromosphere. The main feature to be noted from the figures is the large difference in the radiative loss term between the two geometries. In the radially expanding solution the radiative loss in the transition region is much larger than the enthalpy increase, and conversely for the rapidly expanding solution. In the radial solution essentially all of the downward heat flux is converted into radiation, as is most of the direct transition region heating. This is in agreement with the discussion in

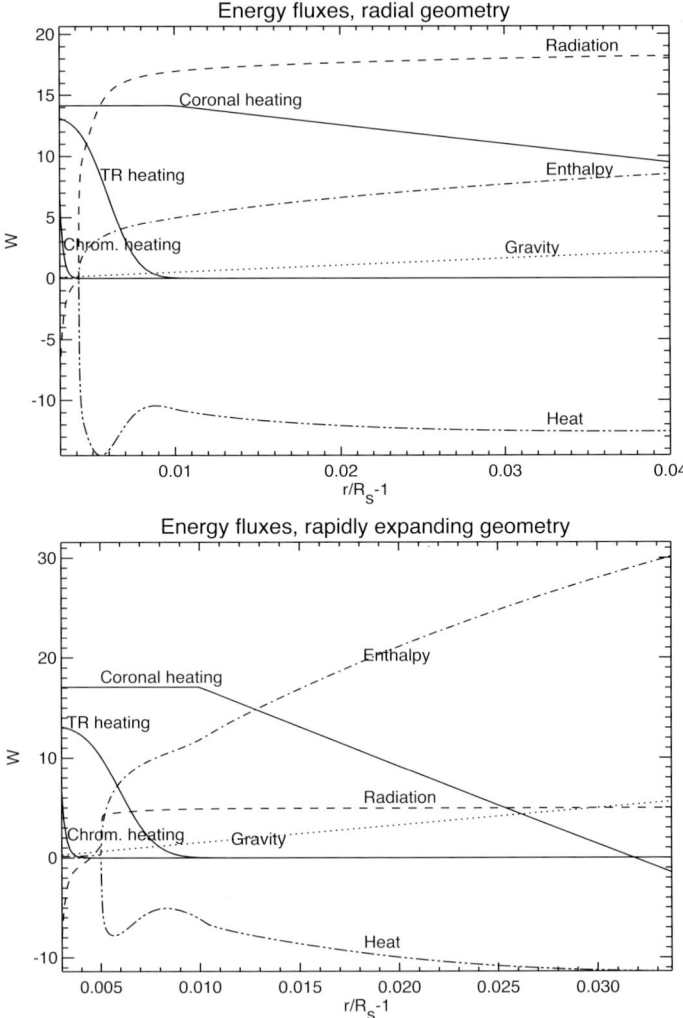

Figure 11.2. Transition region energy budget for the radially and rapidly expanding solutions in Figure 11.1. The solid curves show the various heating terms included in the model ("TR" means the transition region heating proportional to F_t); the dotted curve shows the gravitational potential energy flux; the dash-dotted curve shows the sum of the electron and proton enthalpy fluxes; the dash-dot-dot curve shows $A(q_e + q_p)$ (the total heat flux); and the dashed curve shows the cumulative radiative loss. Arbitrary offsets have been subtracted from the coronal heating and radiative loss fluxes to make the curves fit into the panels.

Sect. 2, where we argued that on closed coronal loops the downward heat flux is balanced by radiative losses, and that this balance sets the transition region pressure. Now we have a solar wind in which most of the energy is lost to the wind; even in the radial solution the downward heat flux of ~ -15 W is much less than the $F_{mp0} = 80$ W supplied to the corona, implying that most is lost to the wind. Still the pressure of the transition region and corona, which is so important for the mass flux, is set by the balance between heat flux and radiation. However, in the rapidly expanding solution this picture is no longer correct. In this case a large fraction of the energy deposited in the transition region is converted into enthalpy flux.

The enthalpy flux of the transition region can be critical for the solar wind mass flux. Say that the proton flux at 1 AU is constrained by observations, with $(n_p u_p)_E \approx 2 \times 10^{12}$ m^{-2} s^{-1} in the fast solar wind (e.g., McComas et al., 2000). Assume for simplicity a common electron and proton temperature T_c in the upper transition region. The increase in the enthalpy flux in the transition region needed to sustain this particle flux is then approximately

$$F_{\text{enth}} = f_{\max}(n_p u_p)_E \left(\frac{r_E}{R_S} \right)^2 5kT_c, \qquad (11.29)$$

where $r_E \approx 215\, R_S$ is the distance from the Sun to Earth. With $f_{\max} = 5$ and the observed particle flux, we find $F_{\text{enth}} \approx 15$ W is needed in order to heat the transition region plasma from chromospheric temperatures up to $T_c = 5 \times 10^5$ K. If this energy cannot be supplied to the transition region somehow, the observed solar wind mass flux cannot be sustained no matter how much energy is available in the corona.

Figure 11.2 shows that for this particular solution it is the energy available below $r = 1.01\, R_S$ (where $T_e \approx 3 \times 10^5$ K) that is most critical, since above this altitude our coronal heating term (proportional to F_{mp0}) is able to heat the upwelling plasma. Closer inspection of this solution shows that of the $F_t = 13$ W deposited as direct heating, 10 W goes into sustaining the enthalpy flux while 3 W is lost as radiation. Of the downward heat flux, ~ 6 W remains at $r = 1.01\, R_S$; of this about 3 W is lost as radiation, about 1 W is used to ionize neutral hydrogen, and hence only about 2 W remains to heat the upwelling plasma. In other words, in this case it is the direct deposition of energy in the transition region, not the downward heat conduction from the corona, that enables us to obtain a solar wind mass flux in agreement with observations. Hence this heating term must have a very large effect on the solar wind properties, as we show below.

Next we therefore consider what happens when we change the amount of energy deposited directly in the transition region. The results of varying F_t while keeping all other parameters identical to the rapidly expanding model of Figure 11.1 are shown in Figure 11.3, where now

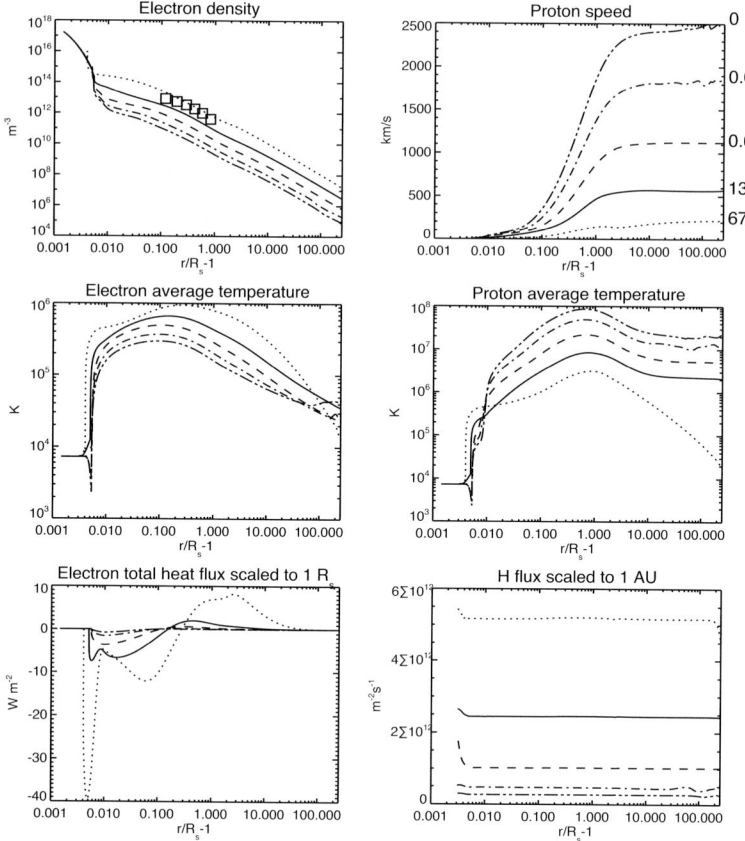

Figure 11.3. Solutions for rapidly expanding geometry with varying transition region heating: $F_t = 13$ W (solid curves), which is identical to the reference model of Figure 11.1; $F_t = 67$ W (dotted curves); $F_t = 0.67$ W (dashed curves); $F_t = 0.07$ W (dash-dotted curves); and $F_t = 0$ (dash-double-dotted curves). The square boxes in the density panel indicate the coronal hole measurements of Koutchmy, 1977.

only the average temperatures are shown, using (11.15). As anticipated from the preceding discussion, the differences are large when F_t is varied, despite that even the maximum value, $F_t = 67$ W, is small compared to the $F_{mp0} = 400$ W deposited in the corona. The solution varies between a massive $((n_p u_p)_E \approx 5 \times 10^{12}$ m^{-2} s$^{-1})$ and slow $(u \approx 200$ km/s) wind for $F_t = 67$ W to an extremely fast wind with a very low flux, $u \approx 2500$ km/s and $(n_p u_p)_E \approx 0.3 \times 10^{12}$ m^{-2} s^{-1} for $F_t = 0$.

Figure 11.3 shows that the changes are even larger than indicated by the discussion above. Even reducing the heating from $F_t = 0.07$ W to 0, which one would have expected to make absolutely no difference for the solution, causes a significant change. The reason for this extreme sensitivity is a nonlinearity of the chromosphere-corona system. In the discussion above we treated the downward heat flux as a given constant, independent of the direct transition region heating proportional to F_t. The solutions in Figure 11.3 demonstrate that this assumption is far from correct; as seen in the lower left panel the downward heat flux is reduced when the direct heating is reduced, so that we get a positive feedback. Hence the reduction in energy to the transition region will be much more than the reduction of direct energy deposition.

The cause of this strong nonlinearity is the collisional coupling between electrons and protons in the corona. When the direct transition region heating is reduced, the coronal density decreases. The energy transfer rate from protons to electrons through collisions therefore decreases. Since we deposit all the coronal energy in the protons, the electron temperature then decreases, as the figure shows, and therefore the downward electron heat flux decreases, too. Since classical heat conduction is $q_e \propto T_e^{5/2} \nabla T_e$, the heat conduction is actually a sensitive function of temperature. The reduction in downward heat flux then reduces the transition region pressure further which in turn decreases the collisional coupling in the corona, and so on.

The radial case is much more "robust" to changes in transition region heating; setting $F_t = 0$ in that case only causes small changes to the solution. The reason is simply that downward heat conduction is so efficient that unless the direct energy deposition in the transition region is large it will not matter.

Hence we find that in solar wind models with proton heating, increasing the expansion factor of the wind (f_{max} in our model) changes the wind from a low-speed wind with high coronal density to a typical high-speed wind (even an "extreme" high-speed wind in the case with no transition region heating) with a low coronal density. The radial geometry is characterized by a high pressure, and a transition region that receives more energy than it needs to sustain the solar wind mass flux, so that the excess energy must by lost by radiation. By contrast, the rapidly expanding flow leads to a low-pressure transition region which is starved of energy, so that the little energy available is used as enthalpy flux and radiative losses are insignificant. In the former case the solar wind mass flux is limited by the amount of coronal energy available to lift particles out of the gravitational field, while in the latter case the mass flux is not limited by coronal heating but rather by the amount of energy available in the transition region.

Why does the geometry have such a large influence on the energy balance of the transition region and hence on solar wind properties? There are two reasons. First of all, in order to drive the solar wind the mechanical energy flux injected at the lower boundary must be much higher in the rapidly expanding model than in the radial model (a factor 5 larger for $f_{max} = 5$), in order to have approximately the same energy flux density at infinity. If the two solutions are to have the same particle flux density at infinity, the enthalpy flux in the transition region has to increase proportionally to the expansion factor, too, as shown by (11.29). However, when only protons are heated in the corona that does not happen; as shown in Figure 11.1, not only is the *fraction* of the coronal energy that ends up as downward heat flux reduced in the rapidly expanding solution, but even its *magnitude* is reduced, despite that the coronal heating rate increased by a factor 5. When protons are heated harder in the rapidly expanding geometry, they tend to become collisionless closer to the Sun and are hence less able to conduct energy down to where it can be transferred to the electrons. Secondly, the magnetic mirror force is much stronger in the rapidly expanding geometry. As long as protons are heated mostly perpendicularly to the magnetic field, they are then reflected by the diverging magnetic field before they get down to an altitude where the density is so high that their energy can be transferred to the electrons.

In the upper left panel of Figure 11.3 we also show the coronal hole electron densities measured by Koutchmy, 1977, which are among the lowest densities observed in the inner corona. We note that most of the modelled densities, except the low-speed, high mass flux wind obtained for $F_t = 67$ W, are well below the observed densities. As we have argued previously, and as argued by Withbroe, 1988, the densities in these types of solar wind models cannot be set, but must result from the heating applied. In other words, if higher densities are sought, a heating mechanism has to be found that can deliver the required amount of energy to the transition region.

This exercise also indicates that if the high-speed solar wind originates from such a rapidly expanding geometry, a mechanism that only heats protons, and only in the perpendicular direction, even if the mechanism is able to deliver the required amount of energy to the solar wind it leads to a mass flux that is much lower than observed (and a correspondingly high speed), and also leads to coronal densities that are orders of magnitude below what is observed. Even using a more physically "realistic" heating model, based on a turbulent cascade of Alfvén waves, we are not able to obtain a reasonable mass flux without forcibly depositing some of the energy directly in the transition region (see Lie-Svendsen et al., 2001).

Apart from direct energy deposition in the transition region, the mass flux can be increased by depositing a significant fraction of the energy in the parallel direction (or having some process that pitch angle scatter protons into the parallel direction). On the other hand the Doppler dimming results (Cranmer et al., 1999) indicate that $T_{p\perp}/T_{p\|} < 2$ in the inner corona (where parallel heating would be most helpful), constraining how much parallel heating we can allow for.

The most effective way of producing a solar wind model in better agreement with observations seems to be explicit heating of coronal electrons; we find that injecting roughly 10% of the coronal energy into electrons leads to a reasonable mass flux. Direct heating of electrons also increases the coronal electron temperature; as mentioned previously, most of the solutions presented here have electron temperatures below what has been inferred from coronal hole observations. Moreover, such low electron temperatures lead to ion charge states that do not agree with *in situ* observations, even for oxygen and carbon (Esser and Edgar, 2001).

Another possible resolution of this problem could be that the downward electron heat conduction is larger than predicted by these models. Because of the steepness of the temperature gradient in the transition region and the relatively low density, electron heat conduction may possibly deviate significantly from classical transport theory (which is inherent in the transport equations we use). However, kinetic models, which solve the Boltzmann equation with the Fokker-Planck collision operator, show that electron heat conduction in the transition region should not deviate significantly from classical transport theory, even for low densities, and when it deviates the heat flux tends to become *smaller* than obtained by classical transport theory, which exacerbates the problem (Shoub, 1983; Lie-Svendsen et al., 1999).

5. Helium in the Corona and Solar Wind

The models discussed in Sect. 4 illustrated how the energy balance of the lower transition region affects the properties of the solar wind itself, and particularly its mass flux. In this section we illustrate how the same region affects the properties of heavier constituents in the corona and solar wind, and particularly their abundances. We shall focus on one species, helium. Not only is helium the second most abundant element in the Sun's atmosphere, with photospheric abundances of order 10%, but also, because of this high abundance, helium may even become a *major* species in the corona with a density comparable to the hydrogen (proton) density, which affects the properties of the solar wind as well.

Modelling of helium is also of interest because observations of the coronal helium abundance have proved to be difficult, particularly for coronal holes. Most observations indicate coronal hole abundances of order 5% or lower (Laming and Feldman, 2001; Laming and Feldman, 2003), which is not more than what is found in the high-speed wind (McComas et al., 2000). Solar energetic particle events, in which coronal particles are accelerated into space by shock waves caused by coronal mass ejections, also provide indirect evidence that the coronal He abundance is of order 5% (Reames, 1998). On the other hand, in the slow wind the α-particle density can be up to 20% of the proton density (Schwenn, 1990), although it is generally much lower (e.g., McComas et al., 2003), and in coronal mass ejections helium can contribute up to 50% of the mass flux (Yermolaev and Stupin, 1997).

The results discussed here are presented in more detail by Lie-Svendsen et al., 2003. However, beware that in that paper the helium *abundance* was defined to be the ratio between the total helium (summed over all charge states) and total hydrogen density, while in this presentation the abundance is defined to be the ratio of the total helium density to the sum of the hydrogen and helium densities, to be more in line with common usage of the word.

Also for helium we shall consider the same two flow geometries as in the preceding section. Initially we choose heating parameters such that we reproduce a fast solar wind with reasonable mass flux, flow speed and He abundance in the rapidly expanding geometry, and a slow speed wind from the radially expanding geometry with the same $\sim 5\%$ He abundance at infinity. In the radial geometry this can be achieved setting $F_{mp0} = 48$ W, $F_{m\alpha 0} = 32$ W, $r_{1p} = r_{1\alpha} = 1.01\ R_S$, deposited over a dissipation scale $H_m = 0.3\ R_S$, and we deposit $F_t = 20$ W as transition region heating. In the rapidly expanding geometry we choose $F_{mp0} = 304$ W, $F_{m\alpha 0} = 76$ W, with $r_{1p} = r_{1\alpha} = 1.03\ R_S$ and $H_m = 0.4\ R_S$. In addition we heat electrons explicitly in the corona with $F_{me0} = 40$ W, starting at $r_{1e} = 1\ R_S$ and damped with $H_m = 0.5\ R_S$, and we heat the transition region with $F_t = 10$ W. As in the electron-proton solar wind models of the previous section, the total energy deposited in the flow tube in the rapidly expanding ($f_{\max} = 5$) model is roughly five times higher than in the radially expanding ($f_{\max} = 1$) model. The energy flux density at 1 AU will thus be roughly the same (when including gravitational potential energy) in the two geometries.

The results are shown in Figure 11.4. The most noticeable feature of these solutions is the large coronal He abundance of the slow wind solution, where the He density becomes close to the hydrogen density so that α-particles carry most of the mass and positive charge of the corona.

Note also the large difference between the radially and rapidly expanding models. In the chromosphere the helium abundance stays close

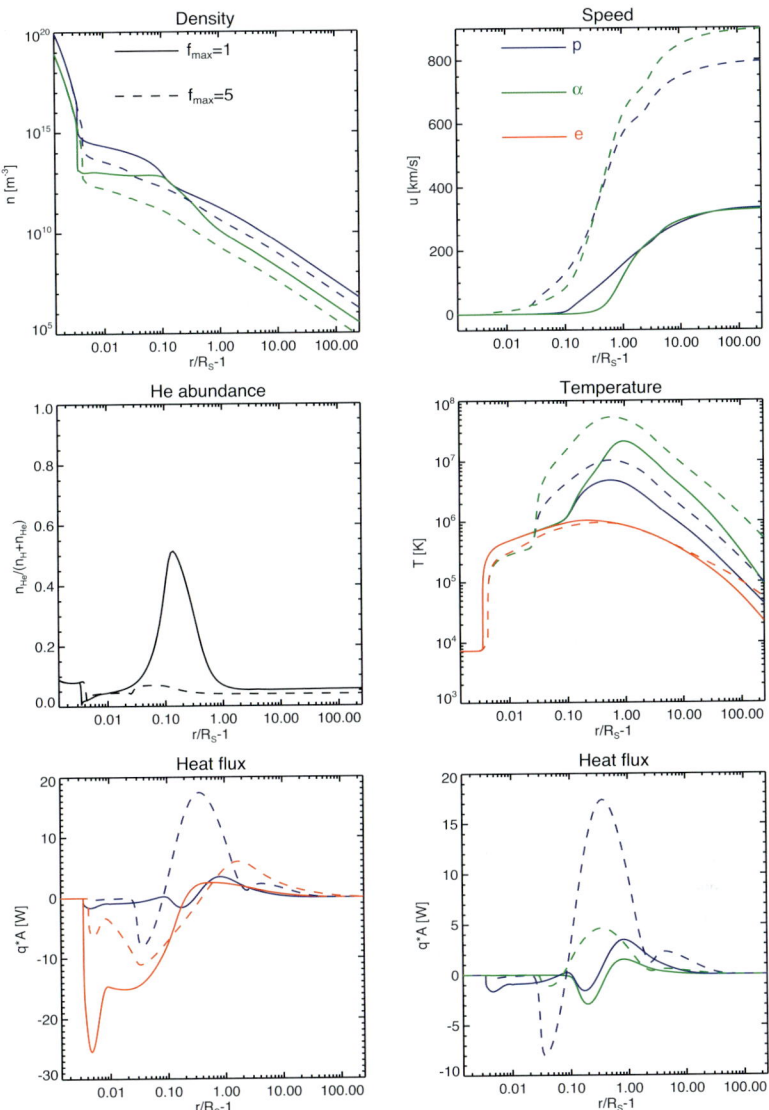

Figure 11.4. Radially expanding ($f_{\max} = 1$) and rapidly expanding ($f_{\max} = 5$) models with helium included, from Lie-Svendsen et al., 2003. The densities are the total (charged plus neutral) H and He densities, and the He abundance is the total (charged plus neutral) helium density divided by sum of the total hydrogen and total helium densities. Blue curves denote hydrogen, green curves helium, and red curves electrons.

to its boundary value of 10%, but as soon as we reach the transition region and particles become ionized the abundance is highly variable with altitude. However, in the rapidly expanding model the He abundance never exceeds its chromospheric value. The densities of both hydrogen and helium throughout the transition region and corona are higher in the radial than in the rapidly expanding model. This is consistent with the discussion of Sect. 4, where we argued that the much higher downward heat flux in the radial geometry must lead to a higher density. As seen in the lower left panel of Figure 11.4, the models with helium included show the same behaviour.

The temperature panel of Figure 11.4 shows that the α-particle temperature in the corona is higher than the proton temperature, reaching a maximum as high as $T_\alpha = 5 \times 10^7$ K. The higher α-particle temperature is not primarily a consequence of a higher heating rate per particle for these, but a consequence of lower energy loss rates for α-particles. Because these particles are heavier than protons, the energy loss by escape into the solar wind is lower unless they become very hot.

The coronal He abundance is influenced by the thermal forces, which are contained in the terms proportional to the heat fluxes in (11.18). In the temperature gradient of the transition region He ions feel a strong upward force caused by collisions with protons and electrons. As a result the heavier helium ions are not pulled through the transition region by friction with protons, but actually stream faster than protons in this region. For instance, in the radial model α-particles stream up to three times faster than protons in the region where $T_e \approx 2 \times 10^5$ K.

The solutions in Figure 11.4, particularly the $f_{max} = 1$ solution, also illustrate that there is no strong reason why the coronal He abundance should be close to the solar wind abundance. Moreover, by choosing the coronal heating "right", one may have a high coronal abundance while still reproducing the observed abundance in the solar wind.

The models presented above demonstrate that it is possible to reproduce the observed solar wind He particle flux (which equals the solar wind He abundance as long as α-particles and protons flow at roughly the same speed) in both types of geometries. However, we would like to get a better understanding of what processes determine this flux. To do so, let us now vary the amount of energy we deposit in α-particles in the corona.

Figures 11.5 and 11.6 show the results from these experiments, where the "standard" cases refer to Figure 11.4. The main result is contained in the bottom panels of the figures. When we change the heating rate in the rapidly expanding geometry, the α-particle coronal abundance and terminal flow speed change, but the α-particle flux does not. That tells us that within the range of chosen heating rates, the flux of helium is not set in the corona, but rather by the supply from the chromosphere and transition region, and since the transition region can readily pull helium

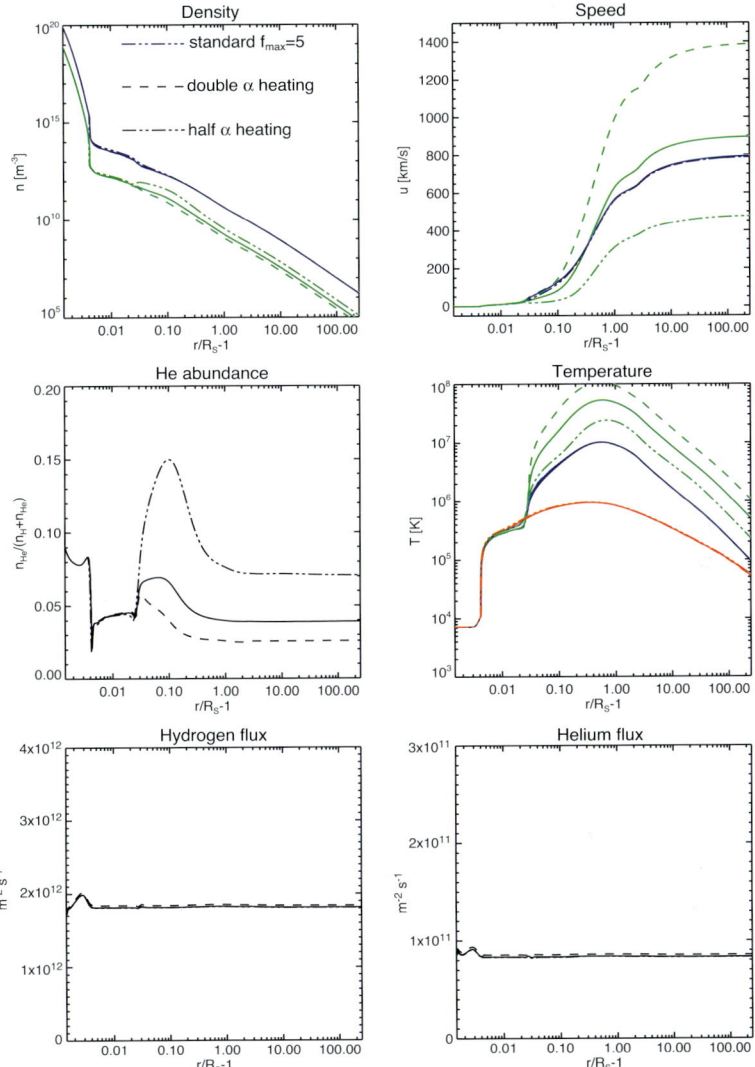

Figure 11.5. Changing the α-particle energy flux $F_{m\alpha 0}$ in the rapidly expanding ($f_{max} = 5$) geometry. The colour coding is the same as in Figure 11.4. The fluxes have been scaled to 1 AU. The "standard" case is also shown in Figure 11.4.

into the corona through the thermal force, the flux must be limited by the supply from the chromosphere and lower transition region where helium is still mostly neutral.

By contrast, the radial solutions in Figure 11.6 show that the corona indeed makes a difference for the helium flux. More surprisingly, the changes for hydrogen are at least as large as for helium, despite that

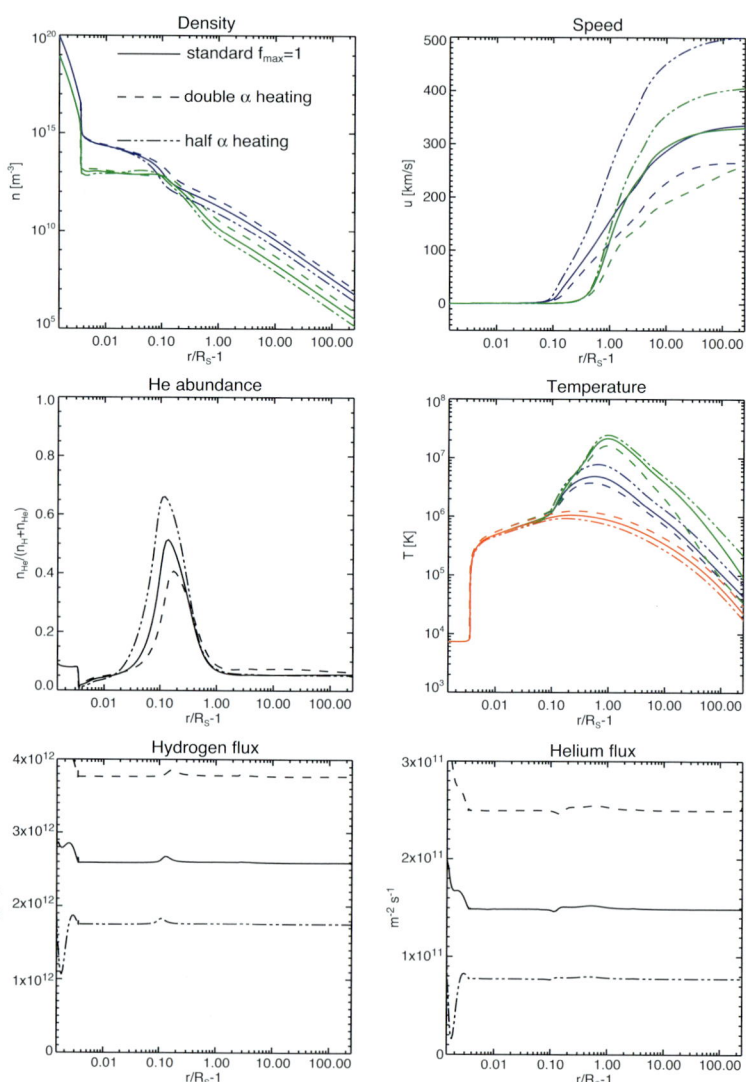

Figure 11.6. Changing the α-particle energy flux $F_{m\alpha 0}$ in the radially expanding $(f_{\max} = 1)$ geometry.

we did not change the proton heating rate. The reason for this is that helium in this case is not a minor species in the corona, and therefore does affect protons and electrons. If the corona becomes helium-rich, the solar wind may even acquire the characteristics of a *polar wind*, in which the light protons are accelerated in the electric field set up by the near static electrons and α-particles.

Protons and electrons feel the changes in helium through collisions. However, Figure 11.6 shows that when, e.g., the heating is doubled, the α-particle and proton temperatures are *reduced*, and this cannot be explained just by the direct collisional coupling in the corona. Instead this example is another illustration of the nontrivial coupling between the chromosphere and corona, and it is mainly through this coupling via the chromosphere and transition region that the coronal protons feel the changes in heating of coronal α-particles: As the heating of α-particles is increased, more energy is being transferred to the electrons, and hence the downward heat flux increases. This increased heat flux then causes the chromosphere-transition region interface to shift to a higher pressure level and hence the density of electrons and protons increase in the corona as well. It is this increase in proton density in the corona that causes the change in proton flux and flow speed. This mechanism also explains why the α-particle temperature decreases when the heating increases: By bringing up more electrons and protons, the increased loss of energy through collisions overcompensates for the increased heating.

Regarding the solar wind helium flux (or, equivalently, the solar wind abundance), the modelling leads to an almost embarrassingly simple picture: If we want to obtain a high-speed solar wind, both for helium and hydrogen, meaning a wind with a terminal speed significantly faster than the gravitational escape speed at the surface, we must heat coronal α-particles so hard that they quickly escape from the corona, resulting in a low coronal abundance (as we found in the $f_{max} = 5$ case above). In that case the helium pressure in the corona is low and therefore does not affect the flow of helium through the chromosphere and transition region. Hence the solar wind helium flux is set by whatever process maintains the helium abundance in the upper chromosphere. As long as there is sufficient energy available in the corona to lift α-particles out of the gravitational field, coronal (and solar wind) processes then have little or no impact on the flux. By contrast, creating a slow wind solution, with the asymptotic speed significantly below the escape speed, is difficult without a low heating rate in the corona. In this case a high coronal helium density results, and helium may even dominate the corona. The helium flux is then determined by the amount of energy available in the corona to heat helium and thus lift helium out of the gravitational potential.

The modelling presented here indicates that it is difficult to create a slow solar wind without also creating a high helium abundance in the corona. This result seems to be at odds with the few observations of helium abundances in the corona which do not indicate high abundances. A weakness of the model used here, based on the transport equations and collision terms developed by Schunk, 1977, and Demars and Schunk, 1979, is that the magnitude of the thermal force may be overestimated. Recently new transport equations have been developed that provide a better description of Coulomb collisions (Killie et al., 2004). The new equations indicate that the thermal force may be overestimated by a factor three or so. It would therefore be of interest to determine whether a better description of the thermal force leads to reduced coronal helium abundances in the slow solar wind models.

We have glossed over another serious problem, regarding the description of the chromosphere. In these models neutral helium is pulled through the chromosphere into the transition region by frictional drag from outflowing hydrogen. However, using accepted values for the He-H elastic cross section, this frictional force is not sufficient to compensate for gravitational settling (Hansteen et al., 1997; Lie-Svendsen et al., 2003). The result is that helium settles in the chromosphere with a scale height that is close to 1/4 of the hydrogen scale height, and hence the helium abundance at the top of the chromosphere is essentially zero. In that case the solar wind helium flux would also be zero, no matter what happens in the corona. In order to obtain a helium abundance at the top of the chromosphere of order 10%, which is necessary for the solar wind helium abundance to be of order 5–10%, we had to artificially increase the elastic cross section by a factor 15 in the radially expanding flow and by a factor 3 in the rapidly expanding flow (the faster hydrogen flow in the chromosphere in the rapidly expanding wind increases the frictional force on helium). Since these cross sections are fairly well known from atomic physics, this means that something is missing from our chromospheric description. One possibility is that the chromosphere is more turbulent than we think, so that mixing maintains the He abundance in the upper chromosphere. In the fast solar wind at least, where the solar wind helium flux seems to be determined by the supply from the chromosphere, it is the unknown process that supplies helium to the upper chromosphere that sets the helium flux. This also illustrates that the elemental composition of the solar wind cannot be understood without understanding chromospheric processes, which is in accordance with studies of the so-called *FIP-effect* (first ionization potential) (e.g., von Steiger and Geiss, 1989; Marsch et al., 1995; Wang, 1996; Peter, 1998).

6.   Summary

Through numerical modelling we have attempted to demonstrate that the solar wind is tightly coupled to the chromosphere, and that this coupling is nontrivial and leads to some surprising properties of the solar wind. The essential part of this coupling is the energy transport between the corona and chromosphere, specifically the fraction of the energy deposited in the corona to drive the wind that ends up as downward heat flux into the transition region. In all models most of the energy deposited in the corona is transferred directly into the solar wind. Hence the magnitude of the solar wind energy flux (which is essentially the sum of the gravitational potential energy and kinetic energy flux), as measured e.g., by the Ulysses spacecraft (McComas et al., 2000), should be close to the actual magnitude of the energy deposited in the corona. However, we have demonstrated how the (small) fraction of the energy transported downwards from the corona can set the pressure of the transition region and corona and influence the mass flux of the solar wind, both for hydrogen and helium. Although the total energy flux of the solar wind is set by the mechanical energy flux heating the corona, the fraction conducted downwards thus influences how the total energy flux is distributed between gravitational potential energy (mass flux) and kinetic energy (flow speed).

The magnetic field configuration near the Sun has a large effect on the magnitude of this downward energy transport and hence the coupling between the chromosphere and corona, particularly in models in which protons receive most of the coronal heating. In a radially expanding geometry downward heat conduction is efficient, resulting in a high coronal pressure and a high coronal helium abundance. In that case the solar wind tends to be slow with a high mass flux whose magnitude is determined by the amount of energy available in the corona. A rapidly expanding geometry, simulating a coronal hole, makes downward heat conduction difficult, resulting in a low coronal density and typically a coronal helium abundance that is close to the chromospheric abundance. In that case the flux of both hydrogen and helium is set below the corona; for hydrogen by the amount of energy available in the transition region to heat the upwelling plasma, and for helium by the amount of helium available for ionization at the top of the chromosphere. Changing the flow geometry therefore naturally leads to a change from one solar wind

	Slow mode	*Fast mode*
geometry	radial	rapidly expanding
wind speed	slow	fast
solar wind mass flux	high	low
coronal density	high	low
proton heat flux	classical	non-classical
transition region pressure	high	low
downward heat flux absorbed by	radiation	enthalpy flux
H mass flux limited by	coronal heating	transition region heating
coronal He abundance	high	low
He mass flux limited by	coronal energy	supply from chromosphere

Table 11.1. The two solar wind modes

"mode" to another, with quite different characteristics and dominating processes, as summarized in Table 11.1.

References

Allen, C. W. (1976). *Astrophysical Quantities.* Athlone, London.

Antonucci, E., Dodero, M. A., and Giordano, S. (2000). *Sol. Phys.*, 197:115–134.

Arnaud, M. and Rothenflug, R. (1985). *Astron. & Astrophys. Suppl.*, 60:425–457.

Cranmer, S. R. et al. (1999). *Astrophys. J.*, 511:481–501.

Cuperman, S., Harten, A., and Dryer, M. (1972). *Astrophys. J.*, 177:555–566.

David, C., Gabriel, A. H., Bely-Dubau, F., Fludra, A., Lemaire, P., and Wilhelm, K. (1998). *Astron. & Astrophys.*, 336:L90–L94.

Demars, H. and Schunk, R. (1979). *J. Phys. D. Appl. Phys.*, 12:1051–1077.

Dorfi, E. A. and Drury, L. O'C. (1987). *J. Comp. Phys.*, 69:175–195.

Doyle, J. G., van den Oord, G. H. J., O'Shea, E., and Banerjee, D. (1998). *Sol. Phys.*, 181:51–71.

Esser, R. and Edgar, R. J. (2001). *Astrophys. J.*, 563:1055–1062.

Habbal, S. R., Esser, R., and Arndt, M. B. (1993). *Astrophys. J.*, 413:435–444.

Hammer, R. (1982a). *Astrophys. J.*, 259:767–778.

Hammer, R. (1982b). *Astrophys. J.*, 259:779–791.

Hansteen, V. H. and Leer, E. (1995). *J. Geophys. Res.*, 100(A11):21577–21593.

Hansteen, V. H., Leer, E., and Holzer, T. E. (1997). *Astrophys. J.*, 482:498–509.

Hartle, R. E. and Sturrock, P. A. (1968). *Astrophys. J.*, 151:1155–1170.

Hollweg, J. V. (1986). *J. Geophys. Res.*, 91:4111–4125.

Hollweg, J. V. and Johnson, W. (1988). *J. Geophys. Res.*, 93:9547–9554.

Holzer, T.E. and Axford, W.I. (1970). *Ann. Rev. Astron. Astrophys.*, 8:31–60.

Judge, P. G. and Meisner, R. (1994). In Hunt, J. J., editor, *The Third Soho Workshop: Solar Dynamic Phenomena and Solar Wind Consequences*, volume ESA SP-373 of *Eur. Space Agency Spec. Publ.*, pages 67–71.

Killie, M. A., Janse, Å. M., Lie-Svendsen, Ø., and Leer, E. (2004). *Astrophys. J.*, 604:842–849.

Kohl, J. L., Esser, R., Cranmer, S. R., Fineschi, S., Gardner, L. D., Panasyuk, A. V., Strachan, L., Suleiman, R. M., Frazin, R. A., and Noci, G. (1999). *Astrophys. J.*, 510:L59–L62.

Kohl, J. L. et al. (1997). *Sol. Phys.*, 175:613–644.

Kopp, R. A. and Holzer, T. E. (1976). *Sol. Phys.*, 49:43–56.

Koutchmy, S. (1977). *Sol. Phys.*, 51:399–407.

Laming, J. M. and Feldman, U. (2001). *Astrophys. J.*, 546:552–558.

Laming, J. M. and Feldman, U. (2003). *Astrophys. J.*, 591:1257–1266.

Landini, M. and Monsignori Fossi, B. C. (1975). *Astron. & Astrophys.*, 42:213–220.

Leer, E. and Axford, W. I. (1972). *Sol. Phys.*, 23:238–250.

Li, X. (1999). *J. Geophys. Res.*, 104:19773–19785.

Li, X., Habbal, S. R., Hollweg, J. V., and Esser, R. (1999). *J. Geophys. Res.*, 104(A2):2521–2535.

Lie-Svendsen, Ø., Hansteen, V. H., and Leer, E. (1997). *J. Geophys. Res.*, 102:4701–4718.

Lie-Svendsen, Ø., Hansteen, V. H., and Leer, E. (2003). *Astrophys. J.*, 596:621–645.

Lie-Svendsen, Ø., Holzer, T. E., and Leer, E. (1999). *Astrophys. J.*, 525:1056–1065.

Lie-Svendsen, Ø., Leer, E., and Hansteen, V. H. (2001). *J. Geophys. Res.*, 106:8217–8232.

Lie-Svendsen, Ø., Leer, E., Hansteen, V. H., and Holzer, T. E. (2002). *Astrophys. J.*, 566:562–576.

Marsch, E., Schwenn, R., Rosenbauer, H., Mühlhäuser, K.-H., Pilipp, W., and Neubauer, F. M. (1982). *J. Geophys. Res.*, 87:52–72.

Marsch, E., von Steiger, R., and Bochsler, P. (1995). *Astron. & Astrophys.*, 301:261–276.

McComas, D. J., Elliott, H. A., Schwadron, N. A., Gosling, J. T., Skoug, R. M., and Goldstein, B. E. (2003). *Geophys. Res. Lett.*, 30(10):1517. doi:10.1029/2003GL017136.

McComas, D. J. et al. (2000). *J. Geophys. Res.*, 105(A5):10419–10433.

McKenzie, J. F., Axford, W. I., and Banaszkiewicz, M. (1997). *Geophys. Res. Lett.*, 24(22):2877–2880.

Munro, R. H. and Jackson, B. V. (1977). *Astrophys. J.*, 213:874–886.

Olsen, E. L. and Leer, E. (1999). *J. Geophys. Res.*, 104(A5):9963–9973.

Peter, H. (1998). *Astron. & Astrophys.*, 335:691–702.

Pilipp, W. G., Miggenrieder, H., Mühlhäuser, K.-H., Rosenbauer, H., Schwenn, R., and Neubauer, F. M. (1987). *J. Geophys. Res.*, 92(A2):1103–1118.

Poletto, G., Suess, S. T., Biesecker, D. A., Esser, R., Gloeckler, G., Ko, Y.-K., and Zurbuchen, T. H. (2002). *J. Geophys. Res.*, 107(A10):1300. doi:10.1029/2001JA000275.

Reames, D. V. (1998). *Space Sci. Rev.*, 85:327–340.

Rosner, R., Tucker, W. H., and Vaiana, G. S. (1978). *Astrophys. J.*, 220:643–665.

Schunk, R. W. (1977). *Rev. Geophys. and Space Physics*, 15(4):429–445.

Schwenn, R. (1990). In Schwenn, Rainer and Marsch, Eckart, editors, *Physics of the Inner Heliosphere I*, volume 20 of *Physics and Chemistry in Space — Space and Solar Physics*, chapter 3, pages 99–181. Springer-Verlag, Berlin Heidelberg.

Shoub, E. C. (1983). *Astrophys. J.*, 266:339–369.

Vernazza, J. E., Avrett, E. H., and Loeser, R. (1981). *Astrophys. J. Suppl. Ser.*, 45:635–725.

von Steiger, R. and Geiss, J. (1989). *Astron. & Astrophys.*, 225:222–238.

Wang, Y.-M. (1996). *Astrophys. J.*, 464:L91–L94.

Wilhelm, K., Marsch, E., Dwivedi, B. N., Hassler, D. M., Lemaire, P., Gabriel, A. H., and Huber, M. C. E. (1998). *Astrophys. J.*, 500:1023–1038.

Withbroe, G. L. (1988). *Astrophys. J.*, 325:442–467.

Yermolaev, Y. I. and Stupin, V. V. (1997). *J. Geophys. Res.*, 102:2125–2136.

Zangrilli, L., Poletto, G., Nicolosi, P., Noci, G., and Romoli, M. (2002). *Astrophys. J.*, 574:477–494.

Chapter 12

ELEMENTAL ABUNDANCES IN THE SOLAR CORONA

John C. Raymond

Harvard-Smithsonian Center for Astrophysics

jraymond@cfa.harvard.edu

Abstract   Elemental abundances in the solar corona and solar wind vary by over an order of magnitude. Two general trends are observed. First, there is an enhancement of elements whose neutral atoms have ionization potentials below about 10 eV (low-FIP) compared with those whose neutral atoms have higher ionization potentials (high-FIP). Second, there is a general depletion of heavier elements relative to hydrogen. This paper summarizes recent observational and theoretical abundance studies.

1.   Introduction

Many analyses of astrophysical observations assume "solar elemental abundances" without recognizing the large abundance variations within the solar atmosphere. However, abundances in the solar corona are modified with respect to those in the photosphere in (at least) two systematic ways. First, the abundances of elements whose neutral species have an ionization potential below about 10 eV are enhanced relative to elements whose atoms have higher ionization potentials. This First Ionization Potential (FIP) effect has been studied for many years, and it is typically a factor of 3 or 4 enhancement of low-FIP elements compared to high-FIP elements in active regions and in the slow solar wind (Meyer 1985a,b). It has also been suggested that the ionization time of the neutral atom, rather than its ionization potential, provides a better correlation with abundance (von Steiger 1998). The inverse of the First Ionization Time (FIT) correlates well with the FIP, but Kr (FIP = 14.0 eV) and Xe (FIP = 12.1 eV) seem to behave like intermediate rather than high-FIP elements in lunar samples (Weiler 1998).

G. Poletto and S.T. Suess (eds.), The Sun and The Heliosphere as an Integrated System, 353–371.

The second effect is a reduction of the abundances of all elements relative to hydrogen. This has been less well studied, because individual *in situ* instruments generally cannot measure both hydrogen and the elements that are several orders of magnitude less abundant, and because the wavelength ranges of many spectroscopic instruments do not include hydrogen lines or continuum. In other cases, the hydrogen lines are optically thick or they are formed in entirely different regions than the other spectral lines. Recent studies of line/continuum ratios in solar flares and of UV line intensities above the solar limb have begun to remedy this problem, and depletions of order a factor of 3-10 are often observed. They are attributed to gravitational settling (Raymond et al. 1997a; Feldman et al. 1999).

Elemental abundances are important as diagnostics in themselves, in that they contain information about the physical processes that heat the corona and accelerate the solar wind. They also provide information about the exchange of material between the chromosphere and corona, and they are a means to connect observations of coronal structures with structures measured in the solar wind. For instance, Borrini et al. (1983) showed that helium abundance differs by up to a factor of 3 between the fast solar wind and the sector boundaries where the solar wind is slow. The FIP enhancement, which is strong in the slow solar wind and in streamers, but weak in the fast solar wind and coronal holes, is an especially good tracer.

Knowledge of the abundances is also required for interpretation of many other observations, such as broad band X-ray measurements (e.g. YOHKOH) or narrow band EUV measurements (e.g. EIT or TRACE). The emission measure derived from these imaging observations is generally inversely proportional to the assumed abundances, and derived temperatures can also depend on the abundances (e.g. Li et al. 1998; Ko et al. 2002).

Elemental abundances also have important physical consequences. The radiative cooling rate is directly proportional to the metal abundances at temperatures up to about 10^7 K, where bremsstrahlung begins to dominate (Raymond, Cox & Smith 1976). The mean particle weight can affect flow dynamics, and Hansteen, Leer and Holzer (1997) have studied the effects of helium on the fast solar wind. Moreover, preferential heating and acceleration of oxygen in the fast solar wind is attributed to cyclotron damping of plasma waves (Kohl et al. 1997, 1998; Cranmer et al. 1999). While the small oxygen abundance implies only a weak effect on the overall dynamics of the wind, analogous preferential heating of helium could be very important.

Two recent books present extensive treatments of elemental abundances in the Sun and heliosphere (Frhlich et al. 1998; Wimmer-Schweingruber 2002). This paper presents an overview of coronal abundances and a summary of recent results.

2. Methods

The measurements of coronal and solar wind abundances are fundamentally different, in that coronal values are derived from remote sensing observations, while solar wind composition is measured *in situ*.

2.1 Coronal Observations

Most coronal abundance determinations rely upon ratios of collisionally excited emission lines. The intensity of such a line, usually given in units of photons/(cm^2 s sr), is

$$ I = \frac{1}{4\pi} \int n_e n_H \frac{N_{elem}}{N_H} \frac{N_i}{N_{elem}} q_{ex} dx \qquad (12.1) $$

where N_i/N_{elem} gives the fraction of the element in the ion observed (usually assumed to be in equilibrium, but see below) and q_{ex} is the excitation rate coefficient for the observed emission line. The integration along the line of sight implies an averaging weighted by the local emissivity. The rapid density falloff with heliocentric distance means that in practice the average is very heavily weighted toward the location where the line of sight passes closest to the Sun.

Both N_i/N_{elem} and q_{ex} are functions of temperature, but one can obtain the relative abundances of two elements by choosing spectral lines such that the product $q_{ex} N_i/N_{elem}$ has the same temperature dependence for both lines. A common example is ratio of the Ne VI and Mg VI multiplets near 400 Å. This set of lines has the additional advantage that the wavelengths are so close that any instrumental radiometric calibration uncertainty cancels out. The derived abundance ratio still depends on the assumed atomic rates, but even this dependence cancels out for comparison of the abundance ratios among different solar features. The ionization fractions of Ne VI and Mg VI are nearly identical on the low temperature side of the peak, but significantly different on the high temperature side. In extreme cases, differences in temperature structure can masquerade as differences in abundance ratio (Del Zanna, Bromage & Mason 2003). Time-dependent ionization might also affect the apparent abundances if flows though a steep temperature gradient are present (e.g. Dupree, Moore & Shaprio 1979), or if impulsive heating occurs, as in nanoflare models for the corona (Raymond 1990; Laming & Feldman 1994; Spadaro et al. 1995).

In the X-ray range, Ne X Lyα and the nearby lines of Fe XVII are formed over similar temperature ranges, and the ratio is used to infer abundance ratios (e.g. Schmelz et al. 1996; Phillips et al. 1997). There is never a perfect match of temperature response, and time-dependent ionization could distort the derived abundances, but in general this method gives reliable relative abundances within the uncertainties in the atomic rates. However, discrepancies between observed and predicted line ratios have created a longstanding controversy regarding the Fe XVII atomic rates, the roles of cascades from highly excited states, resonances in the excitation cross sections, and optical depth effects (Liedahl et al. 1995; Chen & Pradhan 2002; Saba et al. 1999; Wood & Raymond 2000).

To carry the analysis a step further, one can obtain absolute abundances if one of the elements being compared is hydrogen. This can be difficult, in that there are no lines of hydrogen shortward of 912 Å. The Lyman continuum is optically thick, and it is formed in the chromosphere. Therefore it cannot be used for comparison with transition region or coronal emission lines. The Lyα and Lyβ lines in the corona are optically thin, however, and their intensities can be compared with those of other lines. The comparison is complicated by the strong contribution of scattered chromospheric photons to the Lyman line intensities, but it is possible to separate out the collisional contribution to directly compare collisionally excited intensities (Raymond et al. 1997a). At shorter wavelengths, the bremsstrahlung continuum is largely formed by hydrogen, and it can be measured shortward of about 10 Å. Absolute abundances in solar flares can be derived by measuring the equivalent widths of lines such as Ca XIX λ3.178, provided that that ion dominates at the temperature of the flare plasma and that the continuum is not contaminated by fluorescence due to X-rays a shorter wavelengths (Sylwester et al. 1998). At longer wavelengths, the bremsstrahlung brightness can be measured in the radio for comparison with EUV line intensities, though this has the disadvantage of requiring accurate absolute radiometric calibrations of very different instruments (White et al. 2000).

The level of uncertainty in these remote sensing measurements of elemental abundances varies widely. It depends directly on instrumental calibration and the accuracy of the atomic rates used. It also depends upon simplifications such as ionization equilibrium and Maxwellian electron distributions. These approximations can be checked if many spectral lines is available, but are most often untested assumptions. The uncertainties are discussed at length in Raymond et al. (2001). In general, absolute coronal abundances are seldom reliable to better than 20% and in some cases are only good at a factor of 2 level, but relative abundances among different coronal structures or between selected pairs of elements are likely to be better.

2.2   Solar Wind Measurements

Elemental abundances in the solar wind are measured directly by instruments aboard various spacecraft. This paper does not deal with energetic particles (see Schwadron, this volume). Some of the current instruments that measure the abundances of the bulk solar wind are CELIAS/MTOF aboard SOHO (Hovestadt et al. 1995), the SWICS instruments aboard ULYSSES and ACE (Bame et al 1992) and SWEPAM aboard ACE (McComas et al. 1998). The MTOF time of flight sensor provides high mass resolution. The SWICS instruments routinely determine the charges and relative abundances of the 5 or 6 most abundant elements other than hydrogen, while SWEPAM measures the densities of H and He. A major limitation is that absolute abundance measurements often require accurate cross calibration of two separate instruments. While there is no line of sight integration problem for the *in situ* measurements, it is generally necessary to average over a considerable length of time. The abundances in the fast solar wind are quite constant, with a weak FIP effect if any is present at all, while abundances in the slow solar wind show considerable variation and an average FIP enhancement of about 4 (Geiss et al. 1995; von Steiger et al. 2000). This shows up in the striking correlation between solar wind speed and Mg/O abundance ratio. Even in the fast solar wind, short time scale abundance fluctuations are apparent (Raymond et al. 2001). The bimodal nature of the elemental composition, unlike the velocity distribution or the charge state distribution, persists through the solar cycle (Zurbuchen et al. 2002).

3.   FIP Effect

Hnoux (1998) reviewed the models for FIP enhancement available at that time. They included models based on steady state or time-dependent diffusion (Marsch et al. 1995; Wang 1996; Peter 1998) and models that relied upon vertical or horizontal magnetic fields (Vauclair 1996; von Steiger & Giess 1989; Hnoux & Somov 1997). McKenzie, Sukhorukova & Axford (1998) showed that the early diffusion models depend upon artificial boundary conditions, and they went on to consider photoionization of upward flowing chromospheric material. Shock waves in the chromosphere may also play a role in determining the average ionization state and therefore the FIP enhancement. Judge & Peter (1998) discuss the differences in time-averaged ionization states of high- and low-FIP elements in the presence of chromospheric shocks.

More recently, Arge & Mullan (1998) have proposed that low FIP elements are preferentially driven into the current sheet region during reconnection events in the chromosphere. That paper presents detailed calculations of the inflow of neutrals and ions into the current sheet and assumes that the material in the current sheet is ejected into the corona. The overall rates of injection for a nanoflare-heated corona agree with the strong abundance variations and the time scale for the buildup of the FIP enhancement reported by Young & Mason (1997) and Widing & Feldman (2001).

Another model that seeks to explain the global nature of the FIP effect was presented by Schwadron et al. (1999). They followed the evolution of a magnetic loop over time including the heating effects of MHD waves on minor ions. In their model, reconnection between closed loops and open field lines releases FIP-enriched plasma into streamer boundaries and thence into the slow solar wind. Again, the FIP enhancement is expected to increase with a time scale of a few days. This model has the appeal of tying the FIP effect in with solar wind measurements. It predicts that the FIP enhancement is largest for large magnetic loops, and these are the loops that are assumed to reconnect with open field lines to form the slow solar wind.

Laming (2004) has recently extended the ideas of Schwadron et al. with a more detailed calculation of the effects of waves on ions in the chromosphere. He computes the pondermotive force on ions in the chromosphere due to Alfven waves and finds enhancements of the low FIP elements Mg, Si and Fe of 3 to 3.3 in a model based on the empirical chromospheric model from Vernazza, Avrett & Loeser (1981) and a moderate Alfven wave energy density. The model predicts considerably larger enhancements for the very low FIP elements Na and K, and it has the potential for producing a negative FIP bias if Alfven wave reflection leads to cancellation of upward and downward wave fluxes. It also predicts that sulfur, with a FIP of 10.4 eV, has an intermediate FIP enhancement.

The models described above attempt to account for the FIP bias in fairly general terms. The high quality spectra available in recent years have generated many detailed observational studies of the FIP effect. Improvements in the atomic data have also reduced the uncertainty in abundance derivations, and narrow band images from EIT and TRACE have shown the morphologies of the regions being studied. We now consider various types of coronal structures:

Flares.   X-ray observations of flares with SMM and Yohkoh-BCS yield a range of FIP enhancements, for instance a factor of 2 underabundance of high-FIP sulfur and factor of 1.5 overabundance of the low-FIP Ca (Fludra & Schmelz 1995,1999). Sylwester et al. (1998) found significant variations in calcium abundance among different flares, but no correlation between the Ca abundance and other flare parameters. In two events, the Ca abundance increased with time. A recent observation with the RESIK instrument aboard Coronas-F finds a factor of 3 enhancement for the very low FIP element potassium compared to S and Ar (Phillips et al. 2003). Departures from ionization equilibrium could influence the derived abundances, but they are believed to be important only in the very early phase of a flare (Feldman, Doschek & Kreplin 1980).

SMM measurements of narrow γ-ray lines also yield elemental abundances. Share & Murphy (1995) analyzed 19 solar flares observed by SMM. They found clear evidence for a FIP enhancement consistent with an average value of 4, but highly variable. The variation may be attributed to differences in the location of the γ-ray production site, which could be close to the photosphere or considerably higher in the chromosphere or transition region. Murphy et al. (1997) observe abundance variations with time during an individual flare, possibly due to evolution of the height (and ionization state) of the region where the energetic particles deposit their energy in thermal plasma.

Active regions.   Active regions generally show a healthy FIP bias when observed in X-rays (McKenzie & Feldman 1992) or in transition region lines (Mohan et al. 2000). There is some evidence for extreme enhancement of lowest FIP elements, K and Ca (Doschek et al. 1985; Phillips & Feldman 1999; Mohan et al. 2000; Falconer, Davila & Thomas 1997). However CDS observations of two active regions between the solar surface and 1.3 $R_\odot$ by Parenti et al. (2003) showed no enhancement or Al or Ca compared to other low FIP elements. Del Zanna (2003) found no FIP enhancement in quiescent 1 MK active region loops, while Del Zanna & Mason (2003) found a factor of 4 FIP enhancement in other loops. Studies of active regions over time indicate that they form with little FIP enhancement, but the enhancements progressively increase as the active region ages (Widing & Feldman 1997). Sheeley (1995) suggests that material rich in high-FIP elements may be associated with emerging magnetic flux.

At larger heights, the streamers above active regions have been observed by UVCS. Raymond et al. (1997a), Parenti et al. (2003), Uzzo et al. (2004) and Bemporad et al. (2003) observed fairly typical active region streamers, while Ko et al. (2002) and Ciaravella et al. (2002) observed the streamer above a young, extremely unstable and hot active region. These analyses require simultaneous determination of the coronal temperature or Differential Emission Measure to obtain elemental abundances, and they vary in the number of spectral lines available for study. All these analyses found fairly typical FIP enhancements of factors of 2 to 4, except for the two streamers observed by Parenti et al. (2000), where no FIP enhancement was seen.

The UVCS observations of a hot active region by Ko et al. (2002) covered the range from 1.22 to 1.6 $R_\odot$. The FIP bias was constant at about a factor of 4 over this range, though the absolute abundances all declined by about a factor of 2 to 3. Calcium and Potassium behaved like the other low FIP elements, in spite of their unusually low FIPs. Neon appeared to be somewhat overabundant and to vary more dramatically with height compared to the other high-FIP elements. However, the sole neon line available, a Ne IX line at 1248 Å, is extremely sensitive to temperature.

Quiet Sun.   SUMER observations of the quiet Sun between 1.05 and 1.35 $R_\odot$ by Warren (1999) and observations at 1.1 $R_\odot$ by Laming et al. (1999) resulted in FIP biases of 2.3±0.7 and 3-4, respectively. At the base of a coronal streamer at heights between 1.03 and 1.5 $R_\odot$, SUMER observations showed FIP biases of 3-4 (Feldman et al. 1998) based on the abundances of He, N, O, Ne, ,Na, Mg, Si, S, Ca, and Fe. Laming & Feldman (2001) measured the helium abundance at 1.05 $R_\odot$ in the quiet Sun. Their result, 0.052±0.005, is about half the photospheric value, and it is similar to values measured in the slow solar wind.

At solar minimum, a streamer belt associated with the magnetic neutral line encircles the Sun and extends out into the heliosphere to become the slow solar wind. It incorporates active regions, but it is also bright at longitudes where no active regions are present. UVCS observations of quiet Sun streamers at 1.5-1.7 $R_\odot$ showed FIP biases of about a factor of 3 based on the ratios of O, S and Ar to Mg, Al, Si, Ca and Fe (Raymond et al 1997a; Li et al. 1998; Marocchi, Antonucci & Giordano 2001; Uzzo et al. 2003, 2004). Uzzo et al (2003) found the FIP bias to stay constant over several solar rotations in the solar minimum equatorial streamer at 1.7 $R_\odot$.

Coronal Holes.   Observations of coronal holes are more difficult due to their lower brightness at coronal temperatures and to the possible complications of time-dependent ionization balance and non-Maxwellian electron distribution (Ko et al. 1997; Esser, Edgar & Brickhouse 1998;

Edgar & Esser 2000). For heights above about 2 $R_\odot$, the ionization state is frozen in, but too few lines are observable to give a good abundance determination. The spectra at transition region temperatures are nearly indistinguishable from those of the quiet Sun (Vernazza & Reeves 1978; Curdt et al. 2001). SUMER observations by Doschek et al. (1998) and Feldman et al. (1998) below 1.2 $R_\odot$ show essentially no FIP effect. Laming & Feldman (2003) find that the helium abundance below about 0.1 $R_\odot$ varies among several coronal holes, and that the value is at or below the range 0.04-0.05 observed in the fast solar wind. This leads them to conclude that helium is flowing upward faster than hydrogen at these heights in order to match the fluxes observed at 1 AU. Abundances of other elements in polar plumes also appear to be close to photospheric (Teriaca et al. 2001; Antonucci & Giordano 2001).

Prominences. Few abundance determinations are available for prominences. Spicer et al. (1998) used a SUMER spectrum to derive a modest FIP bias of about 2 for one prominence. Mohan et al. (2000) report small scale variations of the FIP bias within a prominence, spanning a range of 1.6 to 8.8. These small scale variations may be related to the small scale, rapid flows seen in prominences (Kucera, Tovar & De Pontieu 2003)

Coronal Mass Ejections. Similarly, there are few abundance determinations available for Coronal Mass Ejections near the Sun. Ciaravella et al. (1997) report a FIP bias near 3 based on doubly ionized species of C and N in one event, indicating that this is prominence material. Ciaravella et al. (2002) and Ko et al. (2003) report normal FIP biases in the current sheets trailing two CMEs, indicating that this higher temperature gas originated in the active region corona.

Average Coronal FIP Bias. A substantial amount of scatter is observed in the FIP bias of different regions, but Figure 1 summarizes typical results. The top panel shows coronal hole abundances relative to the photospheric abundances of Holweger (2001) at a height of 1.1 $R_\odot$ (Feldman et al. 1998; Laming & Feldman 2003). The lower panel shows abundances at the base of a quiet Sun equatorial streamer obtained by SUMER (Feldman et al. 1998), abundances at 1.5 $R_\odot$ in the core of a quiet Sun equatorial streamer obtained with UVCS (Raymond et al. 1997a) and abundances at 1.5 $R_\odot$ in an active region streamer from UVCS (Ko et al. 2002). Dashed lines indicate constant abundances for the coronal hole and factor of 4 FIP biases for the other abundance sets. In summary, the coronal hole shows no FIP bias, while both quiet

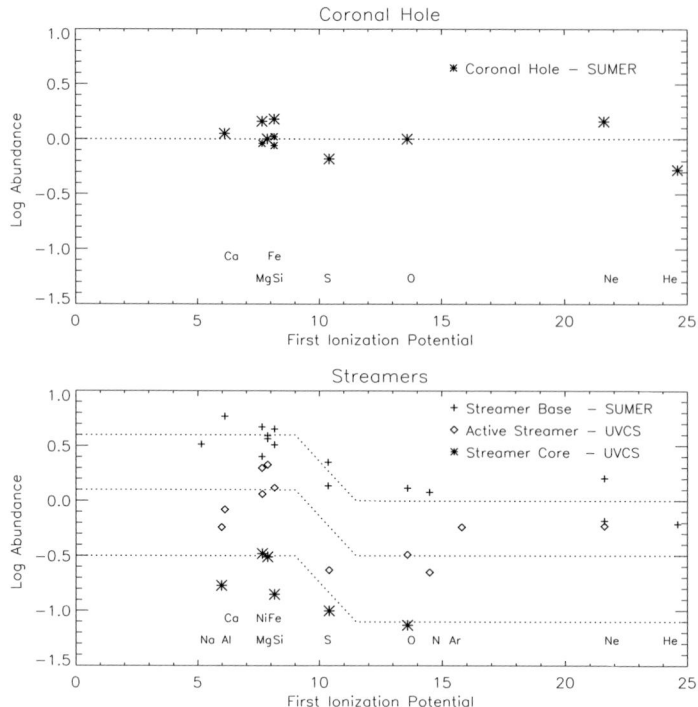

Figure 12.1. Figure 1. Typical coronal abundances in coronal holes, the quiet Sun and active regions.

Sun and active region spectra indicate FIP biases of 3 to 4. There is significant scatter, likely due to observational uncertainty. There is little evidence for the enhanced FIP bias for very low FIP elements predicted by Laming (2004), but some evidence for the intermediate enhancement of S predicted by that model.

Solar Wind.   Solar wind measurements of FIP enhancements show a wide variation, and, as mentioned above, a strong correlation with solar wind speed. The Mg/O abundance ratio is most frequently used (Geiss et al. 1995; von Steiger 1998). Aellig et al. (1999) measured an Fe/O ratio of 0.11±0.03 in generally low speed wind with CELIAS/CTOF. The Fe/O ratio decreased by a factor of 2 when the solar wind speed increased from 350 km/s to 500 km/s. The value at the lowest speed is compatible with the values observed in the equatorial streamer belt at the same time by Raymond et al (1997a). Composition studies of Interplanetary Coronal Mass Ejections (ICMEs) tend to focus on charge state distributions (e.g. Lepri et al. 2001), but Wurz et al. (2001) have derived elemental abundances in 5 magnetic clouds from MTOF measurements. They find variations among the events, but in general oxygen and other low mass elements are depleted relative to hydrogen, while high mass elements are enhanced relative to oxygen. This seems not to be a FIP effect in that argon behaves like calcium and iron. Wurz et al. (1997) present a model for mass fractionation in the pre-CME loops. In an especially well-studied event that was rich in ^{3}He, Gloeckler et al. (1999) report an He/O ratio ten times that of the normal slow solar wind, along with an Fe/O ratio more than twice normal.

The most direct comparisons between coronal and solar wind abundances can be made by observing above the solar limb when ULYSSES is at quadrature, so that one observes the same gas close to the Sun and a few days later at several AU. Parenti et al. (2003) find typical coronal FIP enhancements in the streamer at 1.2 $R_\odot$ from SUMER spectra. Bemporad et al. (2003) find good agreement between Fe/O ratios in the streamers they observed with UVCS at 1.6 and 1.9 $R_\odot$ and in the low speed wind (between 309 and 365 km s^{-1} measured 18 days later by ULYSSES at 1.35 AU, though the uncertainties are large enough to prevent an exact comparison. Ko et al. (2001) attempted to connect UVCS coronal observations with ACE/SWICS solar wind measurements by means of a global solar magnetic field source surface map and an MHD model for the solar wind flow. The comparison proved difficult, perhaps because of line-of-sight effects in the coronal analysis or because of the sensitivity of the traceback from ACE to the solar surface to the details of the solar wind acceleration.

4. Gravitational Settling

The scale height for a metal ion at coronal temperatures is quite small, so the possibility of gravitational settling must be considered. Diffusion models of the transition zone by Roussel-Dupre (1980), Shine, Gerola & Linsky (1975), and Fontenla, Avrett & Loeser (2002) predict very strong changes in absolute abundances resulting from diffusion in the steep temperature gradients of the transition zone. Lenz, Lou & Rosner (1998) modeled the abundance variation along a magnetic loop including diffusion and the effects of the polarization electric field. They found that the Si and Fe abundances near the top of the 10^9 cm loop declined by factors of 10 and 200 relative to abundances at the base if the loop remained in steady state. However, even modest flows along the loop can increase the abundances by orders of magnitude, and models with plasma flow predict relatively constant abundances along the loop (Lenz 2004).

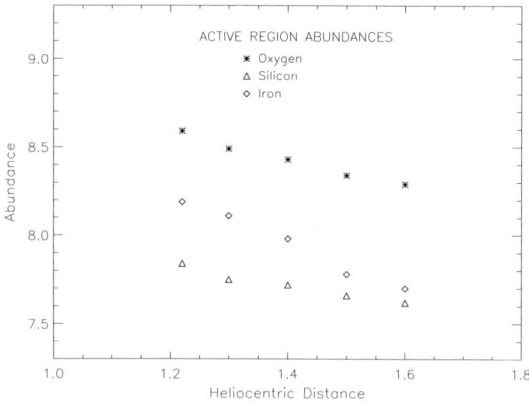

Figure 12.2 Abundances versus height in an active region (Ko et al. 2002).

In flowing plasma of the solar wind, ion drag will tend to carry the heavy elements along with the hydrogen. Ofman (2000) performed multifluid calculations of the flow along open magnetic field lines at the edge of a streamer. He found that Coulomb friction between outflowing protons and oxygen ions maintained an oxygen abundance in the streamer legs several times higher than in the closed field region of the streamer core.

Gravitational settling was proposed as an explanation for the severe depletion observed in the equatorial streamer belt at solar minimum (Raymond et al. 1997a). The UVCS observations at 1.5 $R_\odot$ showed oxygen abundances about 1/3 photospheric along the edges of the streamer, but only about 1/10 photospheric in the streamer core. There was a FIP enhancement of about 3 in low-FIP to high-FIP abundance ratios, but the relative abundances were essentially the same in the streamer core and streamer legs. The change in absolute abundances was attributed to a difference between the open field lines of the streamer legs and the closed field of the streamer core, suggesting that heavy elements in the closed field would settle toward the solar surface. Alternatively, Noci et al. (1997), suggest a multipolar field structure in which the field lines are open near the streamer center. This interpretation requires some other explanation, such as variation in the ion drag as a result of the velocity variation near the center of the streamer, to explain the low abundances in the core.

Elemental abundances appear more normal near the solar surface, in that Feldman et al. (1998) report a normal FIP enhancement and nearly photospheric absolute oxygen abundance based on SUMER spectra. They also find that Fe lines drop off with height more quickly than Si lines, as expected for gravitational settling of the heavier elements. Ko et al. (2002) showed that absolute abundances of all the elements measured (N, O, Ne, Mg, Al, Si, S, Ca, Ar, Fe) declined by about factors of 2 to 5 between 1.2 and 1.6 R$_\odot$ above a bright active region. Parenti et al. (2000) also find depleted abundances at 1.6 R$_\odot$ in several streamers. Figure 2 shows the abundances of O, Si and Fe as functions of heliocentric distance from Ko et al. (2002). The clear decline in abundance with height is qualitatively consistent with the Lenz, Low & Rosner (1998) picture, but the magnitude the the decline is much smaller. This suggests that flows compete with diffusion. As the diffusion settling time is about one day, flows must replenish the coronal gas on about that time scale. There is as yet no clear indication of whether steady flows of the type envisioned by Lenz (2004) or impulsive events of the nanoflare type envisioned by Arge & Mullan (1998) provide the coronal metals. It is conceivable that anisotropic temperature distributions like those in coronal holes also help to support the ions (see Cranmer et al. 1999), but severe broadening of the oxygen lines only sets in at larger heights (Frazin, Cranmer & Kohl 2003).

The low abundance cores seem to be a quite common phenomenon. Streamers that do not show such a core may be fundamentally different, but is is quite likely that they are merely viewed at an unfavorable angle, or that they are observed above the streamer cusp, where no closed field exists. Marocchi et al. (2001) and Uzzo et al. (2003, 2004) have found depletions in streamer cores of roughly a factor of 3 relative to the abundances in the streamer legs.

The low abundance cores of streamers can be identified up to 2.5 or 3 R$_\odot$ in streamers with favorable viewing geometry. This may be a way to identify the magnetic field cusp if the low abundance really does trace the closed magnetic field region. There have been some attempts to determine abundances above 2 R$_\odot$. Zangrilli et al. (2001) and Marocchi et al. (2001) found a continued decline in oxygen abundance at larger heights. It becomes very difficult to separate core and leg contributions much above 1.5 R$_\odot$, however, because projection effects mix the two contributions along the line of sight. Raymond et al (1997b) used an empirical description of the solar minimum equatorial streamer to estimate that emission from the projected legs of the streamer contributed 20% of the observed O V intensity at 1.5 R$_\odot$, 50% at 1.75 R$_\odot$ and 80% at 2 R$_\odot$ for the most favorable case, an arcade viewed end-on. Vasquez, Raymond & van Ballegooijen (2004) find that projection may account for even larger fractions of the core intensities, especially at larger heights. If the streamer is viewed perpendiculate to the magnetic arcade, projection effects will be impossible to disentangle.

Absolute abundance measurements in the solar wind are less common than relative abundance measurements. Wimmer-Schweingruber (1994) obtained H/O = 1890±600 in the slow solar wind and 1590±500 in the fast wind. These values are not significantly different, and they are equal to the photospheric value to within the uncertainties. They are in keeping with the association of fast solar wind with coronal holes, but then one would expect an H/O ratio at least a factor of 2 higher if the slow solar wind originates in the legs of streamers. The discrepancy is surprising, in that the FIP enhancements support these associations (e.g. Geiss et al. 1995). While it is possible that the slow wind originates in open field lines just outside the regions that appear as bright streamers, it is also possible that there is a systematic error in either the coronal or *in situ* value, or that some mixing between fast and slow wind regions occurs in either the coronal measurements (by projection effects) or the *in situ* measurements (by temporal averaging or by actual mixing of solar wind flows). The extremely low oxygen abundance in streamer cores compared with solar wind values implies that plasma from the cores makes at most a small contribution to the slow solar wind (Raymond et al. 1997a).

5. Comparison with Other Stars

Abundance measurements in other stars shed some light on the mechanism for the FIP effect. Drake (2003) provides a review of elemental fractionation in the coronae of other stars based on EUV and X-ray observations. FIP enhancements comparable to those seen in solar active regions have been reported in some stars based on EUVE spectra (ξ Boo, Drake & Kashyap 2001) and grating spectra from XMM and *Chandra* (α Cen, Raassen et al. 2003; AB Dor, Sanz-Forcada, Maggio & Micela 2003), but no FIP effect is seen in other stars (Procyon, Raassen et al. 2002). On the other hand, an inverse FIP effect, with low FIP elements depleted relative to high-FIP elements has been reported in some cases (e.g. HR1099, Drake et al. 2001; AU Mic, Drake 2002). In other cases, the abundances seem to scatter with no clear FIP dependence (II Peg, AR Lac; Huenemoerder et al. 2001, 2003). Drake (2002) suggests that stars with solar-like activity levels show solar-like FIP patterns, while more active stars show Fe depletions of factors of 3 to 10. Audard et al. (2003) report that highly active RS CVn stars show an inverse FIP effect, while intermediate activity RS CVn show a possible solar-like FIP bias.

Clearly these abundances are averages over the entire stellar coronae, but they are strongly weighted toward the highest density, most active emission regions. None of the models proposed to explain the solar FIP effect seems likely to be able to deal with the inverse FIP effect or the more scattered abundance patterns of II Peg or AR Lac except for that of

Laming (2004), which may produce an inverse FIP effect in low gravity stars.

6.   Summary

The pattern of elemental abundances varies considerably among different regions in the solar corona and solar wind, but other stars show other abundance patterns still. These abundances must be considered when deriving other physical quantities from X-ray or EUV observations. The elemental fractionation also presents a potentially powerful means for understanding the heating of stellar chromospheres and coronae. However, there is as yet no clear consensus about the mechanism that drives the FIP effect, and it seems not yet possible to use abundance measurements to learn about the physical processes that heat the corona and drive the solar wind.

References

Aellig, M.R., S., Grnwaldt, H., Bochsler, P., Wurz, P., Ipavich, F.M., & Hovestadt, D. 1999, JGR, 104, 24769

Antonucci, E., & Giordano, S. 2001, in *Solar and Galactic Composition*, R. Wimmer-Schweingruber, ed. (Melville, NY: AIP), p. 77

Arge, A.N., & Mullan, D.J. 1998, Sol. Phys. 182, 293

Audard, M., Gdel, M., Sres, A., Raassen, A.J.J. & Mewe, R. 2003, A&A, 398, 1137

Bame, S.J., et al. 1992, A&A S, 92, 237

Bemporad, A, Poletto, G., Suess, S. T., Ko, Y. K., Parenti, S., Riley, P., Romoli, M., and Zurbuchen, T. 2003, ApJ, 593, 1146

Borrini, G., Gosling, J.T., Bame, S.J., & Feldman, W.C. 1983, Sol. Phys. 83, 367

Chen, G.X., & Pradhan, A.K. 2002, PhRvL, 286, 347

Ciaravella, A., et al. 1997, ApJ, 491, L59

Ciaravella, A., et al. 2002, ApJ, 575, 1116

Cranmer, S.L., et al. 1999, ApJ, 511, 481

Curdt, W., Brekke, P., Feldman, U., Wilhelm, K., Dwivedi, B. N., Schhle, U., Lemaire, P. 2001, A&A, 375, 591

Del Zanna, G. 2003, A&A, 406, L5

Del Zanna, G., Bromage, B.J.I., & Mason, H.E. 2003, A& A, 398, 743

Del Zanna, G. & Mason, H.E. 2003, A& A, 406, 1089

Doschek, G.A., Feldman, U., & Seely, J.F.1985, MNRAS, 217, 317

Doschek, G.A., Laming, J.M., Feldman, U., Wilhelm, K., Lemaire, P., Schhle, U. & Hassler, D.M. 1998, ApJ, 504, 573

Drake, J.J. 2002, in Stellar Coronae in the Chandra and XMM Era, ASP Conf. Ser. vol 277, F. Favata and J.J. Drake, eds (San Francisco: ASP), p. 75

Drake, J.J. 2003, Adv. Sp. Res., 32, 6, 945

Drake, J.J., & Kashyap, V. 2001, ApJ, 547, 428

Drake, J.J., Brickhouse, N.S., Kashyap, V., Laming, J. M, Huenemoerder, D.P., Smith, R. & Wargelin, B.J. 2001, ApJ, 548, L41

Dupree, A.K., Moore, R.T., & Shapiro, P.R. 1979, ApJL, 229, L101

Edgar, R.J., & Esser, R. 2000, ApJL, 538, L167

Esser, R., Edgar, R.J., & Brickhouse, N.A. 1998, ApJ, 498, 448

Falconer, D.A., Davila, J.M., & Thomas, R.J. 1997, ApJ, 482, 1050

Feldman, U., Doschek, G.A., & Kreplin, R.W. 1980, ApJ, 238, 365

Feldman, U., Schhle, U., Widing, K.G., and Laming, J.M. 1998, ApJ, 505, 999

Feldman, U., Doschek, G.A., Schhle, U., & Wilhelm, K. 1999, ApJ, 518, 500

Fludra, A., and Schmelz, J.T. 1995 ApJ, 447, 936

Fludra, A., and Schmelz, J.T. 1999, A& A 348, 286

Fontenla, J.M., Avrett, E.H., & Loeser, R. 2002, ApJ, 572, 636

Frazin, R.A., Cranmer, S.R., & Kohl, J.L. 2003, ApJ, 597, 1145

Frhlich, C., Huber, M.C.E., Solanki, S.K., & von Steiger, R. 1998, *Solar Composition and its Evolution–from Core to Corona* (Dordrecht: Kluwer)

Geiss, J., et al. 1995, Science, 268, 1303

Gloeckler, G., et al. 1999, GRL, 26, 157

Hansteen, V.H., Leer, E., & Holzer, T.E. 1997, ApJ, 482, 498

Hnoux, J.-C. 1998, SSRV, 85, 215

Hnoux, J.-C., & Somov, B.V. 1997, A&A, 318, 947

Holweger, H., in *Solar and Galactic Composition*, R. Wimmer-Schweingruber, ed. (Melville, NY: AIP), p. 23

Hovestadt, D., et al. 1995, Sol. Phys. 162, 441

Huenemoerder, D.P., Canizares, C., & Schulz, N.S. 2001, ApJ, 559, 1135

Huenemoerder, D.P., Canizares, C., Drake, J.J., & Sanz-Forcada, J. 2003, ApJ, 595, 1131

Judge, P.G., & Peter, H. 1998, SSRV, 85, 187

Ko, Y.-K., Fisk, L.A., Geiss, J., Gloeckler, G., & Guhatakurtha, M. 1997, Sol. Phys. 171, 345

Ko, Y.-K., Raymond, J., Li, J., Ciaravella, A., Michels, J., Fineschi, S., & Wu, R. 2002, ApJ, 578, 979

Ko, Y.-K., Raymond, J., Lin, J., Lawrence, G., Li, J., and Fludra, A. 2003, ApJ, 594, 1068

Ko, Y.-K., Zurbuchen, T., Strachan, L., Riley, P., & Raymond, J.C. 2001, in Correlated Phenomena at the Sun, in the Heliosphere and in Geospace, ESA SP-415 (Noordwijk: ESA), p.133

Kohl, J.L., et al. 1997, Sol. Phys. 175, 613

Kohl, J.L., et al. 1998, ApJL, 501, L27

Kucera, T.A., Tovar, M., & De Pontieu, B. 2003, Sol. Phys., 212, 81

Laming, J.M. 2004, preprint

Laming, J.M., & Feldman, U. 1994, ApJ, 426, 414

Laming, J.M., Feldman, U., Drake, J.J., & Lemaire, P. 1999, ApJ, 518, 926

Laming, J.M., & Feldman, U. 2001, ApJ, 546, 552

Laming, J.M., & Feldman, U. 2003, ApJ, 591, 1257

Lenz, D., Lou, Y,-Q, & Rosner, R. 1998, ApJ, 504, 1020

Lenz, D.D. 2004, ApJ, 604, 433

Lepri, S.T., Zurbuchen, T.H., Fisk, L.A., Richardson, I.G., Cane, H.V. & Gloeckler, G. 2001, JGR, 106, 29231

Li, J., et al. et al. 1998, ApJ, 506, 431

Liedahl, D.A., Osterheld, A.L., & Goldstein, W.H. 1995, ApJ Lett., 438, L115

Marocchi, D., Antonucci, E., & Giordano, S. 2001, Annales Geophysicae, 19, 135

Marsch, E., von Steiger, R., & Bochsler, P. 1995, A& A, 301, 261

McComas, D.J., Bame, S.J., Barker, P., Feldman, W.C., Phillips, J.L., Riley, P., & Griffee, J.W. 1998, SSRv, 86, 563

McKenzie, D.L., & Feldman, U. 1992, ApJ, 389, 764

McKenzie, J.F., Sukhorukova, G.V., & Axford, W.I. 1998, A& A, 332, 367

Meyer, J.P. 1985a, ApJ Suppl., 57,151

Meyer, J.P. 1985b, ApJ Suppl., 57, 173

Mohan, A., Landi, E., & Dwivedi, B.N. 2000, A&A, 364, 835

Murphy, R.J., Share, G.H., Grove, J.E., Johnson, W.N., Kinzer, R.L., Kurfess, J.D., Strickman, M.S., & Jung, G.V. 1997, ApJ, 490, 883

Noci, G., et al. 1997, in *The quiescent corona and the slow solar wind, Fifth SOHO workshop*, ESA, SP-404, p. 75

Ofman, L. 2000, GRL, 27, 2885

Parenti ,S., Bromage, B.J.I., Poletto, G., Noci, G., Raymond, J.C., & Bromage, G.E. 2000, A& A, 363, 800

Parenti ,S., Landi, E., & Bromage, B.J.I. 2003, ApJ, 590, 519

Peter, H. 1998, A& A, 335, 691

Phillips, K.J.H., & Feldman, U. 1991, ApJ, 379, 401

Phillips, K.J.H., Greer, C.J., Bhatia, A.K., Coffey, I.H., Barnsley, R., & Keenan, F.P. 1997, A&A 324, 381

Phillips, K. J. H., Sylwester, J., Sylwester, B., & Landi, E. 2003, ApJL 589, 113

Raassen, A.J.J., Mewe, R., Audard, M., Güdel, M., Behar, E., Kaastra, J. S., van der Meer, R.L.J., Foley, C.R. & Ness, J.-U. 2002, A&A, 389, 228

Raassen, A.J.J., Ness, J.-U., Mewe, R., van der Meer, R.L.J., Burwitz, V. & Kaastra, J.S. 2003, A&A, 400, 671

Raymond, J.C. 1990, ApJ, 365, 387

Raymond, J.C., Cox, D.P., & Smith, B.W. 1976, ApJ, 204,290

Raymond, J.C., et al. 1997a, Sol. Phys., 175, 645

Raymond, J., Suleiman, R., van Ballegooijen, A., & Kohl, J. 1997b, in Correlated Phenomena at the Sun, in the Heliosphere and in Geospace, ESA SP-415 (Noordwijk: ESA), p. 383

Raymond, J.C., et al. 2001, in *Solar and Galactic Composition,* R. Wimmer-Schweingruber, ed. (Melville, NY: AIP), p. 491

Roussel-Dupre, R. 1980, ApJ, 241, 402

Saba, J.L.R., Schmelz, J.T., Bhatia, A.K., & Strong, K.T. 1999, ApJ, 510, 1064

Sanz-Forcada, J., Maggio , A. & Micela, g. 2003, A&A, 408, 1087

Schmelz, J.T., Saba, J.L.R., Ghosh, D., & Strong, K.T. 1996, ApJ, 473, 519

Schwadron, N.A., Fisk, L.A., & Zurbuchen, T.H. 1999, ApJ, 521, 859

Share, G.H., & Murphy, R.J. 1995, ApJ, 452, 933

Sheeley, N.R., Jr. 1995, ApJ, 440, 884

Shine, R., Gerola, H., & Linsky, J.L. 1975, ApJL, 202, 101

Spadaro, D., Orlando, S., Peres, G. & Leto, P. 1995, A&A, 302, 285

Spicer, D., Feldman, U., Widing, K.G., & Rilee, M. 1998, ApJ, 494, 450.

Sylwester, J., Lemen, J.R., Bentley, R.D., Fludra, A., & Zolcinski, M.-C. 1998, ApJ, 501, 397

Teriaca, L., Poletto, G., Falchi, A., & Doyle, J.G. 2001, in *Solar and Galactic Composition,* R. Wimmer-Schweingruber, ed. (Melville, NY: AIP), p. 65

Uzzo, M., Ko, Y.-K., Raymond, J.C., Wurz, P., & Ipavich, F.M. 2003, ApJ, 585, 1062

Uzzo, M., Ko, Y.-K., & Raymond, J.C. 2004, ApJ, 603, 760

Vasquez, A., Raymond, J.C., & van Ballegooijen, A. 2004, submitted to ApJ.

Vauclair, S. 1996, A&A, 308, 228

Vernazza, J.E., Avrett, E.A., & Loeser, R. 1981, ApJS, 45, 635

Vernazza, J.E., & Reeves, E.M. 1978, ApJS, 37, 485

von Steiger, R. 1998, SSRv, 85, 407

von Steiger, R., & Geiss, J. 1989, A&A, 225, 222

von Steiger, R., Schwadron, N.S., Fisk, L.A., Geiss, J., Gloeckler, G., Hefti, S., Wilken, B., Wimmer-Schweingruber, R.F., & Zurbuchen, T. 2000, JGR, 105, 27217

Wang, Y.-M. 1996, ApJ, 464, L91

Warren, H.P. 1999, Sol. Phys., 190, 363

Widing, K.G., & Feldman, U. 2001, ApJ, 555, 426

Weiler, R. 1998, SSRV, 85, 303

White, S.M., Thomas, R.J., Brosius, J.W., & Kundu, M.R. 2000, ApJL, 534, L203

Wimmer-Schweingruber, R.F. 1994, PhD Thesis, University of Bern

Wimmer-Schweingruber, R.F. *Solar and Galactic Composition*, R. Wimmer-Schweingruber, ed. (Melville, NY: AIP)

Wood, K., & Raymond, J. 2000, ApJ, 540, 563

Wurz, P., et al. 1997, GRL, 25, 2557

Wurz, P.; Wimmer-Schweingruber, R.F., Issautier, K., Bochsler, P., Galvin, A.B., Paquette, J.A. & Ipavich, F. M. 2001, in *Solar and Galactic Composition*, R. Wimmer-Schweingruber, ed. (Melville, NY: AIP), p. 145

Young, P.R., & Mason, H.E. 1997, Sol. Phys., 175, 523

Zangrilli, L., Poletto, G., Biesecker, D., & Raymond, J.C. 2001, in*Solar and Galactic Composition*, R. Wimmer-Schweingruber, ed. (Melville, NY: AIP), p. 71

Zurbuchen, T.A., Fisk, L.A., Gloeckler, G., & von Steiger, R. 2002, GRL, 29, 66

Chapter 13

THE MAGNETIC FIELD FROM THE SOLAR INTERIOR TO THE HELIOSPHERE

Sami K. Solanki

Max-Planck-Institut für Sonnensystemforschung*,
37191 Katlenburg-Lindau, Germany

solanki@linmpi.mpg.de

Abstract   The Sun's magnetic field permeates all solar layers from the bottom of the convection zone out to the edge of the heliosphere. Through its continuous interaction with the Sun's flow field, the magnetic field evolves, but also influences these flows. In the solar atmosphere the field also affects the thermal state of the gas, producing such diverse phenomena as dark sunspots in the photosphere, plages in the chromosphere and transition region and x-ray loops in the corona, in addition to a rich collection of dynamic and explosive phenomena. Here a brief overview is given of the magnetic field as it threads it way through all these layers. Emphasis is placed on considering the magnetic field in relation to other quantities from one layer to the other. Results of recent investigations are employed to illustrate the magnetic structure and the processes described here.

Keywords: The Sun: magnetic field - convection zone - photosphere - chromosphere - corona - heliosphere

1.   Introduction

The Sun's magnetic field threads its way from the bottom of the convection zone to the edge of the heliosphere. On the way it causes and controls a great variety of phenomena while in turn being plied and

---

*Previously known as Max-Planck-Institut für Aeronomie

G. Poletto and S.T. Suess (eds.), The Sun and The Heliosphere as an Integrated System, 373–395.
© 2004 *Kluwer Academic Publishers. Printed in the Netherlands.*

stretched by such processes as convection, differential rotation and the solar wind.

Due to this interaction and the central role played by the magnetic field this review perforce touches upon a large chunk of solar and heliospheric physics. Limited space (to say nothing of the author's laziness) prevents an in-depth coverage of these topics. Instead I'll briefly touch upon a few aspects and refer to other publications for more details. This overview is structured geometrically, moving from deeper or lower solar layers to outer or higher ones. This has a certain logic to it, since the magnetic field is generated in the solar interior, from where it rises to the solar surface and thence spreads into the Sun's atmosphere. There a fraction of the field lines is pulled out into the heliosphere and 'opened' by the solar wind, i.e. stretched until they reach the local interstellar medium.

In almost all cases, it is the lower layers which affect the higher layers. For example, the magnetic field in the photosphere determines to a large extent the magnetic structure of the corona. Consequently, it makes sense to discuss the lower layers first.

2.   Solar Interior

The generally accepted idea is that a dynamo located at the base of the convection zone is responsible for the magnetic flux appearing at the solar surface in, e.g., active regions (Schmidt 1993; Ossendrijver 2003). In theoretical studies the dynamo is generally placed in the overshoot layer below the convection zone for stability reasons: it is extremely difficult to store sufficiently large amounts of magnetic flux inside the convection zone without this field becoming unstable to buoyancy (Parker 1975; Schüssler 1979), although Nordlund et al. (1992) have argued that convection, in particular rapid downflows, can pump magnetic flux downwards and store significant amounts of flux in the lower convection zone.

The evidence that the dynamo is located near the convection zone's base is strengthened by recent determinations of the Sun's internal rotation. Such studies, based upon data obtained by the Michelson Doppler Interferometer (MDI; Scherrer et al. 1995) and the Global Oscillations Network (GONG); Harvey et al. 1996), have revealed a strong shear layer between the differentially rotating convection zone and the solidly rotating radiative core (Schou et al. 1992, 1998; Tomczyk et al. 1995; Thompson et al. 2003). The differential rotation deduced from MDI data is plotted in Fig. 1. While the nature, the origin and the location of the helical motions that regenerate the magnetic field are still controversially discussed (Schmitt 1987, 2003; Brandenburg and Schmitt 1998; Dikpati and Gilman 2001; Parker 1993; Durney 1995), there is increasing support for the idea that the meridional circulation might play an important role in transporting/redistributing the general magnetic flux (Choudhuri et al. 1995; Durney 1995, 1996, 1997; Dikpati and Charbonneau 1999; Nandy and Choudhuri 2001; Küker et al. 2001; Hathaway et al. 2003). There has also been considerable progress in determining the structure and properties of the meridional circulation, seen as a poleward flow of gas at the solar surface (Hathaway 1996; Duvall and Gizon 2000).

The advantage of storing the field in the overshoot layer is that field strengths much stronger than those in equipartition with the kinetic energy of the convection can be built up and stored before the field succumbs to the Parker instability (Ferriz-Mas and Schüssler 1993, 1995).

Such super-equipartion fields are required in order to reproduce various properties of active regions including the correct emergence latitudes, the tilts relative to the East-West direction of the axes connecting the opposite polarities (Joy's law) (Fan et al. 1993, 1994; D'Silva and Choudhuri 1993; Schüssler et al. 1994; Caligarie et al. 1995, 1998).

Not all the magnetic flux that leaves the overshoot layer in the direction of the surface manages to reach it. If the field is too weak it is susceptible to the influences of the convective flows. These can brake the rise of the field and can distort the flux tube (Schüssler 1979). In extreme cases this can even lead to the 'explosion' and destruction of magnetic flux tubes (Rempel and Schüssler 2001).

3.   Solar Surface

After the magnetic flux has survived the dangers of a passage through the convection zone, it appears at the solar surface as a bipolar magnetic region. At this stage the magnetic field becomes accessible to direct observation. For example, magnetograms mapping the line-of-sight component of the field or the full magnetic vector can be recorded (an example of a full-disk line-of-sight magnetogram is shown in Fig. 2). A greater amount of information can be obtained from spectropolarimetric observations, which allow more reliable determinations of the magnetic vector (Lites et al. 1994; cf. Landi Degl'Innocenti 1992) and also enable the depth dependence of important atmospheric parameters to be determined (e.g. Solanki 1986; Keller et al. 1990; Ruiz Cobo and Del Toro Iniesta 1992; Frutiger et al. 1990; Westendorp Plaza 1997; Socas Navarro et al. 1999).

The largest and most striking bipolar regions are the active regions appearing in the activity belts. These only form the large-scale end of a broad distribution of sizes of bipolar regions of freshly emerged

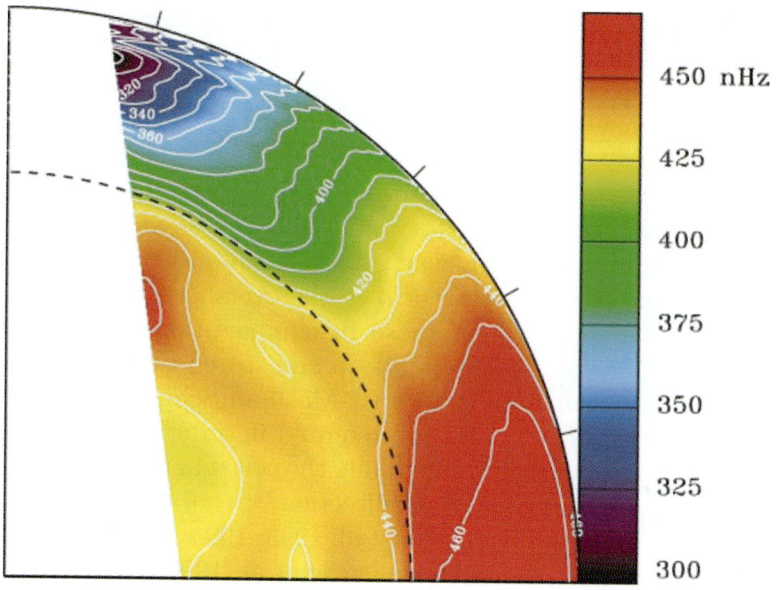

Figure 13.1. Cutaway through the Sun showing rotation frequency (colour coded and iso-frequency contours). The vertical axis is the polar axis, the horizontal axis corresponds to the solar equator, the outer circle (solid) represents the solar surface, the inner circle the bottom of the solar convection zone. The iso-frequency contours running along the convection zone boundary indicate the shear layer (from Thompson et al. 2003, by permission).

flux. Much more common than the active regions are the ephemeral active regions and although each of these individually carries orders of magnitude less flux than an active region, their larger number means that the flux emergence rate in ephemeral regions $(2-4 \times 10^{26} \mathrm{Mx/yr})$ is orders of magnitude larger than in active regions $(3 \times 10^{23} - 3 \times 10^{24} \mathrm{Mx/yr})$ (Harvey 1993). This is compensated somewhat by the fact that the lifetime of the flux emerging in the ephemeral regions is 1–2 orders of magnitude shorter (tens of hours; Harvey 1993; Schrijver et al. 1998; Hagenaar et al. 2003) than of active regions (on the order of months).

At the solar surface the magnetic field is filamented into concentrations having sizes ranging from tens of thousands of km (sunspots; for an overview see Thomas and Weiss 1992; Schmieder et al. 1997; Solanki 2003) to below the spatial resolution (magnetic elements; reviewed by Solanki 1993; Stenflo 1994). Although the magnetic field strength averaged over the solar surface is a few tens of Gauss and can vary by large amounts between the quiet Sun and active regions, individual magnetic concentrations often have kG intrinsic field strengths (Hale 1908; Stenflo

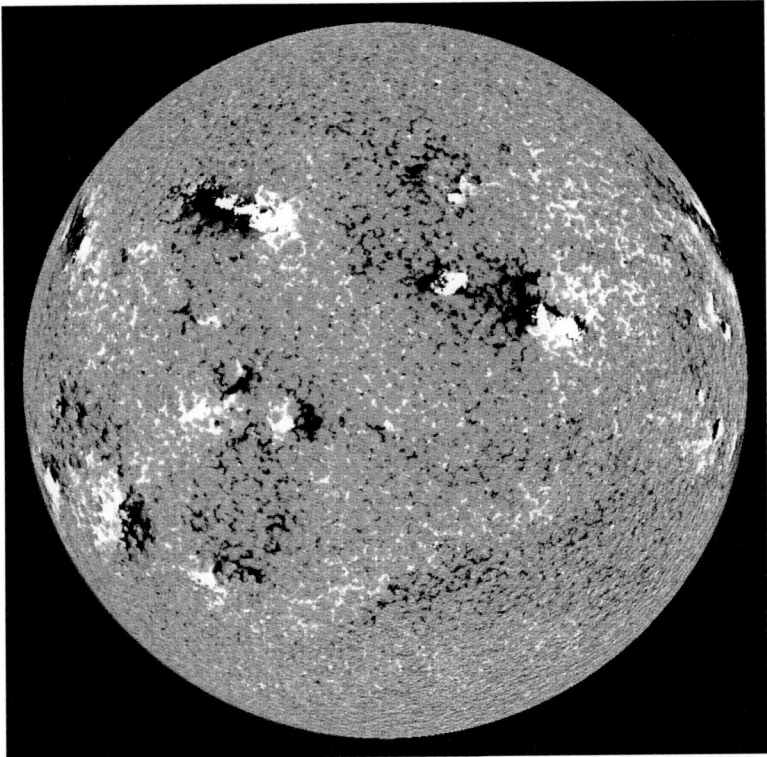

Figure 13.2. Full-disk magnetogram recorded by the Michelson Doppler Interferometer (MDI). Grey areas exhibit little net line-of-sight field, while white and black correspond to opposite polarities. The plotted magnetogram was obtained after integrating over 56 single magnetograms (after compensating for solar rotation) in order to reduce the noise and reveal weaker magnetic features.

1973; Rabin 1992; Rüedi et al. 1992; Solanki 1993). These magnetic concentrations, many of whose global features can be described by simple flux-tube models (e.g. Defouw 1976; Spruit 1976; Deinzer et al. 1984a, b; Jahn and Schmidt 1994; Zayer et al. 1989; Steiner et al. 1986), on closer inspection reveal a highly complex and dynamic (internal) structure. For sunspots this has been deduced directly from observations (e.g., Scharmer et al. 2002; see Sobotka 1997; Solanki 2003 for a review). For the unresolved magnetic elements time series at the highest spatial resolution only give a flavour of the complex internal structure and dynamics (Keller 1992; Berger et al. 1998; Muller et al. 1994). For a more detailed view we have to take recourse to simulations (e.g. Steiner et al.

1998; Gadun et al. 2001; Vögler and Schüssler 2002; Bercik et al. 2003; Vögler 2003).

The latest 3-D simulations of Vögler and co-workers have now reached a level of realism at which they can reproduce sensitive observations with high precision. For example, synthetic G-band images (a spectral region around 430 nm harbouring many CH lines) match the statistical properties and contrasts of observed images after the former have been spatially degraded to the same spatial resolution as the observational data (Schüssler et at. 2003; Shelyag et al. 2004). The comparison is shown in Fig. 3. A detailed analysis reveals that although the temperature at a given height in small magnetic flux concentrations is lower than in the surroundings, they appear brighter (since we observe deeper, hotter layers due to evacuation). Also, it is noticeable that the narrowest magnetic flux sheets are considerably narrower than the intergranular lanes in which they are located (this is visible in both, brightness and velocity maps - the latter not shown in Fig. 3). Thus, almost static gas is co-located with the intense magnetic field, which is bordered on both sides by strong downflows. All these properties are representative of the traditional flux-tube (or flux-sheet; Deinzer et al. 1984a, b) model of magnetic flux concentrations. Other properties predicted by models of thin flux sheets are also seen. These include horizontal pressure equilibrium, the presence of heat influx through the walls and a rapid expansion of the field with height. In addition, the simulations show new features, e.g., in connection with the evolution of magnetic features.

For the largest magnetic structures in the solar photosphere, sunspots, the situation is in some ways reversed. Sunspots are well resolved by observations, which reveal a wealth of detail at all spatial scales. Even the smallest observable scales are important for at least some of the global properties of the sunspot. For example, the complex fine-scale structure of the magnetic and velocity fields in the penumbra (Schmidt et al. 1992; Title et al. 1993; Solanki and Montavon 1993; Westendorp Plaza et al. 1997; Bellot Rubio et al. 2003; Solanki and Rüedi 2003) is responsible for the brightness of the penumbra as a whole (Jahn and Schmidt 1994; Schlichenmaier and Solanki 2003). This makes the self-consistent modelling of sunspots extremely difficult, since a very large range of spatial and temporal scales must be considered simultaneously.

The magnetic field concentrated into sunspots, faculae and the network is also responsible for causing changes in solar irradiance variations. This has been shown convincingly for cycle 23 by Krivova et al. (2003), who employed spectra computed using atmospheric models of these magnetic structures (Unruh et al. 1999) applied to MDI magnetograms in order to reproduce VIRGO (Variability of IRradiance and

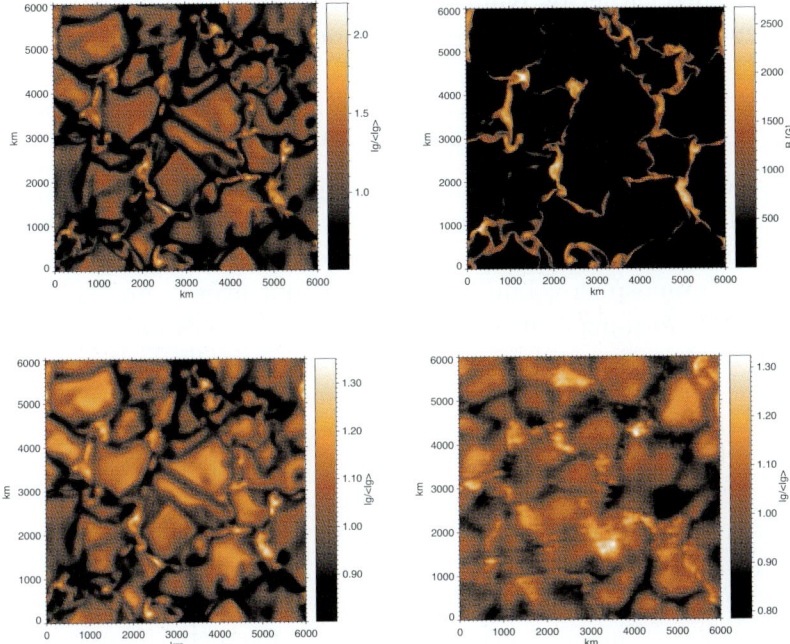

Figure 13.3. Simulated and observed images of the G-band contrast (local G-band brightness divided by spatial average: $I_g/\langle I_g \rangle$) and magnetic structure at the solar surface. Upper left: Synthetic filter image of the simulation area ($6000 \times 6000 \text{ km}^2$) in the G-band spectral region around 430 nm. The extended bright regions are convective upwellings (granules) surrounded by a network of dark downflow lanes. The small brilliant patches in the dark areas coincide with magnetic flux concentrations. Upper right: Gray shading of the magnetic field strength (in units of Gauss) at the same time step as the G-band contrast image. About two-thirds of the vertical magnetic flux penetrating the simulation box has been assembled into flux concentrations with a field strength above 1000 G (100 mT). Lower left: G-band image after spatial smoothing mimicking the diffraction by a telescope of 1m aperture and the image degradation by the Earth's atmosphere. Lower right: Observed G-band image of the same size as the simulated image with about the same area fraction of G-band bright points as in the simulation (subfield of an image that was taken with the SST on La Palma, courtesy of the Royal Swedish Academy of Sciences). Figure from Schüssler et al. (2003).

Gravity Oscillations; Fröhlich et al. 1995) total and spectral irradiance measurements. They found that on time scales up to the solar cycle, surface magnetic features explain over 90% of the measured total irradiance variations. This conclusion has recently been supported by work done by Livingston and Wallace (2003); Walton et al. (2003) and Woodard and Libbrecht (2003). In particular, the latter authors argue that the conclusion reached earlier by Kuhn et al. (1988) that non-magnetic changes are the main cause of irradiance variations over the solar cycle is not tenable.

4.   Chromosphere and Corona

The character of the magnetic field changes with height in the atmosphere. The dominating magnetic structures in the photosphere are best described by the flux-tube (or flux sheet) model. This still holds in the lower chromosphere, but the field fans out ever more rapidly with height, finally forming a magnetic canopy, i.e. a layer of almost horizontal field overlying a nearly field-free atmosphere (Giovanelli and Jones 1982; Jones and Giovanelli 1983; Solanki and Steiner 1990). Although it is agreed that the field from different photospheric sources must merge at some height, there is some controversy about the exact value of this height. The situation is also made more complicated by the fact that many of the field lines connect to nearby regions of opposite magnetic polarity, forming low-lying magnetic loops. In any case, in the upper chromosphere the magnetic field is more homogeneous than in the photosphere below. This is clearly borne out by magnetic maps and scans made in He I 10830 Å (Harvey and Hall 1971; Rüedi et al. 1995; Solanki et al. 2003). By the height at which the He I 10830 Å line is formed, near the top of the chromosphere in standard, plane parallel and time independent atmospheric models (Avrett et al. 1994, cf. Fontenla et al. 1993), the magnetic energy density dominates over the thermal energy density almost everywhere. Above that height the structure of the field depends on the distribution in the photosphere of the magnetic flux and of the magnetic field direction (in particular the polarity). In bipolar regions (e.g. active regions) and in mixed polarity regions the dominant structure in the overlying corona is the magnetic loop. If a single magnetic polarity dominates over a sufficiently large area then a large fraction of the field lines are open, i.e. stretched out into the heliosphere by the solar wind. Figure 4 illustrates the general components of the magnetic structure in the solar atmosphere (adapted from Zwaan and Cram 1989). Note that at any given time only a very small fraction (of the order of a few percent) of the Sun's total magnetic flux is in the form of open field, so that the heliospheric field is not necessarily representative of the Sun's total flux. In other words, almost all of the magnetic field lines leaving the Sun's surface, turn around within a few solar radii and return to the solar interior.

Open and closed magnetic flux manifests itself very differently in its respective radiative properties. Almost all of the EUV and X-ray radia-

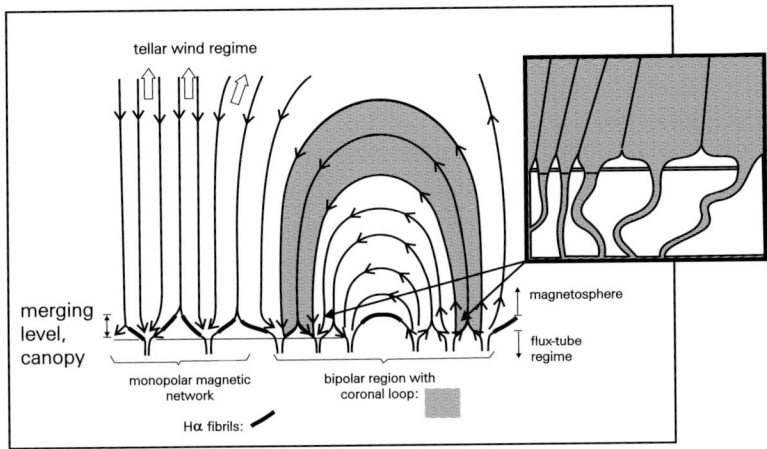

Figure 13.4. Illustration of concepts of the magnetic structure in stellar atmospheres, showing the field rooted in strong flux bundles in the convective zone. The flux-tube domain of discrete flux bundles, separated by nearly field-free plasma, extends upward through the photosphere and into the low chromosphere. In the upper chromosphere the flux tubes merge, creating the magnetic canopy. The magnetosphere, defined as the domain above the merging level, is pervaded by magnetic field. The stellar wind flows along the magnetic field wherever it is (temporarily) open to interstellar space. The inset illustrates that the magnetic field in the photosphere in concentrated in many small fibrils which compose each of the larger elements (e.g. network elements) plotted in the main figure (figure adapted from Zwaan and Cram, 1989).

tion emitted by the Sun comes from closed field lines. I.e., almost all the radiation seen by the SXT instrument on YOHKOH (Acton et al. 1992) and by the channels sampling radiation at above 5×10^5 K in EIT (Delaboudiniére et al. 1995) on SOHO (as well as on TRACE; Handy et al. 1999) comes from closed field regions. A significant fraction of the radiation recorded by the LASCO C2 and C3 coronagraphs (Brückner et al. 1995), however, is emitted by gas located between open field lines. This is mainly due to the relatively large distance to the solar surface sampled by these instruments, since the ratio of open to closed flux increases rapidly with distance from the solar surface. The 'permanently' open field lines, i.e. those remaining open on time scales of multiple weeks or longer, are generally associated with the fast solar wind. In contrast, the slow solar wind is often associated with transiently open magnetic field. The most powerful events during which the field lines open, allowing matter to escape into the heliosphere, are the Coronal Mass Ejections (reviewed e.g., by Webb 2000; Crooker 2000; Gopalswamy et al. 2003).

These different manifestations of closed and open fields are due to a fundamental difference in the way the energy flux input from the solar photosphere (in the form of either wave energy or excess energy in the magnetic field, stored in the form of magnetic tension) is converted into other forms of energy. In a closed field region (i.e. in magnetic loops) no gas can escape across the field lines, so that most of the energy flowing from below is converted into heat and conducted downwards along the field lines, irrespective of the exact mechanism by which the energy is transported and released (wave dissipation, ohmic dissipation of magnetic energy at current sheets or magnetic reconnection; e.g., Narain and Ulmschneider 1996). This hot and dense gas radiates very efficiently.

In open field regions the energy propagating from below is mainly spent accelerating the gas, which then escapes along the field lines in the form of the solar wind. Only a small fraction (estimated to be on the order of 10%) is conducted back down again and radiated away, leading to the fact that coronal holes are much less bright than the quiet Sun (Gabriel 1976).

This simple picture does not explain the fact that at temperatures below approximately 5×10^5 K (e.g. Stucki et al. 2002) coronal holes are equally bright as parts of the quiet Sun containing mainly closed field lines. One proposed solution has been to suggest that many smaller and less hot (transition region) loops are present in both types of regions (Dowdy et al. 1986; Feldman et al. 2000; Feldman 2002; cf. Peter 2002).

The physics of the heating and acceleration processes is complex and, finally, not completely understood. An overview of coronal heating processes is given by Narain and Ulmschneider (1996).

One of the main barriers to progress in this field has been the paucity and limitations of observations of the magnetic field in layers above the solar photosphere. Gyroresonance observations at radio wavelengths have been a major source of our knowledge of the coronal field and have revealed many examples of active regions whose magnetic field is strongly non-potential in the corona (see White 2002 for a review). However, such data (currently) do not have the spatio-temporal resolution needed to resolve the important fine structure. Thus, the structure of magnetic loops has often been derived from either proxy measurements (i.e. brightness structure; Schrijver et al. 1999; Brkovic et al. 2002; Winebarger et al. 2003) or from extrapolations starting from measured photospheric fields (e.g. Lee et al. 1999; Regnier 2002).

Recently, it became possible to deduce the magnetic vector along a set of loops in an emerging flux region. The resulting structure is plotted in Fig. 5 taken from Solanki et al. (2003), cf. Lagg et al. (2003).

At the apex of the highest reconstructed loop, estimated to be roughly 12000–15000 km above the solar surface, the field strength has dropped to roughly 50 G, which is consistent with previous determinations (e.g. from radio observations) of the field strength in coronal loops. These loops are visible in He I 10830 Å because the field along with the gas trapped within it emerged from the solar interior only a short time prior to the observations. Consequently, there had been no time to heat the gas to coronal temperatures. Interestingly, a comparison with extrapolations of field lines starting from the photosphere indicate that the magnetic field was already quite strongly non-potential shortly after emergence. A potential field extrapolation results in a very poor correspondence with the observations, while a constant-α force-free field reproduces the observations already much better (Wiegelmann et al. in preparation). Consequently, significant currents were already present shortly after emergence and must have been built up in the solar interior during the passage of the emerged loop through the convection zone.

Currents are continually fed and supported also later in the life of an active region. It was proposed by Parker (1983, 1988) that the photospheric footpoints of magnetic field lines are continuously shuffled around by convective motions. He argued that these motions lead to the formation of tangential discontinuities or electric current sheets in the corona where energy is dissipated. Strong support for this mechanism as a major heating process in the corona is provided by the work of Gudiksen and Nordlund (2002). Taking a photospheric magnetogram and the potential field extrapolated from it as a starting point, they randomly moved the photospheric magnetic features around with a velocity field exhibiting roughly the horizontal convective velocity spectrum. They then computed the temporal evolution of the resulting coronal magnetic structure, finding that current sheets are continuously being formed, where magnetic energy is dissipated, heating the coronal gas to over a million degrees, with a considerable distribution of temperatures. Interestingly, the picture they obtain for the gas in a limited temperature range (illustrative of the 171 Å channel of TRACE, the Transition Region and Coronal Explorer (Handy et al. 1999)), is much more intermittent and filamented than the magnetic field, in good agreement with the images returned by TRACE in this channel (Schrijver at al. 1999).

Further support for a heating mechanism involving Ohmic dissipation or reconnection at a current sheet comes from the detection of a current sheet using He I 10830 Å spectropolarimetry. This is the first direct detection of an electric current sheet in a layer of the solar atmosphere relevant for coronal and chromospheric heating. The difficulty in mea-

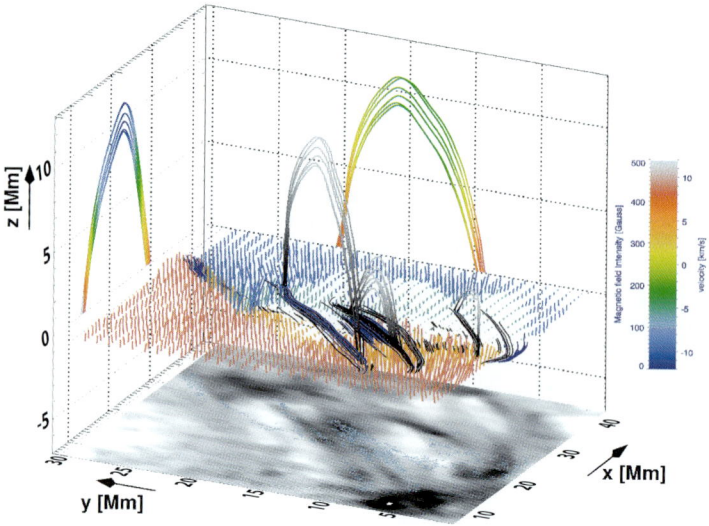

Figure 13.5. Reconstructed magnetic loops in an emerging flux region. The map of the He I 10830 Å lines equivalent width at the bottom is overlaid by a magnetic polarity and connectivity map over which examples of reconstructed loops are plotted. The derived azimuth and inclination angles of the magnetic field in each of the map's pixels are used to trace magnetic field lines. Heavy black lines are projections onto a fixed height of loops that could be successfully reconstructed. These loops trace the equivalent width pattern (shown at the bottom of the plot). Some representative loop field lines are also plotted and for one of the plotted loops they are projected onto the xz- and yz-planes. The color-coding of the loop projections represents the line-of-sight (i.e. basically vertical) velocity (yz-plane) and the magnetic field strength (xz-plane). The loops are rooted in areas with downflowing material, while in the apex upflowing material is observed. This observation, combined with the frozen-in horizontal magnetic field implies that the whole loop is rising. The magnetic field strength decreases with height in both legs from about 390 and 500 Gauss at the two chromospheric footpoints to below 50 Gauss at the apex. Magnetic field vectors at the footpoints of the field lines are shown to visualize the overall magnetic field geometry and strength. Field lines out of (into) the plane parallel to the solar surface are coded in red and orange (blue and green), with orange and green marking field lines for which both footpoints lie within the observed area and emission is seen from the whole length of the loop (see Solanki et al. 2003).

suring such a current sheet lies in the necessity to map the full magnetic vector at chromospheric or coronal heights. Over a distance corresponding to the spatial resolution element of the observations the magnetic vector changes sign within a region of emerging flux in a young active region. This feature is illustrated in Fig. 6, where the current sheet is the narrow valley in the field strength. At the height at which the He I

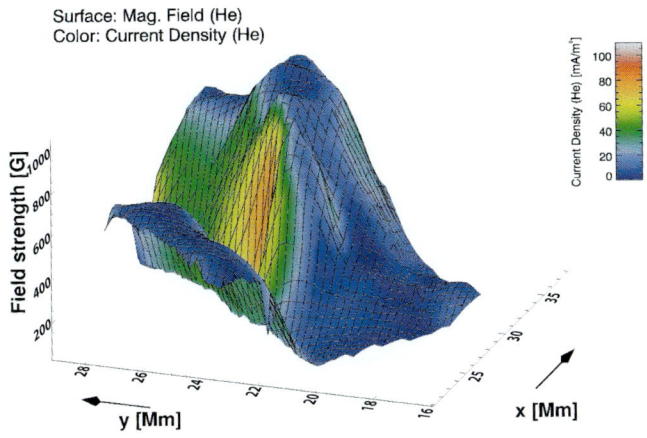

Figure 13.6.   Representation of an electric current sheet near the base of the corona. The elevation of the meshed surface is proportional to the magnetic field strength in the upper chromosphere, as derived from the He I 1083 nm line. A narrow valley of low magnetic field values separates two areas of opposite magnetic polarity, so that an electric current flows along this boundary parallel to the Solar surface. The color-coding in the figure indicates the current density of this current sheet calculated using Ampére's Law. The maximum value of approximately 90 mA/m^2 in the region of the largest magnetic field gradient perpendicular to the current sheet represents only a lower limit of the actual current flowing in this chromospheric current sheet. The width of the valley in the magnetic field strength corresponds to the spatial resolution of 1 Mm, limited by turbulence in the Earth's atmosphere, so that the horizontal gradient of the magnetic field is underestimated (see Solanki et al. 2003).

10830 Å line is formed the field is far more homogeneous in strength than in the photosphere with a magnetic filling factor very close to unity (see description given earlier in this section; cf. Harvey and Hall 1971; Rüedi et al. 1995; Penn and Kuhn 1995). Therefore the gradient between the opposite magnetic polarities is expected to be significantly larger than deduced from the observations (since horizontal gradients are reduced by seeing). Hence the obtained current density, up to 90 mA/m^2 is probably a lower limit. Unfortunately, neither TRACE, EIT, YOHKOH nor BBSO Hα images or movies were available for the day, of these observations, so that it cannot be checked whether this current sheet was responsible for any flare-like episodes.

5.   The Heliosphere

The heliospheric magnetic field is characterized by the fact that higher order multipoles, needed to describe the field at the solar surface, have decayed away beyond a distance of a few solar radii, so that the dipole component dominates the structure.   In addition, beyond the Alfvén radius the energy density of the gas (in particular the kinetic energy density of the solar wind) is far larger than the magnetic energy density (e.g. Weber and Davis 1967). Thus the situation there is reversed relative to the corona. Consequently, the magnetic field is more or less passively dragged along by the solar wind. Logically, only open field lines are present in the heliosphere, if we exclude the magnetic field associated with CMEs and other transient events, which may for some time still be anchored in the photosphere. Due to the combination of the Sun's rotation and the outward motion of the gas the magnetic field follows a spiral pattern (see below).

In recent years very significant progress regarding our knowledge of heliospheric fields has come from the Ulysses spacecraft, which allowed the first measurement of the field outside the ecliptic. The magnetic field strength and solar wind speed during the first Ulysses orbit, carried out under quiet-Sun conditions, is plotted vs. time in Fig. 7, with the heliographic latitude indicated in the top frame (from Balogh and Forsyth 1998). Clearly, at high northern and southern latitudes the field is homogeneous in strength. Although not plotted here, the direction of the magnetic field is also homogeneous at high latitudes, with the azimuth being opposite in both hemispheres, corresponding to opposite magnetic polarities. Note that the gradual increase of the field strength towards the equator in 1994 and 1995 is mainly due to the decreasing distance of the spacecraft to the Sun. This relatively homogeneous field is mainly anchored in the polar coronal holes.

Closer to the ecliptic the field becomes inhomogeneous in both strength and direction, exhibiting multiple jumps of around 180° in azimuth. The magnetic field strength exhibits spikes over which the strength varies by a factor of up to four.

A simple but quite successful model of the large-scale structure of the coronal magnetic field at activity minimum is due to Banaszkiewicz et al. (1998), cf. Forsyth and Marsch (1999). It is composed of dipolar, quadrupolar and equatorial current sheet components. Fig. 8 shows the resulting magnetic structure. Overplotting this on a coronagraphic LASCO C1 brightness image obtained at solar activity minimum exhibits a good agreement (Forsyth and Marsch 1999). According to current thinking the jumps in the magnetic field direction seen by Ulysses

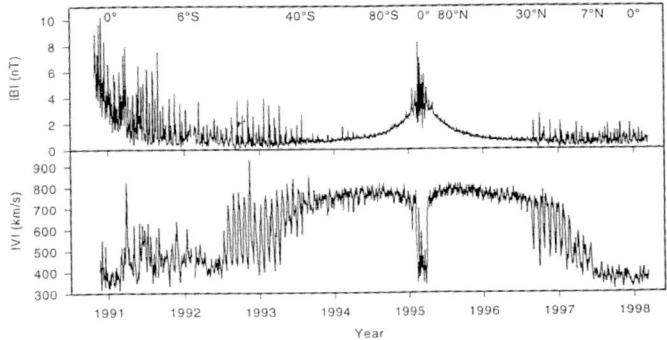

Figure 13.7. The magnitude of the magnetic field and the velocity of the solar wind measured by Ulysses from the beginning of the mission in October 1990 to early March 1998. The heliographic latitude of Ulysses at selected times are indicated in the top frame (figure from Balogh and Forsyth 1998, by permission).

are produced when the spacecraft passes through the current sheet at the equator. This current sheet is warped (appearing like a ballerina's skirt; Schwenn 1990, 1993). Since the Sun completes multiple rotations during the time that the spacecraft crosses the equatorial belt, i.e. the range of latitudes over which the current sheet is warped, the spacecraft can pass through the current sheet multiple times during its passage from pole to pole.

The forces exerted by the solar wind on the field attempt to make it radial, in the absence of solar rotation. When combined with solar rotation, acting through the photospheric footpoints of the field, (which are frozen into the gas) a spiral structure of the field is produced in the equatorial plane (called the Parker spiral; Parker 1958), while outside this plane the field lines follow spirals along cones, with the cones becoming increasingly narrower toward the poles.

Parker's picture has been extended by Fisk (1996) to include the effect of a misalignment between the Sun's axis of rotation and its magnetic dipole axis, as well as of the motion of the magnetic footpoints on the solar surface due to differential rotation (compared to a fixed rotation of the coronal field). This leads to a stronger and less regular displacement of the field lines, in particular at larger distances from the Sun.

Finally, I would like to present an example where the magnetic field at the solar surface directly influences measurements made in the heliosphere by the SWICS instrument (Gloeckler et al. 1992) on Ulysses. In the top panel of Fig. 9 the heliospheric latitude of the Ulysses spacecraft and the sunspot number (as a measure of solar activity) are plotted vs.

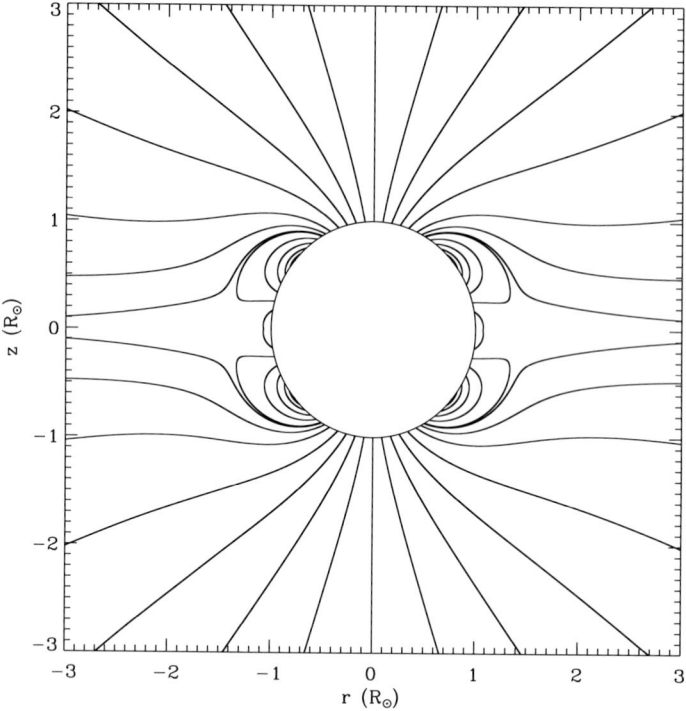

Figure 13.8. The magnetic field lines for the Dipole-Quadrupole-Current-Sheet model of Banaszkiewicz et al. (1998) out to 3 solar radii (solid curves). The model reproduces various observational constraints from SOHO and Ulysses (figure kindly provided by N.-E. Raouafi).

time. In the next panel the solar wind speed and the O^{+7}/O^{+6} abundance ratio measured by SWICS on Ulysses are plotted. It is evident that whereas the solar wind speed is nearly the same while Ulysses is above the southern and the northern coronal holes (marked by the two sets of vertical, dotted lines) the O^{+7}/O^{+6} ratio is significantly different. This ratio is a measure of coronal temperature, so that the southern polar coronal hole (sampled in 1994) was 10–15 % hotter than the northern polar coronal hole in 1995. In the bottom panel of Fig. 9 the spatially averaged field strength at the solar surface in the southern (open circles) and northern polar coronal holes is plotted. It is obvious that the average field strength is higher under the southern polar coronal hole (which is also hotter). This suggests that the average photospheric field strength is closely related to the coronal temperature in coronal holes and the relative abundance of ions in the heliosphere (Zhang et al. 2002).

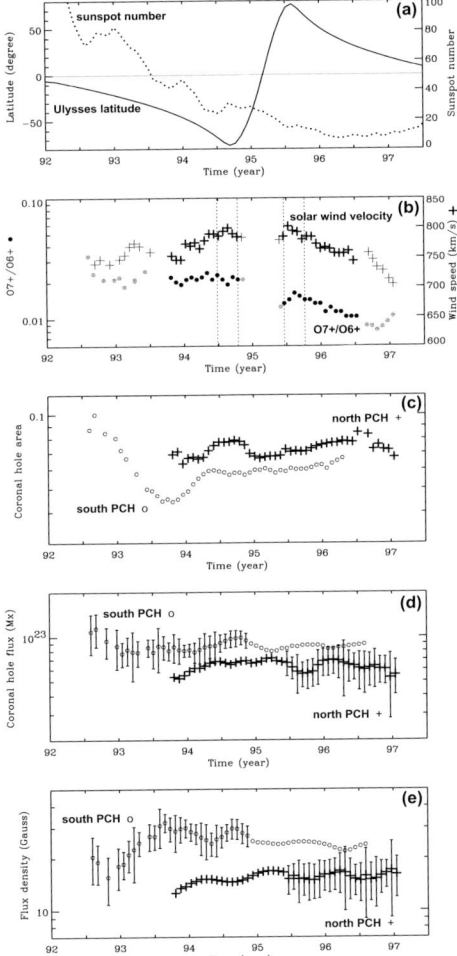

Figure 13.9 Selected solar wind and polar coronal hole parameters between 1992 and 1997. (a) Heliographic latitude of Ulysses and sunspot number; (b) O^{7+}/O^{6+} ratio and solar wind speed as measured by SWICS/Ulysses from 1992 to 1997 (heavy symbols denote pure PCH stream observations, light symbols observations from the mid-latitude fast-slow wind interface regions); (c) area of the south PCH (o) and north PCH (+) estimated from Kitt Peak He I 1083 nm synoptic maps as fraction of the solar surface; (d) magnetic flux of the south (north) PCH derived from Kitt Peak synoptic maps, error bars are included for the south (north) PCH values over the south (north) Ulysses PCH stream intervals; (e) same as in (d) but for magnetic flux density (from Zhang et al. 2002).

6. Conclusion

The magnetic field and its interaction with the solar plasma passes through a variety of regimes from the solar convection zone to the heliosphere. In some layers the magnetic energy density far exceeds the kinetic or thermal energy density, in others it is the opposite, while in yet others the values are similar. The structure and dynamics of the field also change very strongly between convection zone and heliosphere. This variety is one cause of the richness of the types of solar phenomena caused by the magnetic field. More details can be found in the volume edited by Sawaya-Lacoste (2002) as well as the monographs by Parker (1979), Priest (1982), Stenflo (1994) and Zwaan and Schrijver (2000).

Acknowledgements: D. Schmitt critically read through a part of the paper and provided many references. E. Marsch, N.-E. Raouafi, M. Schüssler, A. Vögler and J. Woch provided material and references. My sincere thanks go to all of them.

References

Acton L., Tsuneta S., Ogawara Y., et al., 1992, *Science* **258**, 618.

Avrett E.H., Fontenla J.M., Loeser R., 1994, in *Infrared Solar Physics*, IAU Symp. No. 154, D.M. Rabin et al. (Eds.), Kluwer, Dordrecht, p. 35.

Balogh A. and Forsyth R.J., 1998, The results of the Ulysses mission: A survey of the Heliosphere in three dimensions. In: *Crossroads for European Solan and Heliospheric Physics: Recent Achievements and Future Mission Possibilities*, R.A. Harris (Ed.), ESA SP-417, p. 45.

Banaszkiewicz M., Axford W.I., McKenzie J.F., 1998, *Astron. Astrophys.* **337**, 940.

Bellot Rubio L.R., Balthasar H., Collados M., Schlichenmaier R., 2003, *Astron. Astrophys.* **403**, L47.

Bercik D.J., Nordlund Å., Stein R.F., 2003, Magnetoconvection and micropores. In: *SOHO 12/GONG+ 2002. Local and global helioseismology: the present and future*, H. Sawaya-Lacoste (Ed.), ESA SP-517, p. 201.

Berger T.E., Löfdahl M.G., Shine R.S., Title A.M., 1998, *Astrophys. J.* **495**, 973.

Brandenburg A., and Schmitt D., 1998, *Astron. Astrophys.* **338**, L55.

Brkovic A., Landi E., Landini M., Rüedi I., Solanki S.K., 2002, *Astron. Astrophys.* **383**, 661.

Brueckner G.E., Howard R.A., Koomen M.J. et al., 1995, *Sol. Phys.* **162**, 357.

Caligari P., Moreno-Insertis F., and Schüssler M., 1995, *Astrophys. J.* **411**, 886.

Caligari P., Schüssler M., and Moreno-Insertis F., 1998, *Astrophys. J.* **502**, 481.

Choudhuri A.R., Schüssler M., and Dikpati M., 1995, *Astron. Astrophys.* **303**, L29.

Crooker, N., 2000, *J. Atmosph. Solar Terr. Phys.* **62**, 1071.

Defouw R.J., 1976, *Astrophys. J.* **209**, 266.

Deinzer W., Hensler G., Schüssler M., Weisshaar E., 1984a, *Astron. Astrophys.* **139**, 426.

Deinzer W., Hensler G., Schüssler M., Weisshaar E., 1984b, *Astron. Astrophys.* **139**, 435.

Delaboudiniére J.-P., Artzner G.E., Brunaud J. et al., 1995, *Sol. Phys.* **162**, 291.

Dikpati M., and Charbonneau P., 1999, *Astophys. J.* **518**, 508.

Dikpati M., and Gilman P.A., 2001, *Astophys. J.* **559**, 428.

Dowdy J.F. Jr., Rabin D., Moore R.L., 1986, *Sol. Phys.* **105**, 35.

D'Silva S., Choudhuri A.R., 1993, *Astron. Astrophys.* **272**, 621.

Durney B.R., 1995, *Sol. Phys.* **160**, 213.

Durney B.R., 1996, *Sol. Phys.* **166**, 213.

Durney B.R., 1997, *Astrophys. J.* **486**, 1065.

Duvall T.L., Jr., Gizon L., 2000, *Sol. Phys.* **192**, 177.

Fan Y., Fisher G.H., and Deluca E.E., 1993, *Astrophys. J.* **405**, 390.

Fan Y., Fisher G.H., and Deluca E.E., 1994, *Astrophys. J.* **436**, 907.

Feldman U., 2002, In: *From Solar Min to Max: Half a Solar Cycle with SOHO. Proc. SOHO11 Symp.*, A. Wilson (Ed.), ESA SP-508, p. 531.

Feldman U., Dammasch I., Wilhelm K., 2000, *Space Sci. Rev.* **93**, 411.

Ferriz-Mas A. and Schüssler M., 1993, *Geophys. Astrophys. Fluid Dyn.* **72**, 209.

Ferriz-Mas A. and Schüssler M., 1995, *Geophys. Astrophys. Fluid Dyn.* **81**, 233.

Fisk L., 1996, *J. Geophys. Res.* **101**, 15547.

Fontenla J.M., Avrett E.H., Loeser R., 1993, *Astrophys. J.* **406**, 319.

Forsyth R.J. and Marsch E., 1999, *Space Sci. Rev.* **89**, 7.

Fröhlich C., Romero J., Roth H., et al., 1995, *Sol. Phys.* **162**, 10.

Frutiger C., Solanki S.K., Fligge M., Bruls J.H.M.J., 2000, *Astron. Astrophys.* **358**, 1109.

Gabriel A. H., 1976, *Phil. Trans. Royal Society, Ser. A* **281**, 339.

Gadun A.S., Solanki S.K., Sheminova V.A., Ploner S.R.O., 2001, *Sol. Phys.* **203**, 1.

Giovanelli R.G. and Jones H.P., 1982, *Sol. Phys.* **79**, 267.

Gloeckler et al., 1992, *Astron. Astrophys. Suppl. Ser.* **92**, 267.

Gopalswamy N., Lara A., Yashiro S., Nunes S., Howard R.A., 2003, Coronal mass ejection activity during solar cycle 23. In: *Proc. ISCS 2003 Symposium: Solar Variability as an Input to the Earth's Environment*, ESA-SP-535, p. 403.

Gudiksen B.V. and Nordlund R., 2002, *Astrophys. J.* **572**, 113.

Hagenaar H.J., Schrijver C.J., Title A.M., 2003, *Astrophys. J.* **584**, 1107.

Hale G.E., 1908, *Astrophys. J.* **28**, 315.

Handy B.N., Acton L.W., Kankelborg C.C., et al., 1999, *Sol. Phys.* **187**, 229.

Harvey K.L., 1993, PhD Thesis, Rijksuniv. Utrecht.

Harvey J.W. and Hall D.N.B., 1971, in *Solar Magnetic Fields*, R. Howard (Ed.), IAU Symp. 43, Reidel, Dordrecht, p. 279.

Harvey J.W., Hill F., Hubbard R., et al., 1996, *Science* **272**, 1284.

Hathaway D.H., 1996, *Astrophys. J.* **460**, 1027.

Hathaway D.H., Nandy D., Wilson R.M., Reichmann E.J., 2003, *Astrophys. J.* **589**, 665.

Jahn K. and Schmidt H.U., 1994, *Astron. Astrophys.* **290**, 295.

Jones H.P., Giovanelli R.G., 1983, *Sol. Phys.* **87**, 37.

Keller C.U., 1992, *Nature* **359**, 307.

Keller C.U., Steiner O., Stenflo J.O., Solanki S.K., 1990, *Astron. Astrophys.*, **233**, 583.

Krivova N.A., Solanki S.K., Fligge M., Unruh Y.C., 2003, *Astron. Astrophys.* **399**, L1.

Kuhn J.R., Libbrecht K.G., Dicke R.H., 1989, *Science* **242**, 908.

Küker M., Rüdiger G., and Schultz M., 2001, *Astron. Astrophys.* **374**, 301.

Lagg A., Woch J., Krupp N., Solanki S.K., 2003, *Astron. Astrophys.* in press.

Landi Degl'Innocenti E., 1992, In: *Solar Observations: Techniques and Interpretation*, F. Sánchez, M. Collados, M. Vázquez (Eds.), p. 71.

Lee J., White S.M., Kundu M.R., Mikic Z., McClymont A.N., 1999, *Astrophys. J.* **510**, 413.

Lites B.W., Martínez Pillet V., Skumanich A., 1994, *Sol. Phys.* **155**, 1.

Livingston W., L. Wallace L., 2003, *Sol. Phys.* **212**, 227.

Muller R., Roudhier Th., Vigneau J., Auffret H. 1994, *Astron. Astrophys.* **283**, 232.

Nandy D., and Choudhuri A.R., 2001, *Astrophys. J.* **551**, 576.

Narain U. and Ulmschneider P., 1996, *Space Sci. Rev.* **75**, 453.

Nordlund Å., Brandenburg A., Jennings R.L., et al., 1992, *Astrophys. J.* **392**, 647.

Ossendrijver M., 2003, *Astron. Astrophys. Rev.* **11**, 287.

Parker E.N., 1958, *Astrophys. J.* **128**, 664.

Parker E.N., 1975, *Astrophys. J.* **198**, 205.

Parker E.N., 1979, *Cosmical Magnetic Fields: Their Origin and their Activity*, Clarendon Press, Oxford.

Parker E.N., 1983, *Astrophys. J.* **264**, 642.

Parker E.N., 1988, *Astrophys. J.* **330**, 474.

Parker E.N., 1993, *Astrophys. J.* **408**, 707.

Penn M.J. and Kuhn J.R., 1995, *Astrophys. J.* **441**, 51.

Peter H., 2002, In: *Proc. SOHO 11 Symp., From Solar Min to Max: Half a Solar Cycle with SOHO*, A. Wilson (Ed.), ESA SP-508, p. 237.

Priest E.R., 1982, *Solar Magnetohydrodynamics*, Reidel Publ. Co., Dordrecht.

Rabin D., 1992a, *Astrophys. J.* **390**, L103.

Rabin D., 1992b, *Astrophys. J.* **391**, 832.

Régnier S., Amari T., Kersalé E., 2002, *Astrophys. J.* **392**, 1119.

Rempel M. and Schüssler M., 2001, *Astrophys. J.* **552**, L171.

Rüedi I., Solanki S.K., Livingston W., Stenflo J.O., 1992, *Astron. Astrophys.* **263**, 323.

Rüedi I., Solanki S.K., Livingston W.C., 1995, *Astron. Astrophys.* **293**, 252.

Ruiz Cobo B., Del Toro Iniesta J.C., 1992, *Astrophys. J.* **398**, 375.

Sawaya-Lacoste H. (Ed.), 2002, *SOLMAG 2002: Proc. Magnetic Coupling of the Solar Atmosphere*, Euroconference and IAU Colloquium 188, ESA SP-505.

Scharmer G.B., Gudiksen B.V., Kiselman D., Löfdahl M.G., Rouppe van der Voort L.H.M., 2002, *Nature* **420**, 151.

Scherrer P.H., Bogart R.S., Bush R.I., et al., 1995, *Sol. Phys.* **162**, 129.

Schlichenmaier R. and Solanki S.K., 2003, *Astron. Astrophys.* **411**, 257.

Schmidt D., 1987, *Astron. Astrophys.* **174**, 281.

Schmidt D., 1993, The solar dynamo. In: *IAU Symp. 157: The Cosmic Dynamo*, F. Krause, K.H. Rädler, G. Rüdiger (Eds.), Kluwer, Dordrecht, p. 1.

Schmidt D., 2002, Dynamo action of magnetostrophics waves. In: *Advances in Nonlinear Dynamos*.

Schmidt W., Hofmann A., Balthasar H., Tarbell T.D., Frank Z.A., 1992, *Astron. Astrophys.* **264**, L27.

Schmieder B., del Toro Iniesta J.C., Vázquez M. (Eds.), 1997, *Advances in the Physics of Sunspots*, ASP Conf. Ser. Vol. 118.

Schou J., Christensen-Dalgaard J., and Thompson M.J., 1992, *Astrophys. J.* **385**, L59.

Schou J., Antia H.M., Basu S., 1998, *Astrophys. J.* **505**, 390.

Schou J., Howe R., Basu S., et al., 2002, *Astrophys. J.* **567**, 1234.

Schrijver C.J., Title A.M., Harvey K.L., et al., 1998, *Nature* **394**, 152.

Schrijver C.J., Title A.M., Berger T.E., et al., 1999, *Sol. Phys.* **187**, 261.

Schüssler M., 1979, *Astron. Astrophys.* **71**, 79.

Schüssler M., Caligari P., Ferriz-Mas A., Moreno-Insertis F., 1994, *Astron. Astrophys.* **281**, L69.

Schüssler M., Shelyag S., Berdyugina S., Vögler A., Solanki S.K., 2003, *Astrophys. J.* **597**, L173.

Schwenn R., 1990, Large-scale structure of the interplanetary medium. In: *Physics of the Inner Heliosphere*, Vol. I, Schwenn R., Marsch E. (Eds), Springer, Berlin, p. 99.

Schwenn R., 1993, Interplanetary plasma and magnetic field (solar wind), In: *Landolt-Börnstein Astronomy and Astrophysics New Series VI/3a*

'Instruments, Methods, Solar System', H.H. Voigt (Ed.), Springer, Berlin, Chap. 3.3.5.2, p. 189.

Shelyag S., Schüssler M., Solanki S.K., Berdyugina S., Vögler A., 2004, *Astron. Astrophys.* submitted.

Sobotka M., 1997, in: *Advances in the Physics of Sunspots*, B. Schmieder et al. (Eds.), ASP Conf. Ser. Vol. 118, p. 155.

Socas Navarro H., Trujillo Bueno J., Ruiz Cobo R., 2000, *Science* **288**, 1396.

Solanki S.K., 1986, *Astron. Astrophys.* **168**, 311.

Solanki S.K., 1993, *Space Science Rev.* **63**, 1.

Solanki S.K., 2003, *Astron. Astrophys. Rev.* **11**, 153.

Solanki S.K. and Montavon C.A.P., 1993, *Astron. Astrophys.* **275**, 283.

Solanki S.K. and Rüedi I., 2003, *Astron. Astrophys.* **411**, 249.

Solanki S.K. and Steiner O., 1990, *Astron. Astrophys.* **234**, 519.

Solanki S.K., Lagg A., Woch J., Krupp N., Collados M., 2003, *Nature* **425**, 692.

Spruit H.C., 1976, *Sol. Phys.* **50**, 269.

Stein R.F. and Nordlund Å, 2002, in *SOLMAG 2002: Proc. Magnetic Coupling of the Solar Atmosphere*, Euroconference and IAU Colloquium 188, H. Sawaya-Lacoste (Ed.), ESA SP-505, p.83.

Steiner O., Pneuman G.W., Stenflo J.O., 1986, *Astron. Astrophys.* **170**, 126.

Steiner O., Grossmann-Doerth U., Knölker M., Schüssler M., 1998, *Astrophys. J.* **495**, 468.

Stenflo J.O., 1973, *Sol. Phys.* **32**, 41.

Stenflo J.O., 1994, *Solar Magnetic Fields: Polarized Radiation Diagnostics*, Kluwer, Dordrecht.

Stucki K., Solanki S.K., Pike C.D., et al., 2002, *Astron. Astrophys.* **381**, 653.

Thomas J.H., Weiss N.O. (Eds.), 1992, *Sunspots: Theory and Observations*, Kluwer, Dordrecht.

Thompson M.J., Christensen-Dalsgaard J., Miesch M.S., Toomre J., 2003, *Ann. Rev. Astron. Astrophys.* **41**, 599.

Title A.M., Frank Z.A., Shine R.A., et al., 1993, *Astrophys. J.* **403**, 780.

Tomczyk S., Schou J., and Thompson M.J., 1995, *Astrophys. J.* **448** L57.

Unruh Y.C., Solanki S.K., Fligge M., 1999, *Astron. Astrophys.* **345**, 635.

Vögler A., 2003, PhD Thesis, Univ. Göttingen.

Vögler A. and Schüssler M., 2003, *Astron. Nachr.* **324**, 399.

Walton S.R., Premiger D.G., Chapman G.A., 2003, *Astrophys. J.* **590**, 1088.

Webb, D., 2000, Coronal Mass Ejections: Origins, Evolution, and Role. In: *Space Weather*, IEEE Transactions on Plasma Science, Vol. 28, 1795.

Weber E.J., Davis L., Jr., 1967, *Astrophys. J.* **148**, 217.

Westendrop Plaza C., del Toro Iniesta J.C., Ruiz Cobo B., Martínez Pillet V., Lites B.W., Skumanich A., 1997, *Nature* **389**, 47.

White S.M., 2001, Observations of solar coronal magnetic fields. In: *Magnetic Fields Across the Herzsprung-Russel Diagram*, G. Mathys, S.K. Solanki, D.T. Wickramasinghe (Eds.), ASP Conf. Proc. Vol. 248, p. 67.

Winebarger A., Warren H.P., Mariska J.T., 2003, *Astrophys. J.* **587**, 439.

Woodard M.F. and Libbrecht K.G., 2003, *Sol. Phys.* **212**, 51.

Zayer I., Solanki S.K., Stenflo J.O., 1989, *Astron. Astrophys.*

Zhang J., Woch J., Solanki S.K., von Steiger R., 2002, *Geophys. Res. Lett.* **29**, 77.

Zwaan C. and Cram L.E., 1989, in *FGK Stars and T Tauri Stars*, L.E.. Cram, L.V. Kuhi (Eds.), NASA SP-502, p. 215.

Zwaan C. and Schrijver C.J., 2000, *Solar and Stellar Magnetic Activity*, Cambridge University Press.

Chapter 14

MAGNETIC RECONNECTION

E. R. Priest
and D. I. Pontin
Department of Mathematics
University of St Andrews
St Andrews
KY16 9SS, UK.

Abstract   Magnetic reconnection is a fundamental process in a cosmic plasma such as the solar corona, responsible for heating and for many dynamic phenomena. We here review briefly the basic theory of reconnection in two dimensions and then proceed to describe several features of the way it operates in three dimensions. These include a description of 3D null points, of 3D topology, and of spine, fan and separator reconnection. Also an account of new properties of 3D reconnection at an isolated diffusion region is given, including the fact that a unique field line velocity can no longer be defined and that field line connections change continuously while a field line is traversing the non-ideal region. A summary is given of the various ways in which the corona may be heated, including binary, separator and braiding reconnection, as well as heating by coronal tectonics. Finally, a summary is also given of reconnection processes in the Earth's magnetosphere.

1.   Introduction

In most of the universe magnetic field lines are frozen into the plasma and are carried about with it. However, in tiny current sheets the field lines are able to slip through the plasma, break and re-join. This process of magnetic reconnection is a fundamental one that is responsible for many dynamic phenomena in the cosmos (for a review, see the book by Priest & Forbes, 2000). It can change the topology of the magnetic field, and converts inflowing magnetic energy into heat, bulk kinetic energy and fast particle energy.

G. Poletto and S.T. Suess (eds.), The Sun and The Heliosphere as an Integrated System, 397–422.
© 2004 *Kluwer Academic Publishers. Printed in the Netherlands.*

At the boundary of the Magnetosphere reconnection mediates a transfer of magnetic flux between the solar wind and the Earth's magnetic field. It occurs in both a steady and impulsive way at the front of the Magnetopause (so-called "flux transfer events") and also in the geomagnetic tail (in so-called "geomagnetic substorms"). It is also likely to be a major contributor to solar coronal heating and is certainly the process at the core of a solar flare.

We shall here be discussing reconnection in a magnetohydrodynamic (or MHD) plasma which satisfies the following equations: the plasma velocity $\mathbf{v}$ and magnetic field $\mathbf{B}$ are essentially determined by an equation of motion

$$\rho \frac{d\mathbf{v}}{dt} = -\nabla p + \mathbf{j} \times \mathbf{B} \tag{14.1}$$

and the induction equation

$$\frac{\partial \mathbf{B}}{\partial t} = \nabla \times (\mathbf{v} \times \mathbf{B}) + \eta \nabla^2 \mathbf{B}, \tag{14.2}$$

where the current ($\mathbf{j}$) is given by Ampère's law

$$\mathbf{j} = \frac{\nabla \times \mathbf{B}}{\mu}, \tag{14.3}$$

the pressure (p) by the perfect gas law

$$p = R \rho T, \tag{14.4}$$

the density (ρ) by the equation of continuity

$$\frac{\partial \rho}{\partial t} + \nabla \cdot (\rho \mathbf{v}) = 0, \tag{14.5}$$

and finally the temperature (T) by an energy equation.

In the induction equation (14.2), the ratio of the second term to the third term is the magnetic Reynolds number

$$R_m = \frac{L_0 v_0}{\eta}, \tag{14.6}$$

in terms of the typical plasma velocity (v_0) and scale (L_0) for variations. In most of the universe R_m is extremely large (of order $10^6 - 10^{12}$ typically) and so the first term on the right of (14.2) dominates, which implies that the magnetic field is tied to the plasma and moves with it, hanging onto its energy. The exception is in current singularities where the electric current is extremely large and the gradients of the magnetic field are enormous. They tend to form as sheets, often, although not

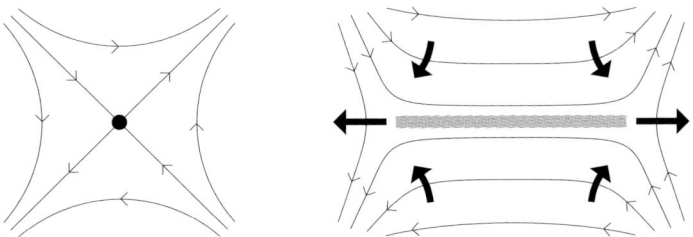

Figure 14.1. (a) The magnetic field lines near an X-type null point, where the magnetic field vanishes. (b) The collapse of this null point to form a current sheet.

only, at null points, where the magnetic field vanishes. It is in these singularities that the second term in the right of (14.2) becomes important and the magnetic field can diffuse through the plasma and reconnect.

In Section 2 we give a summary of reconnection in two dimensions, and then in Section 3 we describe some of the attempts to generalise these ideas to three dimensions. Section 4 gives a new analysis of how 3D reconnection occurs at an isolated diffusion region, highlighting just how different 3D reconnection is from its 2D counterpart. Section 5 summarises some of the ways in which reconnection could be heating the solar corona, including the latest idea of coronal tectonics. Finally, Section 6 describes reconnection processes occurring within the Earth's magnetosphere.

2. Two-Dimensional Reconnection

2.1 X-Collapse

Magnetic reconnection in two dimensions can occur in several ways. It can be driven by external motions, it can occur spontaneously due to the onset of an instability, or it can occur when an X-type null point collapses (Figure 14.1).

The linear field near a potential X-point can be written

$$\mathbf{B} = y\,\hat{\mathbf{x}} + x\,\hat{\mathbf{y}}, \tag{14.7}$$

where the current

$$j_z = \frac{\partial B_y}{\partial x} - \frac{\partial B_x}{\partial y} \tag{14.8}$$

vanishes and so this field is in equilibrium with itself. Now consider a perturbation of this field so that it becomes

$$\mathbf{B} = y\,\hat{\mathbf{x}} + \alpha^2 x\,\hat{\mathbf{y}}. \tag{14.9}$$

Whereas the field lines of (14.7) are rectangular hyperbolae $y^2 - x^2 = const$ with separatrices at right angles, those of (14.9) are hyperbolae with separatrices $y = \pm \alpha x$, which are therefore closed up towards the y-axis if $\alpha > 1$. The resulting magnetic force

$$\mathbf{j} \times \mathbf{B} = \mu^{-1} \left(\alpha^2 - 1 \right) \left(-\alpha^2 x \, \hat{\mathbf{x}} + y \hat{\mathbf{y}} \right)$$

is in such a direction as to continue the perturbation and so, provided the field lines are not anchored and energy can propagate in towards the X-point, it tends to collapse, with α (and so the current)

$$j_z = \frac{\alpha^2 - 1}{\mu}$$

increasing indefinitely.

Linear solutions with magnetic diffusion included (Craig & McClymont, 1991) give collapse on a timescale of $t_A \, log \, (R_m)$ together with a decaying oscillation, where t_A is the Alfvén travel time L_0/v_A, in terms of the Alfvén speed $v_A = B_0/(\mu \rho_0)^{1/2}$. Furthermore, non-linear self-similar solutions to the ideal equations of the form

$$B_x = a(t) \, x + b(t) \, y, \quad B_y = c(t) \, x + d(t) \, y,$$
$$v_x = e(t) \, x + f(t) \, y, \quad v_y = g(t) \, x + h(t) \, y,$$

$$p = i(t) \, x^2 + j(t) \, x \, y + k(t) \, y^2$$

indicate that the collapse occurs in a finite time, although this analysis is only local (Imshennik & Syrovatsky; Klapper, 1998; Bulanov & Olshanetsky, 1984; Mellor et al., 2002).

2.2 Sweet-Parker Reconnection

In two-dimensions reconnection only takes place at an X-point, where the current can become very large and strong dissipation can allow the field lines to break and change their connectivity. In two dimensions the theory is well developed, with several mechanisms possible:

1 Slow Sweet-Parker reconnection (1958);

2 Fast Petschek reconnection (1964);

3 Many other fast regimes, depending on the boundary conditions, such as Almost- Uniform reconnection (Priest & Forbes, 1986) and Non-Uniform reconnection (Priest & Lee, 1990).

Sweet-Parker reconnection considers a simple current sheet of dimension l and L between two regions of uniform but oppositely directed

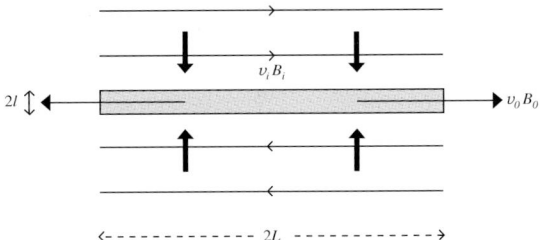

Figure 14.2 Sweet-Parker reconnection.

magnetic field (B_i) and flow (v_i) (Figure 14.2). It is just an order-of-magnitude model. Conservation of mass into and out of the sheet gives

$$L\, v_i \;=\; l\, v_0 \tag{14.10}$$

whereas for a steady state the balance between inwards advection and outwards diffusion gives

$$v_i \;=\; \frac{\eta}{l}. \tag{14.11}$$

Finally, the acceleration of plasma along the sheet by magnetic forces gives roughly

$$v_0 \;=\; v_{Ai}. \tag{14.12}$$

Eliminating l and v_0 between these three equations gives a dimensionless reconnection rate of

$$M_i \;=\; \frac{1}{R_{mi}^{1/2}}, \tag{14.13}$$

where $M_i = v_i/v_{Ai}$ is the Alfvén Mach number for the inflow and $R_{mi} = L\, v_{Ai}/\eta$ is the magnetic Reynolds number based on the sheet length (L) and the inflow Alfvén speed ($v_{Ai} = B_i\,/\,(\mu\rho)^{1/2}$).

In this mechanism the outflow magnetic field (B_0) is

$$B_0 \;=\; \frac{B_i}{R_m^{1/2}} \tag{14.14}$$

and so is much less than the inflow field (B_i) when $R_{mi} \gg 1$. Furthermore the inflow energy is mainly magnetic and half of it is transformed into kinetic energy while the other half is released as ohmic heating. In other words, the mechanism creates hot fast jets of plasma.

2.3   Stagnation-Point Flow Model

The stagnation-point flow model (Sonnerup & Priest, 1975) considers the effect of a stagnation-point flow

$$v_x = -U\,\frac{x}{a}, \qquad v_y = U\,\frac{y}{a} \tag{14.15}$$

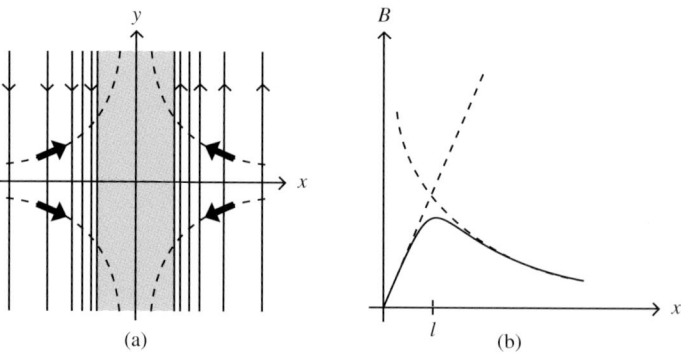

Figure 14.3. (a) Stagnation-point flow model, and (b) the variation with x of the magnetic field (B).

on a uni-directional magnetic field ($B(x)\,\hat{\mathbf{y}}$), as shown in Figure 14.3(a). Ohm's law becomes

$$E - \frac{Ux}{a}B = \eta\frac{dB}{dx},\qquad(14.16)$$

where the electric field $E\,\hat{\mathbf{z}}$ is uniform in space and constant in time. The solution for the magnetic field is

$$B(x) = \frac{2Ea}{v_0 l}\,e^{-x^2/l^2}\int_0^x e^{X^2/l^2}\,dX,\qquad(14.17)$$

where $l^2 = 2\eta a/U$, as shown in Figure 14.3(b). It increases from 0 like Ex/η for small x and decreases like $Ea/(Ux)$ for $x \gg l$. At large distances the magnetic field is carried in towards the origin and is frozen to the plasma. At small distances it diffuses through the plasma in a current sheet of half-width l.

The advantage of the stagnation point flow model is that it is an exact solution of the MHD equations since it also satisfies the equation of motion. Furthermore, it has been generalised to produce so-called "reconnective annihilation" solutions in three dimensions, for both spine and fan geometries (Craig & Fabling, 1996). However, a limitation of the model is that the current sheet is purely one-dimensional and extends to infinity.

2.4 Petschek's Model

The main disadvantage of the Sweet-Parker model is that it is much too slow to explain reconnection in solar flares, and so Petschek (1964) proposed a much faster model in which the Sweet-Parker diffusion region is very much smaller. The diffusion region bifurcates to form two

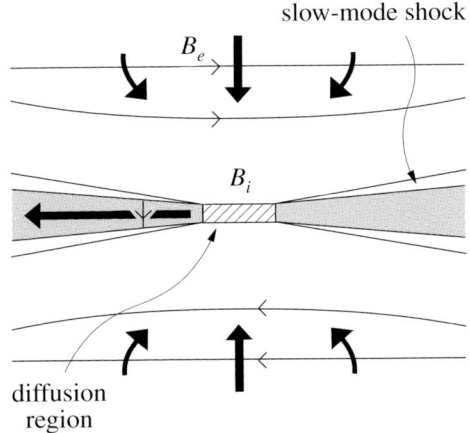

slow-mode shock

diffusion
region

Figure 14.4 Petschek's
model for fast reconnection

pairs of standing slow-mode shock waves at which most of the magnetic
energy conversion takes place (Figure 14.4). Reconnection takes place
at the rate at which it is driven at large distances. As the rate of reconnection increases, so the central diffusion region shrinks in size and the
angle between each pair of shock waves increases. There is, however, a
maximum allowable reconnection rate at which the mechanism chokes
off. This has the form

$$M_{e\,max} = \frac{\pi}{8\,log_e R_{me}}, \tag{14.18}$$

where $M_e = v_e/v_{Ae}$ is the ratio of the inflow speed to Alfvén speed
(v_{Ae}) at large distances and $R_{me} = L_e v_{Ae}/\eta$ is the global magnetic
Reynolds number based on the overall length-scale (L_e). Thus $M_{e\,max}$
varies weakly with R_{me} and is typically 0.01 - 0.1 in magnitude.

2.5   More Recent Fast Mechanisms

The Almost-Uniform family of models (Priest & Forbes, 1986) include
Peschek's model as a special case, but also it has a range of different
regimes, depending on the boundary conditions at the inflow boundary
(Figure 14.5). Regimes are classified by a parameter b, which determines
the inclination of the streamline at the top right-hand corner of the
region. Thus the regimes vary from strongly converging flows (14.5(a))
to strongly diverging flows - so-called "flux pile-up" reconnection (Figure
14.5(f)).

The Almost-Uniform family is characterised by an inflow region containing weakly curved magnetic field lines, but it is also possible for them
to be strongly curved, as in the Non- Uniform family of models (Priest

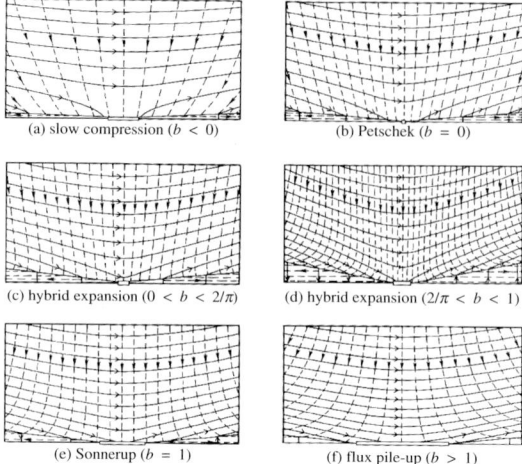

Figure 14.5 Almost uniform reconnection

& Lee, 1990). These exhibit jets of plasma expelled along the separatrices (field lines that pass through the X-point). Furthermore, this model agrees with the numerical experiments of Biskamp (1986) and of Yan et al. (1992,1993), provided the same boundary conditions are adopted (Strachan & Priest, 1994).

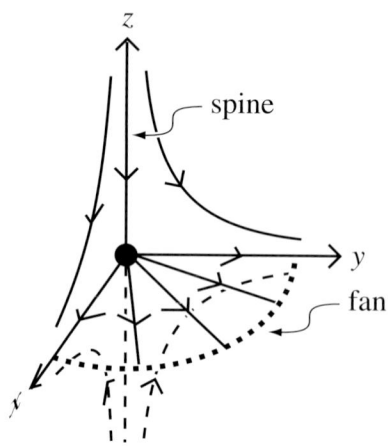

Figure 14.6 A 3D null point.

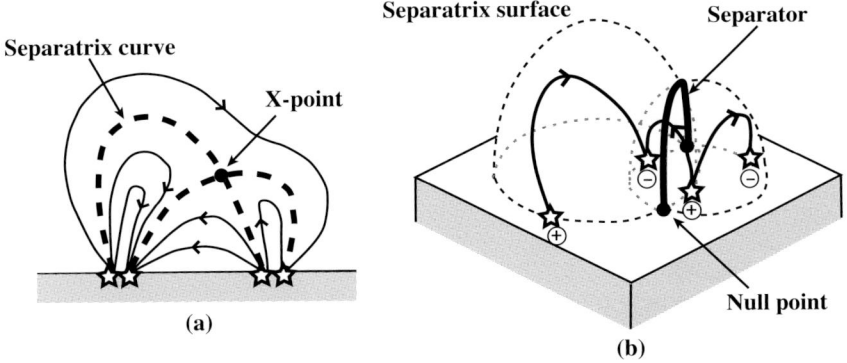

Figure 14.7. Topological structures (a) in 2D and (b) in 3D.

3. Three-Dimensional Reconnection

3.1 Structure of a Null Point

The simplest null point at the origin in three dimensions (Figure 14.6) has a magnetic field with components

$$(B_x, B_y, B_z) = (x, y, -2z),\qquad(14.19)$$

with two families of field lines through the null: an isolated *spine* field line lies along the z-axis and approaches the null from both directions; a surface of *fan* field lines in the xy-plane recedes from the null.

Most generally, the linear field near a null may be written in the form (Parnell et al., 1996)

$$
\begin{aligned}
B_x &= x + \tfrac{1}{2}\left(q - J\right)y \\
B_y &= \tfrac{1}{2}\left(q + J\right)x + py \\
B_z &= jy - (p+1)z
\end{aligned}
$$

in terms of parameters p, q, J and j, where J is the current along the spine and produces a twist in the fan, while j is the current in the fan and determines the angle between the spine and fan.

Just as in 2D, so in 3D a null can collapse to give a current growing along the spine or in the fan (Parnell, 1997).

3.2 Global Topology of Complex Fields

In two dimensions *separatrix curves* (which pass through X-type null points) separate the plane into topologically distinct regions, in the sense that field lines in a particular region all connect from one particular

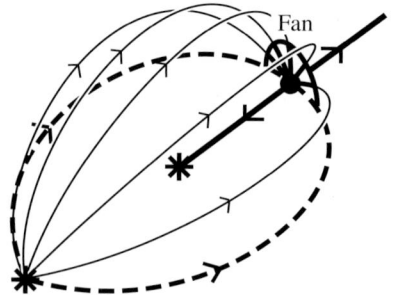

Fan

Figure 14.8 The skeleton of two unbalanced sources.

positive source to one particular negative source (Figure 14.7(a)). In three dimensions the topology of a model solar coronal field due to a series of flux sources lying in a plane is similar, but now the volume above the plane is divided into topologically distinct volumes by separatrix surfaces, which are usually the extensions of fan surfaces from 3D nulls (Figure 14.7(b)).

The way to reveal the structure of a highly complex field is to plot its *skeleton*, namely the set of null points, separatrix surfaces and spine field lines. For example, the skeleton of two unbalanced sources (Figure 14.8) consists of a null, its spine, and its fan, which arches over to form a dome enclosing the weaker source.

In contrast, when there are three magnetic sources, eight different topologies are possible, as categorised by Brown and Priest (1999). You can change from one topology to another by means of a *bifurcation*, either *local* (when nulls are created or destroyed) or *global* (when there is no change in the number of nulls).These include a local separator bifurcation (when a separator joining two nulls is created or destroyed), a global spine-fan bifurcation (when the spine of one null lies in the fan of the other at the moment of bifurcation) and a global separator bifurcation (when two fan surfaces touch at the moment of bifurcation to form a separator). Higher order behaviour due to the presence of four or more sources includes the presence of multiple separators (which can interact, merge or separate) and the appearance of coronal null points which can lift off the plane in a local double-separator bifurcation (Brown & Priest, 2001). If we model the photospheric field as a series of point sources, we find that the number of null points in the plane is roughly equal to the number (N_s) of sources, whereas the number of coronal nulls is about 5- 10% N_s (Longcope et al., 2003).

3.3   3D Reconnection at a Null Point

Priest and Titov (1996) discovered three different types of reconnection at a 3D null, namely *spine reconnection*, when the current concentrates along the spine, *fan reconnection* when it concentrates along the fan and *separator reconnection* when it focuses along the separator joining one null point to another.

Consider steady ideal flow of a magnetic field (14.19), satisfying

$$\mathbf{E} + \mathbf{v} \times \mathbf{B} = \mathbf{0} \tag{14.20}$$

and

$$\nabla \times \mathbf{E} = \mathbf{0}. \tag{14.21}$$

Equation (14.21) implies that $\mathbf{E} = \nabla F$ and so the scalar product of $\mathbf{B}$ with Equation (14.20) gives

$$\mathbf{B} \cdot \nabla F = 0 \tag{14.22}$$

which determines the value of F everywhere in terms of values on the boundary by integrating along field lines. Furthermore, the vector product of $\mathbf{B}$ with Equation (14.20) then determines the plasma velocity normal to the field lines as

$$\mathbf{v}_\perp = \frac{\mathbf{E} \times \mathbf{B}}{B^2}. \tag{14.23}$$

If we impose a flow on the side boundary continuous across the fan, then a singularity arises along the spine, which one can try to resolve by diffusion. If instead a continuous flow is imposed on the top and bottom across the spine, then a singularity arises at the fan.

A separator is a field line joining two 3D nulls and in the generic case it represents the intersection of the two fans of the nulls. In a section across the separator the field topology resembles that of a 2D X-point, and so, just as in 2D an X-point can collapse, so in 3D a separator can collapse to form a current sheet along the separator (Figure 14.9)- and, when the neighbouring field lines are carried into the separator sheet by a locally 2D stagnation point flow, then we find separator reconnection.

Galsgaard and Nordlund (1997) conducted an interesting experiment where their initial configuration contained eight 3D null points and they followed the effect of imposing a shear flow on two of the boundaries. They used a high-order finite-difference scheme on a 100^3 staggered grid with a Reynolds number of 100 and found that a current sheet formed along the separator (Figure 14.10) and produced separator reconnection.

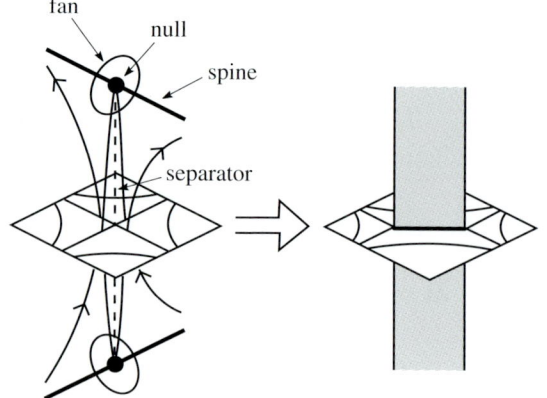

Figure 14.9 Collapse of a separator.

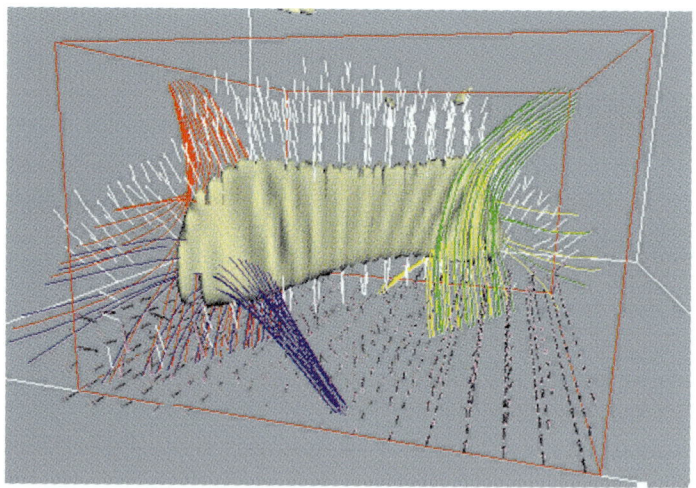

Figure 14.10. Numerical example of separator reconnection (**?**).

4. Three-Dimensional Reconnection at an Isolated Non-Ideal Region

Most astrophysical plasmas are effectively ideal, and non-ideal processes, which allow magnetic field lines to slip through the plasma, only become important on very small length scales. It is thus natural to consider reconnection processes which take place in isolated non-ideal regions. The resulting behaviour in 3D is fundamentally different from that in 2D.

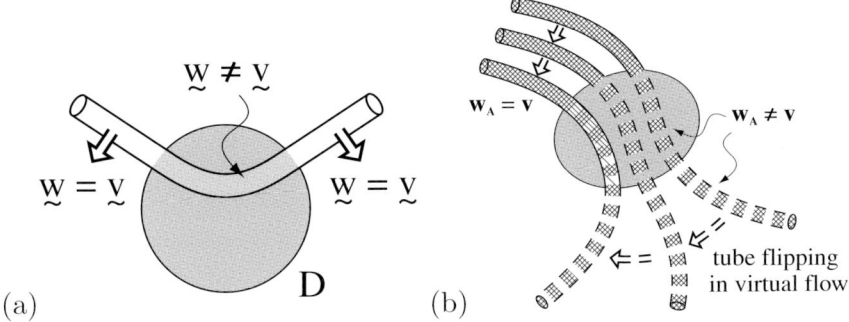

Figure 14.11. The behaviour of a flux tube which is partly within the non-ideal region (shaded) in (a) 2D and (b) 3D.

4.1   Fundamental Properties of 3D Reconnection

In 2D it is always possible to define a flux-conserving velocity ($\mathbf{w}$) which satisfies

$$\mathbf{E} + \mathbf{w} \times \mathbf{B} = \mathbf{0}, \qquad (14.24)$$

and for reconnection to occur, $\mathbf{w}$ must be singular at the X-point. Magnetic flux and field lines move everywhere at the velocity $\mathbf{w}$. By contrast, Priest et al. (2003a) have shown that in general in 3D a unique field line velocity does not exist, and consequently that a number of fundamental properties of 2D reconnection do not follow through when we consider reconnection in 3D. Several new properties of reconnection in 3D are listed below.

1 Since no unique field line velocity exists in 3D, field lines anchored in different regions of the ideal flow move in very different ways. In order to describe fully the motion of the magnetic flux in the volume, it is necessary to split the surface of the non-ideal region, D, into two parts, with magnetic flux entering D through one part and leaving it through the other. The field lines anchored in each of these regions move at completely separate velocities (though field lines which do not thread D move at the ideal plasma velocity everywhere as usual). Consequently, field lines which are followed through and beyond D appear to flip in a virtual flow (Figure 14.11).

2 Due to the non-existence of a unique field line velocity, magnetic field lines continually change their connections as they pass through D.

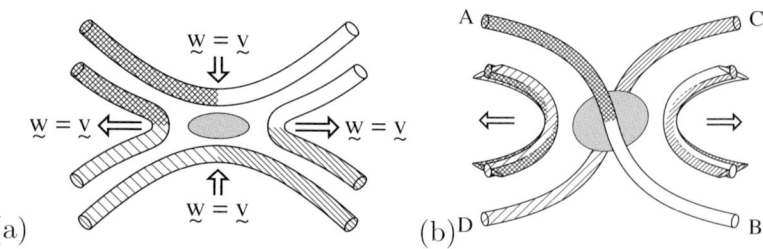

Figure 14.12. (a) Breaking and rejoining of flux tubes in 2D. (b) Breaking of two unique flux tubes in 3D showing partial rejoining, resulting in four distinct flux tubes after reconnection.

3 While the mapping between field line footpoints in 2D changes only at the X-point, where it is discontinuous, the mapping is generally continuous in 3D, the only exception being at separatrices of the field.

4 In 2D, magnetic field lines (and hence flux tubes) which undergo a reconnection process become rejoined in a simple one-to-one fashion. For every field line that is going to reconnect there is a corresponding field line with which it will rejoin perfectly. That is, their footpoints will become pair-wise oppositely connected after the field lines have undergone the reconnection process. The net effect is that before reconnection, we have two unique field lines and after reconnection we again have two unique, though differently connected, field lines (see Figure 14.12(a)).

The situation in 3D is very different. In general, for a given field line which is going to undergo reconnection, there is no corresponding counterpart field line with which its footpoints will become pair-wise oppositely connected after the reconnection process. The same is thus true in general for flux tubes (Figure 14.12(b)).

4.2   Analytical Solutions for 3D Reconnection

The study of reconnection is a highly complex one, which is as yet only in its very early stages. It is thus instructive to seek insight into the basic structure of the process by considering only a reduced set of the MHD equations. A number of analytical three- dimensional solutions, which we shall describe below, have been found. These solutions are kinematic, that is, they satisfy only the induction equation (14.2) as well as Maxwell's equations, but they do not satisfy the equation of motion (14.1). The state is assumed to be steady and a resistive non-ideal term is considered. The equations solved are thus

$$\mathbf{E} + \mathbf{v} \times \mathbf{B} = \eta \mathbf{j}, \tag{14.25}$$

$$\nabla \times \mathbf{E} = \mathbf{0}, \tag{14.26}$$

$$\nabla \cdot \mathbf{B} = 0, \tag{14.27}$$

$$\nabla \times \mathbf{B} = \mu_0 \mathbf{j}, \tag{14.28}$$

where η is assumed to be localised in order to localise the effect of the non-ideal term on the right of (14.25).

In 3D, reconnection can occur either at null points or in their absence (Schindler et al., 1988). Hornig and Priest (2003) solved Equations (14.25)-(14.28) for the situation where there is no null point of the magnetic field. They imposed the steady magnetic field $\mathbf{B} = B_0 (y, kx, b_0)$, a 2D X-point with uniform field in the third direction. The current is then given by $\mathbf{j} = ((k-1)/\mu_0) \hat{\mathbf{z}}$. The imposed resistivity (η) is localised within a region D, centred on the origin. The resulting plasma flow is rotational, with oppositely directed rotation, around the z-axis, above and below D. The nature of the reconnection of magnetic flux is thus rotational as well, with field lines traced from footpoints anchored in the ideal region above D rotating in one sense, while those anchored below rotate in the opposite sense. Hence, in an arbitrarily short period of time, every field line which threads D changes its connection in this rotational fashion. The breakdown of the one-to-one correspondence of reconnecting field lines, and in general flux tubes, is also demonstrated.

Reconnection at a 3D null point with a localised diffusion region centred on the null has been considered by Pontin et al. (2004b) and Pontin et al. (2004a). The nature of the restructuring of magnetic flux is found to be profoundly different depending on the orientation of the electric current with respect to the null.

Pontin et al. (2004b) considered reconnection in the magnetic field $\mathbf{B} = B_0 \left(x - \frac{iy}{2}, y + \frac{jx}{2}, -2z \right)$, so that $\mathbf{j} = (B_0 j/\mu_0) \hat{\mathbf{z}}$ is parallel to the spine of the null. The ideal plasma flows, and thus the restructuring of the magnetic flux, are again found to be rotational. This rotation is once again centred on the axis along which the current is directed, the z-axis, and again has opposite sense above and below the null point. There is no flow across either the spine or fan, which lie along the z-axis and in the $z = 0$ plane respectively.

The case where the current is parallel to the fan plane is described by Pontin et al. (2004a). They consider the magnetic field $\mathbf{B} = B_0(x, y - jz, -2z)$, so that $\mathbf{j} = (B_0 j/\mu_0) \hat{\mathbf{x}}$. The fan of the null again lies in the $z =$

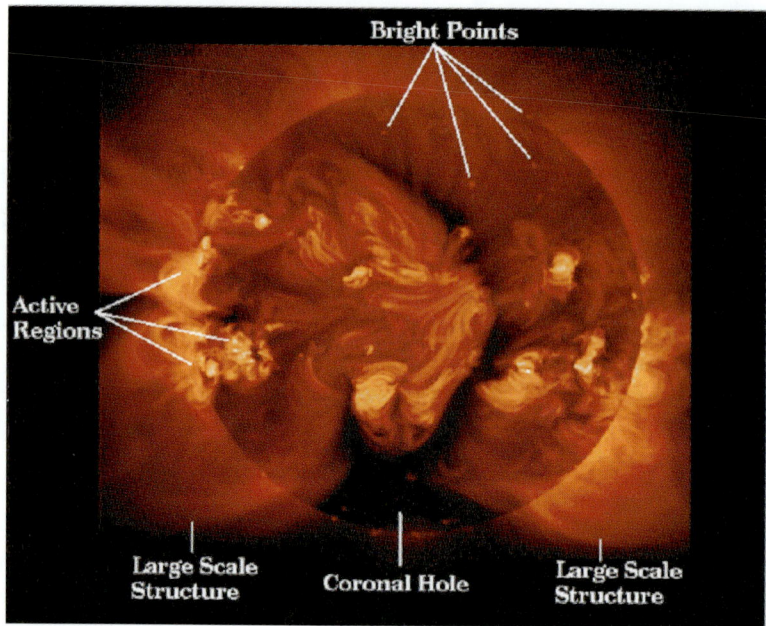

Figure 14.13. The solar corona from the Yohkoh satellite.

0 plane and the spine lies along $x = 0$, $y = jz/3$. This time the behaviour of the field lines is found to be much closer to that seen in the spine and fan reconnection of Priest and Titov (1996). The plasma flow is found to have a roughly stagnation point structure in planes of constant x, crossing both the spine and the fan. Field lines flip in constant succession around the spine line and through the fan plane. Solutions are also found where these two processes can be decoupled, so that the flow crosses only either the spine or fan, and so the field lines behave only either as in spine or fan reconnection. The existence of solutions with fan-crossing flows is particularly important in, for example, coronal fields on the Sun, since a transport of flux across the fan implies a change in the topology of the magnetic field.

5. Heating the Solar Corona by Reconnection

The solar corona at a temperature of a few million degrees consists of X-ray bright points, coronal loops and coronal holes. It is a magnetic world with myriads of magnetic interactions continually taking place (Figure 14.13). Recent space observations have shown examples of low-frequency waves being excited by solar flares (e.g. Nakariakov & Ofman, 2001), but these have too low an amplitude to be heating the corona.

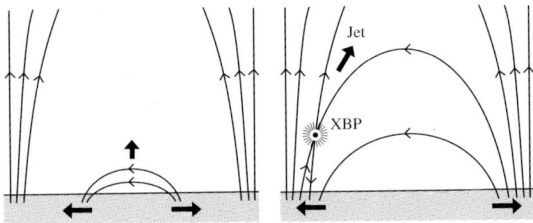

Figure 14.14 Converging flux model for X-ray bright points.

Also, there is a suggestion from broadening of UVCS lines in SOHO that high- frequency waves may be heating the outer corona. However, most of the evidence for heating the low corona is in favour of magnetic reconnection, which provides an elegant explanation for many diverse phenomena.

Yohkoh observations show that the hottest loops are cusps or interacting loops (Yoshida & Tsuneta, 1996). X-ray jets appear to be accelerated by reconnection (Shibata et al., 1992). Large-scale loops appear to have their upper coronal parts heated uniformly, which is consistent with turbulent reconnection (Priest et al., 1998). Furthermore, explosive events observed by SUMER on SOHO reveal that reconnection jets (Innes, Inhester, Axford, & Wilhelm, 1997) and nanoflares too may be caused by turbulent reconnection in myriads of current sheets.

5.1 Converging Flux Model

Reconnection is likely to be heating the corona in a variety of ways. First of all, X-ray bright points appear to be heated mainly by reconnection in cancelling magnetic flux (Parnell et al., 1994). The idea here is that new magnetic flux is continually emerging in supergranule cells as ephemeral regions, and the flux is swept to the edge of a cell, where it reconnects with the network flux and creates an X-ray bright point (Figure 14.14).

5.2 Binary Reconnection

There are many tiny discrete sources of magnetic flux on the solar surface and so it is natural to consider the "binary" interaction due to the relative motion of pairs of magnetic sources. Suppose they are unbalanced and connected - then the skeleton of the magnetic field is as shown in Figure 14.8. As the sources move, heat will be released in a process called *binary reconnection* (Priest et al., 2003b), which has several elements. First of all, the relative motion generates waves which propagate up and dissipate. Secondly, when the motions are slow, a non-linear force-free field is built up below the separatrix dome and then

Figure 14.15. Separator reconnection (Parnell & Galsgaard, 2003).

dissipates by turbulent relaxation to a linear force-free state of lowest energy. Finally, reconnection occurs at the null and at the separatrix dome, both in a driven way and also spontaneously by tearing in the separatrix current sheet.

5.3   Separator Reconnection

The relative motion of two sources that are initially unconnected, although have an overlying field, has been considered by Parnell and Galsgaard (2003). They find that reconnection occurs along a twisted current sheet that forms along a separator. This produces a twisted flux tube joining the two sources, but the subsequent reconnection as the sources separate is rather weak. Various aspects of coronal heating by separator reconnection have also been considered by Longcope (1996, 1998).

5.4   Braiding

Parker (1979) proposed that braiding of the footpoints of coronal field lines can lead to the formation and subsequent dissipation of current sheets at the boundaries between neighbouring flux tubes. In his model he starts with a uniform field stretched between two planes which model the photosphere at two ends of a coronal loop. Also he imposes highly nonlinear complex braiding motions of the footpoints. The reality of the resulting formation of current concentrations has been demonstrated in a numerical experiment by Galsgaard and Nordlund (1996).

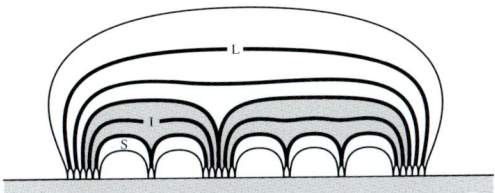

Figure 14.16 Coronal tectonics model.

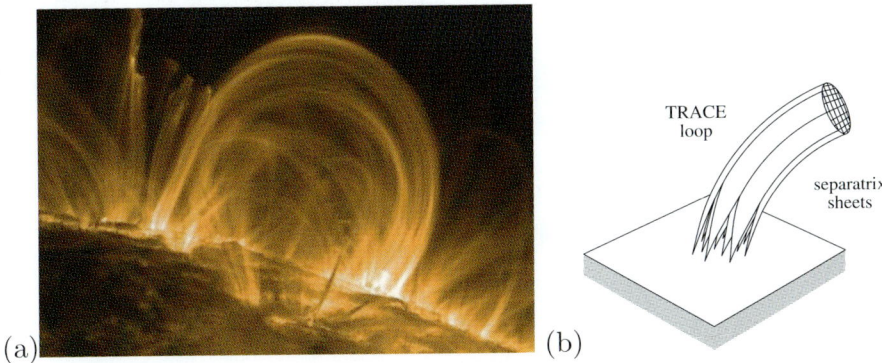

(a)   (b)

Figure 14.17. (a) TRACE coronal loops. (b) A web of separatrices within one loop.

5.5 Coronal Tectonics

In a new Coronal Tectonics Model for coronal heating, Priest, Hey-vaerts and Title (2002) have attempted to model the effect of the *magnetic carpet* (Schrijver et al., 1998) on the corona. The magnetic sources in the photosphere are concentrated and continually move around so that the magnetic flux in the quiet Sun is processed every 14 hours (Hagenaar, 2001). Each observed coronal loop goes down through the surface in many discrete sources, so that the flux from each source is topologically distinct and separated by a network of separatrix surfaces that thread the corona. As the sources move so the coronal fluxes slip past one another and form myriads of separator and separatrix current sheets which heat impulsively by reconnection.

The fundamental flux units are tiny intense flux tubes of field $1200G$, diameter $100km$ and flux $3 \times 10^{17} Mx$. Thus a single X-ray bright point has 100 such sources, and each of the finest TRACE loops (Figure 14.17(a)) consists of 10 finer loops (Figure 14.17(b)), which reach the surface in many footpoints and within which there is a web of separatri-ces.

We have set up a model demonstrating how current sheets form between flux tubes and have estimated the heating. In addition, Mellor et al. (2004) have conducted a numerical experiment. By comparison with

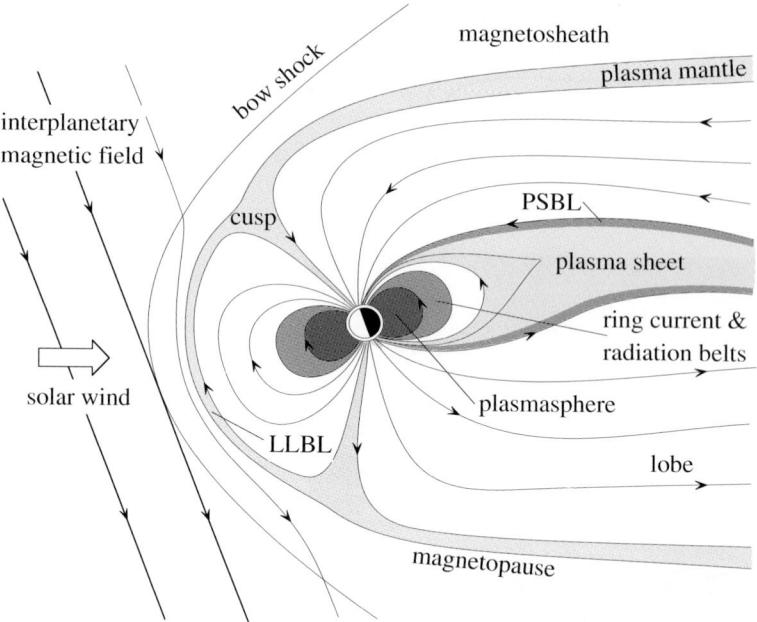

Figure 14.18. Schematic cross-section of the magnetosphere in the noon-midnight plane. LLBL and PSBL are acronyms for low-latitude boundary layer and plasma-sheet boundary layer respectively (after Parks, 1991).

Parker's braiding model, we have an array of coronal flux tubes going down to discrete sources in the photosphere in place of Parker's initial uniform field. Also our current sheets form immediately in response to simple generic motions of all kinds rather than requiring nonlinear braiding.

The results show that heating tends to be rather uniform along each separatrix implying that the elementary sub-telescopic tubes are heated uniformly. However, 90 - 95 % of the photospheric flux closes low down in the magnetic carpet, while the remaining 5 - 10 % forms large-scale connections. This suggests that the carpet should be heated more intensely than the large-scale corona. Also, unresolved observations of coronal loops would show enhanced heating near their feet in the carpet, with the upper parts of coronal loops being heated uniformly and less strongly.

6.   Reconnection in the Magnetosphere

Magnetic reconnection is an important process not only on the Sun, but also closer to home in the Earth's own magnetosphere (see Figure

14.18). This is despite the fact that the plasma conditions are very different, being almost completely collisionless. Reconnection here too may occur in many different ways, which can be described in terms of an 'open' model of the magnetosphere, first proposed by Dungey (1961). It is termed as such because in the model a finite amount of magnetic flux connects the Earth to interplanetary space. In this model there are two neutral points on the dayside and nightside of the magnetopause, and the amount of reconnection which occurs depends strongly on the North-South orientation of the interplanetary magnetic field (IMF) carried by the solar wind. Reconnection can occur in the vicinity of both the dayside and nightside neutral points, as described below.

6.1 Dayside Reconnection

Dayside reconnection was for a long time less studied than its nightside counterpart, since it primarily leads to magnetic energy storage rather than release, and therefore has less obvious direct physical effects (Cowley, 1980). The reconnection is strong when the southward component of the IMF is large at the magnetopause, as this component is anti- parallel to the Earth's magnetic field. Reconnection occurs here in a number of processes.

Firstly, magnetospheric erosion (Aubrey et al., 1970) is a signature of reconnection switching on at the dayside due to a southward turning of the IMF. The erosion happens because there is no outward flow within the magnetosphere to transport magnetic flux towards the reconnection site, and so this site moves earthwards as the reconnection consumes the flux on its earthward side.

Reconnection may occur a quasi-steady way at the magnetopause, since in the open model of the magnetosphere there is a non-zero normal component of the magnetic field there. Different schemes exist for how this may take place such as *component merging*, where reconnection takes place along the magnetopause separator connecting the two neutral points (Sonnerup et al., 1981; Soward, 1982) and *anti-parallel merging* where reconnection occurs only in regions where the magnetospheric and magnetosheath fields are close to anti-parallel (Crooker, 1979).

Also, reconnection occurs on the dayside in localised transient events known as flux transfer events (FTEs). These have been interpreted (Russell & Elphic) as isolated flux tubes interconnecting the IMF and magnetospheric field (see Figure 14.19(a)). Two further models for FTEs that have been proposed are those of Scholer et al. (1988) and Southwood et al. (1988), which produce FTEs by episodic 2D reconnection (Figure 14.19(b)), and that of Lee and Fu (1985), which assumes that an

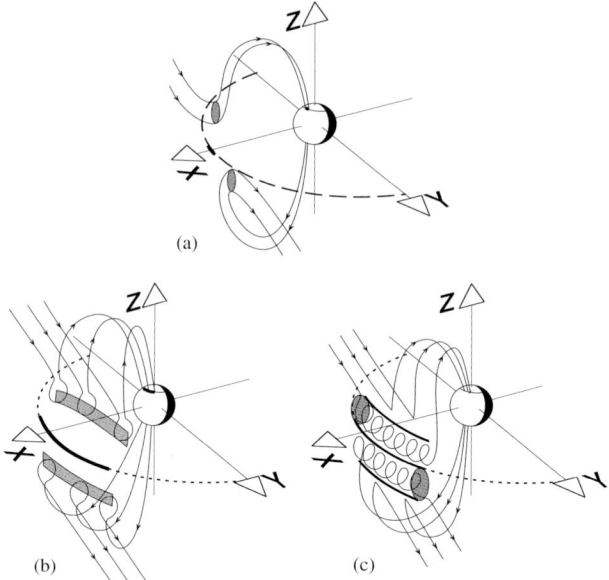

Figure 14.19 Three models for FTEs. The shaded areas indicate the cross-sections of the reconnected field lines at the surface of the magnetopause (after Lockwood, Cowley, & Sandholt, 1990).

FTE is a 2D magnetic island lying on the surface of the magnetopause (Figure 14.19(c)).

6.2   Nightside Reconnection

Reconnection on the nightside is very different. Here it occurs at a distant nightside neutral point, and converts magnetic energy into significant amounts of kinetic and thermal energy. Connected to this neutral point is an elongated current sheet which extends inwards towards the Earth. Importantly, observations have been made in the vicinity of the current sheet of slow-mode shocks (Feldman et al., 1984; Smith et al., 1984), with essentially the configuration predicted by Petschek, described earlier.

As well as occurring in the distant tail, nightside reconnection may also occur closer to the Earth in *magnetospheric substorms*. These are thought to take place when magnetic flux from dayside reconnection builds up in the tail, leading to a narrowing of the current sheet earthward of the nightside neutral point. This leads eventually to the creation of a further near-Earth neutral point. The result is that reconnection occurs, with a plasmoid being ejected down the tail, and earthward field lines relaxing back down to Earth, causing particles to be accelerated into bright aurorae. The phases of a substorm (see e.g. Hones, 1973) are shown in Figure 14.20.

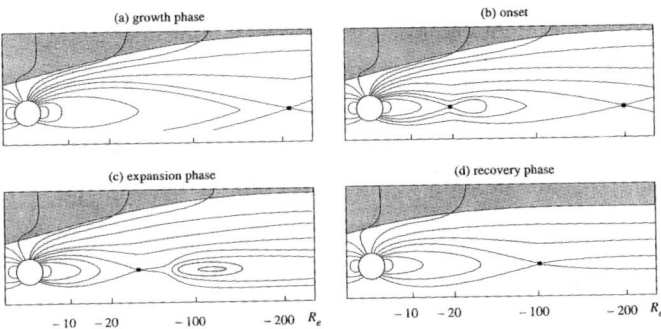

Figure 14.20. The phases of a magnetospheric substorm.

7. Conclusions

The theory of three-dimensional magnetic reconnection is in its infancy, but the indications are that reconnection can occur either in the absence of a null point or at a null point. In the latter case, spine reconnection, fan reconnection and separator reconnection are the main types that have been so far identified. Furthermore, 3D reconnection appears to be completely different from 2D reconnection in several aspects. Whereas field lines in 2D reconnect at a single point (the X-point), in 3D they continually change their connection as they pass through a non-ideal region. Furthermore, in 3D a unique field line velocity cannot be defined, since field lines anchored on one side of a non-ideal region behave differently to those anchored on the other side.

Several proposals have now been made for heating the Sun's corona by magnetic reconnection. X-ray bright points are thought to be due to reconnection driven by the motion of photospheric footpoints, mainly by converging motions. The high corona may possibly be heated by high-frequency waves, but the coronal loops could be heated by reconnection either due to binary reconnection or due to separator reconnection, or due to braiding. Another promising possibility is that coronal loops are driven by turbulent reconnection in myriads of current sheets by coronal tectonics. Furthermore, we have also seen that reconnection occurs in the Earth's magnetosphere, on both the dayside and the nightside.

In future, in order to understand the enigma of coronal heating there is an urgent need to: measure the temperature in coronal loops more accurately; understand fully the effect of the magnetic carpet; determine the effect of complex magnetic topology; unify our understanding of the zoo of coronal transients; test the viability of the Coronal Tectonics

Model; and, with the help of Solar B and Solar Dynamics Observatory, to understand properly the nature of the subtle link between the solar surface and corona.

Acknowledgments

We are grateful to the UK Particle Physics and Astronomy Research Council and the Carnegie Trust for financial support and to Terry Forbes, Gunnar Hornig, Jean Heyvearts, Dana Longcope and Slava Titov for fruitful discussions about the nature of 3D reconnection and coronal heating.

References

Aubrey, M. P., C. T. Russell and M. G. Kivelson (1970), *J. Geophys. Res.*, 75, 7018-7031.

Biskamp, D. (1986), *Phys. Fluids*, 29, 1520-1531.

Brown, D. S., and E. R. Priest (1999), *Proc. R. Soc. Lond. A*, 466, 3931.

Brown, D. S., and E. R. Priest (2001), *Astron. Astrophys.*, 367, 339-346.

Bulanov, S. V., and M. A. Olshanetsky (1984), *Phys. Lett.*, 100, 35-38.

Cowley, S. W. H. (1980), *Space Sci. Rev.*, 26, 217-275.

Craig, I. J. D., & A. N. M. McClymont (1991), *Astrophys. J.*, 371, L41-L44. 5 pages = "L41-L44"

Craig, I. J. D., & R. B. Fabling (1996), *Astrophys. J.*, 462, 969-976. 5 volume = "462",

Crooker, J. U. (1979), *J. Geophys. Res.*, 84, 951-959.

Dungey, J. W. (1961), *Phys. Rev. Lett.*, 6, 47-48.

Feldman, W. C., S. J. Schwartz, S. J. Bame, D. N. Baker, J. Birn, J. T. Gosling, E. W. Hones, Jr., D. J. McComas, J. A. Slavin, E. J. Smith, & R. D. Zwickl (1984), *Geophys. Res. Lett.*, 11, 599-602.

Galsgaard, K., & A. Nordlund (1996), *J. Geophys. Res.*, 101, 13445-13460.

Galsgaard, K., & A. Nordlune (1997), *J. Geophys. Res.*, 102, 231-248.

Hagenaar, H. J. (2001), *Astrophys. J.*, 555, 448-461.

Hones, E. W., Jr. (1973), *Radio Sci.*, 8, 979-990.

Hornig, G., & E. R. Priest (2003), *Physics of Plasmas*, 10(7), 2712-2721.

Imshennik, V. S., & S. I. Syrovatsky (1967), *Sov. Phys. JETP*, 25, 656-664.

Innes, D. E., B. Inhester, W. I. Axford, & K. Wilhelm (1997), *Nature*, 386, 811-813.

Klapper, I. (1998), *Phys. Plasma*, 5, 910-914.

Lee, L. C., & Z. F. Fu (1985), *Geophys. Res. Lett.*, 12, 105-108.

Lockwood, M., S. W. H. Cowley, & P. E. Sandholt (1990), *EOS, Trans. Amer. Geophys. Union*, 71, 719-720.

Longcope, D. W. (1996), *Solar Phys.*, 169, 91-121.

Longcope, D. W. (1998), *Astrophys. J.*, 507, 433-442.

Longcope, D. W., D. S. Brown, & E. R. Priest (2003), *Phys. Plasmas*, 10, 3321-3334.

Mellor, C., V. S. Titov, & E. R. Priest (2002), *J. Plasma Phys.*, 68, 221-235.

Mellor, C., C. Gerrard, K. Galsgaard, A. W. Wood, & E. R. Priest (2004), submitted for publication.

Moffat, H. K. (1978), "Magnetic Field Generation in Electrically Conducting Fluids", Cambdridge Univ. Press: Cambridge, UK.

Nakariakov, V. M., & L. Ofman (2001), *Astron. Astrophys.*, 372, L53-L56.

Parker, E. N. (1979), "Cosmical Magnetic Fields", Clarendon Press: Oxford.

Parnell, C. E., E. R. Priest, & L. Golub (1994), *Solar Phys.*, 151, 57-74.

Parnell, C. E., J. M. Smith, T. Neukirch, & E. R. Priest (1996), *Phys. Plasma*, 3(3), 759-770.

Parnell, C. E., T. Neukirch, J. M. Smith, & E. R. Priest (1997), *Geophys. Astrophys. Fluid Dynamics*, 84, 245-271.

Parnell, C. E., & K. Galsgaard (2003), submitted for publication.

Parks, G. K. (1991), "Physics of Space Plasma", Addison-Wesley: Reading, Massachusetts, USA.

Petschek, H. E. (1964), in "Magnetic Field Annihilation" (W. N. Hess, ed.), 425-439, NASA SP-50, Washington, D.C., USA.

Pontin, D. I., G. Hornig, & E. R. Priest (2004), submitted for publication.

Pontin, D. I., G. Hornig, & E. R. Priest (2004), submitted for publication.

Priest, E. R., & T. G. Forbes (1986), *J. Geophys. Res.*, 91, 5579-5588.

Priest, E. R., & L. C. Lee (1990), *J. Plasma Phys.*, 44, 337-360.

Priest, E. R., & V. S. Titov (1996), *Phil. Trans. R. Soc. Lond. A*, 354, 2951-2992.

Priest, E. R., C. R. Foley, J. Heyvaerts, T. D. Arber, J. L. Culhane, & L. W. Acton (1998), *Nature*, 393, 545-547.

Priest, E. R., & T. G. Forbes (2000), "Magnetic Reconnection: MHD Theory and Applications", Cambridge Univ. Press: Cambridge, UK.

Priest, E. R., J. F. Heyvaerts, & A. M. Title (2002), *Astrophys. J.*, 576, 533-551.

Priest, E. R., G. Hornig, & D. I. Pontin (2003), *J. Geophys. Res.*, 108(A7), SSH6-1.

Priest, E. R., D. W. Longcope, & V. S. Titov (2003), *Astrophys. J.*, 598, 667-677.

Russell, C. T., & R. C. Elphic (1978), *Space Sci. Rev.*, 22, 681-715.

Schindler, K., M. Hesse, & J. Birn (1988), *J. Geophys. Res.*, 93(A6), 5547-5557.

Scholer, M. (1988), *Geophys. Res. Lett.*, 15, 748-751.

Schrijver, C. J., A. M. Title, K. L. Harvey, N. R. Sheeley, Y. -M. Wang, G. H. J. van den Oord, R. A. Shine, T. D. Tarbell, & N. E. Hurlburt (1998), *Nature*, 394, 152-154.

Shibata, K., V. Ishido, L. W. Acton, K. T. Strong, T. Hirayama, Y. Uchida, A. H. McAllister, R. Matsumoto, S. Tsuneta, T. Shimizu, H. Hara, T. Sakurai, K. Ichimoto, V. Nishino, & Y. Ogawara (1992), *Publ. Astron. Soc. Japan*, 44, L173-L179.

Smith, E. J., J. A. Slavin, B. T. Tsurutani, W. C. Feldman, & S. J. Bame (1984), *Geophys. Res. Lett.*, 217, 644-656.

Sonnerup, B. U. Ö., & E. R. Priest (1975), *J. Plasma Phys.*, 14, 283-294.

Sonnerup, B. U. Ö., G. Paschmann, I. Papamastorakis, N. Sckopke, G. Haerendel, S. J. Bame, J. R. Asbridge, J. T. Gosling, & C. T. Russell (1981), *J. Geophys. Res.*, 86, 10049-10067.

Southwood, D. J., C. J. Farrugia, & M. A. Saunders (1988), *Planet. Space Sci.*, 36, 503-508.

Soward, A. M. (1982), *J. Plasma Phys.*, 28, 415-443.

Strachan, N. R., & E. R. Priest (1994), *Gephys. Astrophys. Fluid Dynamics*, 74, 245-274.

Sweet, P. A. (1958), in "Electromagnetic Phenomena in Cosmical Plasma", 123-134, Cambridge Univ. Press: London, UK.

Yan, M., L. C. Lee, & E. R. Priest (1992), *J. Geophys. Res.*, 97, 8277-8293.

Yan, M., L. C. Lee, & E. R. Priest (1993), *J. Geophys. Res.*, 98, 7593-7602.

Yoshida, T., & S. Tsuneta (1996), *Astrophys. J.*, 459, 342-346.

Astrophysics and Space Science Library

Volume 301: *Multiwavelength Cosmology*, edited by Manolis Plionis
Hardbound, ISBN 1-4020-1971-8, March 2004

Volume 300:*Scientific Detectors for Astronomy*, edited by Paola Amico, James
W. Beletic, Jenna E. Beletic
Hardbound, ISBN 1-4020-1788-X, February 2004

Volume 299: *Open Issues in Local Star Fomation,* edited by Jacques Lépine,
Jane Gregorio-Hetem
Hardbound, ISBN 1-4020-1755-3, December 2003

Volume 298: *Stellar Astrophysics - A Tribute to Helmut A. Abt,* edited by
K.S. Cheng, Kam Ching Leung, T.P. Li
Hardbound, ISBN 1-4020-1683-2, November 2003

Volume 297: *Radiation Hazard in Space,* by Leonty I. Miroshnichenko
Hardbound, ISBN 1-4020-1538-0, September 2003

Volume 296: *Organizations and Strategies in Astronomy, volume 4,* edited by
André Heck
Hardbound, ISBN 1-4020-1526-7, October 2003

Volume 295: *Integrable Problems of Celestial Mechanics in Spaces of
Constant Curvature,* by T.G. Vozmischeva
Hardbound, ISBN 1-4020-1521-6, October 2003

Volume 294: *An Introduction to Plasma Astrophysics and
Magnetohydrodynamics,* by Marcel Goossens
Hardbound, ISBN 1-4020-1429-5, August 2003
Paperback, ISBN 1-4020-1433-3, August 2003

Volume 293: *Physics of the Solar System,* by Bruno Bertotti, Paolo Farinella,
David Vokrouhlický
Hardbound, ISBN 1-4020-1428-7, August 2003
Paperback, ISBN 1-4020-1509-7, August 2003

Volume 292: *Whatever Shines Should Be Observed,* by Susan M.P. McKenna-
Lawlor
Hardbound, ISBN 1-4020-1424-4, September 2003

Volume 291: *Dynamical Systems and Cosmology,* by Alan Coley
Hardbound, ISBN 1-4020-1403-1, November 2003

Volume 290: *Astronomy Communication*, edited by André Heck, Claus Madsen
Hardbound, ISBN 1-4020-1345-0, July 2003

Volume 287/8/9: *The Future of Small Telescopes in the New Millennium*, edited by Terry D. Oswalt
Hardbound Set only of 3 volumes, ISBN 1-4020-0951-8, July 2003

Volume 286: *Searching the Heavens and the Earth: The History of Jesuit Observatories*, by Agustín Udías
Hardbound, ISBN 1-4020-1189-X, October 2003

Volume 285: *Information Handling in Astronomy - Historical Vistas*, edited by André Heck
Hardbound, ISBN 1-4020-1178-4, March 2003

Volume 284: *Light Pollution: The Global View*, edited by Hugo E. Schwarz
Hardbound, ISBN 1-4020-1174-1, April 2003

Volume 283: *Mass-Losing Pulsating Stars and Their Circumstellar Matter*, edited by Y. Nakada, M. Honma, M. Seki
Hardbound, ISBN 1-4020-1162-8, March 2003

Volume 282: *Radio Recombination Lines*, by M.A. Gordon, R.L. Sorochenko
Hardbound, ISBN 1-4020-1016-8, November 2002

Volume 281: *The IGM/Galaxy Connection*, edited by Jessica L. Rosenberg, Mary E. Putman
Hardbound, ISBN 1-4020-1289-6, April 2003

Volume 280: *Organizations and Strategies in Astronomy III*, edited by André Heck
Hardbound, ISBN 1-4020-0812-0, September 2002

Volume 279: *Plasma Astrophysics , Second Edition*, by Arnold O. Benz
Hardbound, ISBN 1-4020-0695-0, July 2002

Volume 278: *Exploring the Secrets of the Aurora*, by Syun-Ichi Akasofu
Hardbound, ISBN 1-4020-0685-3, August 2002

Volume 277: *The Sun and Space Weather*, by Arnold Hanslmeier
Hardbound, ISBN 1-4020-0684-5, July 2002

For further information about this book series we refer you to the following web site:
http://www.wkap.nl/prod/s/ASSL

To contact the Publishing Editor for new book proposals:
Dr. Harry (J.J.) Blom: harry.blom@springer-sbm.com